STATISTICAL MECHANICS

Principles and Selected Applications

TERRELL L. HILL

National Institutes of Health
Bethesda, Maryland

DOVER PUBLICATIONS, INC.
New York

Published in Canada by General Publishing Company, Ltd., 30 Lesmill Road, Don Mills, Toronto, Ontario.
Published in the United Kingdom by Constable and Company, Ltd.

This Dover edition, first published in 1987, is an unabridged, unaltered republication of the second printing of the work originally published in the McGraw-Hill Series in Advanced Chemistry by McGraw-Hill Book Company, Inc., New York, in 1956.

Manufactured in the United States of America
Dover Publications, Inc., 31 East 2nd Street, Mineola, N.Y. 11501

Library of Congress Cataloging-in-Publication Data

Hill, Terrell L.
Statistical mechanics.

Includes bibliographies and index.
1. Statistical mechanics. I. Title.
QC174.8.H55 1987 530.1'3 87-589
ISBN 0-486-65390-0 (pbk.)

TO MY PARENTS

Ollie I. Hill

and

George Leslie Hill

FOREWORD

Although several excellent elementary textbooks and a few specialized monographs on the equilibrium theory of statistical mechanics have appeared in recent years, no treatise covering both the basic theory and its applications to the solution of the fundamental problems of physical chemistry and chemical physics has been published since "Statistical Thermodynamics" by Fowler and Guggenheim, which appeared in 1939. Since that time, in spite of interruption during the war years, research in this field has progressed at a rapid rate and a large body of new results has accumulated in the scientific literature. Significant advances have been made in the theory of cooperative phenomena, in the theory of liquids and liquid solutions, as well as in many more specialized areas of application of statistical mechanics. The time is therefore most appropriate for the appearance of a treatise in which the basic theory is reviewed and new results are made accessible in a comprehensive presentation. Dr. Hill's book achieves these objectives in a most satisfactory manner and will, I am certain, be welcomed by all chemists and physicists whose research interests are directed to the interpretation of the macroscopic properties of substances from the standpoint of their structure on the molecular and atomic levels.

John G. Kirkwood
Sterling Chemistry Laboratory
Yale University

PREFACE

There have been a number of important advances in statistical mechanics since the monumental books by Fowler, Tolman, Fowler and Guggenheim, and Mayer and Mayer were published fifteen years or so ago. The primary aim of this volume is to provide a rather detailed account of a selected group of these developments, namely, fluctuation theory (Chap. 4), imperfect gas and condensation theory (Chap. 5), distribution functions (Chap. 6), nearest-neighbor (Ising) statistics (Chap. 7), and free-volume and hole theories of liquids and solids (Chap. 8). These chapters on applications are prefaced, for completeness, by a somewhat condensed treatment of the principles of equilibrium statistical mechanics (Chaps. 1 to 3). From this brief outline it will be clear that a number of important topics have been arbitrarily omitted (e.g., nonequilibrium statistical mechanics, Born–von Kármán theory of solids, many aspects of solution theory, etc.). Certainly more subjects have been left out than included. Thus the book is not intended to be a truly comprehensive treatise on statistical mechanics.

Numerical calculations and comparisons with experiment have not been stressed for the very good reason that these aspects are considered exhaustively in the recent excellent book by Hirschfelder, Curtiss, and Bird. In general, an attempt has been made to emphasize material which is not available in other books.

As for the level of this work, it is neither introductory nor is it by any means the ultimate in sophistication. Rather, the book is primarily for graduate students and research workers in chemistry, physics, and biology who have already some acquaintance with statistical mechanics but who wish to extend their background somewhat. The author hopes that it may prove useful as a text for a second course in statistical mechanics, as a supplement in a first course to a text such as Rushbrooke's, or for self-study or reference.

There are, at various places in the book, modifications, extensions, derivations, points of view, etc., which may be new, but there seems little point in identifying these.

A reasonable number of literature references are provided and, in addition, at the end of each chapter under the heading "General References" there is furnished a list of books, review articles, etc., which can be used for supplementary reading on the material included in the chapter.

Throughout the book, the exp notation is used as follows:

$$\exp\,[a(x + y)]z = e^{a(x+y)}z$$
$$\exp\,(ax)[y(b + z)] = e^{ax}y(b + z)$$

That is, the exponent extends only as far as the closure of the first type of parenthesis used.

The book was written, except for revisions, during the tenure of a John Simon Guggenheim Memorial Fellowship in the Department of Chemistry, Yale University, 1952–1953. The author is greatly indebted to the Guggenheim Foundation for this Fellowship and to Yale University for its kind hospitality. While at Yale, the author benefited from numerous discussions with Professor J. G. Kirkwood and made free use of his published papers and mimeographed lecture notes on statistical mechanics (prepared by J. H. Irving). Specifically, Professor Kirkwood's lecture notes have been used in or have influenced to an appreciable extent parts of Secs. 2, 3, 4, 5, and 7 of Chap. 1, Secs. 16 and 17 of Chap. 3, Sec. 21 of Chap. 4, Secs. 22 and 24 of Chap. 5, and Sec. 48 of Chap. 8.

The author is also indebted to Dr. Elliott Montroll for making available before publication a copy of his review paper with Newell on the Ising model, to Prof. I. Prigogine, Dr. H. N. V. Temperley, Dr. John Ross, Dr. H. M. Peek, and Dr. J. J. Blum for reading and commenting on parts of the manuscript, and especially to Prof. W. G. McMillan, Jr., who read the entire manuscript and made a great many helpful criticisms.

The writing of the book was made a much more pleasant task than it might have been by the constant interest, encouragement, and cooperation of my wife, Laura E. Hill.

Terrell L. Hill

CHAPTER I

PRINCIPLES OF CLASSICAL STATISTICAL MECHANICS

1. Statistical Mechanics and Thermodynamics

Thermodynamics is concerned with the relationships between certain macroscopic properties (the thermodynamic variables or functions) of a system in equilibrium. We must turn to statistical mechanics if our curiosity goes beyond these formal interrelationships and we wish to understand the connection between the observed values of a thermodynamic function and the properties of the molecules making up the thermodynamic system. That is, statistical mechanics provides the *molecular* theory of the macroscopic properties of a thermodynamic system. In current research, both thermodynamics and statistical mechanics are in the process of extension to systems departing from equilibrium, but these developments will not be included here.

In this chapter we summarize the foundations of classical statistical mechanics. The corresponding quantum-mechanical discussion is given in Chap. 2. The relationships between classical statistical mechanics, quantum statistical mechanics, and thermodynamics will then be outlined in Chap. 3.

Consider a thermodynamic system which contains N_1 molecules of component 1, N_2 molecules of component 2, . . . , and N_r molecules of component r, where r is the number of independent components. Also, let there be s external variables $x_1, x_2, \ldots, x_s$ (e.g., volume). Then extensive thermodynamic properties may be considered as functions of $r + s + 1$ independent variables (for example, $T, x_1, x_2, \ldots, x_s, N_1, N_2, \ldots, N_r$). Similarly, intensive properties are functions of $r + s$ independent variables (for example, $T, x_1, x_2, \ldots, x_s, N_2/N_1, N_3/N_1, \ldots, N_r/N_1$). Thus, the thermodynamic state of a system (including its extent) is completely specified by the assignment of values to $r + s + 1$ variables. In contrast to this, the determination of the dynamical state of the same system[1] requires the specification of a very large number of variables, of the order of the total number of molecules $N_1 + N_2 + \cdots + N_r$ (roughly 10^{23} in typical cases). To be specific, if the

[1] We consider only systems which are conservative in the mechanical sense.

system has n degrees of freedom, one usually chooses the $2n$ variables q_1, $q_2, \ldots, q_n$ and $p_1, p_2, \ldots, p_n$, where the q's are (generalized) coordinates and the p's are the associated conjugate momenta defined by

$$p_i = \frac{\partial L}{\partial \dot{q}_i} \qquad i = 1, 2, \ldots, n \tag{1.1}$$

where $$\dot{q}_i = \frac{dq_i}{dt}$$

and $$L = T - U$$

L is the Lagrangian function, $T(\dot{q}_1, \ldots, \dot{q}_n, q_1, \ldots, q_n)$ is the kinetic energy of the system, and $U(q_1, \ldots, q_n)$ is the potential energy of the system. For example, in a one-component system containing N particles with no internal degrees of freedom (rotation, vibration, etc.), $2n = 6N$ since there are three translational degrees of freedom per particle.

From the above it is clear that complete specification of the thermodynamic state leaves the dynamical state undefined. That is to say, there are very many (actually infinitely many in classical mechanics) dynamical ("microscopic") states consistent with a given thermodynamic ("macroscopic") state. The central problem of statistical mechanics, as a molecular theory, is therefore to establish the way in which averages of properties over dynamical states (consistent with the particular thermodynamic state of interest) should be taken in order that they may be put into correspondence with the thermodynamic functions of the system.

2. Phase Space

As suggested by Gibbs, the dynamical state of a system may be specified by locating a point in a $2n$ dimensional space, the coordinates of the point being the values of the n coordinates and n momenta which specify the state. The space is referred to as "phase space" and the point is called a "phase point" or "representative point." It should be noted that with the forces of the system given, assignment of the position of a phase point in phase space at time t actually completely determines the future (and past) trajectory or path of the point as it moves through phase space in accordance with the laws of mechanics. The equations of motion of the phase point are in fact (in "Hamiltonian form")

$$\dot{q}_i = \frac{\partial H}{\partial p_i} \qquad \dot{p}_i = -\frac{\partial H}{\partial q_i} \qquad i = 1, 2, \ldots, n \tag{2.1}$$

where H, the Hamiltonian function, is an expression for the energy of the system as a function of the p's and q's. That is,

$$H = T(p,q) + U(q) \tag{2.2}$$

where we have introduced the usual convention that p and q mean $p_1, p_2, \ldots, p_n$ and $q_1, q_2, \ldots, q_n$, respectively. In principle, this system of $2n$ first-order differential equations may be integrated to give $p(t)$ and $q(t)$. The $2n$ constants of integration would be fixed by knowing the location of the phase point at some time t (i.e., by knowing the coordinates and components of momenta of all molecules at t).

Now the value of a thermodynamic property[1] G is determined by the dynamical state of the system; thus we may write $G = G(p,q)$. Let us consider first the simplest case, that of a perfectly isolated system. At time t_0, let the momenta and coordinates have the values p^0, q^0. The experimental measurement of G, beginning at $t = t_0$, to give G_{obs}, actually involves observation of $G(p,q)$ over a period of time τ (the magnitude of τ is discussed in Sec. 4) with

$$G_{\text{obs}} = \frac{1}{\tau}\int_{t_0}^{\tau+t_0} G(p,q)\,dt \tag{2.3}$$

That is, G_{obs} is a time average. If Eqs. (2.1) are solved using the initial conditions $p(t_0) = p^0$ and $q(t_0) = q^0$, then $G(t)$ in Eq. (2.3) is given by $G(t) = G[p(t), q(t)]$, and G_{obs} may be computed. Thus, the time average indicated in Eq. (2.3) is that average over dynamical states which should in principle be put into correspondence with the value of the thermodynamic function G. Unfortunately, such a purely "mechanical" computation is, of course, quite impossible (even for this perfectly isolated system) owing to the complexity of the problem ($n \cong 10^{23}$ degrees of freedom), and it is for this reason that we must turn (Sec. 4) to another, less direct, method, due to Gibbs, of averaging over dynamical states. The relation between the two methods is considered briefly in Sec. 7.

The above remarks refer to a perfectly isolated system. For a system not in perfect isolation, for example, a system in thermal equilibrium with a heat bath, G_{obs} is still a time average as in Eq. (2.3), but $p(t)$ and $q(t)$ are no longer associated with a single trajectory of a conservative system in phase space. Thus, the mechanical calculation of G_{obs} is even more difficult here; in fact, it can only be carried out, even in principle, by treating the system and surroundings together, as a new, but larger, perfectly isolated system.

[1] To be more specific, we are referring here to thermodynamic properties of a "mechanical" nature, such as energy and pressure. See Sec. 14 for a discussion of "mechanical" and "nonmechanical" thermodynamic properties.

3 Ensembles

Let us imagine that we have a very large number $\mathcal{N}$ of perfectly isolated and therefore independent systems in a variety of dynamical states, but all with the same values of $N_1, \ldots, N_r$ and $x_1, \ldots, x_s$. This collection of systems is called an "ensemble." Whenever it proves useful, we may consider the limit $\mathcal{N} \to \infty$.

An *operational* method of constructing an ensemble would be as follows. Using a real system in a certain macroscopic state as a model, we prepare a collection of $\mathcal{N}$ systems in this same macroscopic state. At some time t' the $\mathcal{N}$ systems are instantaneously removed from their surroundings (if any) and each put in perfect isolation. The result is an ensemble which at t' may certainly be said to be "representative" of the macroscopic state of the real system. We shall actually be interested primarily in ensembles which we attempt to construct *theoretically* in such a way that the distribution of systems of the ensemble over dynamical states is representative of the macroscopic state of some particular real system, in the same sense that the operational ensemble above is representative of such a state at time t'.

In the remainder of this section we discuss general properties of ensembles which hold whether or not the ensemble is supposed to be a representative ensemble in the above sense, and if it is a representative ensemble, whether or not the macroscopic state of interest is an equilibrium state.

Each system in an ensemble has a representative point in a phase space, and if we use the same phase space for all $\mathcal{N}$ phase points, the ensemble itself will appear as a "cloud" of phase points in the phase space.[1] As time passes, each phase point of the cloud pursues its own independent trajectory in phase space.

$\mathcal{N}$ is always taken large enough so that the concept of a continuous density of phase points can be introduced. In fact, it is usually convenient to normalize the density in such a way that it becomes a probability density. To do this, we define a function $f(p_1, \ldots, p_n, q_1, \ldots, q_n; t)$ so that $f(p_1, \ldots, p_n, q_1, \ldots, q_n; t)dp_1 \ldots dp_n\, dq_1 \ldots dq_n$ or, for brevity, $f(p,q;t)\, dp\, dq$ is the fraction of the $\mathcal{N}$ phase points, at time t, in the element of volume $dp_1 \ldots dp_n\, dq_1 \ldots dq_n$ of phase space. Then the number of phase points in this element of volume is $\mathcal{N}f(p,q;t)\, dp\, dq$ and the density of phase points at p,q is $\mathcal{N}f(p,q;t)$. The quantity $f(p,q;t)$ itself is called the probability density (or distribution function), for if a system is chosen at random from the ensemble at time t, the probability that the phase point representative of

[1] Use of the same phase space for all systems is only possible because the ensembles we are considering have the same values of $N_1, \ldots, N_r, x_1, \ldots, x_s$ for all systems. More general ensembles will be discussed in Sec. 6.

its dynamical state is in $dp\,dq$ about the point p, q is $f\,dp\,dq$. The probability density integrates to unity,

$$\int f(p,q;t)\,dp\,dq = 1 \tag{3.1}$$

where the single integral sign means integration over all the p's and q's.

The equations of motion, Eqs. (2.1), determine the trajectory of each phase point, given its location in phase space at some initial time. These equations therefore also determine completely the distribution function $f(p,q;t)$ at any time if the dependence of f on p and q is known at the initial time. It must always be understood, therefore, that the time dependence of f is in accord with the laws of mechanics, and is not arbitrary. This time dependence is discussed below in connection with Liouville's theorem.

The ensemble average of any function $\varphi(p,q)$ of the dynamical state of the system is defined as

$$\bar{\varphi} = \int \varphi(p,q) f(p,q;t)\,dp\,dq \tag{3.2}$$

In view of the fact that the direct computation of the value G_{obs} of a thermodynamic property G from Eq. (2.3) cannot be carried out, Gibbs' alternative suggestion, mentioned at the end of Sec. 2, was that an ensemble average be used in place of a time average. This of course accounts for our interest in ensembles. The particular way in which ensemble averages are employed for this purpose will be discussed in Secs. 4 and 5.

Liouville's Theorem. We derive here a result which is necessary to the further development of our general argument. Suppose that the distribution function $f(p,q;t)$, with its time dependence, is known. Now consider the change df in the value of f at p, q, t as a result of arbitrary infinitesimal changes in p, q, and t:

$$df = \frac{\partial f}{\partial t}\,dt + \sum_{i=1}^{n} \frac{\partial f}{\partial p_i}\,dp_i + \sum_{i=1}^{n} \frac{\partial f}{\partial q_i}\,dq_i \tag{3.3}$$

That is, $f + df$ is the value of f at a point $p + dp$, $q + dq$ near p, q and at a time $t + dt$. Now, instead of using arbitrary quantities dp, dq, let us choose the particular neighboring point $p + dp$, $q + dq$ which the trajectory of the phase point originally at p, q (time t) passes through at $t + dt$. In this case dp and dq cannot be independent variations since along the trajectory $p = p(t)$ and $q = q(t)$. Then, using Eq. (3.3), the change of f with time in the neighborhood of a phase point traveling along its trajectory is

$$\frac{df}{dt} = \frac{\partial f}{\partial t} + \sum_i \left(\frac{\partial f}{\partial p_i}\,\dot{p}_i + \frac{\partial f}{\partial q_i}\,\dot{q}_i \right) \tag{3.4}$$

In contrast to df/dt, $\partial f/\partial t$ gives the change of f with time at a fixed location p, q in phase space. Now Liouville's theorem states that $df/dt = 0$.

To prove the theorem, consider an arbitrary but fixed volume V in phase space whose surface we represent by $\mathscr{A}$. Then the number of phase points in V at t is

$$\mathscr{N}_V = \mathscr{N}\int_V f(p,q;t)\,dp\,dq \tag{3.5}$$

and

$$\frac{d\mathscr{N}_V}{dt} = \mathscr{N}\int_V \frac{\partial f}{\partial t}\,dp\,dq \tag{3.6}$$

The number of phase points passing through the surface $\mathscr{A}$ in unit time gives another expression for $d\mathscr{N}_V/dt$, namely,

$$\begin{aligned}\frac{d\mathscr{N}_V}{dt} &= -\mathscr{N}\int_{\mathscr{A}} f\mathbf{u}\cdot\mathbf{n}\,d\mathscr{A} \\ &= -\mathscr{N}\int_V \nabla\cdot(f\mathbf{u})\,dp\,dq \end{aligned}\tag{3.7}$$

where $\mathbf{n}$ is the usual normal unit vector, and the vectors $\mathbf{u}$ and $f\mathbf{u}$ have components in phase space

$$\dot{p}_1, \ldots, \dot{p}_n, \dot{q}_1, \ldots, \dot{q}_n$$

and

$$f\dot{p}_1, \ldots, f\dot{p}_n, f\dot{q}_1, \ldots, f\dot{q}_n$$

respectively. Because V is arbitrary, on comparing Eqs. (3.6) and (3.7) we obtain

$$\frac{\partial f}{\partial t} + \sum_i \left(\frac{\partial f\dot{p}_i}{\partial p_i} + \frac{\partial f\dot{q}_i}{\partial q_i}\right) = 0 \tag{3.8}$$

or

$$\frac{\partial f}{\partial t} + \sum_i \left(\frac{\partial f}{\partial p_i}\dot{p}_i + \frac{\partial f}{\partial q_i}\dot{q}_i\right) + f\sum_i \left(\frac{\partial \dot{p}_i}{\partial p_i} + \frac{\partial \dot{q}_i}{\partial q_i}\right) = 0 \tag{3.9}$$

Now, from Eq. (2.1),

$$\frac{\partial \dot{p}_i}{\partial p_i} + \frac{\partial \dot{q}_i}{\partial q_i} = -\frac{\partial^2 H}{\partial q_i\,\partial p_i} + \frac{\partial^2 H}{\partial p_i\,\partial q_i} = 0$$

so the second sum in Eq. (3.9) is zero. We therefore have the required result

$$\frac{\partial f}{\partial t} = -\sum_i \left(\frac{\partial f}{\partial p_i}\dot{p}_i + \frac{\partial f}{\partial q_i}\dot{q}_i\right) \tag{3.10}$$

which is equivalent to [see Eq. (3.4)]

$$\frac{df}{dt} = 0 \tag{3.11}$$

or to [see Eq. (2.1)]

$$\frac{\partial f}{\partial t} = -\sum_i \left(\frac{\partial f}{\partial q_i}\frac{\partial H}{\partial p_i} - \frac{\partial f}{\partial p_i}\frac{\partial H}{\partial q_i}\right) \tag{3.12}$$

The sum on the right-hand side has the form of a so-called "Poisson bracket" (of f and H, in this case).

According to Eq. (3.11), in the neighborhood of a phase point following its trajectory, the probability density f remains constant, or, in other words, the "probability fluid" occupying phase space is incompressible. An equivalent statement is that if p, q are the coordinates of a phase point at time $t_0 + s$ which at t_0 were p^0, q^0, then Liouville's theorem states that

$$f(p, q; t_0 + s) = f(p^0, q^0; t_0) \tag{3.13}$$

With the aid of the equations of motion of the phase point, p and q in Eq. (3.13) should be considered functions of the initial conditions p^0, q^0 and of the time. That is,

$$\begin{aligned} p &= p(p^0,q^0,s) \\ q &= q(p^0,q^0,s) \end{aligned} \tag{3.14}$$

Now, let us select a small element of volume at p^0, q^0 and time t_0. At time $t_0 + s$ the phase points originally (t_0) on the surface of this element of volume will have formed a new surface enclosing an element of volume of different shape at p, q. The element of volume at p, q must contain the same number of phase points as the original element of volume at p^0, q^0. This follows because a phase point outside or inside the element of volume can never cross the surface as the element moves through phase space, for otherwise there would be two different trajectories through the same point in phase space. But this is impossible in view of the uniqueness of the solution of the equations of motion of a phase point. As both the density and number of phase points in the element of volume are the same at p^0, q^0 and p, q, we conclude that although the shape of an element of volume is altered as it moves through phase space, its volume remains constant. This fact is expressed mathematically by saying that the Jacobian of the transformation in Eq. (3.14), from p^0, q^0 to p, q, is unity:[1]

$$\frac{\partial(p,q)}{\partial(p^0,q^0)} = 1 \tag{3.15}$$

Equation (3.12) shows explicitly how the value of the distribution function at any fixed point p, q changes with time. In particular, the masses and forces associated with the molecules of the system determine $H(p,q)$ and hence the $\partial H/\partial p_i$ and $\partial H/\partial q_i$, and the initially chosen f at t_0 determines the $\partial f/\partial q_i$ and $\partial f/\partial p_i$ at t_0. Thus Eq. (3.12) can in principle be integrated starting at t_0 to obtain the value of f at any p, q, and t, given $H(p,q)$ and the form of f at t_0.

[1] s is considered as only a parameter of the transformation; that is, different transformations, all with Jacobian unity, may be generated by taking different values of s.

4. Postulate on the Use of Ensemble Averages

In this section and the following one we shall indicate the correspondence used in statistical mechanics between time averages [for example, Eq. (2.3)] and ensemble averages. No completely general and rigorous proof of the validity of this correspondence is available; therefore we shall adopt a postulatory approach. The correctness of the postulates chosen may then be tested by comparing statistical mechanical predictions with experiment. As a matter of fact, many such comparisons have been made with complete success, so that one is justified a posteriori in having virtually complete confidence in the postulates chosen.

As in any logical system based on postulates, there are alternative but equivalent choices as to which ideas are to be considered fundamental. In the present section we introduce Postulate A, which is concerned with correlating time and ensemble averages. This postulate will then be employed to eliminate many possible specific choices of the form of the distribution function to be used in computing ensemble averages. In particular, for a system in equilibrium, we shall find that $f(p,q;t)$ must be independent of t. Postulate B will then complete the specification of the appropriate form of f by indicating its dependence on p and q.

Consider an experimental system in a given thermodynamic state. We measure the average value G_{obs} of a thermodynamic property $G(p,q)$ of this system, from time t_0 to $t_0 + \tau$, as indicated by Eq. (2.3), with the understanding, however, that in general the state is not one of perfect isolation and hence $p(t)$ and $q(t)$ in Eq. (2.3) are not associated with a single trajectory of a conservative system in phase space. Next, we construct at $t'(t' \leqslant t_0)$ an ensemble of systems with the same $N_1, \ldots, N_r$ and $x_1, \ldots, x_s$ as the experimental system, and with a theoretically chosen distribution function which we attempt to make representative at t' of the state of the experimental system.

Now for each (perfectly isolated) system of the ensemble we perform a measurement of the time average of $G(p,q)$ from t_0 to $t_0 + \tau$, giving

$$G_\tau = \frac{1}{\tau}\int_0^\tau G(p,q)\,ds = \frac{1}{\tau}\int_0^\tau G[p(p^0,q^0,s),\,q(p^0,q^0,s)]\,ds \tag{4.1}$$

where p, q is the dynamical state at time $t_0 + s$ of the system whose state was p^0, q^0 at time t_0. The value of G_τ depends of course on the initial state p^0, q^0, but we select τ of sufficient magnitude ("microscopically long") so that G_τ becomes independent of τ through the smoothing out of microscopic fluctuations. For example, instruments designed for the measurement of the pressure automatically record a time average over an interval of the order of magnitude of, say, a microsecond to a second. If the area of the surface on which the normal force is recorded is sufficiently small and the time resolution of the instrument sufficiently fine, fluctuations will of course be observed which are independent of the characteristics of the instrument. With sufficiently fine resolution, the force acting on unit area, therefore, ceases to be a macroscopic property. Thus, when we speak of pressure as a macroscopic property, we mean the normal force per unit area averaged over a macroscopic interval of time of such a magnitude that the recorded average is not sensibly dependent on the magnitude of the interval.

On the theoretical side, a rather general analysis of the proper magnitude of τ can be given[1] in both equilibrium and nonequilibrium cases. However, we merely remark here, as an illustration, that in the case of gases of such low density that only binary collisions are important, it is rather obvious that τ must be long relative to the duration of a representative collision.

We can now state *Postulate A: G_{obs} is identified with the ensemble average of G_τ, $\bar{G}_\tau$.* Explicitly,

$$\bar{G}_\tau(t_0) = \int f(p^0,q^0;t_0)G_\tau(p^0,q^0)\,dp^0\,dq^0 \tag{4.2}$$

where $f(p^0,q^0;t_0)$ is the distribution function at time t_0. We note immediately that it follows from this postulate that, since G_{obs} must be independent of time (i.e., we are concerned with an equilibrium state), $\bar{G}_\tau$ must also be independent of time. But Eq. (4.2) shows that in general $\bar{G}_\tau$ depends on the time t_0 at which the time averaging is commenced and that the necessary and sufficient condition for $\bar{G}_\tau$ to be independent of time is that the distribution function f be independent of time. That is, $\partial f/\partial t = 0$ (for equilibrium). Hence to be representative of any system in equilibrium, the distribution function originally chosen at time t' must be of such form that $\partial f/\partial t = 0$ [for example, it must satisfy Eq. (3.12) with $\partial f/\partial t = 0$]. Also, if the dependence of f on p and q at t' is "representative" of the state of the experimental system, this dependence will also be "representative" at all later times, since it does not change with time.

The validity of the identification of G_{obs} with $\bar{G}_\tau$ depends on the fluctuation in the value of G_τ found from different systems in the ensemble. This is obvious in the (limiting) case of an experimental system which is perfectly isolated, for the experimental system itself may then be regarded as typical of the (perfectly isolated) members of the ensemble. Then if these fluctuations are very small, the probability of any particular G_τ, for example G_{obs}, deviating appreciably from the mean, $\bar{G}_\tau$, is also very small. If the experimental system is not perfectly isolated, it spends, between t_0 and $t_0 + \tau$, varying and "microscopic" periods of time (between interactions with its surroundings) in a very large number of dynamical states, each one of which is essentially the same as that of some member ($\mathcal{N} \to \infty$) of the ensemble at the same time. Thus, G_{obs} is a composite of parts of various G_τ's. But

[1] J. G. Kirkwood, *J. Chem. Phys.*, **14**, 180, 347 (1946). One conclusion is that it is not safe, in general, to go so far as to use the limit $\tau \to \infty$.

again if the fluctuations of G_τ from $\bar{G}_\tau$ are very small, the probability of a composite of G_τ's deviating appreciably from $\bar{G}_\tau$ is also very small, unless the ensemble is very far from representative of the experimental system.[1]

The validity of the identification of G_{obs} with $\bar{G}_\tau$ is thus reduced to the question of whether the fluctuation of G_τ from $\bar{G}_\tau$ would indeed be small. We would expect this to be the case, especially because of the smoothing effect of time averaging. This conclusion is also dependent of course on the ensemble being reasonably representative of the experimental system.

The above argument does not pretend to be rigorous. In fact it is clear from Eq. (4.2) that the value of $\bar{G}_\tau$ to be identified with G_{obs} depends on the particular form of the distribution function. But the preceding paragraphs do lead us to expect that actually $\bar{G}_\tau$ would be quite insensitive to the form of f within certain limits. We shall see later (Chap. 4) that this is indeed the case: average values of thermodynamic properties are somewhat insensitive to f, but there is not this arbitrariness if we are interested in fluctuations in these properties.

Because of the at least formal dependence of $\bar{G}_\tau$ on f just mentioned, it is perhaps preferable to consider Postulates A and B (on the form of f) as two parts of a single fundamental postulate.

We return now to the requirement $\partial f/\partial t = 0$, which, when combined with Liouville's theorem, Eq. (3.13), gives us

$$f(p,q) = f(p^0,q^0) \tag{4.3}$$

That is, if we indicate the trajectory of a phase point by a curve in phase space, at a given time f has the same value at all points on the curve [not required by Eq. (3.13) alone] and this value does not change with time. In other words, f has the *same* constant value in the neighborhood of all phase points moving along the same trajectory. Thus, f depends only on the *trajectory* and is therefore a "constant of the motion."

Equation (4.3) makes possible a simplification in Eq. (4.2). From Eqs. (4.1) and (4.2) we have

$$\bar{G}_\tau = \int f(p^0,q^0;t_0)\left\{\frac{1}{\tau}\int_0^\tau G[p(p^0,q^0,s),q(p^0,q^0,s)]\,ds\right\}dp^0\,dq^0 \tag{4.4}$$

But we can write $f(p^0,q^0;t_0) = f(p,q)$. Also, let us change variables of integration from p^0,q^0 to p,q. The Jacobian of this transformation is unity [Eq. (3.15)], so that $dp^0\,dq^0 = dp\,dq$. Then

$$\bar{G}_\tau = \int f(p,q)\left[\frac{1}{\tau}\int_0^\tau G(p,q)\,ds\right]dp\,dq \tag{4.5}$$

[1] It is clear that the weighting of the various systems of the ensemble, in the composite making up G_{obs}, as $\tau \to$ large, would be the same weighting as that given by the distribution function, if the distribution function were a correct representation of the experimental system.

Now interchanging the order of integration,

$$\bar{G}_\tau = \frac{1}{\tau}\int_0^\tau [\int f(p,q)G(p,q)\, dp\, dq]\, ds = \frac{1}{\tau}\int_0^\tau \bar{G}\, ds = \bar{G} \tag{4.6}$$

since the ensemble average of G, $\bar{G}$, is independent of time in this case (f independent of t). Hence we may replace $\bar{G}_\tau$ by $\bar{G}$ in Postulate A, if desired. Indeed, this identification of G_{obs} with $\bar{G}$ is the standard procedure. Use has been made here of $\bar{G}_\tau$ instead of $\bar{G}$ because the connection between G_{obs} and $\bar{G}_\tau$ is closer, both involving time averages. Also, it is possible to use a similar approach in nonequilibrium statistical mechanics.[1]

5. Postulate on the Form of the Distribution Function

In order to complete the postulatory basis of classical statistical mechanics, particular choices of the distribution function f must be indicated. We have already found in the preceding section that the introduction of ensemble averages in Postulate A together with the stationary value of G_{obs} in an equilibrium state require that f be independent of time [using, for example, Eq. (3.12) with $\partial f/\partial t = 0$ as a criterion]. We now supplement this restriction in possible forms of f by an argument that very strongly suggests further that $f(p,q)$ should be chosen as some function of $H(p,q)$, that is, $f = f[H(p,q)]$.

Since f is a constant of the motion, it may be considered a function of the dynamical invariants, the usual dynamical constants (or integrals) of the motion. Further, for $f(p,q)$ to be single-valued and continuous, it can depend only on the "uniform" integrals of the motion, those which are single-valued, continuous functions of p, q. Such integrals are usually very few, merely energy and perhaps certain momenta or angular momenta. For example, if we have a gas in a circular cylindrical container with smooth walls, then the component of angular momentum about the axis of the cylinder is an integral of the motion, for the external forces (the forces associated with the wall) do not "destroy" this integral (i.e., this component of momentum is conserved in a collision of a molecule with the wall). On the other hand, if the wall is rough or the vessel lacks such symmetry, then each collision with the wall changes the angular momentum of the system about a given axis, so that it is no longer an integral of the motion. Thus, in actual systems confined in containing vessels, the energy is usually the only uniform integral of the motion. We shall therefore postulate, following Gibbs, that f depends only on H in an equilibrium ensemble.

There are many different ways of expressing the state of a thermodynamic system, depending on the particular choice of $r + s + 1$ thermodynamic variables (see Sec. 1). With the type of ensemble (called a *petit* ensemble by Gibbs) we have discussed so far, it is required that the $r + s$ variables

[1] J. G. Kirkwood, *loc. cit.*

$N_1, \ldots, N_r, x_1, \ldots, x_s$ be chosen (see Sec. 3). There remains the selection of the last independent variable, and for each choice a particular form of f is appropriate. We shall consider two cases here.[1]

We now state Postulate B and follow this with a discussion of the postulate.

Postulate B. (*a*) *Microcanonical ensemble.* For a closed, isolated thermodynamic system, i.e., a system with assigned values for the independent variables $E, N_1, \ldots, N_r, x_1, \ldots, x_s$,

$$\begin{aligned} f &= \text{const} \qquad H(p,q) = E \text{ to } E + \delta E \\ &= 0 \qquad\qquad \text{otherwise} \end{aligned} \tag{5.1}$$

where δE is a very small range in E.

(*b*) *Canonical ensemble.* For a closed, isothermal thermodynamic system, i.e., a system with assigned values for the independent variables $T, N_1, \ldots, N_r, x_1, \ldots, x_s$,

$$f = \text{const} \times e^{-\beta H(p,q)} \tag{5.2}$$

where β is a constant.

Equation (5.1), representing the microcanonical ensemble, recognizes the fact that it is impossible for an actual system to be perfectly isolated and hence that the energy can be specified only within some very small range[2] δE. However, in an idealized conservative system in classical mechanics, there is no difficulty in principle in letting $\delta E \to 0$. The situation with regard to this limit is rather different in quantum mechanics (see Chap. 2).

It will be noted that Eq. (5.1) is really the only possible choice for an ensemble each system of which has the energy E, if we accept the argument above leading to the conclusion that in general $f = f(H)$. However, although this argument is rather convincing, it is not rigorous; therefore Eq. (5.1) has the status of a postulate the validity of which remains to be demonstrated.

Although Eq. (5.1), as we have seen, rather obviously represents a closed, isolated system, it is not clear offhand that Eq. (5.2) is appropriate for a closed, isothermal system. However, if we accept Eq. (5.1) as correct, it is then possible to deduce Eq. (5.2). That is, if Eq. (5.1) describes the proper representative ensemble for a closed, isolated system, then Eq. (5.2) describes the representative ensemble for a closed, isothermal system. This deduction is essentially the same in both classical and quantum theory, but is more natural in quantum language. For this reason, and also because classical mechanics can in any case be regarded as a limiting form of quantum mechanics, we defer the derivation of the canonical ensemble from the microcanonical ensemble until Chap. 2. The significance of the constant β will be deduced

[1] A general discussion of the possible choices of independent variables is given in Chap. 3.

[2] Actually, a similar remark can be made about other thermodynamic variables.

through Γ_2, the fraction of time spent in each of these two regions is equal, for $\tau \to$ large.

Next, we assume that if Γ_2 is sufficiently small and Γ_3 is another element of volume located about the center of gravity of Γ_2, then

$$\frac{\text{Fraction of time spent in } \Gamma_2}{\text{Fraction of time spent in } \Gamma_3} = \frac{\text{volume of } \Gamma_2}{\text{volume of } \Gamma_3}$$

independent of the shape of Γ_3. Actually, this would follow from the alternative assumption that the fractional time spent in the neighborhood of a point is a continuous function of the position of that point.

So far we have that for any elements of volume surrounding points on the trajectory of the phase point (whether or not these elements are transforms of each other under motion), the fractions of time spent in these elements are proportional to their volumes. These would include all elements of the energy shell if and only if the trajectory passes arbitrarily close to any point in its energy surface. The hypothesis that this is true is known as the "quasi-ergodic hypothesis," rather general proofs of which have been given by von Neumann and Birkhoff. The older "ergodic hypothesis," that the phase point passes through every point in its energy surface, is certainly false, for the phase path can intersect a $2n - 2$ dimensional "cross section" of the $2n - 1$ dimensional surface at most a denumerable number of times while the number of points in this cross section is nondenumerable.

GENERAL REFERENCES

Fowler, R. H., "Statistical Mechanics" (Cambridge, London, 1936).

Gibbs, J. W., "The Collected Works of J. Willard Gibbs" (Yale, New Haven, 1948), Vol. II.

ter Haar, D., *Revs. Mod. Phys.*, **27**, 289 (1955).

Khinchin, A. I., "Mathematical Foundations of Statistical Mechanics" (Dover, New York, 1949).

Tolman, R. C., "The Principles of Statistical Mechanics" (Oxford, London, 1938).

CHAPTER 2

PRINCIPLES OF QUANTUM STATISTICAL MECHANICS

8. Review of Quantum Mechanics

For the convenience of the reader, we give in this section a brief review of those aspects of quantum mechanics which will be made use of in the remainder of the book. The postulatory foundation (actually, one of many possible choices) of quantum mechanics will be stated first, followed by a summary of deductions (omitting a number of proofs[1]) which can be made from the postulates. We consider conservative systems only, with Hamiltonian $H(p,q)$ (the system is perfectly isolated).

Postulatory Foundation. In quantum mechanics the precise classical specification of dynamical state (i.e., exact values of p and q at t) must be abandoned; instead, the state of the system is described completely by a wave function $\psi(q,t)$. It is not possible to deduce any details about the behavior of the system which are not implicit in $\psi(q,t)$.

Each dynamical variable $G(p,q,t)$ is associated in quantum mechanics with an operator $\mathscr{G}$. Operators must be chosen so that, for any pair of dynamical variables F and G,

$$\mathscr{F}\mathscr{G} - \mathscr{G}\mathscr{F} = -\frac{\hbar}{i}[\mathscr{F},\mathscr{G}] \tag{8.1}$$

where $\hbar = h/2\pi$ and $[\mathscr{F},\mathscr{G}]$ is the operator associated with the classical Poisson bracket [compare Eq. (3.12)],

$$[F,G] = \sum_k \left(\frac{\partial F}{\partial q_k}\frac{\partial G}{\partial p_k} - \frac{\partial F}{\partial p_k}\frac{\partial G}{\partial q_k}\right) \tag{8.2}$$

It can be shown[2] that putting $(\hbar/i)(\partial/\partial q_k)$ in place of p_k everywhere it occurs

[1] For proofs, see, for example, Kemble, Tolman, Rojansky, Schiff, or Bohm (Gen. Ref.).

[2] This may be accomplished by expanding F and G as power series in p. Also a suitable convention [to satisfy Eq. (8.1)] must be adopted to avoid ambiguity in writing the operator for a variable of the form $R(p)S(q)$. See General References.

in F and G will in fact lead to operators $\mathscr{F}[(\hbar/i)\partial/\partial q,q,t]$ and $\mathscr{G}[(\hbar/i)\partial/\partial q,q,t]$ satisfying Eq. (8.1). We adopt this particular recipe (due to Schrödinger) for finding operators.

If a measurement of $G(p,q,t)$ is made on a system in the state $\psi(q,t)$ at t, the only values of G which might be observed are the eigenvalues g_i of the operator $\mathscr{G}$. That is, if a suitable function $\varphi_i(q,t)$ is found which satisfies the equation (in general, a differential equation)

$$\mathscr{G}\left(\frac{\hbar}{i}\frac{\partial}{\partial q},q,t\right)\varphi_i(q,t) = g_i(t)\varphi_i(q,t) \tag{8.3}$$

then (the eigenvalue) $g_i(t)$ is a possible experimental value of G at t and φ_i is called an eigenfunction belonging to the eigenvalue g_i. If two or more independent eigenfunctions belong to the same eigenvalue, the eigenvalue is said to be degenerate. The really important case for our purposes is $G = G(p,q)$, for which

$$\mathscr{G}\left(\frac{\hbar}{i}\frac{\partial}{\partial q},q\right)\varphi_i(q) = g_i\varphi_i(q) \tag{8.4}$$

where g_i is a constant. In the more general equation (8.3) we may adopt the point of view that t is a parameter and an equation of the form (8.4) is to be solved for each value of the parameter t. For simplicity, we shall generally refer below to $G = G(p,q)$ with the understanding that if G is an explicit function of t as well, the treatment may easily be generalized as indicated above.

If we have a number $\mathscr{N}$ of identical systems, with $\mathscr{N} \to \infty$, all in the state $\psi(q,t)$ at t, and if a measurement of $G(p,q)$ is made at t on all of these systems,[1] the average value $\bar{G}$ found is given by

$$\bar{G}(t) = \frac{\int \psi^*(q,t)\mathscr{G}[(\hbar/i)\partial/\partial q,q]\psi(q,t)dq}{\int \psi^*(q,t)\psi(q,t)\,dq} \tag{8.5}$$

where the asterisk means complex conjugate. $\bar{G}$ is usually called the expectation value of G for a system in the state ψ.

The way in which the quantum-mechanical state of the system $\psi(q,t)$ changes with time is determined by

$$\mathscr{H}\psi = i\hbar\frac{\partial\psi}{\partial t} \tag{8.6}$$

where $\mathscr{H}[(\hbar/i)\partial/\partial q,\ q]$ is the operator associated with the Hamiltonian function of the system $H(p,q)$. Thus, the particular dynamical variable $H(p,q)$ plays a special role in quantum mechanics [as it does in classical mechanics; see, for example, Eq. (2.1)].

[1] Note the resemblance to a Gibbsian ensemble. See Sec. 9.

To represent a physically realizable (or "accessible") state, the wave function $\psi(q,t)$ must be of such form that if all the coordinates associated with two identical molecules of the system are interchanged in $\psi(q,t)$, (1) $\psi(q,t)$ changes sign but is otherwise unaffected if each of the two molecules is made up of an odd number of neutrons + protons + electrons, and (2) $\psi(q,t)$ is unaffected, even with regard to sign, if each of the two molecules is made up of an even number of neutrons + protons + electrons. In other words, the wave function must be symmetrical in the coordinates of identical molecules of type 2 and antisymmetrical in the coordinates of molecules of type 1. The Pauli exclusion principle is a special case (for a system of electrons) of type 1.

Deductions from the Postulates. For convenience, we use below normalized wave functions ψ; that is,

$$\int \psi^*(q,t)\psi(q,t)\, dq = 1 \tag{8.7}$$

For a system confined to a container (as in conventional statistical mechanical applications), ψ will always be normalizable (or "quadratically integrable"). In cases where ψ is not normalizable (e.g., a "free" particle), the same finite limits are used in both integrals of Eq. (8.5); then, to obtain $\bar{G}$, the limit of the quotient is taken as the limits on the integrals are allowed simultaneously to become infinite.

On physical grounds $\bar{G}$ must be real for arbitrary ψ. It can be proved[1] that to ensure the reality of $\bar{G}$, $\mathcal{G}$ must be an Hermitian operator. That is, $\mathcal{G}$ must satisfy

$$\int \psi^*(\mathcal{G}\varphi)\, dq = \int \varphi(\mathcal{G}\psi)^*\, dq \tag{8.8}$$

where ψ and φ are arbitrary. Then it follows that the eigenvalues of $\mathcal{G}$ are real, as is also required physically. Thus, using the notation of Eq. (8.4),

$$\begin{aligned} \int \varphi_i^* \mathcal{G} \varphi_i\, dq &= g_i \int \varphi_i^* \varphi_i\, dq = g_i \\ &= \int \varphi_i (\mathcal{G}\varphi_i)^*\, dq = g_i^* \end{aligned} \tag{8.9}$$

Since $g_i^* = g_i$, g_i is real.

Let φ_i and φ_j be two eigenfunctions of $\mathcal{G}$:

$$\mathcal{G}\varphi_i = g_i\varphi_i \qquad \mathcal{G}\varphi_j = g_j\varphi_j \tag{8.10}$$

Then

$$\begin{aligned} \int \varphi_j^* \mathcal{G} \varphi_i\, dq &= g_i \int \varphi_j^* \varphi_i\, dq \\ &= \int \varphi_i (\mathcal{G}\varphi_j)^*\, dq = g_j \int \varphi_j^* \varphi_i\, dq \end{aligned} \tag{8.11}$$

That is,

$$(g_i - g_j)\int \varphi_j^* \varphi_i\, dq = 0 \tag{8.12}$$

Hence, if φ_i and φ_j belong to different eigenvalues ($g_i \neq g_j$), the two functions are orthogonal:

$$\int \varphi_j^* \varphi_i\, dq = 0 \tag{8.13}$$

[1] See General References.

Also, it is always possible[1] to choose (in an infinite number of ways) linear combinations of independent eigenfunctions belonging to the same eigenvalue in such a way that the linear combinations are orthogonal. It will thus be understood below that the eigenfunctions of $\mathscr{G}$ form an orthonormal set (except in nonnormalizable cases). It can be shown[2] further that an arbitrary normalizable wave function can be expanded as an infinite series in such an orthonormal set of functions [see, for example, Eq. (8.21) below].

The eigenfunctions associated with the variables q, p, and $H(p,q)$ are of particular importance. For a coordinate q_k, Eq. (8.4) becomes

$$q_k\psi = q'_k\psi \tag{8.14}$$

where ψ is an eigenfunction and q'_k an eigenvalue of q_k. Hence $(q_k - q'_k)\psi = 0$. The solution of this equation is the Dirac δ function,

$$\psi = \delta(q_k - q'_k) = \begin{cases} 0 & q_k \neq q'_k \\ \infty & q_k = q'_k \end{cases} \tag{8.15}$$

with

$$\int_{-\infty}^{+\infty} \delta(q_k - q'_k)\, dq_k = 1 \tag{8.16}$$

as part of the definition of $\delta(q_k - q'_k)$. Thus, *any* real number q'_k is a nondegenerate eigenvalue of q_k and the eigenfunction belonging to q'_k is $\delta(q_k - q'_k)$. For a component of momentum p_k, Eq. (8.4) reads

$$\frac{\hbar}{i}\frac{\partial\psi}{\partial q_k} = p'_k\psi \tag{8.17}$$

where p'_k is an eigenvalue. The solution of Eq. (8.17) is

$$\psi(q_k) = Ae^{ip'_kq_k/\hbar} \tag{8.18}$$

which is a suitable wave function for *any* real number p'_k (a nondegenerate eigenvalue). A is a constant; the momentum eigenfunctions can be normalized by the use of special methods.[3]

Thus quantum mechanics leads to a continuous "spectrum" of possible experimental values of q and p, as in classical mechanics. On the other hand, the possible experimental values of the energy, the energy "levels" E_i, are often "discrete" or "quantized." The E_i are the eigenvalues of the operator $\mathscr{H}$,

$$\mathscr{H}\psi_i = E_i\psi_i \tag{8.19}$$

When written out in the form of a partial differential equation, Eq. (8.19) is, of course, the well-known Schrödinger equation. In addition to the possible

[1] See Tolman (Gen. Ref.), p. 250.
[2] See Kemble (Gen. Ref.), Chap. 4.
[3] See Schiff (Gen. Ref.), pp. 49–51.

discreteness of eigenvalues already mentioned, energy eigenvalues and eigenfunctions differ from those of coordinates and momenta in another important respect: the solution of Eq. (8.19) depends on the "nature" of the system, that is, on the number and masses of the particles of the system and on the potential energy expression $U(q)$. Hence, unlike Eqs. (8.14) and (8.17), each problem, as defined by $H(p,q)$, requires special treatment.

Suppose, at $t = 0$, a system is in an energy eigenstate (i.e., a state represented by an energy eigenfunction) $\psi_i(q)$ belonging to the eigenvalue E_i. How does the state of this system change with time? Consider the function

$$\psi_i(q,t) = \psi_i(q)e^{-iE_it/\hbar} \tag{8.20}$$

$\psi_i(q,t)$ reduces to $\psi_i(q)$ at $t = 0$ and satisfies Eq. (8.6) for all t. Equation (8.20) therefore gives the desired time dependence. We may note also that $\psi_i(q,t)$ satisfies Eq. (8.19) for all t, and hence a (perfectly isolated) system in an energy eigenstate always remains in an energy eigenstate belonging to the same energy eigenvalue. Next, consider a system in a state $\psi(q)$ at $t = 0$ which is not necessarily an energy eigenstate. The time dependence is more complicated in this more general case. First, we expand[1] $\psi(q)$ in the orthonormal energy eigenfunctions $\psi_i(q)$:

$$\psi(q) = \sum_i c_i\psi_i(q) \tag{8.21}$$

$$c_i = \int \psi_i^*(q)\psi(q)\,dq \tag{8.22}$$

Then it is easy to verify that the required time-dependent wave function which reduces to Eq. (8.21) at $t = 0$ and satisfies Eq. (8.6) for all t is

$$\psi(q,t) = \sum_i c_i\psi_i(q)e^{-iE_it/\hbar} \tag{8.23}$$

Thus, the way in which the state of a quantum-mechanical system changes with time is determined by an initial state ($\psi(q)$) and by the nature of the Hamiltonian function for the system, that is, the ψ_i and E_i follow from $H(p,q)$. Incidentally, if $\psi(q)$ is normalized, $\psi(q,t)$ is also normalized for any t, for

$$\begin{aligned}\int \psi^*(q,t)\psi(q,t)\,dq &= \int \Big(\sum_i c_i^*\psi_i^*e^{iE_it/\hbar}\Big)\Big(\sum_j c_j\psi_je^{-iE_jt/\hbar}\Big)\,dq \\ &= \sum_i c_i^*c_i \\ &= \int \psi^*(q)\psi(q)\,dq = 1\end{aligned} \tag{8.24}$$

Our postulates state that the only possible results of an experimental

[1] It will always be understood that $\sum_i$ is to be replaced by $\int di$ for those values of i where the eigenvalue spectrum is continuous.

measurement of $G(p,q)$ are the eigenvalues g_k of $\mathscr{G}$, and that $\bar{G}(t)$, which must be an appropriately weighted average of the g_k, is given by Eq. (8.5) for a system in the state $\psi(q,t)$. We now ask the more detailed question: for a system in the state $\psi(q,t)$ at t, if a measurement of G is made, what is the *probability* of observing some particular eigenvalue g_k? If this probability is $P_k(t)$ we have, for example,

$$\bar{G}(t) = \sum_k P_k(t) g_k \tag{8.25}$$

In fact, for any positive integer m,

$$\overline{G^m}(t) = \sum_k P_k(t) g_k{}^m \tag{8.26}$$

By using suitable expansions, we shall now obtain a more explicit expression of the same form as Eq. (8.26), which by comparison with Eq. (8.26), will give us $P_k(t)$. Let us first expand $\psi(q,t)$ as in Eqs. (8.22) and (8.23). Now suppose the eigenfunctions of $\mathscr{G}$ are the $\varphi_k(q)$; that is, $\mathscr{G}\varphi_k(q) = g_k\varphi_k(q)$. Then, $\psi(q,t)$ can be expanded in these functions as follows:

$$\psi(q,t) = \sum_k A_k(t)\varphi_k(q) \tag{8.27}$$

$$\begin{aligned} A_k(t) &= \int \varphi_k^*(q)\psi(q,t)\, dq = \int \varphi_k^*\Big(\sum_i c_i\psi_i e^{-iE_it/\hbar}\Big)\, dq \\ &= \sum_i c_i a_{ki} e^{-iE_it/\hbar} \end{aligned} \tag{8.28}$$

where

$$a_{ki} = \int \varphi_k^*(q)\psi_i(q)\, dq \tag{8.29}$$

In this way [Eq. (8.27)] any time-dependent wave function can be expanded in the eigenfunctions associated with an arbitrary dynamical variable $G(p,q)$. Of course, if $G = G(p,q,t)$, the g_k and a_{ki} are functions of time, and $\varphi_k = \varphi_k(q,t)$. The state $\psi(q,t)$ is said to be represented here as a superposition of G states. Finally, we use Eq. (8.27) to obtain $\overline{G^m}(t)$:

$$\begin{aligned} \overline{G^m}(t) &= \int \psi^*(q,t)\mathscr{G}^m\psi(q,t)\, dq \\ &= \int \Big(\sum_j A_j^*\varphi_j^*\Big)\mathscr{G}^m\Big(\sum_k A_k\varphi_k\Big)\, dq \\ &= \int \Big(\sum_j A_j^*\varphi_j^*\Big)\Big(\sum_k A_k g_k{}^m\varphi_k\Big)\, dq \\ &= \sum_k A_k^*(t)A_k(t)g_k{}^m = \sum_k |A_k(t)|^2 g_k{}^m \end{aligned} \tag{8.30}$$

Thus we have the result

$$P_k(t) = A_k^*(t)A_k(t) = |A_k(t)|^2 \tag{8.31}$$

where $A_k(t)$ is given by Eq. (8.28). Naturally, the probability of observing a degenerate eigenvalue at t is the sum over $|A_k(t)|^2$ for all φ_k belonging to the eigenvalue.

In summary, we find in quantum mechanics that only the eigenvalues of the operator associated with a dynamical variable can be observed experimentally, and further that in general which particular eigenvalue will actually be observed at t cannot be predicted with certainty. Instead, one can only state that there is a *probability* $P_k(t)$ of observing the eigenvalue g_k. The procedure in calculating $P_k(t)$ for a variable G is to expand the state of the system $\psi(q,t)$ in the eigenfunctions of $\mathscr{G}$ [Eq. (8.27)] and use the coefficients in the expansion to find $P_k(t)$ [Eq. (8.31)].

It should be observed from Eq. (8.28) that $P_k(t)$ depends on $\psi(q)$ at $t = 0$ through the c_i, on $H(p,q)$ for the system through the c_i, the E_i, and the a_{ki}, and of course on the dynamical variable $G(p,q)$ being discussed through the a_{ki}.

There are a number of important special consequences of Eq. (8.31) which are easy to deduce. (1) If $\psi(q)$ at $t = 0$ is an eigenfunction of $\mathscr{G}$, say $\psi(q) = \varphi_k(q)$, then $P_k(0) = 1$. That is, if the wave function representing the state of the system at the time of measurement is an eigenfunction associated with the dynamical variable being measured, then the result of the measurement is certain to be the eigenvalue to which the eigenfunction belongs. In this special case, the probability distribution function is "sharp," at one particular eigenvalue. However, for $t > 0$ (or $t < 0$), $P_k(t)$ is in general [see (2)] no longer sharp; hence other eigenvalues may be observed if the measurement is made at $t > 0$ instead of $t = 0$. (2) If $\psi(q)$ at $t = 0$ is an energy eigenfunction, say $\psi(q) = \psi_i(q)$, the energy probability distribution function is sharp at E_i not only at $t = 0$ but for all t. The probability distribution functions $P_k(0)$ for other dynamical variables are in general (see below) not sharp at $t = 0$, but they are independent of time [unless $G = G(p,q,t)$]: $P_k(0) = P_k(t)$. Since probability distribution functions for all variables $G(p,q)$ are stationary in time in this special case, energy eigenstates are often called "stationary states." (3) If $\psi(q)$ at $t = 0$ is an arbitrary wave function, the energy probability distribution function is in general not sharp at $t = 0$, but it is independent of time. This is the quantum-mechanical analogue for a conservative system of the conservation of energy in classical mechanics. If we use averages, $\bar{H} = \bar{T} + \bar{U} =$ constant. The bars are omitted in classical mechanics. (4) If the state of a system is $\psi(q,t)$ at t, what is the probability distribution function in some coordinate, say q_j? We give a few details here because this case is less straightforward than the others. First consider a dynamical variable F which is a function of only one coordinate and its conjugate momentum, say q_1 and p_1. Let the eigenfunctions of $\mathscr{F}$ be the $u_k(q_1)$, normalized according to

$$\int u_k^* u_k \, dq_1 = 1 \tag{8.32}$$

For a system in the state $\psi(q_1, q_2, \ldots, q_n, t)$, we have the expansion

$$\psi(q_1, q_2, \ldots, q_n, t) = \sum_k A_k(q_2, \ldots, q_n, t)u_k(q_1) \tag{8.33}$$

$$A_k = \int u_k^* \psi \, dq_1$$

where ψ is normalized according to

$$\int \psi^* \psi \, dq_1 \, dq_2 \cdots dq_n = 1 \tag{8.34}$$

That is, from Eq. (8.33),

$$\sum_k \int A_k^* A_k \, dq_2 \cdots dq_n = 1 \tag{8.35}$$

Now

$$\begin{aligned} \overline{F^m}(t) &= \int \psi^* \mathscr{F}^m \psi \, dq_1 \cdots dq_n \\ &= \sum_k f_k{}^m \int A_k^* A_k \, dq_2 \cdots dq_n \end{aligned} \tag{8.36}$$

therefore in this case

$$P_k(t) = \int A_k^* A_k \, dq_2 \cdots dq_n \tag{8.37}$$

In particular, let us choose $F = q_1$. Then Eq. (8.33) becomes

$$\psi(q_1, q_2, \ldots, q_n, t) = \int A(q_1'; q_2, \ldots, q_n, t)\delta(q_1 - q_1') \, dq_1' \tag{8.38}$$

where summation over k has been replaced by integration over the continuous spectrum of eigenvalues q_1'. The right-hand side of Eq. (8.38) is just $A(q_1; q_2, \ldots, q_n, t)$ so that

$$A(q_1; q_2, \ldots, q_n, t) = \psi(q_1, q_2, \ldots, q_n, t) \tag{8.39}$$

The probability of observing q_1 between q_1 and $q_1 + dq_1$ is, from Eq. (8.37), dropping the prime on q_1',

$$P(q_1,t) \, dq_1 = \int \psi^* \psi \, dq_2 \cdots dq_n \tag{8.40}$$

A similar result can of course be obtained for any coordinate q_j, so it is clear that

$$\psi^* \psi \, dq_1 \, dq_2 \cdots dq_n = \psi^* \psi \, dq \tag{8.41}$$

is the probability of *simultaneously* finding q_1 between q_1 and $q_1 + dq_1$, . . . , and q_n between q_n and $q_n + dq_n$ on measuring all coordinates at t. This probability description of the configuration of a dynamical system replaces the precise description of configuration possible in classical mechanics.

The state of a system at t is described by a wave function $\psi(q,t)$, from which we can obtain the configuration probability distribution $\psi^*\psi$. If, instead of knowing ψ, our only information about the system is in the form of the configuration probability distribution, $\psi^*\psi = |A|^2$, at t_0, the state $\psi(q,t_0)$ of the system is still unspecified to the extent of an arbitrary phase function $\gamma(q)$. That is,

$$\psi(q,t_0) = |A(q)|e^{i\gamma(q)} \tag{8.42}$$

If $\psi(q,t)$ is expanded in the eigenfunctions $\varphi_k(q)$ of some arbitrary $\mathscr{G}$ as in Eq. (8.27), the set of functions $A_k(t)$ furnish an alternative and completely equivalent description of the quantum-mechanical state $\psi(q,t)$. If our only information about the system is in the form of the probability distribution function in G, $|A_k|^2$, at t_0, this information can be expressed by writing for the state ψ,

$$\psi(q,t_0) = \sum_k |A_k|e^{i\gamma_k}\varphi_k(q) \tag{8.43}$$

where the γ_k are arbitrary phase factors.

Even when G is a function of t, $G = G(p,q,t)$, it is of course still possible to expand $\psi(q,t)$ in a set of time-independent orthonormal eigenfunctions of $\mathscr{G}$, for example, the $\varphi_k(q,0)$. Thus

$$\psi(q,t) = \sum_k B_k(t)\varphi_k(q,0) \tag{8.44}$$

instead of

$$\psi(q,t) = \sum_k A_k(t)\varphi_k(q,t) \tag{8.45}$$

where the *full* time dependence in Eq. (8.44) is contained in the $B_k(t)$. Here, $P_k(t) = |A_k(t)|^2 \neq |B_k(t)|^2$.

It has been mentioned above that precise knowledge (a "sharp" probability distribution) concerning the value of some variable G at t is available if the state of the system is an eigenstate of G at t. It is then natural to ask whether it is possible to have such precise knowledge concerning two variables F and G simultaneously. Suppose the state of the system is $\psi(q,t_0)$ at t_0. For the probability distributions in both F and G to be sharp, $\psi(q,t_0)$ must be a "simultaneous eigenfunction" of both $\mathscr{F}$ and $\mathscr{G}$. That is,

$$\mathscr{F}\psi(q,t_0) = f'\psi(q,t_0) \qquad \mathscr{G}\psi(q,t_0) = g'\psi(q,t_0) \tag{8.46}$$

where f' and g' are eigenvalues. Then

$$\mathscr{G}\mathscr{F}\psi = \mathscr{F}\mathscr{G}\psi = f'g'\psi$$

or

$$\mathscr{G}\mathscr{F} - \mathscr{F}\mathscr{G} = 0 \tag{8.47}$$

Two operators satisfying the above operator equation are said to commute. The possibility of simultaneous precise information about two dynamical variables therefore implies that the associated operators commute. The converse theorem can also be proved.[1] In particular, it is simple to verify that

[1] See General References.

the pairs of operators associated with q_j and q_k, and with p_j and p_k, commute, while the operators associated with q_k and p_j commute if $k \neq j$ but do not commute if $k = j$. Precise simultaneous knowledge of both q_k and p_k is therefore impossible. This statement is, of course, closely related to the familiar Heisenberg uncertainty relation, which can be derived from the postulates adopted here.[1]

If, at a certain time t_0, the precise configuration $q_1', \ldots, q_n'$ of the system is known, the wave function expressing this is a simultaneous eigenfunction of $q_1, \ldots, q_n$:

$$\psi(q_1, \ldots, q_n, t_0) = \delta(q_1 - q_1')\delta(q_2 - q_2') \cdots \delta(q_n - q_n') \tag{8.48}$$

If, on the other hand, the components of momenta are known to have the values p_k', then

$$\psi(q_1, \ldots, q_n, t_0) = \text{const} \times e^{i(p_1'q_1 + \cdots + p_n'q_n)/\hbar} \tag{8.49}$$

where, in this case, ψ is a simultaneous eigenfunction of all the momentum operators.

Vector and Matrix Language. We indicate very briefly here how the above discussion may be recast into the equivalent and frequently used language of (infinite) vectors and matrices. Consider the wave function $\psi(q,t)$ expanded in the eigenfunctions $\varphi_k(q)$ associated with an arbitrary dynamical variable $G(p,q)$,

$$\psi(q,t) = \sum_k A_k(t)\varphi_k(q) \tag{8.50}$$

$$A_k(t) = \int \varphi_k^* \psi \, dq$$

Now ψ may be regarded as a "vector" in function (or Hilbert) space, where the φ_k play the role of "unit" or "basic" vectors and the A_k are the "components" of ψ along the various "axes" of function space. The same vector ψ will of course be represented by a different set of components A_k if an alternative choice of basic vectors is made. A vector ψ is usually represented by a column symbol:

$$\begin{pmatrix} A_1 \\ A_2 \\ \cdot \\ \cdot \\ \cdot \end{pmatrix}$$

Now let $\mathscr{F}$ be the operator associated with the dynamical variable $F(p,q)$. The wave function (or vector) $\psi(q,t)$ is transformed into a new function (or vector), say $\chi(q,t)$, by $\mathscr{F}$; that is, $\mathscr{F}\psi = \chi$. Let us now find the components

[1] See General References.

C_k of χ in terms of the original set of basic vectors φ_k. We have, from Eq. (8.50),

$$\mathscr{F}\psi = \chi = \sum_j A_j \mathscr{F}\varphi_j \tag{8.51}$$

But we may expand $\mathscr{F}\varphi_j$ in the set of functions φ_k:

$$\mathscr{F}\varphi_j = \sum_k F_{kj}\varphi_k$$

where
$$F_{kj} = \int \varphi_k^* \mathscr{F}\varphi_j \, dq \tag{8.52}$$

The F_{kj} will be functions of t if either F or G depends on t. Then Eq. (8.51) becomes

$$\begin{aligned} \mathscr{F}\psi = \chi &= \sum_j A_j \sum_k F_{kj}\varphi_k \\ &= \sum_k \Big(\sum_j A_j F_{kj}\Big)\varphi_k \end{aligned} \tag{8.53}$$

Thus the components of χ are

$$C_k(t) = \sum_j A_j(t) F_{kj} \tag{8.54}$$

But the linear transformation exhibited in Eq. (8.54) has just the form encountered in the multiplication of a column symbol (vector) by a matrix to give a second column symbol:

$$\begin{pmatrix} F_{11} & F_{12} & F_{13} & \cdots \\ F_{21} & F_{22} & F_{23} & \cdots \\ F_{31} & F_{32} & F_{33} & \cdots \\ \cdot & \cdot & \cdot & \cdots \end{pmatrix} \begin{pmatrix} A_1 \\ A_2 \\ A_3 \\ \cdot\,\cdot \end{pmatrix} = \begin{pmatrix} A_1F_{11} + A_2F_{12} + \cdots \\ A_1F_{21} + A_2F_{22} + \cdots \\ A_1F_{31} + A_2F_{32} + \cdots \\ \cdots\cdots\cdots \end{pmatrix} \tag{8.55}$$

Equation (8.55) states that $\mathscr{F}\psi = \chi$; therefore if the wave functions ψ and χ are regarded as vectors, the operator $\mathscr{F}$ is represented by a matrix with elements given by Eq. (8.52), $\mathscr{F} = \|F_{kj}\|$.

The same operator $\mathscr{F}$ may of course be represented by different matrices, depending on the choice of basic vectors [see Eq. (8.52)]. To illustrate this, suppose we change the basic vectors above from the φ_k to the u_k. We employ the expansions

$$\varphi_i = \sum_k \alpha_{ki} u_k \tag{8.56}$$

$$u_k = \sum_i \beta_{ik}\varphi_i \tag{8.57}$$

where
$$\alpha_{ki} = \int u_k^* \varphi_i \, dq \qquad \beta_{ik} = \int \varphi_i^* u_k \, dq = \alpha_{ki}^*$$

If we use Eq. (8.57) for u_k in Eq. (8.56), we find the useful relation

$$\delta_{ni} = \sum_k \alpha_{ki}\alpha^*_{kn} \tag{8.58}$$

Equation (8.58) represents the product of two matrices $\alpha = \|\alpha_{ki}\|$ and $\beta = \|\beta_{nk}\| \equiv \|\alpha^*_{kn}\|$ to give the unit matrix: $\beta\alpha = \mathbf{1}$. By definition, then, β is the "inverse" of α, $\beta = \alpha^{-1} = \|\alpha_{nk}{}^{-1}\|$. Hence, we may write $\alpha^*_{kn} = \alpha_{nk}{}^{-1}$. Let F'_{kj} be the kjth element of $\mathscr{F}$ with the u_k as basic vectors:

$$F'_{kj} = \int u_k^* \mathscr{F} u_j \, dq \tag{8.59}$$

From Eq. (8.57),

$$\begin{aligned} F'_{kj} &= \int \Big(\sum_i \alpha_{ki}\varphi_i^*\Big)\mathscr{F}\Big(\sum_n \alpha_{nj}{}^{-1}\varphi_n\Big)\, dq \\ &= \sum_{i,n} \alpha_{ki}F_{in}\alpha_{nj}{}^{-1} = \sum_i \alpha_{ki}(\mathscr{F}\alpha^{-1})_{ij} \\ &= (\alpha\mathscr{F}\alpha^{-1})_{kj} \end{aligned} \tag{8.60}$$

where, in Eq. (8.60), $\mathscr{F} = \|F_{kj}\|$. Equation (8.60) shows the desired relation between the original and transformed matrices.

Let us consider the special case of Eq. (8.55) in which ψ is an eigenfunction of $\mathscr{F}$ belonging to the eigenvalue λ. Then $\mathscr{F}\psi = \lambda\psi$. That is $\chi = \lambda\psi$, and $C_k = \lambda A_k$. In this case, in vector-matrix language, ψ is called an eigenvector of the matrix $\mathscr{F}$ belonging to the eigenvalue λ. The linear transformation [Eq. (8.54)] represented by the matrix $\mathscr{F}$ has the effect of converting the eigenvector ψ of the matrix $\mathscr{F}$ into a new vector χ lying on the same ray in function space as ψ but differing in length from ψ by a factor λ. It is possible in principle to find the eigenvalues and eigenvectors of a matrix (and therefore the eigenvalues and eigenfunctions of the associated operator) by use of Eq. (8.54), putting $C_k = \lambda A_k$. We have

$$\begin{aligned} (F_{11} - \lambda)A_1 + F_{12}A_2 + \cdots &= 0 \\ F_{21}A_1 + (F_{22} - \lambda)A_2 + \cdots &= 0 \\ \cdots\cdots\cdots\cdots\cdots \end{aligned} \tag{8.61}$$

This set of homogeneous linear equations in the unknowns A_k has nontrivial solutions if and only if

$$\begin{vmatrix} F_{11} - \lambda & F_{12} & F_{13} & \cdots \\ F_{21} & F_{22} - \lambda & F_{23} & \cdots \\ F_{31} & F_{32} & F_{33} - \lambda & \cdots \\ \cdots & \cdots & \cdots & \cdots \end{vmatrix} = 0 \tag{8.62}$$

Equation (8.62) is called the secular equation of the matrix $\mathscr{F}$. Each root λ_i of Eq. (8.62) is an eigenvalue of the matrix. If we substitute the λ_i successively in Eq. (8.61), the corresponding eigenvectors (with components A_k) may be obtained. Normalization is accomplished as usual by requiring that

$$\sum_k A_k^* A_k = 1$$

If a root λ_i is repeated, this eigenvalue is degenerate.

The eigenvalues and eigenvectors of a matrix are also the eigenvalues and eigenfunctions of the associated operator. These properties of the operator, of course, are independent of the choice of basic vectors. Therefore we may conclude (and verify algebraically, if desired) that the eigenvalues and eigenvectors of a matrix are also independent of the choice of basic vectors, although the actual numerical values of the *components* of the eigenvectors will depend on this choice (that is, different sets of basic vectors correspond to different sets of "axes" in function space, but merely changing the reference axes does not change the actual vector).

If the basic vectors φ_k happen to be eigenfunctions of $\mathscr{F}$, then Eq. (8.52) leads to $F_{kj} = 0$ if $k \neq j$. Thus, the matrix $\|F_{kj}\|$ is diagonal, and Eq. (8.62) becomes

$$\begin{vmatrix} F_{11} - \lambda & 0 & 0 & \cdots \\ 0 & F_{22} - \lambda & 0 & \cdots \\ 0 & 0 & F_{33} - \lambda & \cdots \\ \cdots & \cdots & \cdots & \cdots \end{vmatrix} = 0 \tag{8.63}$$

The eigenvalues are $\lambda_k = F_{kk}$.

In the general case where F and G both depend on t, $F_{kj} = F_{kj}(t)$ and $\lambda = \lambda(t)$ in Eq. (8.62), and it is implicit that this equation is to be solved, in principle, for each t, using basic vectors $\varphi_k(q,t)$ [Eq. (8.45)] which themselves depend on t. Thus, there is a different set of basic vectors for each t. However, since the eigenvalues and eigenvectors of $\mathscr{F}$ at any t are actually independent of the choice of basic vectors, a single set of basic vectors, say $\varphi_k(q,0)$, could in fact be used for all values of t. Then

$$F_{kj}(t) = \int \varphi_k^*(q,0) \mathscr{F}\left(\frac{\hbar}{i}\frac{\partial}{\partial q}, q, t\right) \varphi_j(q,0)\, dq \tag{8.64}$$

But it should be noted that if the $\varphi_k(q,t)$ happen to be eigenfunctions of $\mathscr{F}[(\hbar/i)\partial/\partial q, q, t]$, the matrix $\|F_{kj}\|$ using the $\varphi_k(q,0)$ as basic vectors would be diagonal only at $t = 0$.

The trace of a matrix, for example $||F_{kj}||$, is defined as

$$\text{trace}\,||F_{kj}|| = \sum_k F_{kk} \tag{8.65}$$

From the form of Eq. (8.62) as a polynomial in λ it is easy to see that

$$\text{trace}\,||F_{kj}|| = \lambda_1 + \lambda_2 + \lambda_3 + \cdots \tag{8.66}$$

Now the eigenvalues of $\mathscr{F}$, and hence also their sum, are invariant under a change in basic vectors. Therefore, from Eq. (8.66), trace $\mathscr{F}$ is independent of the choice of basic vectors. This is easy to verify algebraically.

The operators of quantum mechanics are Hermitian operators [Eq. (8.8)], and therefore

$$\int \varphi_k^* \mathscr{F} \varphi_j \, dq = \int \varphi_j (\mathscr{F} \varphi_k)^* \, dq \tag{8.67}$$

The left-hand side of Eq. (8.67) is F_{kj} while the right-hand side is F_{jk}^*. Thus, if $\mathscr{F}$ is an Hermitian operator, the matrix $||F_{kj}||$ has the property

$$||F_{kj}|| = ||F_{jk}^*|| \tag{8.68}$$

Any matrix having this property [whether or not the elements F_{kj} are obtained in the manner of Eq. (8.52)] is called an Hermitian matrix. Hermitian matrices have real eigenvalues only.

Time Dependence of Quantum-mechanical Systems. We conclude this brief synopsis of quantum mechanics by a further discussion of the time dependence of quantum-mechanical systems. In classical mechanics the state of a system is specified by definite values of p and q, and the way in which the state of the system changes with time is expressed by $p(t)$ and $q(t)$, where the time dependence may be deduced (at least in principle) from the equations of motion, Eq. (2.1). In quantum mechanics, the state of a system is described by a wave function $\psi(q,t)$. By expanding this wave function in the eigenfunctions of any dynamical variable (e.g., coordinates, momenta, energy, etc.), we may state what the probability is of observing any particular value of the variable if the variable is measured at t (the probability is zero except for eigenvalues). The change of the state of the system $\psi(q,t)$ with time is governed by Eq. (8.6), which is therefore a quantum-mechanical equation of motion. We have already seen that an explicit expression exhibiting the time dependence in $\psi(q,t)$ is given by Eq. (8.23),

$$\psi(q,t) = \sum_k A_k(t)\psi_k(q) \tag{8.69}$$

$$A_k(t) = c_k e^{-iE_k t/\hbar} \qquad c_k = \int \psi_k^*(q)\psi(q)\, dq$$

where the ψ_k are energy eigenfunctions and $\psi(q,0) = \psi(q)$. Actually, Eq. (8.69) is a special case of the more general Eq. (8.27) in which $\psi(q,t)$ is expanded in the eigenfunctions φ_k of an arbitrary dynamical variable $G(p,q)$ instead of those of $H(p,q)$. In general, then, the description of the way in which the state $\psi(q,t)$ of the system changes with time is contained in the time dependence of the coefficients in the expansion of $\psi(q,t)$ in the eigenfunctions of $\mathscr{G}$; the relative contributions to $\psi(q,t)$ of the various terms in the series change with time, as would be observed experimentally through $P_k(t) = |A_k(t)|^2$.

It is clear from the above discussion that an alternative quantum-mechanical equation of motion, analogous to Eq. (8.6), but more general in a sense, would be an expression for dA_k/dt. This is found by putting Eq. (8.27) in Eq. (8.6):

$$\mathscr{H}\sum_k A_k\varphi_k = i\hbar\sum_k \frac{dA_k}{dt}\varphi_k \tag{8.70}$$

If we multiply both sides of Eq. (8.70) by φ_j^* and integrate over q, the result is

$$\sum_k A_k(t)H_{jk} = i\hbar\frac{dA_j(t)}{dt} \tag{8.71}$$

where the matrix element H_{jk} is given by

$$H_{jk} = \int\varphi_j^*\mathscr{H}\varphi_k\,dq \tag{8.72}$$

Equation (8.28) must satisfy Eq. (8.71), as is not very difficult to confirm.

In addition to discussing the change in the state of the system with time, it is of interest to consider the quantum-mechanical analogue of the change in the value of some dynamical variable $F(p,q,t)$ with time. In classical mechanics, using Eq. (2.1),

$$\frac{dF}{dt} = \frac{\partial F}{\partial t} + \sum_{i=1}^{n}\left(\frac{\partial F}{\partial q_i}\dot{q}_i + \frac{\partial F}{\partial p_i}\dot{p}_i\right) \tag{8.73}$$

$$= \frac{\partial F}{\partial t} + \sum_{i=1}^{n}\left(\frac{\partial F}{\partial q_i}\frac{\partial H}{\partial p_i} - \frac{\partial H}{\partial q_i}\frac{\partial F}{\partial p_i}\right) \tag{8.74}$$

$$= \frac{\partial F}{\partial t} + [F,H] \tag{8.75}$$

The first term on the right-hand side of Eq. (8.75) is due to the explicit dependence of F on t, while the second term, the Poisson bracket of F and H, arises from the dependence of F on p and q, since $p = p(t)$ and $q = q(t)$ in accordance with the classical equations of motion of the system. In quantum mechanics, dF/dt itself has no meaning in general, but we can discuss the time dependence of $\bar{F}$ or of the matrix or operator $\mathscr{F}$. For this purpose, consider any two wave functions $\psi(q,t)$ and $\varphi(q,t)$ representing possible states of the same system, both with suitable time dependence (that is, satisfying Eq. (8.6)). Consider the integral

$$w(t) = \int\psi^*(q,t)\mathscr{F}\left(\frac{\hbar}{i}\frac{\partial}{\partial q},q,t\right)\varphi(q,t)\,dq \tag{8.76}$$

Then $$\frac{dw}{dt} = \int \psi^* \mathscr{F} \frac{\partial \varphi}{\partial t} \, dq + \int \psi^* \frac{\partial \mathscr{F}}{\partial t} \varphi \, dq + \int \frac{\partial \psi^*}{\partial t} \mathscr{F} \varphi \, dq \tag{8.77}$$

From Eq. (8.6) and the fact that $\mathscr{H}$ is an Hermitian operator, we find

$$\frac{d}{dt} \int \psi^* \mathscr{F} \varphi \, dq = \int \psi^* \frac{\partial \mathscr{F}}{\partial t} \varphi \, dq - \frac{i}{\hbar} \int \psi^* (\mathscr{F}\mathscr{H} - \mathscr{H}\mathscr{F}) \varphi \, dq \tag{8.78}$$

If we take $\varphi = \psi$, the change of $\bar{F}$ with time for the state ψ is given by

$$\frac{d\bar{F}}{dt} = \overline{\frac{\partial F}{\partial t}} + \overline{[F,H]} \tag{8.79}$$

Also, because of the arbitrariness of ψ^* and φ, Eq. (8.78) may be regarded as an operator or matrix equation:

$$\frac{d\mathscr{F}}{dt} = \frac{\partial \mathscr{F}}{\partial t} - \frac{i}{\hbar} (\mathscr{F}\mathscr{H} - \mathscr{H}\mathscr{F}) \tag{8.80}$$

$$= \frac{\partial \mathscr{F}}{\partial t} + [\mathscr{F},\mathscr{H}] \tag{8.81}$$

For example, if $\varphi = \chi_k(q,t)$ and $\psi = \chi_j(q,t)$, we have for the matrix element $F_{jk}(t)$,

$$\frac{dF_{jk}}{dt} = \left(\frac{\partial F}{\partial t}\right)_{jk} + [F,H]_{jk} \tag{8.82}$$

The resemblance between the classical equations (8.74) and the quantum-mechanical equations (8.79) and (8.81) is very striking.

If we take, in Eq. (8.79), $F = H(p,q)$, then $d\bar{H}/dt = 0$, which is one way of stating the quantum-mechanical law of conservation of energy in a conservative system. If $F = F(p,q)$ is any dynamical variable whose operator $\mathscr{F}$ commutes with $\mathscr{H}$, then $d\bar{F}/dt = 0$ also. In particular, $\mathscr{H}^m$ commutes with $\mathscr{H}$ (m a positive integer); thus $d\overline{H^m}/dt = 0$. If $F = f(H)$, where the function $f(H)$ can be expanded in a power series in H, then again $d\overline{f(H)}/dt = 0$. In all of these cases, each matrix element is also stationary in time. For example,

$$\frac{dH_{jk}}{dt} = 0 \qquad \text{and} \qquad \frac{d[f(H)]_{jk}}{dt} = 0 \tag{8.83}$$

9. Ensembles and Ensemble Averages in Quantum Statistical Mechanics

Since we know that for dynamical systems on an atomic scale classical mechanics must be replaced by quantum mechanics, it follows that a proper formulation of statistical mechanics, as a molecular theory, must be in quantum-mechanical language. Having accomplished such a formulation, one may then investigate the conditions under which classical statistical mechanics, as a limiting case, may be employed in applications. It turns out, as we shall see, that there are in fact very many problems, or parts of problems, that can quite correctly be treated classically.

The general argument involved in the development of the principles of quantum statistical mechanics parallels that of the classical treatment (Chap. 1) very closely. Differences arise primarily because the dynamical states of a quantum-mechanical system are expressed by wave functions instead of by the exact specification of p and q. There are a very large number of quantum-mechanical ("microscopic") states consistent with a given thermodynamic ("macroscopic") state of a system, so that the central problem is again one of indicating how averages over microscopic states are to be taken in order to calculate macroscopic properties of the system. Because the exact specification of both p and q is not possible in quantum mechanics, phase space cannot be used in the same way as in classical mechanics. However, the concept of a representative ensemble of perfectly isolated systems is carried over to quantum statistics and used in the same way. That is, it is even more hopeless than in the classical case to attempt to compute the change of a quantum-mechanical state $\psi(q,t)$ with time when the number of degrees of freedom is extremely large. Ensemble averages must therefore be resorted to again and are put in correspondence with the observed values of thermodynamic properties. It is to be noted that the use of ensemble averages introduces averaging and probability concepts at a *second* level in quantum statistics, as these concepts are already present, as we have seen, even in the discussion of a single quantum-mechanical system.

In accordance with the above remarks, we may anticipate the introduction of ensemble averages in Postulate A below, and define here precisely what we mean by an ensemble average in quantum statistics. This definition will be seen to introduce automatically the so-called density matrix which plays a role analogous to the distribution function $f(p,q;t)$ in classical phase space. Some of the general properties of the density matrix will then be considered.

An ensemble, here, is a collection of a very large number of perfectly isolated (conservative) and therefore independent systems in a variety of quantum-mechanical states $\psi(q,t)$, but all with the same values of $\mathbf{N} = N_1, \ldots, N_r$ and $\mathbf{x} = x_1, \ldots, x_s$. Our interest will eventually be in ensembles which we attempt to construct theoretically in such a way as to be "representative" of the thermodynamic state of a real system with $\mathbf{N}$ and $\mathbf{x}$. The operational method of constructing an ensemble, mentioned in Chap. 1, applies here also and may again be used to define what is meant by "representative," with the modification that a distribution of systems over different quantum-mechanical states $\psi(q,t)$ is obtained in this case.

Let us first consider a special case. Suppose we have an ensemble of $\mathcal{N}'$ systems all in the *same* state $\psi(q,t)$. With $\mathcal{N}' \to \infty$, the ensemble average $\overline{\overline{F}}(t)$ (we use two bars for ensemble averages) of $F(p,q)$ is appropriately[1]

[1] See Eq. (8.5) and the detailed discussion in Kemble (Gen. Ref.), pp. 53–55.

defined here as just the expectation value itself for the state $\psi(q,t)$,

$$\bar{\bar{F}}(t) = \bar{F}(t) = \int \psi^*(q,t) \mathscr{F} \psi(q,t)\, dq \tag{9.1}$$

This ensemble is representative of a real system known to be in the state $\psi(q,t)$. We may note that if we expand $\psi(q,t)$ in the eigenfunctions $\varphi_k(q)$ of $\mathscr{G}$, as in Eq. (8.27), Eq. (9.1) becomes

$$\bar{\bar{F}}(t) = \bar{F}(t) = \sum_{k,l} A_k^*(t) A_l(t) F_{kl} \tag{9.2}$$

$$F_{kl} = \int \varphi_k^* \mathscr{F} \varphi_l\, dq \tag{9.3}$$

In the general case, let the ensemble contain systems in states, $\psi^{(i)}(q,t)$, $i = 1, 2, \ldots, \mathscr{N}$, not necessarily all different, each one of these $\mathscr{N}$ states occurring $\mathscr{N}'$ times in the ensemble, so that there are altogether $\mathscr{N}'' = \mathscr{N}\mathscr{N}'$ systems in the ensemble. As usual, $\mathscr{N} \to \infty$ and $\mathscr{N}' \to \infty$. The (partial ensemble) average of $F(p,q)$ over those $\mathscr{N}'$ systems in the state $\psi^{(i)}(q,t)$ is $\bar{F}^{(i)}(t)$ as in Eqs. (9.1) and (9.2). The complete ensemble average $\bar{\bar{F}}(t)$ is then the mean of the $\mathscr{N}$ quantities $\bar{F}^{(i)}(t)$:

$$\bar{F}^{(i)}(t) = \sum_{k,l} A_k^{(i)*}(t) A_l^{(i)}(t) F_{kl} \tag{9.4}$$

$$\bar{\bar{F}}(t) = \sum_{k,l} \overline{\overline{A_k^*(t) A_l(t)}} F_{kl} \tag{9.5}$$

where

$$\overline{\overline{A_k^* A_l}} = \frac{1}{\mathscr{N}} \sum_{i=1}^{\mathscr{N}} A_k^{(i)*} A_l^{(i)} \tag{9.6}$$

A simpler and completely equivalent point of view, which, however, does not stress the "ensemble average" nature of the expectation value itself, is to state that the ensemble consists of $\mathscr{N}$ systems in the states $\psi^{(i)}(q,t)$, $i = 1, 2, \ldots, \mathscr{N}$. The ensemble average $\bar{\bar{F}}$ of $F(p,q)$ is then *defined* as the average of the $\mathscr{N}$ expectation values $\bar{F}^{(i)}$ of $F(p,q)$, as in Eqs. (9.4) and (9.5). We adopt this usual second point of view for simplicity.

To shorten notation, let us write

$$\overline{\overline{A_k^*(t) A_l(t)}} = f_{lk}(t) \tag{9.7}$$

$$\bar{\bar{F}}(t) = \sum_{k,l} f_{lk}(t) F_{kl} \tag{9.8}$$

Equation (9.8) for the ensemble average of $F(p,q)$ may be compared with the classical equation (3.2) for the same quantity,

$$\bar{F}(t) = \int f(p,q;t) F(p,q)\, dp\, dq \tag{9.9}$$

We see that an integration over p and q has been replaced by a double sum over quantum states, and the matrix $\mathfrak{f} = \|f_{lk}\|$ plays a role analogous to that of the distribution function or probability density $f(p,q;t)$. For this reason $\mathfrak{f}$ is called the "density matrix." According to the rules of matrix multiplication, Eq. (9.8) can also be written

$$\bar{\bar{F}}(t) = \sum_l [\mathfrak{f}(t)\mathscr{F}]_{ll} \tag{9.10}$$

$$= \text{trace } \mathfrak{f}(t)\mathscr{F} \tag{9.11}$$

Hence, the alternative statement can be made that integration of the product fF over p and q has been replaced by the sum of the diagonal elements of the matrix product $\mathfrak{f}\mathscr{F}$.

A further analogy of this type concerns normalization: we replace the matrix $\mathscr{F}$ in Eq. (9.8) by the unit matrix $\mathbf{1} = ||\delta_{kl}||$, and substitute unity for $F(p,q)$ in Eq. (9.9). Then

$$\sum_{k,l} f_{lk}(t)\delta_{kl} = \sum_l f_{ll} = \text{trace } \mathfrak{f}$$

$$= \sum_l \overline{\overline{A_l^*(t)A_l(t)}} = 1 \tag{9.12}$$

and
$$\int f(p,q;t)\, dp\, dq = 1 \tag{9.13}$$

where the right-hand side of Eq. (9.12) is equal to unity because of the normalization of the $\psi^{(i)}$ and φ_k.

The diagonal element $f_{kk}(t)$ of the density matrix is the ensemble average $\bar{\bar{P}}_k$ of $P_k^{(i)}$ and is therefore also the probability of observing the eigenvalue g_k of $\mathscr{G}$ at t if a measurement of $G(p,q)$ is made on a system chosen at random from the ensemble. Thus, for the ensemble average of $G(p,q)$,

$$\bar{\bar{G}}(t) = \sum_{k,l} f_{lk}(t)G_{kl}$$

$$= \sum_{k,l} f_{lk}(t)\delta_{kl}g_k$$

$$= \sum_k f_{kk}(t)g_k = \sum_k \bar{\bar{P}}_k(t)g_k \tag{9.14}$$

We have introduced ensemble averages and the density matrix above using a particular but arbitrary choice of basic vectors, the φ_k. Now suppose we change the basic vectors from the φ_k to another set, the u_k, using the notation of Eqs. (8.56) to (8.60). Then

$$\psi^{(i)} = \sum_k A_k^{(i)}\varphi_k = \sum_k A_k^{(i)}\left(\sum_j \alpha_{jk}u_j\right)$$

$$= \sum_j B_j^{(i)}u_j \tag{9.15}$$

where
$$B_j^{(i)} = \sum_k A_k^{(i)}\alpha_{jk} \tag{9.16}$$

The new density matrix elements are by definition

$$f'_{lk} = \overline{\overline{B_k^* B_l}} \tag{9.17}$$

From Eq. (9.16),

$$\begin{aligned} f'_{lk} &= \sum_{j,n} \alpha_{ln} f_{nj} \alpha_{jk}^{-1} \\ &= (\alpha \mathfrak{f} \alpha^{-1})_{lk} \end{aligned} \tag{9.18}$$

This relation between original and transformed density matrices is the same as in Eq. (8.60).

It is, of course, essential, for the ensemble average to be useful, that $\overline{\overline{F}}$ be invariant under a change of basic vectors. This property is easy to verify, using Eqs. (8.58), (8.60), and (9.18):

$$\begin{aligned} \overline{\overline{F}}' &= \sum_{k,l} f'_{lk} F'_{kl} = \sum_{k,l,j,n,i,p} (\alpha_{ln} f_{nj} \alpha_{jk}^{-1})(\alpha_{ki} F_{ip} \alpha_{pl}^{-1}) \\ &= \sum_{j,n} f_{nj} F_{jn} = \overline{\overline{F}} \end{aligned} \tag{9.19}$$

Also, the normalization is invariant:

$$\sum_k f'_{kk} = \sum_{k,j,n} \alpha_{kn} f_{nj} \alpha_{jk}^{-1} = \sum_n f_{nn} = 1 \tag{9.20}$$

Analogous invariance properties under canonical transformations $(p,q \to p',q')$ hold in classical mechanics.[1]

Several analogies between $\mathfrak{f}$ and $f(p,q;t)$ have been pointed out above. In addition to these, we may now show that $\mathfrak{f}$ satisfies a quantum-mechanical version of Liouville's theorem. From Eqs. (8.71) and (9.7), we have

$$\begin{aligned} \frac{\partial f_{lk}}{\partial t} &= \overline{\overline{A_k^* \frac{dA_l}{dt}}} + \overline{\overline{\frac{dA_k^*}{dt} A_l}} \\ &= -\frac{i}{\hbar} \sum_j f_{jk} H_{lj} + \frac{i}{\hbar} \sum_j f_{lj} H_{jk} \\ &= \frac{i}{\hbar} [\mathfrak{f}\mathscr{H} - \mathscr{H}\mathfrak{f}]_{lk} \end{aligned} \tag{9.21}$$

or

$$\frac{\partial \mathfrak{f}}{\partial t} = -[\mathfrak{f},\mathscr{H}] \tag{9.22}$$

This result is strictly analogous to the classical equation (3.12) [see Eq. (8.75)].

If we are willing to assume at this point that the density matrix is some function of the various matrices associated with dynamical variables of the system (and is therefore itself associated with some dynamical variable—

[1] See Tolman (Gen. Ref.).

which may be complicated), then Eq. (8.81) applies and there results the matrix equation

$$\frac{d\mathfrak{f}}{dt} = 0 \tag{9.23}$$

which is to be compared with the classical equation (3.11). This result is not required in our argument, however, and it is mentioned essentially as a digression. But it may be noted in this connection that the above assumption (which will later, in a more specific form, be a deduction from Postulate *A*, below) is actually consistent with several properties of the density matrix. In the first place, we see from Eqs. (8.60) and (9.18) that $\mathfrak{f}$ transforms under a change of basic vectors in the same way that the matrix associated with a dynamical variable transforms. Second, trace $\mathfrak{f}$ is invariant under a change of basic vectors, as is trace $\mathscr{F}$, where $F(p,q)$ is a dynamical variable [Eq. (8.66)]. Incidentally, the sum of the eigenvalues of the matrix $\mathfrak{f}$ is unity [Eq. (9.20)]. A third important property, consistent with the association of $\mathfrak{f}$ with a dynamical variable, is that $\mathfrak{f}$ is an Hermitian matrix, as can be seen from Eq. (9.7).

We conclude this section by considering the form of the density matrix in the special case that $\psi^{(i)}$ is expanded in the simultaneous eigenfunctions of the coordinates [Eq. (8.48)]. This will give the density matrix in "coordinate language." Let $\delta(q - q')$ represent the product

$$\delta(q_1 - q_1')\delta(q_2 - q_2') \cdot \cdot \cdot \delta(q_n - q_n')$$

Then

$$\begin{aligned}\psi^{(i)}(q,t) &= \int A^{(i)}(q',t)\delta(q - q')\, dq' \\ &= A^{(i)}(q,t)\end{aligned} \tag{9.24}$$

and we may write the density matrix as

$$\begin{aligned}f(q',q'';t) &= \overline{\overline{A^*(q'',t)A(q',t)}} \\ &= \overline{\overline{\psi^*(q'',t)\psi(q',t)}}\end{aligned} \tag{9.25}$$

where $q = q'$ and $q = q''$ are particular but arbitrary configurations. Equation (9.12) becomes

$$\int f(q',q';t)\, dq' = \int \overline{\overline{\psi^*(q',t)\psi(q',t)}}\, dq' = 1 \tag{9.26}$$

The quantity $f(q',q';t)\, dq'$, a "diagonal element," is the probability of observing a system chosen at random from the ensemble in a configuration between q' and $q' + dq'$ [compare Eq. (8.41)], at t.

If we expand $\psi^{(i)}$ in Eq. (9.25) in the usual arbitrary basic vectors φ_k, we obtain a relation, which will prove to be useful, between $f(q',q'';t)$ and the matrix elements $f_{lk}(t)$ associated with the basic vectors φ_k:

$$f(q',q'';t) = \sum_{k,l} \varphi_k^*(q'')\varphi_l(q')f_{lk}(t) \tag{9.27}$$

10. Postulate on the Use of Ensemble Averages

In this section we shall state the quantum-mechanical form of Postulate A, the purpose of which postulate, as before, is to indicate the association to be used between F_{obs} and ensemble averages. The argument is practically identical with that in Chap. 1, to which reference should be made, except for several points included in the discussion below.

We may begin by remarking that, in quantum mechanics, it is meaningless to discuss the experimental value F_{obs}, of a thermodynamic property $F(p,q)$ for a *perfectly* isolated real system. This is because the measurement itself (of a macroscopic property), carried out over a period of time τ (see Chap. 1), destroys the perfect isolation of the system even if it existed before the time of the measurement. In the classical view, the interaction between experimental system and measuring instrument can in principle be reduced to zero during the period τ, but this is not the case in quantum mechanics because of the uncertainty principle. On the other hand, the systems of the ensemble, being only mental representatives of a real system, are defined and may be considered as perfectly isolated.

The ith system of the ensemble is in the state $\psi^{(i)}(q,t)$ at t and the expectation value of $F(p,q)$ at t is $\bar{F}^{(i)}(t)$ [Eq. (9.4)]. The time average of this expectation value, between t_0 and $\tau + t_0$, is

$$F_\tau^{(i)}(t_0) = \frac{1}{\tau}\int_{t_0}^{\tau+t_0} (\sum_{k,l} A_k^{(i)*}(t)A_l^{(i)}(t)F_{kl})\, dt \tag{10.1}$$

Unlike the classical discussion, this time averaging is not to be considered a measuring process on the ith system of the ensemble, as all systems of the ensemble are perfectly isolated. The period τ is chosen long enough so that $F_\tau^{(i)}(t_0)$ may be assumed independent of τ.

We now state *Postulate A: F_{obs} is identified with the ensemble average of $F_\tau^{(i)}$, $\bar{\bar{F}}_\tau$, where*

$$\bar{\bar{F}}_\tau(t_0) = \frac{1}{\mathscr{N}}\sum_{i=1}^{\mathscr{N}} F_\tau^{(i)}(t_0) = \frac{1}{\tau}\int_{t_0}^{\tau+t_0} (\sum_{k,l} f_{lk}(t)F_{kl})\, dt \tag{10.2}$$

We note immediately that since F_{obs} must be independent of t_0 for an equilibrium state, $\bar{\bar{F}}_\tau$ must also be independent of t_0, and hence f_{lk} must be independent of t. Putting f_{lk} = constant, Eq. (10.2) becomes[1]

$$\bar{\bar{F}}_\tau = \sum_{k,l} f_{lk}F_{kl} = \bar{\bar{F}} \tag{10.3}$$

[1] It may be noted that the physically reasonable assumption above, of $F_\tau^{(i)}$ independent of τ, is not necessary, since putting f_{lk} = constant removes the dependency, if any, of $\bar{\bar{F}}_\tau$ on both t_0 and τ. Analogous remarks hold in the classical case.

We can, therefore, replace $\bar{\bar{F}}_\tau$ in Postulate A by $\bar{\bar{F}}$, if desired, as is in fact customary. We have introduced time averaging here primarily in order to maintain the analogy with the classical argument in Chap. 1.

As before, F_{obs} may be regarded as a composite of parts of many different F_τ's, since during the period τ the real system makes transitions from one state ψ to another ψ' on every interaction with the measuring instrument and surroundings. The remarks in Chap. 1 (nonperfectly isolated case) concerning the justification of the identification of F_{obs} with the ensemble average of the time average of F apply here as well, and need not be repeated.

In summary: we postulate that F_{obs} may be calculated theoretically, from a representative ensemble, as the ensemble average F; the ensemble average depends on the density matrix $\mathfrak{f}$ of the ensemble, the elements of which matrix must be independent of t if the ensemble is to be representative of a system in thermodynamic equilibrium.

11. Postulate on the Form of the Density Matrix

From Postulate A we have deduced that $\partial f_{lk}/\partial t = 0$ for an equilibrium ensemble. In this section we complete the specification (Postulate B) of the form of the density matrix for closed, isolated and closed, isothermal thermodynamic systems. We give first an argument that makes our postulate on the closed, isolated case (microcanonical ensemble) very plausible. The postulate for the closed, isothermal case (canonical ensemble) is stated here without a supporting argument. But in Sec. 13 a complete discussion of the derivation of other ensembles from the microcanonical ensemble will be given, as already promised in Chap. 1.

From $\partial f_{lk}/\partial t = 0$ and Eq. (9.21) we see that $\mathfrak{f}$ and $\mathscr{H}$ commute. Therefore, $\mathfrak{f}$ is a matrix associated with some constant of the motion of the system (by definition, F is a "constant of the motion" if $\partial F/\partial t = 0$ and $\mathscr{F}$ and $\mathscr{H}$ commute). One class of possibilities is that $\mathfrak{f}$ is some function of $\mathscr{H}$, $\mathfrak{f} = f(\mathscr{H})$, where $f(\mathscr{H})$ can be expanded in a power series in $\mathscr{H}$. On the basis of the classical argument on this point (Chap. 1) and on the established analogy between $f(p,q)$ and $\mathfrak{f}$, we restrict ourselves to this choice. The meaning of this assumption in terms of matrix elements is that

$$f_{lk} = \int \varphi_k^* f(\mathscr{H}) \varphi_l \, dq \tag{11.1}$$

In the special case that the energy eigenfunctions ψ_k are chosen as basic vectors (the "energy language"), Eq. (11.1) becomes

$$f_{lk} = \int \psi_k^* f(\mathscr{H}) \psi_l \, dq \tag{11.2}$$

If we consider $f(\mathscr{H})$ as a power series in $\mathscr{H}$,

$$f(\mathscr{H})\psi_l = f(E_l)\psi_l$$

and

$$f_{lk} = \delta_{lk} f(E_k) \tag{11.3}$$

Thus, in the energy language, any $f(\mathscr{H})$ is a diagonal matrix. Also, $f(E_k)$ is the probability of observing the eigenvalue E_k if a measurement of the energy is made on a system chosen at random from the ensemble:

$$f(E_k) = \bar{\bar{P}}_k \tag{11.4}$$

Now suppose we want to select an $\mathfrak{f}$ to be representative of a thermodynamic system (with $\mathbf{N}$ and $\mathbf{x}$) known to have a definite energy E within a small range δE. In the first place, because of the special role played by the energy in this definition of the thermodynamic state, we should certainly expect the energy language to be particularly appropriate in this case. In the second place, for a degenerate level,[1] say with $k = k'$ to $k = k''$,

$$\bar{\bar{P}}_{k'} = \bar{\bar{P}}_{k'+1} = \cdots = \bar{\bar{P}}_{k''} = f(E_{k'}) \tag{11.5}$$

That is, since $\bar{\bar{P}}_k$ depends only on E_k, all basic states ψ_k belonging to the same eigenvalue have the same probability $\bar{\bar{P}}_k$. Now the state of a system known to have an energy E, within a range δE very small compared to E, must be represented by a superposition of basic states ψ_k with E_k such that $E \leqslant E_k \leqslant E + \delta E$. Since f_{kk} depends on E_k only and δE is very small, $f_{kk} = f(E_k)$ is for practical purposes constant in the range δE and hence all basic states ψ_k with eigenvalues E_k falling in this range may be considered as belonging to an effective expanded "degenerate" energy level E. Thus, as for the "pure" degenerate level discussed above, if k runs from k' to $k' + \Omega - 1$ for the Ω basic states ψ_k with $E \leqslant E_k \leqslant E + \delta E$,

$$\bar{\bar{P}}_{k'} = \bar{\bar{P}}_{k'+1} = \cdots = \bar{\bar{P}}_{k'+\Omega-1} = f(E) \tag{11.6}$$

We are thus led by the above argument, which is completely analogous to that employed in Chap. 1, to state the following postulate.

Postulate B. (*a*) *Microcanonical ensemble.* For a closed, isolated thermodynamic system, i.e., a system with assigned values for the independent variables $E, N_1, \ldots, N_r, x_1, \ldots, x_s$, using energy eigenfunctions as basic vectors,

$$f_{lk} = \delta_{lk}\bar{\bar{P}}_k \tag{11.7}$$

and

$$\begin{aligned} \bar{\bar{P}}_k &= \frac{1}{\Omega} \qquad E \leqslant E_k \leqslant E + \delta E \\ &= 0 \qquad \text{(otherwise)} \end{aligned} \tag{11.8}$$

where δE is a very small range in E, and Ω is the number of basic states ψ_k belonging to eigenvalues E_k in the range $E \leqslant E_k \leqslant E + \delta E$. The constant $1/\Omega$ results from the normalization equation, $\sum_k \bar{\bar{P}}_k = 1$.

Some uncertainty δE in E will arise, as already indicated, because the

[1] In systems made up of very large numbers of molecules, the energy levels will in general be highly degenerate.

measurement of the thermodynamic property E implies interaction of the experimental system with the measuring instrument over the microscopically long period of measurement τ. Thus, even if somehow the system had been prepared to have exactly the energy E at the time t_0 of commencement of the measurement, a small variation in E, δE, would still be introduced. But, actually, also because of the uncertainty principle, there will be an uncertainty $\delta E'$ already present at $t = t_0$, related to the period δt, prior to $t = t_0$, during which the system is known to be in perfect isolation, by

$$\delta E' \delta t \geqslant a\hbar \tag{11.9}$$

where a is a number of order unity. We can have $\delta E' = 0$ only under the physically impossible condition of perfect isolation of the experimental system for an infinite period of time prior to commencement of the measurement at $t = t_0$.

Let $||f'_{lk}||$ be the microcanonical density matrix for arbitrary basic vectors φ_k, whereas $||f_{lk}||$ [Eq. (11.7)] is in the energy language. It is easy to see that $||f'_{lk}||$ is in general not in a very simple form. Let [compare Eq. (8.56)]

$$\psi_i = \sum_k \beta_{ki} \varphi_k \tag{11.10}$$

Then, from Eq. (9.18),

$$\begin{aligned} f'_{lk} &= \sum_{j,n} \beta_{ln} \beta^*_{kj} f_{nj} = \sum_n \beta_{ln} \beta^*_{kn} \bar{\bar{P}}_n \\ &= \frac{1}{\Omega} \sum_{n=k'}^{k'+\Omega-1} \beta_{ln} \beta^*_{kn} \end{aligned} \tag{11.11}$$

$||f'_{lk}||$ is thus a rather complicated nondiagonal matrix, in contrast to $||f_{lk}||$.

We saw above that Postulate $B(a)$ may be deduced for a closed, isolated system, if we assume $\mathfrak{f} = f(\mathscr{H})$. Part of this postulate, Eq. (11.8), states that for a system known to have an energy between E and $E + \delta E$, all basic (energy) states ψ_k belonging to an E_k in this range have the same probability $\bar{\bar{P}}_k$. This is usually called the postulate of "equal a priori probabilities" and we emphasize here that it is included in our Postulate $B(a)$. It should also be stressed that Postulate $B(a)$ contains as well the usual postulate of "random phases." To see this, we note that $\psi^{(i)}$, the state of the ith system of the microcanonical ensemble, is given by the superposition of energy states,

$$\psi^{(i)} = \sum_{k=k'}^{k'+\Omega-1} A_k^{(i)} \psi_k \tag{11.12}$$

where the complex number $A_k^{(i)}$ can be written as

$$A_k^{(i)} = r_k^{(i)} e^{i\gamma_k^{(i)}} \tag{11.13}$$

with $r_k^{(i)}$ and $\gamma_k^{(i)}$ real and $r_k^{(i)} \geqslant 0$. Then

$$f_{lk} = \overline{\overline{r_k r_l e^{i(\gamma_l - \gamma_k)}}} \tag{11.14}$$

Now in order for Eq. (11.14) to agree with Eqs. (11.7) and (11.8), we must have

$$\begin{aligned} r_k^2 &= \frac{1}{\Omega} \qquad k' \leqslant k \leqslant k' + \Omega - 1 \\ &= 0 \qquad \text{otherwise} \end{aligned} \tag{11.15}$$

and also

$$\overline{\overline{r_k r_l e^{i(\gamma_l - \gamma_k)}}} = 0 \begin{cases} k \neq l \\ k' \leqslant k,l \leqslant k' + \Omega - 1 \end{cases} \tag{11.16}$$

for $\mathfrak{f}$ will not be a diagonal matrix, as in Eq. (11.7), unless Eq. (11.16) holds. In the absence of any information concerning the phases $\gamma_k^{(i)}$, the only reasonable way to make Eq. (11.16) true is to assume that, for each k, values of $\gamma_k^{(i)}$ are assigned completely at random among the members of the ensemble. Thus, in summary, Eq. (11.8) is essentially the postulate of "equal a priori probabilities" and Eq. (11.7) is essentially the postulate of "random phases."

We may summarize the above discussion of the microcanonical ensemble as follows: Postulate $B(a)$ completes our specification of the density matrix representative of a closed, isolated thermodynamic system; an argument is given which makes the postulate plausible, but actual verification must rest on the comparison of predictions made from the postulate with experiment.

Canonical Ensemble. We next state Postulate $B(b)$ for a closed system in thermal equilibrium with its surroundings, without any attempt, until Sec. 13, to make the postulate physically reasonable.

Postulate B. (*b*) *Canonical ensemble.* For a closed, isothermal thermodynamic system, i.e., a system with assigned values for the independent variables, T, N_1, . . . , N_r, x_1, . . . , x_s, using arbitrary basic vectors φ_k,

$$\mathfrak{f} = \text{const} \times \exp(-\beta\mathscr{H}) \tag{11.17}$$

where β is a constant. That is,

$$f_{lk} = \text{const} \times \int \varphi_k^* \exp(-\beta\mathscr{H})\varphi_l \, dq \tag{11.18}$$

We may recall that $f_{kk} = \overline{\overline{P}}_k$ is the probability of observing g_k if a measurement of G is made on a system chosen at random from the ensemble. Then

$$\sum_k f_{kk} = 1 = \text{const} \times \sum_k \int \varphi_k^* \exp(-\beta\mathscr{H})\varphi_k \, dq \tag{11.19}$$

$$\text{const} = \frac{1}{Q} \tag{11.20}$$

and

$$\mathfrak{f} = \frac{\exp(-\beta\mathscr{H})}{Q} \tag{11.21}$$

where

$$Q = \sum_k \int \varphi_k^* \exp(-\beta\mathscr{H})\varphi_k \, dq \tag{11.22}$$

$Q(\beta,\mathbf{x},\mathbf{N})$ is called the partition function. We see, on comparing Eqs. (5.4) and (11.22), that integration over p in classical mechanics is replaced here by a summation over quantum states. Q may be written in the alternative ways

$$Q = \sum_k [\exp(-\beta\mathscr{H})]_{kk} = \text{trace} \exp(-\beta\mathscr{H}) \tag{11.23}$$

$$= \text{sum of eigenvalues of} \exp(-\beta\mathscr{H}) \tag{11.24}$$

Q is obviously invariant under a change of basic vectors.

The most important special choice of basic vectors is the set of energy eigenfunctions ψ_k. In this case,

$$f_{lk} = \frac{1}{Q}\int \psi_k^* \exp(-\beta\mathscr{H})\psi_l \, dq \tag{11.25}$$

$$= \frac{1}{Q}\int \psi_k^* \psi_l e^{-\beta E_l} \, dq \tag{11.26}$$

$$= \frac{\delta_{lk} e^{-\beta E_k}}{Q} = \delta_{lk}\bar{\bar{P}}_k \tag{11.27}$$

where $\bar{\bar{P}}_k$ is the probability of observing E_k if the energy is measured on a system selected at random from the canonical ensemble. Alternative forms for $\bar{\bar{P}}_k$ are then

$$\bar{\bar{P}}_k = \frac{1}{Q}\int \psi_k^* \exp(-\beta\mathscr{H})\psi_k \, dq \tag{11.28}$$

$$= \frac{1}{Q}\int |\psi_k|^2 e^{-\beta E_k} \, dq \tag{11.29}$$

$$= \frac{1}{Q} e^{-\beta E_k} \tag{11.30}$$

Now we know that $|\psi_k|^2 \, dq$ is the probability of the system in the energy eigenstate ψ_k being observed to have a configuration between q and $q + dq$. It follows then that

$$\frac{e^{-\beta E_k}}{Q} \cdot |\psi_k|^2 \, dq$$

is the probability of the system being simultaneously observed in the energy eigenstate ψ_k and in a configuration between q and $q + dq$. If we sum over all energy eigenstates [see Eq. (9.25)],

$$\frac{1}{Q}\sum_k |\psi_k|^2 e^{-\beta E_k} \, dq = \frac{1}{Q}\sum_k \psi_k^* \exp(-\beta\mathscr{H})\psi_k \, dq = f(q,q) \, dq \tag{11.31}$$

is then the probability of a configuration between q and $q + dq$ (irrespective of the energy state). Equation (11.31) is a special case of Eq. (9.27), is closely related to the "Slater sum" (see Chap. 3), and is analogous to the classical expression

$$dq \int f(p,q)\, dp \tag{11.32}$$

The partition function, in the energy language, takes on the various forms

$$Q = \sum_k \int \psi_k^* \exp(-\beta\mathscr{H})\psi_k\, dq \tag{11.33}$$

$$= \sum_k \int |\psi_k|^2 e^{-\beta E_k}\, dq \tag{11.34}$$

$$= \sum_k e^{-\beta E_k} \tag{11.35}$$

where Eq. (11.35) is by far the most widely used of the alternatives.

Finally, returning to arbitrary basic vectors, we carry the discussion of Eq. (9.27) somewhat further. This equation, it will be recalled, relates the density matrix in coordinate language to the density matrix in an arbitrary language. We have

$$f(q',q'') = \sum_{k,l} \varphi_k^*(q'')\varphi_l(q')f_{lk} \tag{11.36}$$

where $\mathfrak{f} = \|f_{lk}\| = \exp(-\beta\mathscr{H})/Q$. Now $\mathfrak{f}\varphi_k(q')$ is the product of a matrix $\mathfrak{f}$ and a vector φ_k. The components of φ_k, when expanded in the basic vectors φ_l, are δ_{lk}. The components of the new vector $\mathfrak{f}\varphi_k(q')$, also in terms of the basic vectors φ_l, are [Eq. (8.54)]

$$\sum_n \delta_{nk} f_{ln} = f_{lk} \tag{11.37}$$

That is,

$$\mathfrak{f}\varphi_k(q') = \sum_l f_{lk}\varphi_l(q') \tag{11.38}$$

We now use this relation in Eq. (11.36), obtaining

$$f(q',q'') = \frac{1}{Q} \sum_k \varphi_k^*(q'') \exp(-\beta\mathscr{H})\varphi_k(q') \tag{11.39}$$

and

$$f(q',q') = \frac{1}{Q} \sum_k \varphi_k^*(q') \exp(-\beta\mathscr{H})\varphi_k(q') \tag{11.40}$$

Equation (11.40) is a more general expression of the same type as Eq. (11.31).

12. Grand Ensembles

This discussion of grand ensembles can be rather brief because of the close correlation with Sec. 6.

A grand ensemble, as introduced in Chap. 1, is a collection of petit ensembles. For each petit ensemble, we have, as in Eqs. (9.5) and (9.8),

$$\bar{\bar{F}}(\mathbf{N},t) = \sum_{k,l} \overline{\overline{A_k^*(\mathbf{N},t)A_l(\mathbf{N},t)}} F_{kl}(\mathbf{N}) \tag{12.1}$$

$$= \sum_{k,l} f_{lk}(\mathbf{N},t) F_{kl}(\mathbf{N}) \tag{12.2}$$

where $\mathbf{N}$ is the composition of each of the systems in the petit ensemble. Now let $\theta(\mathbf{N})$ be the fraction of systems of the grand ensemble with composition $\mathbf{N}$. Then the grand ensemble average (three bars) of $F(p,q)$ is [see Eq. (9.8)]

$$\bar{\bar{\bar{F}}} = \sum_{\mathbf{N} \geqslant \mathbf{0}} \theta(\mathbf{N}) \bar{\bar{F}}(\mathbf{N},t) \tag{12.3}$$

$$= \sum_{\mathbf{N},k,l} \theta(\mathbf{N}) f_{lk}(\mathbf{N},t) F_{kl}(\mathbf{N}) \tag{12.4}$$

Also [see Eq. (9.14)],

$$\bar{\bar{\bar{G}}} = \sum_{\mathbf{N},k} \theta(\mathbf{N}) f_{kk}(\mathbf{N},t) g_k(\mathbf{N}) \tag{12.5}$$

$$= \sum_{\mathbf{N},k} \theta(\mathbf{N}) \bar{\bar{P}}_k(\mathbf{N},t) g_k(\mathbf{N}) \tag{12.6}$$

The normalization equations here are

$$\sum_k f_{kk}(\mathbf{N},t) = 1 \tag{12.7}$$

$$\sum_{\mathbf{N}} \theta(\mathbf{N}) = 1 \tag{12.8}$$

$$\sum_{\mathbf{N},k} \theta(\mathbf{N}) f_{kk}(\mathbf{N},t) = 1 \tag{12.9}$$

where the probability of selecting at random from the grand ensemble a system with composition $\mathbf{N}$ and of obtaining $g_k(\mathbf{N})$ from a measurement of G on this system is $\theta(\mathbf{N}) f_{kk}(\mathbf{N},t)$.

Each system of the grand ensemble, for example, the ith, is closed (fixed $\mathbf{N}$) and perfectly isolated so that the time average of $\bar{F}^{(i)}(t)$ between t_0 and $t_0 + \tau$, $F_\tau^{(i)}(\mathbf{N},t_0)$, is still given by Eq. (10.1). Then we define the (grand) ensemble average of $F_\tau^{(i)}$ as

$$\bar{\bar{\bar{F}}}_\tau(t_0) = \sum_{\mathbf{N}} \theta(\mathbf{N}) \bar{\bar{F}}_\tau(\mathbf{N},t_0) \tag{12.10}$$

where $\bar{\bar{F}}_\tau$ is given by Eq. (10.2). We may now state *Postulate A: F_{obs} is identified with the ensemble average of $F_\tau^{(i)}$, $\bar{\bar{\bar{F}}}_\tau(t_0)$.*

F_{obs} is independent of time for an equilibrium system; therefore $\bar{\bar{\bar{F}}}_\tau$ cannot depend on t_0. The necessary and sufficient condition for this is that $f_{lk}(\mathbf{N})$ be independent of t at equilibrium. In this case $\bar{\bar{F}}_\tau(\mathbf{N}) = \bar{\bar{F}}(\mathbf{N})$, and by

Eqs. (12.3) and (12.10), $\overline{\overline{\overline{F}}}_\tau = \overline{\overline{\overline{F}}}$. Hence, $\overline{\overline{\overline{F}}}_\tau$ may be replaced by $\overline{\overline{\overline{F}}}$ in Postulate *A*.

As before, we conclude that the density matrix $\mathfrak{f}(\mathbf{N}) = f(\mathscr{H}(\mathbf{N}))$. Then we have *Postulate B. Grand canonical ensemble.* For an open, isothermal thermodynamic system, i.e., a system with assigned values for the independent variables $T, \mu_1, \ldots, \mu_r, x_1, \ldots, x_s$,

$$\theta(\mathbf{N})\mathfrak{f}(\mathbf{N}) = \text{const} \times \exp[-\beta\mathscr{H}(\mathbf{N})] \exp(-\boldsymbol{\gamma} \cdot \mathbf{N}) \tag{12.11}$$

where $\boldsymbol{\gamma} \cdot \mathbf{N}$ means $\gamma_1 N_1 + \cdots + \gamma_r N_r$ and β and the γ's are constants. The postulate concerns only the product of (the scalar) θ and (the matrix) $\mathfrak{f}$, as only the product is involved in calculating ensemble averages which are to be compared with experiment. Equation (12.11) means that

$$\theta(\mathbf{N})f_{lk}(\mathbf{N}) = \text{const} \times \exp(-\boldsymbol{\gamma} \cdot \mathbf{N}) \int \varphi_k^*(\mathbf{N},q) \exp[-\beta\mathscr{H}(\mathbf{N})]\varphi_l(\mathbf{N},q)\, dq \tag{12.12}$$

If we use Eq. (12.12) in Eq. (12.9), we find

$$\text{const} = \frac{1}{\Xi} \tag{12.13}$$

and

$$\theta(\mathbf{N})\mathfrak{f}(\mathbf{N}) = \frac{\exp[-\beta\mathscr{H}(\mathbf{N})] \exp(-\boldsymbol{\gamma} \cdot \mathbf{N})}{\Xi} \tag{12.14}$$

where

$$\Xi(\beta,\mathbf{x},\boldsymbol{\gamma}) = \sum_{\mathbf{N} \geqslant 0} Q(\beta,\mathbf{x},\mathbf{N}) \exp(-\boldsymbol{\gamma} \cdot \mathbf{N}) \tag{12.15}$$

Ξ is the "grand partition function." Any of the forms of Q in the preceding section may be used in Eq. (12.15).

For $\theta(\mathbf{N})$ we find, from Eqs. (12.12) and (12.13),

$$\theta(\mathbf{N}) = \sum_k \theta(\mathbf{N}) f_{kk}(\mathbf{N}) \tag{12.16}$$

$$= \frac{\exp(-\boldsymbol{\gamma} \cdot \mathbf{N})Q(\mathbf{N})}{\Xi} \tag{12.17}$$

and hence, from Eq. (12.14),

$$\mathfrak{f}(\mathbf{N}) = \frac{\exp[-\beta\mathscr{H}(\mathbf{N})]}{Q(\mathbf{N})} \tag{12.18}$$

as in the canonical ensemble.

13. Derivation of Generalized Ensembles from the Microcanonical Ensemble

Up to this point, in both classical and quantum statistics, we have introduced postulates relating to closed, isolated systems (microcanonical ensemble), closed, isothermal systems (canonical ensemble), and open, isothermal systems (grand canonical ensemble). The choice of the microcanonical ensemble for a closed, isolated system has been justified to a considerable extent in our discussion. We now wish to show that actually the canonical and grand canonical ensembles (and other ensembles) may be deduced from the microcanonical ensemble, so that from the point of view of logical structure, the postulates on the canonical and grand canonical ensemble could have been omitted in our prior treatment. We have not omitted these postulates partly for historical reasons and partly to be able to indicate at the outset something of the scope of statistical mechanics through the introduction of ensembles other than the microcanonical. But at the conclusion of this section, we may definitely adopt the position that Postulates A and B for a closed, isolated system (microcanonical ensemble) form the entire postulatory foundation of statistical mechanics. This refers specifically to the quantum-mechanical version of the microcanonical ensemble since the classical treatment is included as a special case of this formulation (Chap. 3). For completeness, however, we shall indicate briefly at the end of this section how the classical canonical and grand canonical ensembles follow from the classical microcanonical ensemble.

A closed, isothermal system is in thermal contact with its surroundings. An open, isothermal system is in thermal and material contact with its surroundings. A closed, isothermal, isobaric (including pressure and/or other generalized forces associated with the external variables $\mathbf{x}$) system is in thermal and mechanical contact with its surroundings—that is, the values of the external variables (e.g., volume) can fluctuate, just as the numbers of molecules do in a system in material contact with its surroundings. The above "environments" for thermodynamic systems are common; the argument of the present section (below) will provide us with the appropriate statistical mechanical representative ensemble to apply to each situation. For example, from the proper representative ensemble we shall be able to calculate the probability of observing a given energy and given numbers of molecules in an open, isothermal system.

There is an additional, more general, "environment" that must be considered for completeness, although this case is of only trivial interest as far as applications of statistical mechanics are concerned. That is, it is possible for a system to be in thermal, material, *and* mechanical contact with its surroundings. It should certainly be part of the task of statistical mechanics to determine the representative ensemble for this situation also. Thus, we should like to know: What is the probability of observing such a system with a given energy, given numbers of molecules, and given values of the external variables? There are certain complications relating to the assignment of thermodynamic "independent" variables here (see Chap. 3), but this does not alter the fact that statistical mechanics ought to be able to provide a definite answer to the above question (see the end of Appendix 2 for an illustration).

Examples of systems of this type are shown in Figs. 2 and 3. In Fig. 2, a crystal (the system) is in contact with a large amount of its equilibrium vapor (the surroundings). In Fig. 3 we have a more general situation (the system is not restricted to conditions under which two phases exist in equilibrium). A movable, weightless, and frictionless piston separates the system (bottom portion of the container) from its surroundings.

The simplest procedure, and the one we adopt, is to treat the most general ensemble (thermal, material, and mechanical contact) explicitly and then obtain the required results for less general ensembles as special cases. If the

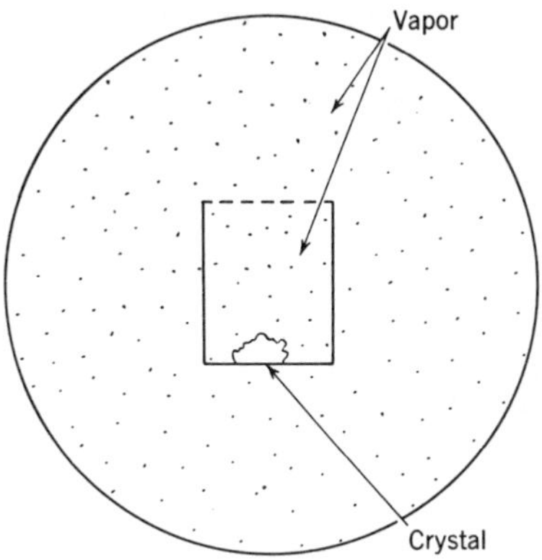

FIG. 2. Crystal in thermal, mechanical, and material contact with surroundings.

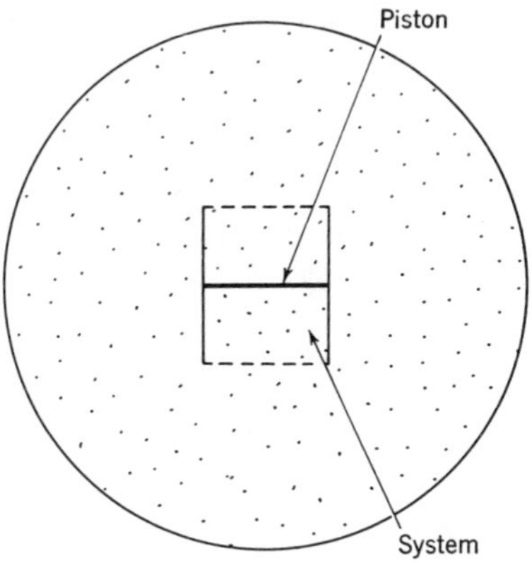

FIG. 3. System in thermal, mechanical, and material contact with surroundings.

reader prefers to avoid the most general ensemble, the argument below goes through separately, with obvious changes, for an open, isothermal system [omit Eq. (13.6)] and for a closed, isothermal, isobaric system [omit Eq. (13.7)]. A closed, isothermal system may then be regarded as a special case of either of these. The parts of Chaps. 3 and 4 relating to the most general ensemble may also be passed over, at the reader's option.

We turn, then, to the consideration of a perfectly isolated system of fixed volume V_t, number of molecules N_t, and energy E_t. For ease of visualization and to simplify notation, we have chosen a one-component system, and the volume V as the only external parameter. The extension to the general case $\mathbf{N}$, $\mathbf{x}$ is very straightforward.[1] The system consists of a collection of $\mathscr{N}$

[1] For each component there is an equation of type (13.7) and for each external variable there is an equation of type (13.6).

distinguishable (i.e., labeled) subsystems[1] ($\mathcal{N} \to \infty$) separated from each other by walls which allow heat transfer from one subsystem to another, which are permeable to molecules, and which are freely movable (so that the volume of each subsystem can fluctuate, though the total volume V_t of the system is constant). Aside from this means of establishing thermal, material, and mechanical contact between all the subsystems in the system, the subsystems are independent of each other (in the usual terminology, the subsystems are only weakly interacting).

Since the system is closed and perfectly isolated, our fundamental postulates [Eqs. (10.2), (11.7), and (11.8)] tell us that we may deduce equilibrium properties of the system from a microcanonical ensemble of systems. The particular equilibrium properties of the system so obtained will then prove to be completely equivalent to our canonical and grand canonical postulates.

In order to be able to use a microcanonical ensemble, our first task is to list the basic (energy) states or vectors of the system. For those of the $\mathcal{N}$ subsystems in the system with particular values of N and V (at any time t), the basic (energy) states on the subsystem level are the $\psi_k(N,V)$, where

$$\mathscr{H}(N,V)\psi_k(N,V) = E_k(N,V)\psi_k(N,V) \tag{13.1}$$

Now if we have two independent, distinguishable quantum-mechanical subsystems a and b and if

$$\mathscr{H}(a)\psi(a) = E(a)\psi(a)$$

$$\mathscr{H}(b)\psi(b) = E(b)\psi(b)$$

then the energy eigenfunctions of the combined subsystem $a + b$ are products of the energy eigenfunctions of the separate subsystems and the energy eigenvalues are sums of the separate eigenvalues:

$$\mathscr{H}(a + b) = \mathscr{H}(a) + \mathscr{H}(b)$$

$$\begin{aligned}\mathscr{H}(a + b)\psi(a)\psi(b) &= \psi(b)\mathscr{H}(a)\psi(a) + \psi(a)\mathscr{H}(b)\psi(b)\\ &= [E(a) + E(b)]\psi(a)\psi(b)\end{aligned} \tag{13.2}$$

By this same kind of argument, we see that the basic (energy) states of the system are of the form

$$\psi = \psi(1)\psi(2)\cdots\psi(l)\cdots\psi(\mathcal{N}) \tag{13.3}$$

where $\psi(l)$ is a basic (energy) state for the lth subsystem. For example, if the lth subsystem happens to have (at any particular time t) $N = N'$ and $V = V'$, then $\psi(l)$ would be one of the $\psi_k(N',V')$. Of course, the actual state $\psi^{(l)}$ of the lth subsystem would in general be a superposition of these basic states:

$$\psi^{(l)} = \sum_k A_k\psi_k(N',V') \tag{13.4}$$

[1] The physical system of interest is equivalent to a subsystem here.

At another time, with $N = N''$ and $V = V''$, $\psi(l)$ would be one of the $\psi_k(N'',V'')$. Clearly, each $\psi(l)$ $(l = 1, 2, \ldots, \mathcal{N})$ in Eq. (13.3) might in principle be any one of the $\psi_k(N,V)$ where $0 \leqslant N \leqslant N_t$ and $0 \leqslant V \leqslant V_t$, and the basic states of the system [Eq. (13.3)] must be enumerated with all these possibilities in mind.

For a particular basic state $\psi = \psi(1) \cdots \psi(\mathcal{N})$, let $n_k(N,V)$ be the number of times $\psi_k(N,V)$ occurs as a factor. In order to be a basic state, ψ must be constructed in such a way that

$$\sum_{k,N,V} n_k(N,V) = \mathcal{N} \tag{13.5}$$

$$\sum_{k,N,V} n_k(N,V)V = V_t \tag{13.6}$$

$$\sum_{k,N,V} n_k(N,V)N = N_t \tag{13.7}$$

$$\sum_{k,N,V} n_k(N,V)E_k(N,V) = E_t \tag{13.8}$$

where summation over V (instead of integration) is used for simplicity of notation.[1] Equation (13.8) must hold in order for ψ to be an energy eigenfunction of the system [see Eq. (13.2)].

A "distribution" n is defined as a particular assignment of a numerical value to each of the $n_k(N,V)$ *consistent with Eqs.* (13.5) *to* (13.8). We observe that there are Ω_n different basic states all with the same distribution n, where

$$\Omega_n = \frac{\mathcal{N}!}{\prod_{k,N,V} n_k^{(n)}(N,V)!} \tag{13.9}$$

and $n_k^{(n)}(N,V)$ represents the value of $n_k(N,V)$ in the distribution n. In Eq. (13.9), $\mathcal{N}!$ is the total number of ways of assigning the $\mathcal{N}$ factors in ψ to the $\mathcal{N}$ subsystems. But a new basic state is not obtained by an exchange of the same $\psi_k(N,V)$ between two subsystems. Hence, we must correct $\mathcal{N}!$ as in Eq. (13.9) to obtain Ω_n. The total number of basic states Ω is then the sum of Ω_n over *all* distributions n:

$$\Omega = \sum_n \Omega_n \tag{13.10}$$

Let the Ω basic (energy) states of the system, each of the form $\psi(1) \cdots \psi(\mathcal{N})$, be denoted by $\psi[1], \psi[2], \ldots, \psi[\Omega]$. Then the state Ψ of the system, may be represented by

$$\Psi = \sum_{j=1}^{\Omega} A[j]\psi[j] \tag{13.11}$$

[1] Integration over V is discussed in Chap. 3.

That is, the system is known to be in the degenerate energy state E_t (degeneracy Ω) so that the expansion in Eq. (13.11) is limited to the Ω energy eigenfunctions belonging to this eigenvalue [see Eq. (11.12)].

With this background, let us find an expression for the probability $p_k(N,V)$ defined as follows: if we select at random a subsystem from the system and make an observation to determine the basic (energy) state of the selected subsystem,[1] then $p_k(N,V)$ is the probability of observing the state $\psi_k(N,V)$. To actually calculate $p_k(N,V)$, our postulates state that the system here may be represented by a microcanonical ensemble of systems and that $p_k(N,V)$ may be taken as the average of the probability defined above over all systems in the microcanonical ensemble.

Let the state of the ith system of the ensemble be

$$\Psi^i = \sum_{j=1}^{\Omega} A^i[j]\psi[j] \tag{13.12}$$

If an observation is made to determine the basic state of the ith system, the probability of observing $\psi[j]$ is $|A^i[j]|^2$. When we say that the ith system is observed in some basic state $\psi(1) \cdots \psi(\mathcal{N})$, this of course means that subsystem 1 is observed in the basic state (on the subsystem level) $\psi(1)$, 2 in $\psi(2)$, etc. Let $n_k^{[j]}(N,V)$ be the value of $n_k(N,V)$ in the product making up $\psi[j]$. Then if the ith system is observed in the state $\psi[j]$, the fraction of subsystems in the ith system observed in the state $\psi_k(N,V)$ is $n_k^{[j]}(N,V)/\mathcal{N}$. This is also the probability that a subsystem selected at random from the ith system would be one of those observed in the state $\psi_k(N,V)$. The *simultaneous* probability, then, of observing the ith system in the state $\psi[j]$ and also a random subsystem selected from the ith system in the state $\psi_k(N,V)$ is the product

$$|A^i[j]|^2 \cdot \frac{n_k^{[j]}(N,V)}{\mathcal{N}}$$

The (total) probability of observing the random subsystem in the state $\psi_k(N,V)$ when the basic state of the ith system is observed, irrespective of the basic state of the ith system, is the sum of the above product over basic states,

$$P_k^i(N,V) = \sum_{j=1}^{\Omega} |A^i[j]|^2 \frac{n_k^{[j]}(N,V)}{\mathcal{N}} \tag{13.13}$$

The ensemble average of $P_k^i(N,V)$, $\bar{\bar{P}}_k(N,V)$, is then the quantity to be set equal to $p_k(N,V)$. Now

$$\bar{\bar{P}}_k(N,V) = \sum_{j=1}^{\Omega} \overline{\overline{|A[j]|^2}} \frac{n_k^{[j]}(N,V)}{\mathcal{N}} \tag{13.14}$$

[1] For example, N,V and the energy are observed, in the presence of a perturbation to split degeneracy, if necessary.

and
$$\overline{\overline{|A[j]|^2}} = \frac{1}{\Omega} \tag{13.15}$$

according to Eq. (11.8) for a microcanonical ensemble; thus

$$\bar{\bar{P}}_k(N,V) = \frac{\sum_{j=1}^{\Omega} n_k^{[j]}(N,V)}{\mathcal{N}\Omega} \tag{13.16}$$

$$= \frac{\sum_n \Omega_n n_k^{(n)}(N,V)}{\mathcal{N}\Omega} \tag{13.17}$$

$$= \frac{\bar{n}_k(N,V)}{\mathcal{N}} = p_k(N,V) \tag{13.18}$$

since all Ω_n basic states with the same distribution n have the same value of $n_k(N,V)$, $n_k^{(n)}(N,V)$. The quantity $\bar{n}_k(N,V)$ is the mean value of $n_k(N,V)$ over all Ω basic states.

We next obtain an explicit expression for $\bar{n}_k(N,V)$. The well-known Darwin and Fowler application of the method of steepest descents can be employed for this purpose.[1] However, the same result is obtained, and can be justified, by the much simpler procedure which we adopt here. Namely, we find first the value of $n_k(N,V)$, $n_k^*(N,V)$, in the most probable distribution n^* (i.e., the distribution with maximum Ω_n), and then show that $\bar{n}_k(N,V) \rightarrow n_k^*(N,V)$ as $\mathcal{N} \rightarrow \infty$.

It is more convenient to use $\ln \Omega_n$ than Ω_n here; therefore we write Eq. (13.9) in the form

$$\ln \Omega_n = \mathcal{N} \ln \mathcal{N} - \mathcal{N} - \sum_{k,N,V} [n_k^{(n)}(N,V) \ln n_k^{(n)}(N,V) - n_k^{(n)}(N,V)] \tag{13.19}$$

The introduction of the Stirling approximation is legitimate here since we shall be concerned only with distributions in the neighborhood of the most probable distribution, in the limiting case $\mathcal{N} \rightarrow \infty$ where (as we shall see) $n_k^*(N,V) \rightarrow \infty$. That is, $n_k^{(n)}(N,V)$ in Eq. (13.19) will always be a very large number. Let us expand $\ln \Omega_n$ in a Taylor series about n^*:

$$\ln \Omega_n = \ln \Omega_{n^*} + \sum_{k,N,V} \left(\frac{\partial \ln \Omega_n}{\partial n_k^{(n)}(N,V)} \right)_{n=n^*} \delta n_k^{(n)}(N,V) + \frac{1}{2} \sum_{k,N,V} \left(\frac{\partial^2 \ln \Omega_n}{\partial n_k^{(n)}(N,V)^2} \right)_{n=n^*} \delta n_k^{(n)}(N,V)^2 + \cdots \tag{13.20}$$

where $\delta n_k^{(n)}(N,V) = n_k^{(n)}(N,V) - n_k^*(N,V)$. Second-order cross-product terms do not appear in Eq. (13.20) because cross derivatives vanish [see Eq.

[1] See Fowler and Guggenheim (Gen. Ref.).

(13.19)]. The variations $\delta n_k^{(n)}(N,V)$ in Eq. (13.20) are not all independent, of course, because of the four relations, Eqs. (13.5) to (13.8), which must be maintained for all distributions as already discussed. In view of these four relations, the most probable distribution is not found in the usual way from

$$\frac{\partial \ln \Omega_n}{\partial n_k^{(n)}(N,V)} = 0 \qquad \text{all } k,N,V \tag{13.21}$$

but rather by the Lagrange method of undetermined multipliers. The value of $n_k^*(N,V)$ is given by

$$\frac{\partial}{\partial n_k^{(n)}(N,V)}(\ln \Omega_n - \alpha\mathcal{N} - \nu V_t - \gamma N_t - \beta E_t) = 0 \qquad \text{all } k,N,V \tag{13.22}$$

where α, ν, γ, and β are constants. On substitution of Eqs. (13.5) to (13.8) and (13.19) into Eq. (13.22), the result is

$$n_k^*(N,V) = e^{-\alpha}e^{-\nu V}e^{-\gamma N}e^{-\beta E_k(N,V)} \tag{13.23}$$

Equation (13.23) shows the form of the dependence of $n_k^*(N,V)$ on N, V, and $E_k(N,V)$, but the significance of the constants is unspecified at this point. However, we may eliminate α immediately using Eq. (13.5):

$$\sum_{k,N,V} n_k^*(N,V) = \mathcal{N} = e^{-\alpha} \sum_{k,N,V} e^{-\nu V}e^{-\gamma N}e^{-\beta E_k(N,V)} \tag{13.24}$$

or

$$n_k^*(N,V) = \mathcal{N}\frac{e^{-\nu V}e^{-\gamma N}e^{-\beta E_k(N,V)}}{\sum\limits_{k,N,V} e^{-\nu V}e^{-\gamma N}e^{-\beta E_k(N,V)}} \tag{13.25}$$

Equations (13.22) ensure that the sum of linear terms in $\delta n_k^{(n)}(N,V)$ vanishes in Eq. (13.20), taking into account the four relations between the $\delta n_k^{(n)}(N,V)$. Then Eq. (13.20) becomes (to second-order terms)

$$\ln \Omega_n = \ln \Omega_{n^*} + \frac{1}{2}\sum_{k,N,V}\left(\frac{\partial^2 \ln \Omega_n}{\partial n_k^{(n)}(N,V)^2}\right)_{n=n^*} \delta n_k^{(n)}(N,V)^2 \tag{13.26}$$

or

$$\Omega_n = \Omega_{n^*} \exp\left\{-\frac{1}{2}\sum_{k,N,V}\left[\frac{\delta n_k^{(n)}(N,V)}{n_k^*(N,V)}\right]^2 n_k^*(N,V)\right\} \tag{13.27}$$

where again it is understood that the set of $\delta n_k^{(n)}(N,V)$ used must be consistent with Eqs. (13.5) to (13.8).

Equations (13.25) and (13.27), together, lead to a most important conclusion. According to the first equation, $n_k^*(N,V)$ increases linearly with $\mathcal{N}$ (which we can take as large as we please[1]). Now, with $\mathcal{N}$ a certain value, consider a distribution n, in the neighborhood of n^*, having $n_k^{(n)}(N,V) = n_k^*(N,V) + \delta n_k^{(n)}(N,V)$ (for the various k, N, V). If we substitute these

[1] V_t, N_t, and E_t must be increased proportionally with $\mathcal{N}$.

$\delta n_k^{(n)}(N,V)$ in Eq. (13.27) we can calculate Ω_n/Ω_{n*}. Now let $\mathcal{N} \to \infty$, increasing each $n_k^*(N,V)$ and $\delta n_k^{(n)}(N,V)$ in proportion to $\mathcal{N}$. That is, as we let $\mathcal{N} \to \infty$, we consider a series of distributions all with precisely the same *relative* deviation from the most probable distribution as that of the original distribution n. Thus, in Eq. (13.27), $\delta n_k^{(n)}(N,V)/n_k^*(N,V)$ is constant as $\mathcal{N} \to \infty$. But $n_k^*(N,V) \to \infty$, and $\Omega_n/\Omega_{n*} \to 0$. Hence, the contribution, for example in Eq. (13.17), of a distribution n with a given relative deviation, however small, from the most probable distribution, becomes insignificant as $\mathcal{N} \to \infty$. In the limit, then, only the most probable distribution itself need be considered and Eq. (13.18) becomes

$$p_k(N,V) = \frac{\bar{n}_k(N,V)}{\mathcal{N}} = \frac{n_k^*(N,V)}{\mathcal{N}} \tag{13.28}$$

This simple and rather remarkable result is a direct consequence of the form of Eq. (13.19), leading to $\partial^2 \ln \Omega_n/\partial n_k^{(n)}(N,V)^2 = -1/n_k^{(n)}(N,V)$.

Distributions with deviations from the most probable distribution of order higher than the second, are, of course, even less important than those already discussed and disposed of.

Generalizing the notation at this point to $\mathbf{x}$ and $\mathbf{N}$, our conclusion is then that

$$\frac{\bar{n}_k(\mathbf{N},\mathbf{x})}{\mathcal{N}} = p_k(\mathbf{N},\mathbf{x}) = \frac{\exp(-\boldsymbol{\nu}\cdot\mathbf{x})\exp(-\boldsymbol{\gamma}\cdot\mathbf{N})\exp[-\beta E_k(\mathbf{N},\mathbf{x})]}{\sum\limits_{k,\mathbf{N},\mathbf{x}} \exp(-\boldsymbol{\nu}\cdot\mathbf{x})\exp(-\boldsymbol{\gamma}\cdot\mathbf{N})\exp[-\beta E_k(\mathbf{N},\mathbf{x})]} \tag{13.29}$$

where $p_k(\mathbf{N},\mathbf{x})$, it will be recalled, is the probability of observing the state $\psi_k(\mathbf{N},\mathbf{x})$ if an observation is made to determine the basic (energy) state of a subsystem selected at random from a perfectly isolated collection (the system) of $\mathcal{N}$ subsystems in thermal, material, and mechanical contact with each other. The final step in our argument is to point out that if the $\mathcal{N}$ subsystems (of the system) are instantaneously separated from each other and each put in perfect isolation, $p_k(\mathbf{N},\mathbf{x})$ is unaffected in view of the initial independence (weak interaction) of the subsystems. Also, after the separation we have an ensemble of $\mathcal{N}$ subsystems which is certainly representative of a physical system in thermal, material, and mechanical contact with its surroundings. This follows because each (sub)system in the ensemble is "prepared" by being put in contact with the other $\mathcal{N} - 1$ (sub)systems (its surroundings). In other words, this ensemble is prepared "operationally" (see Chap. 2) and is therefore a representative ensemble.

In summary, by applying the microcanonical postulates, we have deduced that the generalized ensemble representative of a system in thermal, material, and mechanical contact with its surroundings has the property that $p_k(\mathbf{N},\mathbf{x})$ [Eq. (13.29)] is the probability of observing the state $\psi_k(\mathbf{N},\mathbf{x})$ if an observation is made to determine the basic (energy) state of a system selected at random from the ensemble.

To obtain the grand canonical ensemble as a special case, we select from the generalized ensemble only those systems with the same (constant) $\mathbf{x}$. The grand canonical ensemble is therefore representative of a system in thermal and material contact with its surroundings. Equation (13.29) becomes

$$p_k(\mathbf{N}) = \frac{\exp(-\boldsymbol{\nu}\cdot\mathbf{x})\exp(-\boldsymbol{\gamma}\cdot\mathbf{N})\exp[-\beta E_k(\mathbf{N})]}{\sum_{k,\mathbf{N}}\exp(-\boldsymbol{\nu}\cdot\mathbf{x})\exp(-\boldsymbol{\gamma}\cdot\mathbf{N})\exp[-\beta E_k(\mathbf{N})]}$$

$$= \frac{\exp(-\boldsymbol{\gamma}\cdot\mathbf{N})\exp[-\beta E_k(\mathbf{N})]}{\sum_{k,\mathbf{N}}\exp(-\boldsymbol{\gamma}\cdot\mathbf{N})\exp[-\beta E_k(\mathbf{N})]} \tag{13.30}$$

This equation is the same as Eq. (12.14) for $\theta(\mathbf{N})f_{kk}(\mathbf{N})$ in the energy language.

Similarly, the canonical ensemble, representative of a closed system in thermal (contact) equilibrium, is the result if we select from the generalized ensemble only those systems with the same (constant) $\mathbf{N}$ and $\mathbf{x}$:

$$p_k = \frac{\exp(-\boldsymbol{\nu}\cdot\mathbf{x})\exp(-\boldsymbol{\gamma}\cdot\mathbf{N})\,e^{-\beta E_k}}{\sum_k \exp(-\boldsymbol{\nu}\cdot\mathbf{x})\exp(-\boldsymbol{\gamma}\cdot\mathbf{N})\,e^{-\beta E_k}}$$

$$= \frac{e^{-\beta E_k}}{\sum_k e^{-\beta E_k}} \tag{13.31}$$

in agreement with Eqs. (11.30) and (11.35).

Classical Mechanics. In conclusion, we indicate very briefly here how the type of argument used above can also be applied to derive the corresponding classical generalized ensemble (and therefore the classical canonical and grand canonical ensembles) from the classical microcanonical ensemble.

The phase space of each of the $\mathcal{N}$ subsystems in the system is divided up into energy shells of width δE, where δE is very small and constant. We denote the value of the energy in the kth shell by E_k, where $(k-1)\delta E \leqslant E_k \leqslant k\delta E$. Let $\omega_k(N,V)$ be the volume in phase space of the kth energy shell for a subsystem with N and V. At an arbitrary time t, let $n_k(N,V)$ be the number of subsystems in the system with E_k, N, and V. A "distribution" is a set of values for the $n_k(N,V)$ consistent with Eqs. (13.5) to (13.8), replacing $E_k(N,V)$ by E_k. A "configuration" of the system is a definite assignment of values of E_k, N, and V to each subsystem in the system. Because of the weak interaction between subsystems, the volume in the phase space of the system associated with a given configuration is the product of $\mathcal{N}$ factors of the type $\omega_k(N,V)$, one for each subsystem. The system is represented by a microcanonical ensemble of systems which includes every possible configuration and gives a weight, in forming ensemble averages, to each configuration proportional to its volume in the phase space of the system.

The number of configurations Ω_n with the same distribution n is given by Eq. (13.9). Then the volume in the phase space of the system, associated with these Ω_n configurations, is

$$\Omega_n' = \Omega_n \prod_{k,N,V} \omega_k(N,V)^{n_k^{(n)}(N,V)} \tag{13.32}$$

Let us write

$$\Omega' = \sum_n \Omega_n' \tag{13.33}$$

Then, because of the microcanonical postulates, the probability of a system selected at random from the ensemble having the distribution n is Ω_n'/Ω'. Also, the probability of a subsystem selected at random from this system having E_k, N, and V is $n_k^{(n)}(N,V)/\mathcal{N}$. The total probability of finding E_k, N and V, irrespective of the distribution of the system, is then $p_k(N,V)$ in Eqs. (13.17) and (13.18) (with Ω_n and Ω'). By the same argument as before, we find

$$p_k(N,V) = \frac{e^{-\nu V}e^{-\gamma N}\omega_k(N,V)e^{-\beta E_k}}{\sum\limits_{k,N,V} e^{-\nu V}e^{-\gamma N}\omega_k(N,V)e^{-\beta E_k}} \tag{13.34}$$

Now we let $\delta E \to dE$ and replace summation by integration. Then

$$P(E,N,V)\,dE = \frac{e^{-\nu V}e^{-\gamma N}\omega(E,N,V)e^{-\beta E}\,dE}{\sum\limits_{N,V} e^{-\nu V}e^{-\gamma N}\int\omega(E,N,V)e^{-\beta E}\,dE} \tag{13.35}$$

is the probability that a (sub)system selected at random from the generalized ensemble (formed from the system by separation and perfect isolation of the $\mathcal{N}$ subsystems) will have N, V, and E between E and $E + dE$. Finally, since $\omega(E,N,V)\,dE$ is just the volume in phase space [of a (sub)system with N and V] between E and $E + dE$, the integration can be carried out by summing over elements of volume $dp\,dq$ instead of over energy shells. Thus,

$$\frac{P(E,N,V)}{\omega(E,N,V)}\,dp\,dq = \frac{e^{-\nu V}e^{-\gamma N}e^{-\beta E(p,q)}\,dp\,dq}{\sum\limits_{N,V} e^{-\nu V}e^{-\gamma N}\int e^{-\beta E(p,q)}\,dp\,dq} \tag{13.36}$$

is the probability of the selected (sub)system having N and V and a dynamical state between p, q and $p + dp$, $q + dq$, where $E(p,q) = H(p,q)$. This is the desired result [see Eqs. (5.3) and (6.7)].

GENERAL REFERENCES

Quantum Mechanics

Bohm, D., "Quantum Theory" (Prentice-Hall, New York, 1951).

Kemble, E. C., "The Fundamental Principles of Quantum Mechanics" (McGraw-Hill, New York, 1937).

Pauling, L., and E. B. Wilson, Jr., "Introduction to Quantum Mechanics" (McGraw-Hill, New York, 1935).
Rojansky, V., "Introductory Quantum Mechanics" (Prentice-Hall, New York, 1938).
Schiff, L. I., "Quantum Mechanics" (McGraw-Hill, New York, 1949).
Tolman, R. C., "The Principles of Statistical Mechanics" (Oxford, London, 1938).

Statistical Mechanics

Fowler, R. H., "Statistical Mechanics" (Cambridge, London, 1936).
Fowler, R. H., and E. A. Guggenheim, "Statistical Thermodynamics" (Cambridge, London, 1939).
ter Haar, D., *Revs. Mod. Phys.*, **27**, 289 (1955).
Mayer, J. E., and M. G. Mayer, "Statistical Mechanics" (Wiley, New York, 1940).
Schrodinger, E., "Statistical Thermodynamics" (Cambridge, London, 1948).
Tolman, R. C., "The Principles of Statistical Mechanics" (Oxford, London, 1938).

CHAPTER 3

STATISTICAL MECHANICS AND THERMODYNAMICS

In this chapter we consider interrelationships between quantum statistical mechanics, classical statistical mechanics, and thermodynamics. We first indicate (Sec. 14) the associations between thermodynamic and quantum statistical mechanical quantities for the most important ensembles, and then supplement this (Sec. 15) with a brief summary of ensembles of possible interest. The connection between classical statistical mechanics and thermodynamics is established by regarding classical statistical mechanics as a limiting form of quantum statistics. The transition from the quantum to the classical case is carried out explicitly in Sec. 16 for the canonical ensemble. Finally, the relation between entropy and the classical Boltzmann $\mathfrak{H}$ theorem is discussed briefly in Sec. 17.

14. Association of Thermodynamic Variables with Statistical Mechanical Quantities

Generalized Ensemble. The formal treatment of the generalized ensemble[1] introduced in Sec. 13 is actually simpler and more direct than in the other cases; therefore we consider it first. There are, however, certain additional features to be discussed, not encountered with the other ensembles, which arise from the fact that thermodynamically $\boldsymbol{\nu}$, β and $\boldsymbol{\gamma}$ cannot all be independent. We therefore carry the analysis here only to the point of establishing the associations between thermodynamic and statistical mechanical quantities, and return to the additional features mentioned after considering other ensembles.

It is perhaps appropriate to restate at the outset that the generalized ensemble is unlikely to be of much value in applications. We include a discussion of this ensemble primarily for completeness and to make possible a fuller understanding of the relationships between the various ensembles of

[1] From this point on we use the term "generalized ensemble" to refer specifically to this most general case, rather than to *any* ensemble (e.g., the grand canonical ensemble) more general than the canonical ensemble.

statistical mechanics. Also, physical systems *can* exist for which the present ensemble is representative (thermal, mechanical, and material equilibrium). Several elementary applications of the generalized ensemble are carried out in Appendixes 2 to 4 to show that it is possible to obtain correct thermodynamic results by this method.[1]

We may associate immediately (Postulate A) the thermodynamic internal energy E with the ensemble average $\bar{E}$ (we use a single bar for all ensemble averages from this point on unless otherwise noted):

$$E \leftrightarrow \bar{E} \tag{14.1}$$

$$\bar{E} = \sum_{k,\mathbf{N},\mathbf{x}} p_k(\mathbf{N},\mathbf{x}) E_k(\mathbf{N},\mathbf{x}) \tag{14.2}$$

where
$$p_k(\mathbf{N},\mathbf{x}) = \frac{\exp(-\boldsymbol{\nu}\cdot\mathbf{x})\exp(-\boldsymbol{\gamma}\cdot\mathbf{N})\exp[-\beta E_k(\mathbf{N},\mathbf{x})]}{\Upsilon} \tag{14.3}$$

$$\Upsilon = \sum_{k,\mathbf{N},\mathbf{x}} \exp(-\boldsymbol{\nu}\cdot\mathbf{x})\exp(-\boldsymbol{\gamma}\cdot\mathbf{N})\exp[-\beta E_k(\mathbf{N},\mathbf{x})] \tag{14.4}$$

To obtain further associations of type (14.1), we write Eq. (14.2) in differential form,

$$d\bar{E} = \sum_{k,\mathbf{N},\mathbf{x}} dp_k(\mathbf{N},\mathbf{x}) E_k(\mathbf{N},\mathbf{x}) \tag{14.5}$$

where the $E_k(\mathbf{N},\mathbf{x})$ are constants (the sum is over k, $\mathbf{N}$, $\mathbf{x}$). We now use Eq. (14.3) to eliminate $E_k(\mathbf{N},\mathbf{x})$ in Eq. (14.5), with the result

$$d\bar{E} = -\frac{1}{\beta}\sum_{k,\mathbf{N},\mathbf{x}} [\boldsymbol{\nu}\cdot\mathbf{x} + \boldsymbol{\gamma}\cdot\mathbf{N} + \ln p_k(\mathbf{N},\mathbf{x}) + \ln \Upsilon]\, dp_k(\mathbf{N},\mathbf{x}) \tag{14.6}$$

Since
$$\bar{N}_i = \sum_{k,\mathbf{N},\mathbf{x}} p_k(\mathbf{N},\mathbf{x}) N_i \tag{14.7}$$

and
$$\bar{x}_i = \sum_{k,\mathbf{N},\mathbf{x}} p_k(\mathbf{N},\mathbf{x}) x_i \tag{14.8}$$

Equation (14.6) becomes

$$\begin{aligned} d\bar{E} &= -\frac{1}{\beta}\sum_{k,\mathbf{N},\mathbf{x}} [\ln p_k(\mathbf{N},\mathbf{x})\, dp_k(\mathbf{N},\mathbf{x})] - \frac{1}{\beta}\boldsymbol{\nu}\cdot d\bar{\mathbf{x}} - \frac{1}{\beta}\boldsymbol{\gamma}\cdot d\bar{\mathbf{N}} \\ &= -\frac{1}{\beta}\, d\left\{\sum_{k,\mathbf{N},\mathbf{x}} [p_k(\mathbf{N},\mathbf{x}) \ln p_k(\mathbf{N},\mathbf{x})]\right\} - \frac{1}{\beta}\boldsymbol{\nu}\cdot d\bar{\mathbf{x}} - \frac{1}{\beta}\boldsymbol{\gamma}\cdot d\bar{\mathbf{N}} \end{aligned} \tag{14.9}$$

where we have employed $\sum dp_k(\mathbf{N},\mathbf{x}) = 0$. With the further associations

[1] Guggenheim first introduced this ensemble [see *J. Chem. Phys.*, **7**, 103 (1939), and Fowler and Guggenheim (Gen. Ref.), Chap. 6]. It has been criticized recently by Prigogine [*Physica*, **16**, 133 (1950)]. The author believes that the present treatment removes Prigogine's objections. Professor Prigogine has authorized the author to state that he agrees with the treatment of the generalized ensemble presented in this chapter and in the Appendixes.

(Postulate A) $x_i \leftrightarrow \bar{x}_i$ and $N_i \leftrightarrow \bar{N}_i$, Eq. (14.9) is seen to be just the statistical mechanical version of the thermodynamic equation

$$dE = T\,dS - \mathbf{X}\cdot d\mathbf{x} + \boldsymbol{\mu}\cdot d\mathbf{N} \tag{14.10}$$

where X_i is the generalized force appropriate to x_i. By comparing Eqs. (14.9) and (14.10), we have further

$$\mu_i \leftrightarrow -\frac{\gamma_i}{\beta} \tag{14.11}$$

and

$$X_i \leftrightarrow \frac{\nu_i}{\beta} \tag{14.12}$$

Also

$$dS \leftrightarrow -\frac{1}{\beta T}\, d\left\{ \sum_{k,\mathbf{N},\mathbf{x}} [p_k(\mathbf{N},\mathbf{x}) \ln p_k(\mathbf{N},\mathbf{x})]\right\} \tag{14.13}$$

The thermodynamic entropy S depends only on the thermodynamic state of the system and therefore dS, on the left-hand side of (14.13), is an exact differential. The right-hand side of (14.13) can also be made an exact differential by putting $1/\beta T = \text{constant} = \mathbf{k}$ (called the Boltzmann constant). The numerical value of $\mathbf{k}$ may be obtained, for example, from the perfect gas thermometer (see Appendix 1). Thus, we adopt the associations

$$\frac{1}{\mathbf{k}T} \leftrightarrow \beta \tag{14.14}$$

and

$$S \leftrightarrow -\mathbf{k} \sum_{k,\mathbf{N},\mathbf{x}} p_k(\mathbf{N},\mathbf{x}) \ln p_k(\mathbf{N},\mathbf{x}) \tag{14.15}$$

where a constant of integration c has been set equal to zero in Eq. (14.15). This constant is independent of the thermodynamic state of the system. Now in thermodynamics only entropy *differences* are defined. When we refer to the entropy as an extensive thermodynamic property we actually mean, for example, that if we double the extent of a system (keeping intensive properties constant) in two states A and B, then $S_B - S_A$ is doubled. If we use Eq. (14.15) for S, the constant c is not doubled but it cancels in taking the difference $S_B - S_A$. It is therefore not necessary, as might at first be supposed, to set $c = 0$ in order for the entropy to be an extensive quantity.[1] However, as a matter of convenience, since c is arbitrary and devoid of any thermodynamic significance, we set $c = 0$.

Equation (14.3) may now be written

$$p_k(\mathbf{N},\mathbf{x}) = \frac{\exp(-\mathbf{X}\cdot\mathbf{x}/\mathbf{k}T)\exp(\mathbf{N}\cdot\boldsymbol{\mu}/\mathbf{k}T)\exp[-E_k(\mathbf{N},\mathbf{x})/\mathbf{k}T]}{\Upsilon} \tag{14.16}$$

where

$$\Upsilon = \sum_{k,\mathbf{N},\mathbf{x}} \exp\left(-\frac{\mathbf{X}\cdot\mathbf{x}}{\mathbf{k}T}\right)\exp\left(\frac{\mathbf{N}\cdot\boldsymbol{\mu}}{\mathbf{k}T}\right)\exp\left[-\frac{E_k(\mathbf{N},\mathbf{x})}{\mathbf{k}T}\right] \tag{14.17}$$

[1] See Schrödinger (Gen. Ref.).

If we substitute Eq. (14.16) in Eq. (14.15), we find

$$\mathbf{k}T \ln \Upsilon = TS - \mathbf{X} \cdot \bar{\mathbf{x}} + \bar{\mathbf{N}} \cdot \boldsymbol{\mu} - \bar{E} = 0 \tag{14.18}$$

and, using Eq. (14.10),

$$d(\mathbf{k}T \ln \Upsilon) = S\, dT - \bar{\mathbf{x}} \cdot d\mathbf{X} + \bar{\mathbf{N}} \cdot d\boldsymbol{\mu} = 0 \tag{14.19}$$

This expresses the fact that not all of the intensive properties T, $\mathbf{X}$, $\boldsymbol{\mu}$ can be independent variables; there must be one relation between them.

If we let $f(T,\mathbf{X},\boldsymbol{\mu}) = 0$ represent the thermodynamic relation between T, $\mathbf{X}$, and $\boldsymbol{\mu}$, then Eq. (14.18) can be written more explicitly as

$$\lim_{f \to 0} \mathbf{k}T \ln \Upsilon = TS - \mathbf{X} \cdot \bar{\mathbf{x}} + \bar{\mathbf{N}} \cdot \boldsymbol{\mu} - \bar{E} = 0$$

That is, Υ may be computed by carrying out the sums in Eq. (14.17) with arbitrary values of T, $\mathbf{X}$, and $\boldsymbol{\mu}$; in the thermodynamic limit $f \to 0$, we must have $\ln \Upsilon = 0$. The condition $\ln \Upsilon = 0$ does not imply that $\Upsilon = 1$, but rather only that $\mathbf{k}T \ln \Upsilon$ is of negligible order compared to $\bar{N}\mathbf{k}T$, where $\bar{N}$ is the average number of molecules in the system, and therefore also that $\mathbf{k}T \ln \Upsilon$ is negligible compared to the other terms in Eq. (14.18). For example, Υ might be of order $\bar{N}, (\bar{N})^2$, etc.

It might appear that if Υ were calculated with arbitrary T, $\mathbf{X}$, and $\boldsymbol{\mu}$, then the condition $\ln \Upsilon = 0$ could be used to deduce $f(T,\mathbf{X},\boldsymbol{\mu}) = 0$. This procedure is not applicable in general, however, because $f = 0$ is a sufficient but not a necessary condition foı $\ln \Upsilon = 0$. The actual method of deducing $f = 0$ from the generalized ensemble is discussed later.

It will be observed that if we differentiate the right-hand side of Eq. (14.17), which is a function of T, $\mathbf{X}$, and $\boldsymbol{\mu}$, with respect to T, X_i, and μ_i, and compare Eqs. (14.2), (14.7), (14.8), and (14.18), we obtain formal expressions for S, $\bar{x}_i$, and $\bar{N}_i$. These formal relations have thermodynamic significance, however, only when the parameters T, $\mathbf{X}$, and $\boldsymbol{\mu}$ satisfy the thermodynamic equation $f(T,\mathbf{X},\boldsymbol{\mu}) = 0$. Thus, we have

$$S = \lim_{f \to 0} \mathbf{k}T \left(\frac{\partial \ln \Upsilon}{\partial T}\right)_{\boldsymbol{\mu},\mathbf{X}} \tag{14.20}$$

$$-\bar{x}_i = \lim_{f \to 0} \mathbf{k}T \left(\frac{\partial \ln \Upsilon}{\partial X_i}\right)_{T,\boldsymbol{\mu},X} \tag{14.21}$$

$$\bar{N}_i = \lim_{f \to 0} \mathbf{k}T \left(\frac{\partial \ln \Upsilon}{\partial \mu_i}\right)_{T,\mu_j,\mathbf{X}} \tag{14.22}$$

The application of these equations is considered further below and in Appendixes 2 to 4.

Superficially, there is some arbitrariness in Υ and S connected with the summation over $\mathbf{x}$. To illustrate, let the only external parameter be the volume V, and let ΔV be the interval between successive values of V used in the summation over[1] V in Υ. ΔV must be small enough so that the result of the summation over V is independent of ΔV, except for a constant factor (see below). A reasonable choice for ΔV would be a volume of the order of the volume per molecule (close packed, or at the density of interest), although another choice, based on the uncertainty principle and the mean kinetic energy at temperature T, would be Λ^3 where[2]

$$\Lambda = \frac{h}{(2\pi m\mathbf{k}T)^{1/2}} \tag{14.23}$$

and m is the mass of a typical molecule of the system.

It is obvious that if we calculate, from Eq. (14.15), S_1 using ΔV_1 and S_2 using ΔV_2, then

$$S_2 = S_1 + \mathbf{k} \ln \frac{\Delta V_1}{\Delta V_2} \tag{14.24}$$

But S_1 is of order $\bar{N}\mathbf{k}$ while $S_2 - S_1$ is only of order $\mathbf{k}$; thus this arbitrariness in ΔV is inconsequential. Similarly, from Eq. (14.17),

$$\Upsilon_2 = \frac{\Upsilon_1 \Delta V_1}{\Delta V_2}$$

$$\ln \Upsilon_2 = \ln \Upsilon_1 + \ln \frac{\Delta V_1}{\Delta V_2} \tag{14.25}$$

In Eq. (14.25), although Υ_1 and Υ_2 differ, if $\ln \Upsilon_1$ is of negligible order compared to $\bar{N}$ then $\ln \Upsilon_2$ is also, and this is all that is necessary in Eq. (14.18).

We conclude that the choice of ΔV is arbitrary for thermodynamic purposes, provided only that it is sufficiently small (say, a volume per molecule).

If we wish to replace summation in Eq. (14.17) by integration,

$$\Upsilon = \frac{\sum\limits_{k,\mathbf{N},\mathbf{x}} \Delta x_1 \cdots \Delta x_s \exp(-\mathbf{X}\cdot\mathbf{x}/\mathbf{k}T)\exp(\mathbf{N}\cdot\boldsymbol{\mu}/\mathbf{k}T)\exp[-E_k(\mathbf{N},\mathbf{x})/\mathbf{k}T]}{\Delta x_1 \cdots \Delta x_s}$$

$$= \frac{\int \cdots \int dx_1 \cdots dx_s \sum\limits_{k,\mathbf{N}} \exp(-\mathbf{X}\cdot\mathbf{x}/\mathbf{k}T)\exp(\mathbf{N}\cdot\boldsymbol{\mu}/\mathbf{k}T)\exp\left[-\frac{E_k(\mathbf{N},\mathbf{x})}{\mathbf{k}T}\right]}{x_1^* \cdots x_s^*}$$

[1] In some problems the volume is "discrete" rather than continuous so that summation is natural. See Chap. 7 and Appendixes 4 and 5.

[2] That is, Λ^3 is of the order of the smallest volume within which a molecule of mass m can definitely be located at T. See Appendix 6 for a discussion of the order of magnitude of Λ^3.

where $x_i^* = \Delta x_i$. That is,

$$\Upsilon = \int \cdots \int d\left(\frac{x_1}{x_1^*}\right) \cdots d\left(\frac{x_s}{x_s^*}\right) \sum_{k,\mathbf{N}} \exp\left(-\frac{\mathbf{X}\cdot\mathbf{x}}{\mathbf{k}T}\right) \exp\left(\frac{\mathbf{N}\cdot\boldsymbol{\mu}}{\mathbf{k}T}\right) \exp\left[-\frac{E_k(\mathbf{N},\mathbf{x})}{\mathbf{k}T}\right] \tag{14.26}$$

The limits on the sums and integrals for this ensemble, as well as for the other ensembles to be considered, are discussed later.

Grand Canonical Ensemble. It is not possible, as might be expected off-hand, to give a complete discussion of an open isothermal system by selecting in Eq. (14.6), above, only those systems in the generalized ensemble with specified $\mathbf{x}$. For if we do this, the external work term [as in Eq. (14.10)] will be missing.

To introduce external work in an ensemble of systems with the same $\mathbf{x}$, we consider an *infinitely slow* change dx_j in one of the external parameters, x_j, for all of the $\mathcal{N}$ systems of the ensemble, keeping each system otherwise perfectly isolated and closed. The resulting change in the expectation value[1] of the energy for the ith system of the ensemble, $dE_{(j)}{}^{(i)}$, must then be equal (conservation of energy) to the thermodynamic work done by the outside world on the ith system, $-dW_j{}^{(i)} = -X_j{}^{(i)}dx_j$, where $X_j{}^{(i)}$ is the generalized force associated with (in fact, defined by) $dW_j{}^{(i)}$ and dx_j. If the ith system has a composition $\mathbf{N}$, let the expectation value of the energy before the variation in x_j be

$$\hat{E}^{(i)} = \sum_k |A_k{}^{(i)}|^2 E_k(\mathbf{N};\mathbf{x}) \tag{14.27}$$

in the notation of Sec. 8 (energy language) except that $(\mathbf{N};\mathbf{x})$ emphasizes that $\mathbf{x}$ is a parameter here. It can be proved[2] that the $|A_k{}^{(i)}|^2$ remain constant if changes are made in the external parameters of a quantum-mechanical system at an infinitely slow rate ("adiabatic principle").[3] We are therefore justified in writing

$$d\hat{E}_{(j)}{}^{(i)} = \sum_k |A_k{}^{(i)}|^2 \frac{\partial E_k(\mathbf{N};\mathbf{x})}{\partial x_j} dx_j = -dW_j{}^{(i)} = -X_j{}^{(i)}\,dx_j \tag{14.28}$$

The ensemble average of $X_j{}^{(i)}$, $\bar{X}_j$, is then (Postulate A) to be associated with the thermodynamic external force X_j:

$$X_j \leftrightarrow \bar{X}_j \tag{14.29}$$

[1] We use $\hat{G}$ for expectation value in this section in order to continue the use of $\bar{G}$ for ensemble average.

[2] See Tolman (Gen. Ref.), pp. 412–414.

[3] It should be added that, for a degenerate energy level that "splits" on varying external parameters, the basic energy $\psi_k(\mathbf{N},\mathbf{x})$'s for the degenerate level must be so chosen that they go over into the proper (split) energy eigenfunctions when the parameters are changed.

where, on taking ensemble averages in Eq. (14.28), $\bar{X}_j$ is given by

$$-\bar{X}_j = \sum_{k,\mathbf{N}} p_k(\mathbf{N};\mathbf{x}) \frac{\partial E_k(\mathbf{N};\mathbf{x})}{\partial x_j} \tag{14.30}$$

and [Eq. (13.30)]

$$p_k(\mathbf{N};\mathbf{x}) = \frac{\exp(-\boldsymbol{\gamma}\cdot\mathbf{N})\exp[-\beta E_k(\mathbf{N};\mathbf{x})]}{\Xi} \tag{14.31}$$

$$\Xi = \sum_{k,\mathbf{N}} \exp(-\boldsymbol{\gamma}\cdot\mathbf{N})\exp[-\beta E_k(\mathbf{N};\mathbf{x})] \tag{14.32}$$

It should be noted in our present notation that $E_k(\mathbf{N};\mathbf{x}) = E_k(\mathbf{N},\mathbf{x})$ but $p_k(\mathbf{N};\mathbf{x}) \neq p_k(\mathbf{N},\mathbf{x})$ [compare Eqs. (14.3) and (14.31)] in general. The complete external work term ($j = 1, 2, \ldots, s$) is then

$$-\bar{\mathbf{X}}\cdot d\mathbf{x} = \sum_{k,\mathbf{N}} p_k(\mathbf{N};\mathbf{x})\, dE_k(\mathbf{N};\mathbf{x}) = -dW \tag{14.33}$$

where

$$dE_k(\mathbf{N};\mathbf{x}) = \sum_{j=1}^{s} \frac{\partial E_k(\mathbf{N};\mathbf{x})}{\partial x_j} dx_j \tag{14.34}$$

The association

$$E \leftrightarrow \bar{E} \tag{14.35}$$

is again adopted, where

$$\bar{E}(\boldsymbol{\gamma},\beta,\mathbf{x}) = \sum_{k,\mathbf{N}} p_k(\mathbf{N};\mathbf{x})E_k(\mathbf{N};\mathbf{x}) \tag{14.36}$$

Then, from Eqs. (14.32) and (14.33),

$$\begin{aligned} d\bar{E} &= \sum_{k,\mathbf{N}} p_k(\mathbf{N};\mathbf{x})\, dE_k(\mathbf{N};\mathbf{x}) + \sum_{k,\mathbf{N}} E_k(\mathbf{N};\mathbf{x})\, dp_k(\mathbf{N};\mathbf{x}) \\ &= -\bar{\mathbf{X}}\cdot d\mathbf{x} - \frac{1}{\beta}\sum_{k,\mathbf{N}}[\boldsymbol{\gamma}\cdot\mathbf{N} + \ln p_k(\mathbf{N};\mathbf{x}) + \ln \Xi]\, dp_k(\mathbf{N};\mathbf{x}) \\ &= -\frac{1}{\beta} d[\sum_{k,\mathbf{N}} p_k(\mathbf{N};\mathbf{x}) \ln p_k(\mathbf{N};\mathbf{x})] - \bar{\mathbf{X}}\cdot d\mathbf{x} - \frac{1}{\beta}\boldsymbol{\gamma}\cdot d\bar{\mathbf{N}} \end{aligned} \tag{14.37}$$

where

$$\bar{N}_i(\boldsymbol{\gamma},\beta,\mathbf{x}) = \sum_{k,\mathbf{N}} p_k(\mathbf{N};\mathbf{x})N_i \tag{14.38}$$

Again, comparing Eqs. (14.10) and (14.37), we have by the same arguments as before,

$$N_i \leftrightarrow \bar{N}_i \tag{14.39}$$

$$\mu_i \leftrightarrow -\frac{\gamma_i}{\beta} \tag{14.40}$$

$$\frac{1}{\mathbf{k}T} \leftrightarrow \beta \tag{14.41}$$

$$S(\boldsymbol{\gamma},\beta,\mathbf{x}) \leftrightarrow -\mathbf{k}\sum_{k,\mathbf{N}} p_k(\mathbf{N};\mathbf{x}) \ln p_k(\mathbf{N};\mathbf{x}) \tag{14.42}$$

With the above associations introduced, Eq. (14.31) becomes

$$p_k(\mathbf{N};\mathbf{x}) = \frac{\exp(\mathbf{N}\cdot\boldsymbol{\mu}/\mathbf{k}T)\exp[-E_k(\mathbf{N};\mathbf{x})/\mathbf{k}T]}{\Xi} \tag{14.43}$$

$$\Xi(T,\boldsymbol{\mu},\mathbf{x}) = \sum_{k,\mathbf{N}} \exp\left(\frac{\mathbf{N}\cdot\boldsymbol{\mu}}{\mathbf{k}T}\right)\exp\left[-\frac{E_k(\mathbf{N};\mathbf{x})}{\mathbf{k}T}\right] \tag{14.44}$$

$$= \sum_{\mathbf{N}} Q(\mathbf{N},\mathbf{x},T)\exp\left(\frac{\mathbf{N}\cdot\boldsymbol{\mu}}{\mathbf{k}T}\right) \tag{14.45}$$

where Q is given (below) by Eq. (14.84). If Eq. (14.43) is substituted into Eq. (14.42), the result is

$$\mathbf{k}T \ln \Xi = TS + \bar{\mathbf{N}}\cdot\boldsymbol{\mu} - \bar{E} = \mathbf{x}\cdot\bar{\mathbf{X}} \tag{14.46}$$

and [Eq. (14.10)]

$$d(\mathbf{k}T \ln \Xi) = \bar{\mathbf{X}}\cdot d\mathbf{x} + S\,dT + \bar{\mathbf{N}}\cdot d\boldsymbol{\mu} = d(\mathbf{x}\cdot\bar{\mathbf{X}}) \tag{14.47}$$

All the thermodynamic properties of the system may then be deduced from $\Xi(T,\boldsymbol{\mu},\mathbf{x})$ using Eq. (14.46) and

$$\bar{X}_i(T,\boldsymbol{\mu},\mathbf{x}) = \mathbf{k}T\left(\frac{\partial \ln \Xi}{\partial x_i}\right)_{T,\boldsymbol{\mu},x_j} \tag{14.48}$$

$$S(T,\boldsymbol{\mu},\mathbf{x}) = \mathbf{k}\ln \Xi + \mathbf{k}T\left(\frac{\partial \ln \Xi}{\partial T}\right)_{\mathbf{x},\boldsymbol{\mu}} \tag{14.49}$$

$$\bar{N}_i(T,\boldsymbol{\mu},\mathbf{x}) = \mathbf{k}T\left(\frac{\partial \ln \Xi}{\partial \mu_i}\right)_{T,\mathbf{x},\mu_j} \tag{14.50}$$

Equations (14.48) to (14.50) follow directly from Eq. (14.47), or by differentiating Eq. (14.44).

A number of applications of the grand canonical ensemble, which is representative of a system in thermal and material equilibrium with its surroundings (independent variables T, $\boldsymbol{\mu}$, $\mathbf{x}$), will appear in other chapters. See also Appendixes 2 to 5.

Ensemble for Isothermal-Isobaric System. We consider here a closed system in thermal and mechanical equilibrium with its surroundings. We see as a special case of Eq. (13.29) that

$$p_k(\mathbf{x};\mathbf{N}) = \frac{\exp(-\boldsymbol{\nu}\cdot\mathbf{x})\exp[-\beta E_k(\mathbf{x};\mathbf{N})]}{\Delta} \tag{14.51}$$

is the appropriate probability here, where

$$\Delta = \sum_{k,\mathbf{x}} \exp(-\boldsymbol{\nu}\cdot\mathbf{x})\exp[-\beta E_k(\mathbf{x};\mathbf{N})] \tag{14.52}$$

The notation $(\mathbf{x};\mathbf{N})$ means that $\mathbf{N}$ is a parameter (as was $\mathbf{x}$ in the grand canonical ensemble). Of course $E_k(\mathbf{x};\mathbf{N}) = E_k(\mathbf{N},\mathbf{x})$, but in general $p_k(\mathbf{x};\mathbf{N}) \neq p_k(\mathbf{N};\mathbf{x}) \neq p_k(\mathbf{N},\mathbf{x})$ [compare Eqs. (14.3), (14.31), and (14.51)].

There is nothing like an "adiabatic principle" in quantum mechanics[1] to handle changes in the parameter $\mathbf{N}$; therefore we establish associations with thermodynamics for $\mathbf{N}$ constant and then use a thermodynamic definition to extend the treatment to variations in $\mathbf{N}$.

The procedure is the same as for the generalized ensemble above. We write ($\mathbf{N}$ constant)

$$E \leftrightarrow \bar{E} \tag{14.53}$$

$$\bar{E}(\boldsymbol{\nu},\beta,\mathbf{N}) = \sum_{k,\mathbf{x}} p_k(\mathbf{x};\mathbf{N}) E_k(\mathbf{x};\mathbf{N}) \tag{14.54}$$

$$d\bar{E} = \sum_{k,\mathbf{x}} E_k(\mathbf{x};\mathbf{N})\, dp_k(\mathbf{x};\mathbf{N}) \tag{14.55}$$

$$= -\frac{1}{\beta}\sum_{k,\mathbf{x}} [\boldsymbol{\nu}\cdot\mathbf{x} + \ln p_k(\mathbf{x};\mathbf{N}) + \ln \Delta]\, dp_k(\mathbf{x};\mathbf{N}) \tag{14.56}$$

$$= -\frac{1}{\beta} d[\sum_{k,\mathbf{x}} p_k(\mathbf{x};\mathbf{N}) \ln p_k(\mathbf{x};\mathbf{N})] - \frac{1}{\beta}\boldsymbol{\nu}\cdot d\bar{\mathbf{x}} \tag{14.57}$$

where

$$\bar{x}_i(\boldsymbol{\nu},\beta,\mathbf{N}) = \sum_{k,\mathbf{x}} p_k(\mathbf{x};\mathbf{N}) x_i \tag{14.58}$$

Then, as before,

$$X_i \leftrightarrow \frac{\nu_i}{\beta} \tag{14.59}$$

$$\frac{1}{\mathbf{k}T} \leftrightarrow \beta \tag{14.60}$$

$$S(\boldsymbol{\nu},\beta,\mathbf{N}) \leftrightarrow -\mathbf{k}\sum_{k,\mathbf{x}} p_k(\mathbf{x};\mathbf{N}) \ln p_k(\mathbf{x};\mathbf{N}) \tag{14.61}$$

Equations (14.51) and (14.52) become

$$p_k(\mathbf{x};\mathbf{N}) = \frac{\exp(-\mathbf{X}\cdot\mathbf{x}/\mathbf{k}T) \exp[-E_k(\mathbf{x};\mathbf{N})/\mathbf{k}T]}{\Delta} \tag{14.62}$$

$$\Delta(T,\mathbf{X},\mathbf{N}) = \sum_{k,\mathbf{x}} \exp\left(-\frac{\mathbf{X}\cdot\mathbf{x}}{\mathbf{k}T}\right) \exp\left[-\frac{E_k(\mathbf{x};\mathbf{N})}{\mathbf{k}T}\right] \tag{14.63}$$

$$= \sum_{\mathbf{x}} Q(\mathbf{N},\mathbf{x},T) \exp\left(-\frac{\mathbf{X}\cdot\mathbf{x}}{\mathbf{k}T}\right) \tag{14.64}$$

[1] Particles cannot be added infinitely slowly to a system. They are discrete or "quantized," the minimum variation being a single particle.

where Q is given by Eq. (14.84). From Eqs. (14.10), (14.61), and (14.62) we find

$$\mathbf{k}T \ln \Delta = TS - \mathbf{X} \cdot \bar{\mathbf{x}} - \bar{E} = -F \tag{14.65}$$

and

$$d(\mathbf{k}T \ln \Delta) = S\,dT - \bar{\mathbf{x}} \cdot d\mathbf{X} - \boldsymbol{\mu} \cdot d\mathbf{N} = -dF \tag{14.66}$$

where μ_i is defined thermodynamically by

$$\mu_i = \left(\frac{\partial F}{\partial N_i}\right)_{T,\mathbf{X},N_j} \tag{14.67}$$

Besides Eq. (14.65), the other relations connecting thermodynamic functions directly with Δ are

$$-\mu_i(T,\mathbf{X},\mathbf{N}) = \mathbf{k}T \left(\frac{\partial \ln \Delta}{\partial N_i}\right)_{T,\mathbf{X},N_j} \tag{14.68}$$

$$S(T,\mathbf{X},\mathbf{N}) = \mathbf{k} \ln \Delta + \mathbf{k}T \left(\frac{\partial \ln \Delta}{\partial T}\right)_{\mathbf{X},\mathbf{N}} \tag{14.69}$$

$$-\bar{x}_i(T,\mathbf{X},\mathbf{N}) = \mathbf{k}T \left(\frac{\partial \ln \Delta}{\partial X_i}\right)_{T,\mathbf{N},X_j} \tag{14.70}$$

Equations (14.68) to (14.70) follow from Eq. (14.66); Eqs. (14.69) and (14.70) may also be obtained by differentiating Eq. (14.63).

There is again, as in the generalized ensemble, some arbitrariness in Δ and S owing to the choice of, for example, ΔV. But, as before, the magnitudes involved are thermodynamically insignificant.

Equation (14.63) can, of course, be written in integral form:

$$\Delta = \int \cdots \int d\left(\frac{x_1}{x_1^*}\right) \cdots d\left(\frac{x_s}{x_s^*}\right) \sum_k \exp\left(-\frac{\mathbf{X} \cdot \mathbf{x}}{\mathbf{k}T}\right) \exp\left[-\frac{E_k(\mathbf{x};\mathbf{N})}{\mathbf{k}T}\right] \tag{14.71}$$

Applications of this ensemble are included in Appendixes 2 to 4 and in Chap. 7.

Canonical Ensemble. We are concerned here with a closed system, having external variables $\mathbf{x}$, in thermal equilibrium with its surroundings. The external work term can be introduced as in the grand canonical ensemble and variations in composition as in the preceding ensemble.

Equations (14.27) and (14.28) apply to the ith system of the canonical ensemble except that we shall use the notation $E_k(;\mathbf{N},\mathbf{x})(= E_k(\mathbf{N},\mathbf{x}))$ to indicate that both $\mathbf{N}$ and $\mathbf{x}$ are parameters. On taking ensemble averages,

$$X_j \leftrightarrow \bar{X}_j \tag{14.72}$$

$$-\bar{X}_j = \sum_k p_k(;\mathbf{N},\mathbf{x}) \frac{\partial E_k(;\mathbf{N},\mathbf{x})}{\partial x_j} \tag{14.73}$$

where
$$p_k(;\mathbf{N},\mathbf{x}) = \frac{\exp[-\beta E_k(;\mathbf{N},\mathbf{x})]}{Q} \tag{14.74}$$

$$Q = \sum_k \exp[-\beta E_k(;\mathbf{N},\mathbf{x})] \tag{14.75}$$

and in general $p_k(;\mathbf{N},\mathbf{x}) \neq p_k(\mathbf{N};\mathbf{x}) \neq p_k(\mathbf{x};\mathbf{N}) \neq p_k(\mathbf{x},\mathbf{N})$.

Also, with $\mathbf{N}$ constant,

$$E \leftrightarrow \bar{E} \tag{14.76}$$

$$\bar{E}(\beta,\mathbf{N},\mathbf{x}) = \sum_k p_k(;\mathbf{N},\mathbf{x})E_k(;\mathbf{N},\mathbf{x}) \tag{14.77}$$

$$d\bar{E} = \sum_k p_k(;\mathbf{N},\mathbf{x})\, dE_k(;\mathbf{N},\mathbf{x}) + \sum_k E_k(;\mathbf{N},\mathbf{x})\, dp_k(;\mathbf{N},\mathbf{x}) \tag{14.78}$$

$$= -\bar{\mathbf{X}} \cdot d\mathbf{x} - \frac{1}{\beta}\sum_k [\ln p_k(;\mathbf{N},\mathbf{x}) + \ln Q]\, dp_k \tag{14.79}$$

$$= -\frac{1}{\beta} d[\sum_k p_k(;\mathbf{N},\mathbf{x}) \ln p_k(;\mathbf{N},\mathbf{x})] - \bar{\mathbf{X}} \cdot d\mathbf{x} \tag{14.80}$$

so that
$$\frac{1}{\mathbf{k}T} \leftrightarrow \beta \tag{14.81}$$

$$S(\beta,\mathbf{N},\mathbf{x}) \leftrightarrow -\mathbf{k}\sum_k p_k(;\mathbf{N},\mathbf{x}) \ln p_k(;\mathbf{N},\mathbf{x}) \tag{14.82}$$

Equation (14.74), using these associations, is then

$$p_k(;\mathbf{N},\mathbf{x}) = \frac{\exp[-E_k(;\mathbf{N},\mathbf{x})/\mathbf{k}T]}{Q} \tag{14.83}$$

$$Q(T,\mathbf{N},\mathbf{x}) = \sum_k \exp\left[-\frac{E_k(;\mathbf{N},\mathbf{x})}{\mathbf{k}T}\right] \tag{14.84}$$

Equations (14.10), (14.82), and (14.83) then lead to

$$\mathbf{k}T \ln Q = TS - \bar{E} = -A \tag{14.85}$$

and
$$d(\mathbf{k}T \ln Q) = S\, dT + \bar{\mathbf{X}} \cdot d\mathbf{x} - \boldsymbol{\mu} \cdot d\mathbf{N} = -dA \tag{14.86}$$

where μ_i is defined thermodynamically by

$$\mu_i = \left(\frac{\partial A}{\partial N_i}\right)_{T,\mathbf{x},N_j} \tag{14.87}$$

In addition to Eq. (14.85), the relations between Q and thermodynamic properties are

$$-\mu_i(T,\mathbf{N},\mathbf{x}) = \mathbf{k}T\left(\frac{\partial \ln Q}{\partial N_i}\right)_{T,\mathbf{x},N} \tag{14.88}$$

$$S(T,\mathbf{N},\mathbf{x}) = \mathbf{k} \ln Q + \mathbf{k}T \left(\frac{\partial \ln Q}{\partial T}\right)_{\mathbf{x},\mathbf{N}} \tag{14.89}$$

$$\bar{X}_i(T,\mathbf{N},\mathbf{x}) = \mathbf{k}T \left(\frac{\partial \ln Q}{\partial x_i}\right)_{T,\mathbf{N},x_j} \tag{14.90}$$

Equations (14.88) to (14.90) follow from Eq. (14.86); Eqs. (14.89) and (14.90) also result on differentiating Eq. (14.84).

As will be seen in Chap. 4, for the usual thermodynamic systems containing very large numbers of particles, fluctuations about mean values are extremely small.[1] In such systems, so long as one is interested only in mean values (as is ordinarily the case), there is no need to be conscientious about using the particular ensemble appropriate to the actual experimental arrangement of the system of interest. The same answers are in fact obtained from all ensembles; which one is adopted is purely a matter of mathematical convenience. But if fluctuations are of interest or if the number of particles in the system is not very large, then the proper ensemble must, of course, be used. We mention these points here by way of explaining that the canonical ensemble, primarily for reasons of mathematical convenience, is by far the most extensively used ensemble in statistical mechanics. However, in the past decade or two, there has been an increasing awareness of the utility of other ensembles, particularly the grand canonical ensemble, for some purposes and problems.

Microcanonical Ensemble. We obtain the desired results for a microcanonical ensemble, representative of a closed, isolated system, as a limiting form of the canonical ensemble.

Suppose we are interested in a closed, isolated system with energy in the narrow range between E and $E + \delta E$, where the basic energy states in this interval are given by $E \leqslant E_k(;\mathbf{N},\mathbf{x}) \leqslant E + \delta E$ and $k' \leqslant k \leqslant k' + \Omega - 1$. For practical purposes, we may regard the Ω basic (energy) states as belonging to a single degenerate energy level, $E_{k'}(;\mathbf{N},\mathbf{x}) = E$. A canonical ensemble would be representative of the state of this system provided we imagine that somehow all basic (energy) states with k outside the above range are inaccessible. With this restriction, we take over all of the associations established in the canonical case. In particular, Eqs. (14.74) and (14.82) become

$$\begin{aligned} p_k(;E,\mathbf{N},\mathbf{x}) &= \frac{\exp[-\beta E_k(;\mathbf{N},\mathbf{x})]}{\Omega \exp[-\beta E_k(;\mathbf{N},\mathbf{x})]} \qquad k' \leqslant k \leqslant k' + \Omega - 1 \\ &= \frac{1}{\Omega} \qquad k' \leqslant k \leqslant k' + \Omega - 1 \\ &= 0 \qquad \text{otherwise} \end{aligned} \tag{14.91}$$

[1] Except in special circumstances such as phase transitions and critical points.

where the notation $(;E,\mathbf{N},\mathbf{x})$ is analogous to that already used, and

$$S \leftrightarrow -\mathbf{k} \sum_{k=k'}^{k'+\Omega-1} \frac{1}{\Omega} \ln \frac{1}{\Omega} = \mathbf{k} \ln \Omega \tag{14.92}$$

In place of Eq. (14.86), we have, using Eq. (14.10),

$$d(\mathbf{k} \ln \Omega) = dS = \frac{dE}{T} + \frac{1}{T} \bar{\mathbf{X}} \cdot d\mathbf{x} - \frac{1}{T} \boldsymbol{\mu} \cdot d\mathbf{N} \tag{14.93}$$

The following equations, in addition to Eq. (14.92), then relate $\Omega(E,\mathbf{N},\mathbf{x})$ directly to thermodynamic properties (as functions of the variables E, $\mathbf{N}$, and $\mathbf{x}$):

$$\frac{1}{\mathbf{k}T} = \left(\frac{\partial \ln \Omega}{\partial E}\right)_{\mathbf{x},\mathbf{N}} \tag{14.94}$$

$$\frac{\bar{X}_i}{\mathbf{k}T} = \left(\frac{\partial \ln \Omega}{\partial x_i}\right)_{E,\mathbf{N},x_j} \tag{14.95}$$

$$-\frac{\mu_i}{\mathbf{k}T} = \left(\frac{\partial \ln \Omega}{\partial N_i}\right)_{E,\mathbf{x},N_j} \tag{14.96}$$

$\bar{X}_i$ retains the meaning of an ensemble average [Eqs. (14.73) and (14.91)], because, although $E_k(;\mathbf{N},\mathbf{x})$ has (virtually) the same value E for all Ω values of k, this is not in general true of $\partial E_k(;\mathbf{N},\mathbf{x})/\partial x_i$ on account of the possibility of "splitting" of the degeneracy.

Because of the difficulty in practice of finding $\Omega(E,\mathbf{N},\mathbf{x})$, the microcanonical ensemble is not extensively used in statistical mechanical applications.

It is rather interesting that of all the ensembles considered in this section, the microcanonical ensemble bears the most direct relation to mechanics (as we have seen in Chaps. 1 and 2) but is the most difficult ensemble to relate to thermodynamics. For this reason we have avoided trying to establish the connection directly and have instead treated the microcanonical ensemble as a special case of the canonical ensemble.

The effect of the range δE on Ω and on thermodynamic properties derived from Ω will be discussed in Chap. 4.

Further Discussion of the Generalized Ensemble. Some of the results above are collected in Table 1.

All partition functions except Υ give a thermodynamic "potential," or fundamental thermodynamic function. Also, in all cases but Υ, at least one extensive variable is included in the set of independent variables. Thus, in these cases, the finite extent of the system is fixed explicitly. In applications of thermodynamics we are, in fact, always interested in systems of finite extent. This is true no less of systems in thermal, mechanical, and material equilibrium (Υ) than in systems with other types of equilibrium. Since an ensemble must be representative of the experimental system on which it is modeled, Υ, the partition function for the generalized ensemble, must contain the extent of the system implicitly, even though it does not appear explicitly in the "independent" variables $\mathbf{X}$, $\boldsymbol{\mu}$, T.

We mentioned in Sec. 13 that an example of a system in thermal, mechanical, and material contact with its surroundings is a crystal in contact with a large amount of its equilibrium vapor. The maximum extent of the crystal (the system in this case) is determined by the size of the container (Fig. 2). The maximum extent of the system shown in Fig. 3 is also determined by the

TABLE 1. SUMMARY OF ENSEMBLES

Independent variables	Types of contact with surroundings	Partition function	Fundamental thermodynamic function	Name of ensemble
1. $\mathbf{X},\boldsymbol{\mu},T$	Mechanical, material, thermal	$\Upsilon = \sum_{k,\mathbf{N},\mathbf{x}} \exp\left[\frac{(-\mathbf{X}\cdot\mathbf{x} + \mathbf{N}\cdot\boldsymbol{\mu} - E_k)}{\mathbf{k}T}\right]$	$-\mathbf{k}T\ln\Upsilon = 0$	Generalized
2. $\mathbf{x},\boldsymbol{\mu},T$	Material, thermal	$\Xi = \sum_{k,\mathbf{N}} \exp\left[\frac{(\mathbf{N}\cdot\boldsymbol{\mu} - E_k)}{\mathbf{k}T}\right]$	$-\mathbf{k}T\ln\Xi = -\mathbf{x}\cdot\mathbf{X}$	Grand canonical
3. $\mathbf{X},\mathbf{N},T$	Mechanical, thermal	$\Delta = \sum_{k,\mathbf{x}} \exp\left[\frac{(-\mathbf{X}\cdot\mathbf{x} - E_k)}{\mathbf{k}T}\right]$	$-\mathbf{k}T\ln\Delta = F = \mathbf{N}\cdot\boldsymbol{\mu}$	Isothermal–isobaric
4. $\mathbf{x},\mathbf{N},T$	Thermal	$Q = \sum_{k} e^{-E_k/\mathbf{k}T}$	$-\mathbf{k}T\ln Q = A$	Canonical
5. $\mathbf{x},\mathbf{N},E$	None	$\Omega = \sum_{k} 1$	$-\mathbf{k}T\ln\Omega = -TS$	Microcanonical

size of the container. In general, a system in thermal, mechanical, and material contact with its surroundings is fully characterized by (self-consistent) values of $\mathbf{X}$, $\boldsymbol{\mu}$, and T *and* by the maximum extent of the system (in terms, say, of one of the $\mathbf{x}$).

In using the partition functions Q, Δ, and Ξ, the proper limits to employ are $1 \leqslant k \leqslant \infty$, $0 \leqslant N_i \leqslant \infty$ and $0 \leqslant x_i \leqslant \infty$. These limits are appropriate in view of: (1) the derivation in Sec. 13 (with $\mathcal{N} \to \infty$); and (2) the circumstance that, in any case, fluctuations away from the mean values of any of the $\mathbf{N}$ or $\mathbf{x}$ which are free to fluctuate are so small in general[1] that the limits are immaterial (see Chap. 4). The situation is different with respect to Υ, but before we go into this we have to digress briefly to introduce two

[1] Except for phase transitions and critical points.

partition functions, on the one hand intermediate between Υ and Ξ and, on the other, between Υ and Δ. Incidentally, these partition functions are not without interest in their own right.

We omit detailed derivations but obtain the correct results by analogy with Table 1. For this purpose we observe, in Table 1, that the sum of the thermodynamic functions in the exponential of the partition function, when added to the "fundamental thermodynamic function," is always $-TS$. For example, for Q, $-E + A = -TS$ and for Ξ, $\mathbf{N} \cdot \boldsymbol{\mu} - E - \mathbf{x} \cdot \mathbf{X} = -TS$. Now consider the two partition functions

$$\Gamma_{N_i} = \sum_{k,N_{j \neq i},\mathbf{x}} \exp\left(-\frac{\mathbf{X} \cdot \mathbf{x}}{\mathbf{k}T}\right) \exp\left(\frac{\sum_{j \neq i} N_j \mu_j}{\mathbf{k}T}\right) \exp\left[-\frac{E_k(\mathbf{N},\mathbf{x})}{\mathbf{k}T}\right] \tag{14.97}$$

$$\Gamma_{x_i} = \sum_{k,\mathbf{N},x_{j \neq i}} \exp\left(-\frac{\sum_{j \neq i} X_j x_j}{\mathbf{k}T}\right) \exp\left(\frac{\mathbf{N} \cdot \boldsymbol{\mu}}{\mathbf{k}T}\right) \exp\left[-\frac{E_k(\mathbf{N},\mathbf{x})}{\mathbf{k}T}\right] \tag{14.98}$$

Γ_{N_i} and Γ_{x_i} are the same as Υ [Eq. (14.17)] *except* that the sum over N_i is omitted in Γ_{N_i} and the sum over x_i is omitted in Γ_{x_i}. Γ_{N_i} is the appropriate partition function for a system in thermal and mechanical contact with its surroundings, and also in material contact with respect to all species except $\boldsymbol{i}$. By analogy with Table 1, the thermodynamic "potential" is

$$N_i \mu_i = -\mathbf{k}T \ln \Gamma_{N_i}(N_i, \mu_{j \neq i}, T, \mathbf{X}) \tag{14.99}$$

This partition function has been used by Stockmayer[1] in investigating the relation between light scattering and composition fluctuations (Chap. 4). Similarly,

$$-\bar{X}_i x_i = -\mathbf{k}T \ln \Gamma_{x_i}(\boldsymbol{\mu}, T, X_{j \neq i}, x_i) \tag{14.100}$$

Other related partition functions can be generated by omitting summation over two or more of the $\mathbf{N}$, $\mathbf{x}$. For both Γ_{N_i} and Γ_{x_i}, the extent of the system is fixed by one extensive variable, and the limits on k, $\mathbf{N}$, and $\mathbf{x}$ are those specified above for Q, Δ, and Ξ.

Let us return now to Υ. We note first that Υ can be written as

$$\Upsilon = \sum_{\mathbf{x}} \exp\left(-\frac{\mathbf{X} \cdot \mathbf{x}}{\mathbf{k}T}\right) \Xi\,(\mathbf{x}, \boldsymbol{\mu}, T) \tag{14.101}$$

$$= \sum_{\mathbf{N}} \exp\left(\frac{\mathbf{N} \cdot \boldsymbol{\mu}}{\mathbf{k}T}\right) \Delta(\mathbf{X}, \mathbf{N}, T) \tag{14.102}$$

$$= \sum_{x_i} e^{-X_i x_i / \mathbf{k}T} \Gamma_{x_i} \tag{14.103}$$

$$= \sum_{N_i} e^{N_i \mu_i / \mathbf{k}T} \Gamma_{N_i} \tag{14.104}$$

[1] W. H. Stockmayer, *J. Chem. Phys.*, **18**, 58 (1950).

Only the final summation in Υ causes any difficulty; therefore we may confine our attention here to Eqs. (14.103) and (14.104). According to Eqs. (14.99) and (14.100), we have

$$\Gamma_{N_i} = C_{N_i} e^{-N_i\mu_i/\mathbf{k}T} \tag{14.105}$$

$$\Gamma_{x_i} = C_{x_i} e^{\bar{X}_i x_i/\mathbf{k}T} \tag{14.106}$$

where $\ln C_{N_i}$ and $\ln C_{x_i}$ are thermodynamically negligible quantities which we take as constants (see Appendix 9). Equations (14.103) and (14.104) become, then,

$$\lim_{f\to 0} \Upsilon = C_{x_i} \int d\,\frac{x_i}{x_i^*} \tag{14.107}$$

$$= C_{N_i} \sum_{N_i} 1 \tag{14.108}$$

The physical significance of a constant integrand in Eq. (14.107) is that every value of x_i from $x_i = 0$ to the maximum value of x_i ("maximum extent of the system") has equal probability of occurrence [see Eq. (14.8)]. If $\bar{x}_i$ denotes the average value of x_i, the maximum value of x_i must then be $2\bar{x}_i$ (or if $x_{i\,(\max)}$ is the maximum value of x_i, $\bar{x}_i = x_{i\,(\max)}/2$). Thus Eq. (14.103) can be written more explicitly as

$$\Upsilon = \int_0^{2\bar{x}_i/x_i^*} d\left(\frac{x_i}{x_i^*}\right) e^{-X_i x_i/\mathbf{k}T}\Gamma_{x_i} \tag{14.109}$$

$$\lim_{f\to 0} \Upsilon = 2C_{x_i}\frac{\bar{x}_i}{x_i^*} \tag{14.110}$$

Similar remarks apply to N_i so we have also

$$\Upsilon = \sum_{N_i=0}^{2\bar{N}_i} e^{N_i\mu_i/\mathbf{k}T}\Gamma_N \tag{14.111}$$

$$\lim_{f\to 0} \Upsilon = 2C_{N_i}\bar{N}_i \tag{14.112}$$

As regards $p_k(\mathbf{N},\mathbf{x})$ [see Eq. (14.3)], Eqs. (14.109) and (14.111) imply that there must be a "cutoff" in the most probable distribution at $2\bar{x}_i$ and $2\bar{N}_i$ in order to satisfy the restraints, Eqs. (14.7) and (14.8). This is illustrated in Appendix 2.

In the grand ensemble, one of Eqs. (14.48) is redundant since Eq. (14.46) for the sum $\mathbf{x}\cdot\bar{\mathbf{X}}$ is already available. Similarly, one of Eqs. (14.68) in the isothermal-isobaric ensemble, one of Eqs. (14.88) and (14.90) in the canonical ensemble, and one of Eqs. (14.95) and (14.96) in the microcanonical ensemble are redundant. In the generalized ensemble, however, since there is no thermodynamic "potential" [Eq. (14.18) and Table 1], all of Eqs. (14.21) and (14.22) provide new information. One of these equations is unique, though:

if the final summation in Υ is over x_i [Eq. (14.109)], then Eq. (14.21) for $\bar{x}_i$ gives us $f(T,\mathbf{X},\boldsymbol{\mu}) = 0$ as the necessary and sufficient condition in order to obtain the identity $\bar{x}_i = \bar{x}_i$; and if the final summation is over N_i [Eq. (14.111)], then Eq. (14.22) for $\bar{N}_i$ provides $f(T,\mathbf{X},\boldsymbol{\mu}) = 0$. This, then, is the way in which the thermodynamic relation between $T,\mathbf{X}$ and $\boldsymbol{\mu}$ is deduced from the generalized ensemble.

Examples are included in Appendixes 2 to 4. See also the discussion of fluctuations in Chap. 4.

Entropy in Statistical Mechanics. In each of the ensembles above we have found the entropy S to be given by an expression of the form

$$S = -\mathbf{k}\sum_j p(j) \ln p(j) \tag{14.113}$$

where $p(j)$ is a probability, $p(j) \leqslant 1$ with $\sum_j p(j) = 1$, and j represents k and possibly $\mathbf{N}$ or $\mathbf{x}$ or both. We might begin by summarizing here the expressions given above for $p(j)$ in a way which shows a certain formal similarity between the different cases.[1]

1. "Independent" (i.e., assigned) variables $\mathbf{X}$, $\boldsymbol{\mu}$, T; fluctuations in $\mathbf{x}$, $\mathbf{N}$, E:

$$p_k(\mathbf{N},\mathbf{x}) = \exp\left(-\frac{\mathbf{X}\cdot\mathbf{x}}{\mathbf{k}T}\right)\exp\left(\frac{\mathbf{N}\cdot\boldsymbol{\mu}}{\mathbf{k}T}\right)\exp\left[-\frac{E_k(\mathbf{N},\mathbf{x})}{\mathbf{k}T}\right] \tag{14.114}$$

2. Independent variables $\mathbf{x}$, $\boldsymbol{\mu}$, T; fluctuations in $\mathbf{X}$, $\mathbf{N}$, E:

$$p_k(\mathbf{N};\mathbf{x}) = \exp\left(-\frac{\bar{\mathbf{X}}\cdot\mathbf{x}}{\mathbf{k}T}\right)\exp\left(\frac{\mathbf{N}\cdot\boldsymbol{\mu}}{\mathbf{k}T}\right)\exp\left[-\frac{E_k(\mathbf{N},\mathbf{x})}{\mathbf{k}T}\right] \tag{14.115}$$

3. Independent variables $\mathbf{X}$, $\mathbf{N}$, T; fluctuations in $\mathbf{x}$, E:

$$p_k(\mathbf{x};\mathbf{N}) = \exp\left(-\frac{\mathbf{X}\cdot\mathbf{x}}{\mathbf{k}T}\right)\exp\left(\frac{\mathbf{N}\cdot\boldsymbol{\mu}}{\mathbf{k}T}\right)\exp\left[-\frac{E_k(\mathbf{N},\mathbf{x})}{\mathbf{k}T}\right] \tag{14.116}$$

4. Independent variables $\mathbf{x}$, $\mathbf{N}$, T; fluctuations in $\mathbf{X}$, E:

$$p_k(;\mathbf{N},\mathbf{x}) = \exp\left(-\frac{\bar{\mathbf{X}}\cdot\mathbf{x}}{\mathbf{k}T}\right)\exp\left(\frac{\mathbf{N}\cdot\boldsymbol{\mu}}{\mathbf{k}T}\right)\exp\left[-\frac{E_k(\mathbf{N},\mathbf{x})}{\mathbf{k}T}\right] \tag{14.117}$$

5. Independent variables $\mathbf{x}$, $\mathbf{N}$, E; fluctuations in $\mathbf{X}$:

$$p_k(;E,\mathbf{N},\mathbf{x}) = \exp\left(-\frac{\bar{\mathbf{X}}\cdot\mathbf{x}}{\mathbf{k}T}\right)\exp\left(\frac{\mathbf{N}\cdot\boldsymbol{\mu}}{\mathbf{k}T}\right)e^{-E/\mathbf{k}T} \tag{14.118}$$

$$= e^{-S/\mathbf{k}} = \frac{1}{\Omega} \tag{14.119}$$

[1] Omitting, for simplicity, thermodynamically negligible quantities such as Υ, $C_{\mathbf{N}}$, $C_{\mathbf{x}}$, etc. [see Eqs. (14.105) and (14.106)].

Each term in the sum in Eq. (14.113) is negative unless there is only one term with $p(j) = 1$. The entropy is clearly larger the greater the number of states over which the probabilities $p(j)$ are spread. This is the origin of the general qualitative correlation between entropy and "randomness" of molecular motion and configuration in a system.

It will be recalled that Eqs. (14.115) to (14.118) are special cases of Eq. (14.114), obtained by selecting only certain systems from the generalized ensemble. Depending on the particular case, this amounts to selecting certain of the $p_k(\mathbf{N},\mathbf{x})$ in Eq. (14.114) and renormalizing these quantities to unity. This procedure necessarily gives a sum in Eq. (14.113) less negative than the original sum with all terms present. Thus, as an illustration of the relation between entropy and "spread" in the $p(j)$, we may say that if a system is arranged to have, successively, the kind of contact with its surroundings appropriate to each of cases (1) to (5) above, with the same values or mean values of E, $\mathbf{N}$, $\boldsymbol{\mu}$, $\mathbf{x}$, $\mathbf{X}$, and T in each instance, the entropy will be largest in case (1) and smallest[1] in case (5). Physically, as is clear from the procedure already mentioned of generating special cases from Eq. (14.114), the large entropy in case (1) is associated with the possibility of fluctuations in E, $\mathbf{N}$, and $\mathbf{x}$, while in case (5), E, $\mathbf{N}$, and $\mathbf{x}$ are fixed. However, as has already been suggested in another connection and as will be shown in Chap. 4, the fluctuations from mean values in systems with a very large number of particles are so small[2] that the entropy differences mentioned above are completely negligible in practice. Thus, all ensembles lead to virtually the same value of the entropy. It will also be seen in Chap. 4 that, when the N_i are very large, even the magnitude of δE in the microcanonical ensemble does not have an appreciable effect on S.

The smallest possible "spread" in $p(j)$ and therefore the minimum possible value of the entropy, $S = 0$, occurs when, say, $p(j') = 1$ and $p(j) = 0$ for $j \neq j'$. This is a limiting form of the microcanonical (or canonical) ensemble, with $\Omega = 1$. The ensemble is in this case representative of a system known to be in a nondegenerate energy eigenstate, say $\psi_{k'}(\mathbf{N},\mathbf{x})$, where $\mathbf{N}$ and $\mathbf{x}$ are fixed values. From Eq. (14.116) for the canonical ensemble it is clear that this situation arises when $T \rightarrow 0$, provided the lowest ("ground state") energy eigenvalue ($k' = 1$) of the system is nondegenerate. This result is the statistical mechanical form of the third law of thermodynamics, which has

[1] Cases (2) and (3) are essentially equivalent in this respect.

[2] The generalized ensemble is exceptional in that it has large fluctuations in extensive properties (see Chap. 4). However, the same entropy is obtained, for example, for $\bar{\mathbf{x}}$, T, and $\boldsymbol{\mu}$ in the generalized ensemble as for $\mathbf{x}$ ($= \bar{\mathbf{x}}$), T, and $\boldsymbol{\mu}$ in the grand canonical ensemble (and other ensembles—see Appendixes 2 to 4). That is, the additional "spread" of $p(j)$ over x_i from 0 to $2\bar{x}_i$ or over N_i from 0 to $2\bar{N}_i$ in the generalized ensemble is unimportant for the entropy since the added term in the entropy is only of the order of $\mathbf{k} \ln (2\bar{x}_i/x_i^*)$ or $\mathbf{k} \ln (2\bar{N}_i)$.

been stated as follows:[1] "If the entropy of each element in some crystalline state be taken as zero, then every substance has a finite positive entropy, but at the absolute zero of temperature the entropy may become zero, and does so become zero in the case of a perfect crystalline [pure] substance." In view of work done since the above formulation by Lewis and Randall, certain restrictions on this statement must be pointed out.[2] For example, as already mentioned above, the crystal must have a nondegenerate ground state (electronic, rotational, vibrational, translational), and substances must be excluded in which several possible molecular orientations persist to $T = 0$. Even if we pass over these exceptions, there is still a certain amount of arbitrariness in the selection of the zero of entropy[3] and therefore in so called "absolute entropies," quite aside from putting the integration constant $c = 0$ above. This situation is due to the fact that, in enumerating basic energy states $\psi_k(\mathbf{N},\mathbf{x})$ in any given example, an implicit or explicit assumption is made that matter is structureless beyond a certain point in the subatomic realm. Thus, in a typical case, $\psi_k(\mathbf{N},\mathbf{x})$ might be written as a function of the coordinates associated with the usual translational, rotational, internal vibrational, and electronic degrees of freedom of the molecules making up the system. But something is known about nuclear energy levels and degeneracies; therefore to be quite proper we should certainly take the nuclear ground state degeneracy into account in setting the zero of entropy. However, we have no way of knowing how many more stages of ultrastructure will eventually be discovered; hence any attempt to fix an absolute entropy zero is clearly illusory on this account as well as on account of our arbitrary choice of $c = 0$. Because of this arbitrariness, and as a matter of convenience, it is customary in many applications *not* to include nuclear degeneracy in Ω, as nuclear changes do not occur in thermodynamic processes ordinarily studied. Of course, as already mentioned, only *changes* in entropy are experimentally measurable and have thermodynamic significance.

The special position of energy states in connection with the zero of entropy should be pointed out here. We have seen that for a system known to be in a definite energy state, $\psi_{k'}(\mathbf{N},\mathbf{x})$, $S = 0$. What if the system is known to be in some other state $\psi(q,t)$ [see Eqs. (9.1) to (9.3)], not an energy eigenstate? That is, every system in the representative ensemble is in the state $\psi(q,t)$. In this case, if we expand $\psi(q,t)$ in the energy eigenfunctions ψ_k,

$$\psi = \sum_k A_k \psi_k$$

there will be more than one nonzero A_k and p_k, and hence $S > 0$. But actually, this situation can never arise in a system in equilibrium. All

[1] G. N. Lewis and M. Randall, "Thermodynamics" (McGraw-Hill, New York, 1923).

[2] See Fowler and Guggenheim (Gen. Ref.) for details.

[3] As implied in the Lewis and Randall statement: ". . . be taken as . . . ."

elements of the density matrix must be independent of time for any choice of basic vectors φ_k, and therefore all the B_k's in

$$\psi = \sum_k B_k \varphi_k$$

must be independent of time for any set of φ_k's. The necessary and sufficient condition for this is that ψ be an energy eigenfunction (stationary state). Thus, the only "pure" state an equilibrium system can be in is an energy eigenstate, for which $S = 0$.

It is appropriate in this discussion of the entropy to make some general comments on the nature of the various thermodynamic variables or functions. In our statistical mechanical postulates we provided explicitly, in terms of ensemble averages, for thermodynamic variables with a "mechanical" origin. One of these "mechanical" thermodynamic variables has a well-defined value for each separate system in an ensemble, and we average this "mechanical" quantity over the ensemble to find the mean value to be associated with the experimental system. Variables of this type are E, $\mathbf{N}$, $\mathbf{X}$, and $\mathbf{x}$. Also, a familiar composite variable of the same sort is the heat content, $H = E + \mathbf{X} \cdot \mathbf{x}$. Of course, in certain ensembles some of these variables have the same value in all systems of the ensemble, in which case the ensemble average "bar" is omitted.

On the other hand, other thermodynamic variables have been introduced[1] which do not have such a "mechanical" nature. These are S, T and μ_i, and the derivative functions $F = \mathbf{N} \cdot \boldsymbol{\mu}$ and $A = E - TS$. S and μ_i can be put in the form of averages with weights $p(j)$; for example, in the canonical ensemble

$$S = \sum_k p_k(;\mathbf{N},\mathbf{x}) \left[\mathbf{k} \ln \frac{1}{p_k(;\mathbf{N},\mathbf{x})}\right] \tag{14.120}$$

and

$$\mu_i = \sum_k p_k(;\mathbf{N},\mathbf{x}) \left[\frac{\partial E_k(;\mathbf{N},\mathbf{x})}{\partial N_i}\right] \tag{14.121}$$

where Eq. (14.121) follows from Eq. (14.88). The meaning of the derivative is

$$\frac{\partial E_k(;\mathbf{N},\mathbf{x})}{\partial N_i} = E_k(;N_{j\neq i}, N_i + 1, \mathbf{x}) - E_k(;\mathbf{N},\mathbf{x}) \tag{14.122}$$

where k has the same numerical value for N_i and $N_i + 1$ (that is, the lowest energy state is $k = 1$ in both cases, and we recognize the fact that there will be a completely new set of states when $N_i \to N_i + 1$). However, unlike

[1] These were not introduced through the postulates of Chap. 1 and 2. Hence, limiting the postulates to "mechanical" variables was in fact not a real restriction, as might have seemed to be the case.

similar equations for E, $\mathbf{N}$, $\mathbf{X}$, and $\mathbf{x}$, Eqs. (14.120) and (14.121) cannot be recast as ensemble averages of the form

$$\bar{\varphi} = \frac{1}{\mathcal{N}} \sum_{i=1}^{\mathcal{N}} \varphi^{(i)} \tag{14.123}$$

and therefore do not have the real significance of ensemble averages.[1] This conclusion with respect to Eq. (14.121) results from the fact that an "adiabatic principle" does not apply for variations in N_i [compare Eq. (14.28)]. Addition of a single particle to a system of the ensemble disturbs the system in an unpredictable way, adds new coordinates and new basic vectors, etc. Equation (14.120) involves an average over a function of $p_k(;\mathbf{N},\mathbf{x})$, a quantity with no "mechanical" significance whatever. The entropy is in fact peculiarly a property of a probability distribution only, which in turn is an indivisible property of the *entire* ensemble. Physically, this means that the entropy is a measure of the possible dispersion or spread (as *represented* by the ensemble) of the actual system over different states rather than an average of some dynamical variable over these states.

15. Summary of Ensembles[2]

Table 1 and Eqs. (14.114) to (14.119) summarize a number of important properties of the principal ensembles we have considered. In addition, two "hybrid" ensembles [Eqs. (14.97) and (14.98)] have been mentioned briefly. A further "hybrid" ensemble of some physical interest is that representative of a system in thermal and *partial* material contact with its surroundings [that is, a partially open system with $\mathbf{x}$ fixed instead of $\mathbf{X}$ as in Eq. (14.97)]. Suppose only the ith species cannot pass in and out of the system. The independent variables are then $\mathbf{x}$, T, $\mu_{j\neq i}$, N_i. By analogy with Table 1,

$$-\mathbf{x}\cdot\bar{\mathbf{X}} + N_i\mu_i = -\mathbf{k}T \ln \sum_{k,N_{j\neq i}} \exp\left(\frac{\sum_{j\neq i} N_j\mu_j}{\mathbf{k}T}\right) \exp\left[-\frac{E_k(\mathbf{N},\mathbf{x})}{\mathbf{k}T}\right] \tag{15.1}$$

Certain other choices of independent variables suggest themselves immediately to complete the set in Table 1: (a) $\mathbf{X}$, $\mathbf{N}$, E; (b) $\mathbf{x}$, $\boldsymbol{\mu}$, E; and (c) $\mathbf{X}$, $\boldsymbol{\mu}$, E. These are of no real interest, however, because of the rather artificial

[1] It may be well to remind the reader that a given system in the ensemble is not in any particular state $\psi_k(\mathbf{N},\mathbf{x})$, but rather in a superposition of such states, and $p_k(;\mathbf{N},\mathbf{x})$ is itself an ensemble average [Eqs. (9.6) and (9.14)]. $p_k(;\mathbf{N},\mathbf{x})$ is thus not (1) the fraction of systems in the ensemble in the state $\psi_k(\mathbf{N},\mathbf{x})$, but rather (2) the probability of observing the state $\psi_k(\mathbf{N},\mathbf{x})$ if the basic (energy) state of a system chosen at random from the ensemble is determined. Hence an average with weights $p_k(;\mathbf{N},\mathbf{x})$ is not necessarily an *ensemble* average (it is for E, $\mathbf{N}$, $\mathbf{X}$, and $\mathbf{x}$). The first interpretation, (1) above, of $p_k(;\mathbf{N},\mathbf{x})$ (the one usually adopted in textbooks), though wrong in principle, leads to correct results in applications.

[2] See also Fowler and Guggenheim (Gen. Ref.).

conditions which must be satisfied by the experimental system to make these variables appropriate. Specifically, from

$$dE = dq - \mathbf{X} \cdot d\mathbf{x} + \boldsymbol{\mu} \cdot d\mathbf{N}$$
$$dq = T\,dS \tag{15.2}$$

we see that the system must absorb heat according to:

(a) $dq = \mathbf{X} \cdot d\mathbf{x}$ (b) $dq = -\boldsymbol{\mu} \cdot d\mathbf{N}$ (c) $dq = \mathbf{X} \cdot d\mathbf{x} - \boldsymbol{\mu} \cdot d\mathbf{N}$

for each fluctuation $d\mathbf{x}$ or $d\mathbf{N}$ in order to keep E constant in the different cases.

The presence or absence of thermal contact is associated naturally with the pair of variables T,S rather than T,E, so that in addition to the eight sets of independent variables already listed, we should also mention: (d) $\mathbf{X}$, $\mathbf{N}$, S (mechanical contact); (e) $\mathbf{x}$, $\boldsymbol{\mu}$, S (material contact); and (f) $\mathbf{X}$, $\boldsymbol{\mu}$, S (mechanical and material contact). Case 5, in Table 1, is the same as $\mathbf{x}$, $\mathbf{N}$, S, of course [Eq. (15.2)]. Cases (e) and (f) are not realizable in practice[1] because of the difficulty of having a transfer of molecules in and out of a system without heat transfer. Case (d) is experimentally attainable but not very important.

16. Transition from Quantum to Classical Statistics

The object here is to obtain the main results of Chap. 1 (on classical statistical mechanics) as limiting expressions in a quantum-mechanical treatment. This derivation is of more than incidental interest, however, since, as we shall see, some important alterations must be made if the classical partition functions of Chap. 1 are to have the same formal relations to thermodynamic functions as the quantum-mechanical partition functions in Secs. 14 and 15. For example, although in the classical limit Eq. (5.4) is a perfectly legitimate definition of a quantity Q, we shall find that this Q is not related to thermodynamics by $A = -\mathbf{k}T \ln Q$, as in Sec. 14, unless it is multiplied by an appropriate factor.

We may confine the discussion to the canonical ensemble, as more general ensembles differ formally in classical and quantum statistics only as regards the energy distribution, while the microcanonical ensemble may be considered a special case of the canonical ensemble in both classical and quantum statistics. To simplify the argument we restrict ourselves to a one-component system of N monatomic molecules of mass m. No new fundamental features are involved in a more general treatment.[2] We choose cartesian coordinates for the analysis; a canonical transformation to other generalized coordinates can be made on the final result.

[1] Except possibly in liquid helium II, where the set of variables $\boldsymbol{\mu}$, S, E is even more appropriate.

[2] The discussion of the quantum-classical transition for internal degrees of freedom (rotation, vibration) in polyatomic molecules is standard (see General References) and will not be included here.

Let the position vector of the jth particle be $\mathbf{r}_j$ and its momentum vector be $\mathbf{p}_j$. For future use we define $\boldsymbol{\varkappa}_j$ by $\hbar\boldsymbol{\varkappa}_j = \mathbf{p}_j$. We let $\mathbf{r}$ represent the vector with $3N$ components $x_1, y_1, z_1, \ldots, x_N, y_N, z_N$, and define in an analogous way $\mathbf{p}$ and $\boldsymbol{\varkappa}$. Also,

$$\nabla^2 = \frac{\partial^2}{\partial x_1^2} + \cdots + \frac{\partial^2}{\partial z_N^2}$$

$$\nabla = \mathbf{i}_{1x}\frac{\partial}{\partial x_1} + \cdots + \mathbf{i}_{Nz}\frac{\partial}{\partial z_N}$$

where the $\mathbf{i}$'s are unit vectors.

We begin with Eq. (11.22) for Q,

$$Q = \sum_k \int \varphi_k^* \exp(-\beta\mathscr{H})\varphi_k \, d\mathbf{r} \tag{16.1}$$

where, as we found in Chap. 3, $\beta = 1/\mathbf{k}T$, and $d\mathbf{r} = dx_1 \cdots dz_N$. The Hamiltonian operator is

$$\mathscr{H} = \mathscr{K} + U(\mathbf{r}) \tag{16.2}$$

where $\mathscr{K}$ is the kinetic energy operator. The kinetic energy K is given by

$$K = \frac{1}{2m}(p_{x1}^2 + \cdots + p_{zN}^2) \tag{16.3}$$

and thus

$$\mathscr{K} = -\frac{\hbar^2}{2m}\nabla^2 \tag{16.4}$$

For the φ_k in Eq. (16.1) we wish to introduce the momentum eigenfunctions [Eq. (8.49)], $\exp(i\boldsymbol{\varkappa}\cdot\mathbf{r})$, since this will lead (see below) to an expression for Q involving integration over coordinates and momenta, as in classical mechanics. However, these functions require special normalization; therefore we cannot use Eq. (16.1) without some modification. We therefore proceed by writing

$$Q = \sum_m \int \psi_m^* \exp(-\beta\mathscr{H})\psi_m \, d\mathbf{r} \tag{16.5}$$

where the ψ_m are energy eigenfunctions, and expand ψ_m explicitly in the $\exp(i\boldsymbol{\varkappa}\cdot\mathbf{r})$. That is, we write

$$\psi_m(\mathbf{r}) = \int A_m(\boldsymbol{\varkappa}) \exp(i\boldsymbol{\varkappa}\cdot\mathbf{r}) \, d\boldsymbol{\varkappa} \tag{16.6}$$

Thus $\psi_m(\mathbf{r})$ is the Fourier transform of $A_m(\boldsymbol{\varkappa})$. According to the inversion formula for the Fourier transform,

$$A_m(\boldsymbol{\varkappa}) = \frac{1}{(2\pi)^{3N}} \int \psi_m(\mathbf{r}) \exp(-i\boldsymbol{\varkappa}\cdot\mathbf{r}) \, d\mathbf{r} \tag{16.7}$$

The limits in Eqs. (16.6) and (16.7) are of course $-\infty$ to $+\infty$ for each component. Now we substitute Eq. (16.6) for ψ_m in Eq. (16.5), giving

$$Q = \sum_m \int_{\mathbf{r}} \int_{\boldsymbol{\varkappa}} \psi_m^*(\mathbf{r}) A_m(\boldsymbol{\varkappa}) \exp(-\beta \mathscr{H}) \exp(i\boldsymbol{\varkappa} \cdot \mathbf{r})\, d\boldsymbol{\varkappa}\, d\mathbf{r} \tag{16.8}$$

and then employ Eq. (16.7) (denoting the variable of integration as $\mathbf{r}'$ instead of $\mathbf{r}$) to eliminate $A_m(\boldsymbol{\varkappa})$ from Eq. (16.8):

$$Q = \frac{1}{(2\pi)^{3N}} \int_{\mathbf{r}} \int_{\boldsymbol{\varkappa}} \int_{\mathbf{r}'} [\sum_m \psi_m^*(\mathbf{r}) \psi_m(\mathbf{r}')] \exp(-i\boldsymbol{\varkappa} \cdot \mathbf{r}') \exp(-\beta \mathscr{H}) \times \exp(i\boldsymbol{\varkappa} \cdot \mathbf{r})\, d\mathbf{r}'\, d\boldsymbol{\varkappa}\, d\mathbf{r} \tag{16.9}$$

where the order of summation and integration has been reversed. This equation can be simplified in view of the relation[1]

$$\sum_m \psi_m^*(\mathbf{r}) \psi_m(\mathbf{r}') = \delta(\mathbf{r} - \mathbf{r}') \tag{16.10}$$

The $\mathbf{r}'$ integration in Eq. (16.9) is then simply

$$\int_{\mathbf{r}'} \delta(\mathbf{r} - \mathbf{r}') \exp(-i\boldsymbol{\varkappa} \cdot \mathbf{r}')\, d\mathbf{r}' = \exp(-i\boldsymbol{\varkappa} \cdot \mathbf{r}) \tag{16.11}$$

Equation (16.9) therefore becomes

$$Q = \frac{1}{(2\pi)^{3N}} \int_{\mathbf{r}} \int_{\boldsymbol{\varkappa}} \exp(-i\boldsymbol{\varkappa} \cdot \mathbf{r}) \exp(-\beta \mathscr{H}) \exp(i\boldsymbol{\varkappa} \cdot \mathbf{r})\, d\boldsymbol{\varkappa}\, d\mathbf{r} \tag{16.12}$$

which is thus seen to be the proper form of Eq. (16.1) for our present purposes.[2]

Equation (16.12) is an integral over the p's ($\boldsymbol{\varkappa}$) and q's ($\mathbf{r}$) as required in the classical expression, Eq. (5.4), but the classical Q has $e^{-\beta H}$ instead of the operator $\exp(-\beta\mathscr{H})$ in the integrand. This suggests[3] the following substitution in Eq. (16.12):

$$\exp(-\beta \mathscr{H}) \exp(i\boldsymbol{\varkappa} \cdot \mathbf{r}) = \exp[-\beta H(\boldsymbol{\varkappa},\mathbf{r})] \exp(i\boldsymbol{\varkappa} \cdot \mathbf{r}) w(\boldsymbol{\varkappa},\mathbf{r},\beta) \tag{16.13}$$

$$= F(\boldsymbol{\varkappa},\mathbf{r},\beta) \tag{16.14}$$

[1] Equation (16.10) may be derived by expanding $\delta(\mathbf{r} - \mathbf{r}')$ in the set of functions $\psi_m^*(\mathbf{r})$:

$$\delta(\mathbf{r} - \mathbf{r}') = \sum_m \alpha_m(\mathbf{r}') \psi_m^*(\mathbf{r})$$

$$\alpha_m(\mathbf{r}') = \int \psi_m(\mathbf{r}) \delta(\mathbf{r} - \mathbf{r}')\, d\mathbf{r} = \psi_m(\mathbf{r}')$$

Using this result for $\alpha_m(\mathbf{r}')$ in the expansion, we find Eq. (16.10).

[2] Writing $\varphi(\boldsymbol{\varkappa}) = (2\pi)^{-3N/2} \exp(i\boldsymbol{\varkappa} \cdot \mathbf{r})$ instead of $\exp(i\boldsymbol{\varkappa} \cdot \mathbf{r})$ is obviously all that is necessary for this case to be included in Eq. (16.1) without modification.

[3] J. G. Kirkwood, *Phys. Rev.*, **44**, 31 (1933); **45**, 116 (1934). See also E. Wigner, *Phys. Rev.*, **40**, 749 (1932).

These equations define the functions w and F. When Eq. (16.13) is introduced in Eq. (16.12), the result is

$$Q = \frac{1}{(2\pi)^{3N}} \int_{\mathbf{r}} \int_{\varkappa} \exp\left[-\beta H(\varkappa,\mathbf{r})\right] w(\varkappa,\mathbf{r},\beta)\, d\varkappa\, d\mathbf{r} \tag{16.15}$$

so that the quantum correction to the classical equation (5.4) is contained in w/h^{3N} (since $\varkappa = \mathbf{p}/\hbar$). We therefore turn to a consideration of w.

One can proceed in a straightforward way by carrying out explicitly the operation on the left-hand side of Eq. (16.13), and comparing the result with the right-hand side to find w. However, a much more satisfactory approach is the following. First, we differentiate Eq. (16.14) with respect to β, and arrive at the Bloch differential equation

$$\begin{aligned} \frac{\partial F}{\partial \beta} &= \frac{\partial}{\partial \beta}\left(1 - \beta \mathscr{H} + \frac{1}{2!}\beta^2 \mathscr{H}^2 - \cdots\right) \exp(i\varkappa \cdot \mathbf{r}) \\ &= -\mathscr{H} F \end{aligned} \tag{16.16}$$

which has the boundary condition [Eq. (16.14)]

$$F(\beta = 0) = \exp(i\varkappa \cdot \mathbf{r}) \tag{16.17}$$

Now we substitute $e^{-\beta H} \exp(i\varkappa \cdot \mathbf{r}) w$ for F in Eq. (16.16), and find

$$\begin{aligned} &\exp(i\varkappa \cdot \mathbf{r}) \left(e^{-\beta H} \frac{\partial w}{\partial \beta} - H w e^{-\beta H}\right) \\ &= -Ue^{-\beta H} \exp(i\varkappa \cdot \mathbf{r}) w + \frac{\hbar^2}{2m}[e^{-\beta H} \exp(i\varkappa \cdot \mathbf{r}) \nabla^2 w + e^{-\beta H} w \nabla^2 \exp(i\varkappa \cdot \mathbf{r}) \\ &\quad + \exp(i\varkappa \cdot \mathbf{r}) w \nabla^2 e^{-\beta H} + 2 \exp(i\varkappa \cdot \mathbf{r}) \nabla w \cdot \nabla e^{-\beta H} + 2e^{-\beta H} \nabla w \cdot \nabla \\ &\qquad \times \exp(i\varkappa \cdot \mathbf{r}) + 2w \nabla e^{-\beta H} \cdot \nabla \exp(i\varkappa \cdot \mathbf{r})] \end{aligned} \tag{16.18}$$

But we have

$$\nabla e^{-\beta H} = e^{-\beta K} \nabla e^{-\beta U} = -\beta e^{-\beta H} \nabla U \tag{16.19}$$

$$\nabla^2 e^{-\beta H} = e^{-\beta K} \nabla^2 e^{-\beta U} = -\beta e^{-\beta H} \nabla^2 U + \beta^2 e^{-\beta H} (\nabla U)^2 \tag{16.20}$$

$$\nabla \exp(i\varkappa \cdot \mathbf{r}) = \frac{i}{\hbar} \exp(i\varkappa \cdot \mathbf{r}) \mathbf{p} \tag{16.21}$$

$$\nabla^2 \exp(i\varkappa \cdot \mathbf{r}) = -\frac{2m}{\hbar^2} K \exp(i\varkappa \cdot \mathbf{r}) \tag{16.22}$$

so that Eq. (16.18) reduces to the differential equation in w,

$$\frac{\partial w}{\partial \beta} = \frac{\hbar^2}{2m}[\nabla^2 w - \beta w \nabla^2 U - 2\beta \nabla w \cdot \nabla U + \beta^2 w (\nabla U)^2] + \frac{i\hbar}{m}(\nabla w \cdot \mathbf{p} - \beta w \nabla U \cdot \mathbf{p}) \equiv M(\beta) \quad (16.23)$$

The boundary condition here, from Eq. (16.13), is $w(\beta = 0) = 1$.

In quantum theory, the standard technique ("WKB method") for investigating the relation of classical mechanics to quantum mechanics is to make use of an expansion in powers of $\hbar$, with the classical limit corresponding to $\hbar \to 0$. We follow a similar procedure here. Let us write

$$w(\mathbf{p},\mathbf{r},\beta) = \sum_{l=0}^{\infty} \hbar^l w_l(\mathbf{p},\mathbf{r},\beta) \quad (16.24)$$

and substitute this series for w in

$$w = 1 + \int_0^\beta M(t)\, dt \quad (16.25)$$

which is just Eq. (16.23) rewritten as an integral equation (satisfying the boundary condition at $\beta = 0$). The function $M(t)$ is defined in Eq. (16.23). On equating coefficients of successive powers of $\hbar$, we find

$$w_0 = 1 \quad (16.26)$$

$$w_1 = \int_0^\beta \frac{i}{m}[\nabla w_0(t) \cdot \mathbf{p} - t w_0(t) \nabla U \cdot \mathbf{p}]\, dt = -\frac{i\beta^2}{2m} \nabla U \cdot \mathbf{p} \quad (16.27)$$

$$w_2 = \int_0^\beta \left\{ \frac{1}{2m}[\nabla^2 w_0(t) + t^2 w_0(t)(\nabla U)^2 - t w_0(t) \nabla^2 U - 2t \nabla w_0(t) \cdot \nabla U] + \frac{i}{m}[\nabla w_1(t) \cdot \mathbf{p} - t w_1(t) \nabla U \cdot \mathbf{p}] \right\} dt$$

$$= -\frac{1}{2m}\left\{ \frac{\beta^2}{2} \nabla^2 U - \frac{\beta^3}{3}\left[(\nabla U)^2 + \frac{1}{m}(\mathbf{p} \cdot \nabla)^2 U\right] + \frac{\beta^4}{4m}(\nabla U \cdot \mathbf{p})^2 \right\} \quad (16.28)$$

etc.

We now put our series for w in Eq. (16.15) with the result [compare Eq. (5.4)]

$$Q = \frac{1}{h^{3N}} \int_{\mathbf{r}} \int_{\mathbf{p}} \exp[-\beta H(\mathbf{p},\mathbf{r})]\, d\mathbf{p}\, d\mathbf{r} + O(\Lambda^2) \quad (16.29)$$

where, as in Eq. (14.23),

$$\Lambda = \frac{h}{(2\pi m \mathbf{k} T)^{1/2}} \quad (16.30)$$

Λ is approximately the de Broglie wavelength corresponding to the energy $\mathbf{k}T$. The term in Λ (from $\hbar w_1$ in the series for w) vanishes on integrating over the momenta in Eq. (16.15) because the integrand is an odd function of each component of $\mathbf{p}$. The terms of order Λ^2 (from $\hbar^2 w_2$) are integrals over phase space (with weighting $e^{-\beta H}$) of quantities of the type $[\Lambda\partial(U/\mathbf{k}T)/\partial x_1]^2$, which involve the intermolecular forces. If $U = 0$ (perfect gas), only the leading integral in Eq. (16.29) remains, *without* taking the limit $\Lambda \to 0$.

Unfortunately, the derivation of Eq. (16.29) is not complete, for we have ignored the quantum-mechanical postulate on the symmetry of wave functions for a system of identical particles. This accounts for the surprising (and incorrect) implication above that the classical phase integral (except for a multiplicative constant), is generally valid in quantum statistical mechanics when $U = 0$.

However, Eq. (16.29) is instructive in showing how the classical Q must be corrected irrespective of symmetry requirements. It also shows that the volume in classical phase space corresponding to a single quantum state is h^{3N} [see also Eq. (16.58)]:

$$h^{3N} = \Delta p_{x_1}\,\Delta x_1 \cdot \cdot \cdot \Delta p_{z_N}\,\Delta z_N \tag{16.31}$$

This becomes obvious if we divide phase space into equal regions $(\Delta\mathbf{p}\,\Delta\mathbf{r})_i$ of volume h^{3N}, where H is substantially constant in each region, having the value H_i in the ith region. Then, from Eq. (16.29),

$$Q = \sum_i \frac{e^{-\beta H_i}(\Delta\mathbf{p}\,\Delta\mathbf{r})_i}{h^{3N}} = \sum_i e^{-\beta H_i} \tag{16.32}$$

which is just the form of the quantum-mechanical sum over states. From the point of view of the uncertainty principle, Eq. (16.31) states that the position of a point in phase space, specifying the classical dynamical state (p's and q's) of a system, cannot be located more closely than within a "cell" of volume h^{3N}.

Symmetry. We must now take symmetry considerations into account. We shall limit ourselves here to systems with wave functions which are either symmetric ("Bose-Einstein statistics") or antisymmetric ("Fermi-Dirac statistics") in the $3N$ *spatial* coordinates already introduced. This is not the most general case because of the neglect of spin coordinates.

For any symmetric wave function $\psi(\mathbf{r}_1, \ldots, \mathbf{r}_N)$,

$$P_r\psi(\mathbf{r}_1, \ldots, \mathbf{r}_N) = \psi(\mathbf{r}_1, \ldots, \mathbf{r}_N) \tag{16.33}$$

where P_r is an operator which permutes the $\mathbf{r}_j$'s. There are, of course, altogether $N!$ different possible permutations of the $\mathbf{r}_j$'s, and Eq. (16.33) states that ψ is unchanged by any of these permutations. For an antisymmetric ψ,

$$P_r\psi(\mathbf{r}_1, \ldots, \mathbf{r}_N) = (-1)^{|P_r|}\psi(\mathbf{r}_1, \ldots, \mathbf{r}_N)$$

or

$$(-1)^{|P_r|}P_r\psi(\mathbf{r}_1, \ldots, \mathbf{r}_N) = \psi(\mathbf{r}_1, \ldots, \mathbf{r}_N) \tag{16.34}$$

where $|P_r|$ is the number of exchanges of pairs of $\mathbf{r}_j$'s into which the permutation P_r can be decomposed. Each exchange contributes a factor -1 in Eq. (16.34). $|P_r|$ is not unique for a given P_r, but it will be either always even (P_r an "even permutation") or always odd (P_r an "odd permutation"). Thus, $(-1)^{|P_r|} = +1$ for an even permutation and $(-1)^{|P_r|} = -1$ for an odd permutation. Equations (16.33) and (16.34) may be summarized by

$$(\pm 1)^{|P_r|} P_r \psi(\mathbf{r}_1, \ldots, \mathbf{r}_N) = \psi(\mathbf{r}_1, \ldots, \mathbf{r}_N) \tag{16.35}$$

where the upper sign is for the symmetric case (Bose-Einstein) and the lower sign for the antisymmetric case (Fermi-Dirac). Since there are $N!$ different permutations, it follows from Eq. (16.35) that

$$\psi(\mathbf{r}_1, \ldots, \mathbf{r}_N) = \frac{1}{N!} \sum_{P_r} (\pm 1)^{|P_r|} P_r \psi(\mathbf{r}_1, \ldots, \mathbf{r}_N) \tag{16.36}$$

Now as in Eq. (16.5) the energy eigenfunction ψ_m may be expanded as

$$\psi_m(\mathbf{r}_1, \ldots, \mathbf{r}_N) = \int a_m(\boldsymbol{\varkappa}_1, \ldots, \boldsymbol{\varkappa}_N) \exp\,(i \sum_j \boldsymbol{\varkappa}_j \cdot \mathbf{r}_j)\, d\boldsymbol{\varkappa} \tag{16.37}$$

with the inversion

$$a_m(\boldsymbol{\varkappa}_1, \ldots, \boldsymbol{\varkappa}_N) = \frac{1}{(2\pi)^{3N}} \int \psi_m(\mathbf{r}_1, \ldots, \mathbf{r}_N) \exp\,(-i \sum_j \boldsymbol{\varkappa}_j \cdot \mathbf{r}_j)\, d\mathbf{r} \tag{16.38}$$

where $1 \leqslant j \leqslant N$. If we substitute this expansion of ψ_m into the right-hand side of Eq. (16.36), writing $\psi = \psi_m$, we have

$$\psi_m(\mathbf{r}_1, \ldots, \mathbf{r}_N) = \frac{1}{(N!)^{1/2}} \int a_m(\boldsymbol{\varkappa}_1, \ldots, \boldsymbol{\varkappa}_N) \varphi(\boldsymbol{\varkappa}, \mathbf{r})\, d\boldsymbol{\varkappa} \tag{16.39}$$

where

$$\varphi(\boldsymbol{\varkappa}, \mathbf{r}) = \frac{1}{(N!)^{1/2}} \sum_{P_r} (\pm 1)^{|P_r|} P_r \exp\,(i \sum_j \boldsymbol{\varkappa}_j \cdot \mathbf{r}_j)$$

$$= \frac{1}{(N!)^{1/2}} \sum_{P_\kappa} (\pm 1)^{|P_\kappa|} P_\kappa \exp\,(i \sum_j \boldsymbol{\varkappa}_j \cdot \mathbf{r}_j) \tag{16.40}$$

The operator P_κ permutes the $\boldsymbol{\varkappa}_j$'s. Equation (16.40) follows because of the symmetry with respect to $\boldsymbol{\varkappa}_j$ and $\mathbf{r}_j$ in $\sum_j \boldsymbol{\varkappa}_j \cdot \mathbf{r}_j$.

If $F(\boldsymbol{\varkappa}_1, \ldots, \boldsymbol{\varkappa}_N)$ is any function such that in the integral

$$I = \int F(\boldsymbol{\varkappa}_1, \ldots, \boldsymbol{\varkappa}_N)\, d\boldsymbol{\varkappa}_1 \cdots d\boldsymbol{\varkappa}_N \tag{16.41}$$

it is valid to change the order of integration, then

$$I = \int [P_\kappa F(\boldsymbol{\varkappa}_1, \ldots, \boldsymbol{\varkappa}_N)]\, d\boldsymbol{\varkappa}_1 \cdots d\boldsymbol{\varkappa}_N \tag{16.42}$$

since after performing the permutation, changing notation, and interchanging the order of integration one arrives back at the original form, Eq. (16.41). From this result, we may write

$$\int F(\varkappa_1, \ldots, \varkappa_N)\, d\varkappa = \frac{1}{N!} \int [\sum_{P_\kappa} P_\kappa F(\varkappa_1, \ldots, \varkappa_N)]\, d\varkappa \tag{16.43}$$

We now apply Eq. (16.43) to the integral in Eq. (16.39):

$$\psi_m(\mathbf{r}_1, \ldots, \mathbf{r}_N) = \frac{1}{(N!)^{3/2}} \int \sum_{P_\kappa} P_\kappa [a_m(\varkappa_1, \ldots, \varkappa_N) \varphi(\varkappa, \mathbf{r})]\, d\varkappa \tag{16.44}$$

According to the form of Eq. (16.40), φ is symmetric or antisymmetric in the $\varkappa_j$'s depending on whether the upper or lower sign is used (which in turn depends on the symmetry of ψ_m). Therefore

$$P_\kappa \varphi(\varkappa, \mathbf{r}) = (\pm 1)^{|P_\kappa|} \varphi(\varkappa, \mathbf{r}) \tag{16.45}$$

and, in Eq. (16.44),

$$P_\kappa [a_m(\varkappa_1, \ldots, \varkappa_N) \varphi(\varkappa, \mathbf{r})] = (\pm 1)^{|P_\kappa|} \varphi(\varkappa, \mathbf{r}) P_\kappa a_m(\varkappa_1, \ldots, \varkappa_N) \tag{16.46}$$

Thus, Eq. (16.44) can be written

$$\psi_m(\mathbf{r}_1, \ldots, \mathbf{r}_N) = \frac{1}{N!} \int B_m(\varkappa_1, \ldots, \varkappa_N) \varphi(\varkappa, \mathbf{r})\, d\varkappa \tag{16.47}$$

where
$$B_m(\varkappa_1, \ldots, \varkappa_N) = \frac{1}{(N!)^{1/2}} \sum_{P_\kappa} (\pm 1)^{|P_\kappa|} P_\kappa a_m(\varkappa_1, \ldots, \varkappa_N)$$
$$= \frac{1}{(2\pi)^{3N}} \int \psi_m(\mathbf{r}_1, \ldots, \mathbf{r}_N) \varphi^*(\varkappa, \mathbf{r})\, d\mathbf{r} \tag{16.48}$$

having used Eq. (16.38) for a_m. This is the inversion of Eq. (16.47).

We now follow our earlier procedure, beginning with the substitution of Eq. (16.47) into Eq. (16.5), where now it is understood that the sum is taken over accessible states, that is, over the complete set of independent energy eigenfunctions with proper symmetry. We have

$$Q = \frac{1}{N!} \sum_m \int_{\mathbf{r}} \int_{\varkappa} \psi_m^*(\mathbf{r}) B_m(\varkappa) \exp(-\beta \mathscr{H}) \varphi(\varkappa, \mathbf{r})\, d\varkappa\, d\mathbf{r} \tag{16.49}$$

$$= \frac{1}{N!(2\pi)^{3N}} \int_{\mathbf{r}} \int_{\varkappa} \int_{\mathbf{r}'} [\sum_m \psi_m^*(\mathbf{r}') \psi_m(\mathbf{r}')] \varphi^*(\varkappa, \mathbf{r}') \exp(-\beta \mathscr{H}) \varphi(\varkappa, \mathbf{r})\, d\mathbf{r}'\, d\varkappa\, d\mathbf{r} \tag{16.50}$$

$$= \frac{1}{N!(2\pi)^{3N}} \int_{\mathbf{r}} \int_{\varkappa} \varphi^*(\varkappa, \mathbf{r}) \exp(-\beta \mathscr{H}) \varphi(\varkappa, \mathbf{r})\, d\varkappa\, d\mathbf{r} \tag{16.51}$$

which is the analogue of Eq. (16.12). The result of the above analysis has therefore been to replace $\exp(i\varkappa \cdot \mathbf{r})$ in Eq. (16.12) by φ, which is a linear combination [Eq. (16.40)] of functions of this type but with proper symmetry.

To proceed further, we put the explicit form of φ, Eq. (16.40), into Eq. (16.51):

$$Q = \frac{1}{(N!)^2(2\pi)^{3N}} \sum_{P_\kappa'} \sum_{P_\kappa''} (\pm 1)^{|P_\kappa'|+|P_\kappa''|} \int_{\mathbf{r}} \int_{\varkappa} [P_\kappa'' \exp(-i\sum_j \boldsymbol{\varkappa}_j \cdot \mathbf{r}_j)]$$
$$\times \exp(-\beta\mathscr{H})[P_\kappa' \exp(i\sum_j \boldsymbol{\varkappa}_j \cdot \mathbf{r}_j)]\, d\boldsymbol{\varkappa}\, d\mathbf{r} \quad (16.52)$$

By the theorem of Eqs. (16.41) and (16.42), the integral of Eq. (16.52) is unaltered if we operate on the integrand with $(P_\kappa')^{-1}$, the inverse[1] of P_κ'. Then

$$Q = \frac{1}{(N!)^2(2\pi)^{3N}} \sum_{P_\kappa'} \sum_{P_\kappa''} (\pm 1)^{|P_\kappa|} \int_{\mathbf{r}} \int_{\varkappa} [P_\kappa \exp(-i\sum_j \boldsymbol{\varkappa}_j \cdot \mathbf{r}_j)]$$
$$\times \exp(-\beta\mathscr{H}) \exp(i\sum_j \boldsymbol{\varkappa}_j \cdot \mathbf{r}_j)\, d\boldsymbol{\varkappa}\, d\mathbf{r} \quad (16.53)$$

where $$P_\kappa = (P_\kappa')^{-1} P_\kappa''$$

and we have used

$$|P_\kappa| = |(P_\kappa')^{-1}| + |P_\kappa''|$$
$$= |P_\kappa'| + |P_\kappa''|$$

The double sum in Eq. (16.53) over P_κ' and P_κ'' together with the definition of P_κ generates each of the $N!$ different permutations, P_κ, $N!$ times. That is,

$$\sum_{P_\kappa'} \sum_{P_\kappa''} = N! \sum_{P_\kappa}$$

Hence,

$$Q = \frac{1}{N!(2\pi)^{3N}} \sum_{P_\kappa} (\pm 1)^{|P_\kappa|} \int_{\mathbf{r}} \int_{\varkappa} [P_\kappa \exp(-i\sum_j \boldsymbol{\varkappa}_j \cdot \mathbf{r}_j)] \exp(-\beta\mathscr{H})$$
$$\times \exp(i\sum_j \boldsymbol{\varkappa}_j \cdot \mathbf{r}_j)\, d\boldsymbol{\varkappa}\, d\mathbf{r} \quad (16.54)$$

Now $\exp(-\beta\mathscr{H}) \exp(i\boldsymbol{\varkappa} \cdot \mathbf{r})$ in Eq. (16.54) has already been discussed in Eqs. (16.13) to (16.28). Therefore,

$$Q = \frac{1}{N!\,h^{3N}} \int_{\mathbf{r}} \int_{\mathbf{p}} \exp[-\beta H(\mathbf{p},\mathbf{r})](1 + w_1\hbar + w_2\hbar^2 + \cdots)\, d\mathbf{p}\, d\mathbf{r}$$
$$+ \frac{1}{N!\,(2\pi)^{3N}} \sum_{P_\kappa \neq 1} (\pm 1)^{|P_\kappa|} \int_{\mathbf{r}} \int_{\varkappa} [P_\kappa \exp(-i\sum_j \boldsymbol{\varkappa}_j \cdot \mathbf{r}_j)] \exp(i\sum_j \boldsymbol{\varkappa}_j \cdot \mathbf{r}_j)$$
$$\times \exp[-\beta H(\boldsymbol{\varkappa},\mathbf{r})](1 + w_1\hbar + w_2\hbar^2 + \cdots)\, d\boldsymbol{\varkappa}\, d\mathbf{r} \quad (16.55)$$

[1] That is, if the permutation P_κ' is followed by $(P_\kappa')^{-1}$, one arrives back at the original order.

where the identity permutation has been written separately as the first integral. The first integral is the same as Eq. (16.29), except for the factor $1/N!$. The other permutations lead to new terms. For example, consider the permutation that exchanges $\varkappa_1$ and $\varkappa_2$. The corresponding term in Q is

$$\pm \frac{1}{N!h^{3N}} \int_{\mathbf{r}} \int_{\mathbf{p}} \exp\left[\frac{i\mathbf{p}_1 \cdot (\mathbf{r}_1 - \mathbf{r}_2)}{\hbar}\right] \exp\left[\frac{i\mathbf{p}_2 \cdot (\mathbf{r}_2 - \mathbf{r}_1)}{\hbar}\right] \times e^{-\beta K(\mathbf{p})} e^{-\beta U(\mathbf{r})} (1 + w_1\hbar + w_2\hbar^2 + \cdots)\, d\mathbf{p}\, d\mathbf{r}$$

When only the leading term in the series for w is retained, the integration over $\mathbf{p}$ yields $3N - 6$ factors h/Λ and six factors such as (from p_{x_1})

$$\frac{h}{\Lambda} e^{-\pi(x_1 - x_2)^2/\Lambda^2}$$

Thus, the second integral in Eq. (16.55) vanishes for each $P_\kappa \neq 1$ as $\Lambda \to 0$. As pointed out in connection with Eq. (16.29), for $U = 0$ (perfect gas) and $\Lambda \nrightarrow 0$ only the leading term in the series for w remains. However, because of the nonidentity permutations [which were ignored in deriving Eq. (16.29)], the classical phase integral will not be the only term left in Eq. (16.55). The perfect Bose-Einstein and Fermi-Dirac gases are the well-known consequences.

We have seen above that in the limit $\Lambda \to 0$, only the classical phase integral remains:

$$Q = \frac{1}{N!h^{3N}} \int_{\mathbf{r}} \int_{\mathbf{p}} e^{-\beta H(\mathbf{p},\mathbf{r})}\, d\mathbf{p}\, d\mathbf{r} \tag{16.56}$$

The conditions (temperature, density, etc.) under which it is legitimate to use this limiting expression are examined in Appendix 6. The new factor $1/N!$ [compare Eq. (16.29)] has arisen from our consideration of the symmetry of wave functions. In turn, the quantum-mechanical postulate on symmetry of wave functions has its origin in the experimental indistinguishability of identical particles. Gibbs had the intuitive foresight, before the advent of quantum mechanics, to insert this factor $1/N!$.

As in Eq. (16.32), let us write Eq. (16.56) in the form

$$Q = \sum_j \frac{e^{-\beta H_j}(\Delta\mathbf{p}\, \Delta\mathbf{r})_j}{N!h^{3N}} = \sum_j e^{-\beta H_j} \tag{16.57}$$

where

$$(\Delta\mathbf{p}\, \Delta\mathbf{r})_j = N!(\Delta\mathbf{p}\, \Delta\mathbf{r})_i = N!h^{3N} \tag{16.58}$$

The meaning of Eqs. (16.57) and (16.58) is the following. The volume in phase space associated with a single quantum-mechanical state of a system of N identical and indistinguishable particles is $N!h^{3N}$. The factor h^{3N} represents the uncertainty principle limit of accuracy in assigning the coordinates and momenta of the N particles in the neighborhood of a particular point in phase space, while $N!$ is the number of such cells of volume h^{3N} in phase space which differ *only* in the exchange of values of coordinates and momenta between identical particles.

The generalization of this result to a multicomponent system of polyatomic molecules is the following. If f_i degrees of freedom per molecule for species i are separable[1] and can be treated classically, then the volume in the phase space of these degrees of freedom associated with a single quantum state of these degrees of freedom is

$$\prod_{i=1}^{r} N_i!h^{f_iN_i} \tag{16.59}$$

Quantum and Classical Partition Functions. We have established, in Sec. 14, the connection between thermodynamics and the *quantum*-mechanical partition function Q as

$$A = -\mathbf{k}T \ln Q$$

On the other hand, in the completely classical (all degrees of freedom) treatment of Chap. 1 [Eqs. (5.3) and (5.4)],

$$f = \frac{e^{-H(p,q)/\mathbf{k}T}}{Q_c} \tag{16.60}$$

$$Q_c = \int e^{-H(p,q)/\mathbf{k}T}\, dp\, dq \tag{16.61}$$

If it is indeed legitimate to treat all degrees of freedom classically, (16.59) states that

$$Q = \frac{Q_c}{\prod_{i=1}^{r} N_i!h^{f_iN_i}} \tag{16.62}$$

where f_i here includes all degrees of freedom of a molecule of species i. Then

$$f = \frac{e^{-H(p,q)/\mathbf{k}T}}{Q\prod_{i=1}^{r} N_i!h^{f_iN_i}} \tag{16.63}$$

$$A = -\mathbf{k}T \ln \frac{Q_c}{\prod_{i=1}^{r} N_i!h^{f_iN_i}} \tag{16.64}$$

Similar alterations must of course be made if Ξ_c of Eq. (6.9) is to be related to its thermodynamic "potential" as in Sec. 14.

[1] That is, the Hamiltonian can be written as a sum of two parts, one part containing the coordinates and momenta associated with the N_if_i degrees of freedom for all species i and only these coordinates and momenta. The f_i degrees of freedom always include, if anything, translation of the center of mass of the molecule, so that each $f_i \geqslant 3$.

When only certain (separable) degrees of freedom can be treated classically, as discussed in connection with (16.59),

$$H = H_{\text{class.}} + H_{\text{quant.}}$$

$$Q = Q_{\text{class.}} Q_{\text{quant.}} \tag{16.65}$$

$$= \frac{Q_{\text{quant.}}}{\prod_{i=1}^{r} N_i ! h^{f_i N_i}} \int e^{-H_{\text{class.}}/\mathbf{k}T}\, dp_{\text{class.}}\, dq_{\text{class.}} \tag{16.66}$$

where f_i is the number of classical degrees of freedom of a molecule of species i, and Eq. (16.65) follows from Eqs. (11.35) and (13.2).

Configuration Integral and Slater Sum. We return to a one-component monatomic gas. The integration over momenta in Eq. (16.56) can be carried out immediately yielding

$$Q = \frac{1}{N!\Lambda^{3N}} \int e^{-U(\mathbf{r})/\mathbf{k}T}\, d\mathbf{r} \tag{16.67}$$

The integral in Eq. (16.67) is known as the configuration integral (often denoted by Z). The quantity $e^{-U(\mathbf{r})/\mathbf{k}T}$ is proportional to the probability of a closed system in thermal equilibrium with its surroundings having the configuration $\mathbf{r}$.

It is of some interest to find the quantum-mechanical analogue of $e^{-U/\mathbf{k}T}$. From [Eq. (11.22)]

$$Q = \sum_k \int \varphi_k^* \exp\left(-\frac{\mathscr{H}}{\mathbf{k}T}\right) \varphi_k\, d\mathbf{r} \tag{16.68}$$

which is also an integral over $\mathbf{r}$, it is obvious that the desired analogue of $e^{-U/\mathbf{k}T}$ is [see Eq. (11.31)]

$$[S] = N!\Lambda^{3N} \sum_k \varphi_k^* \exp\left(-\frac{\mathscr{H}}{\mathbf{k}T}\right) \varphi_k \tag{16.69}$$

$$= N!\Lambda^{3N} \sum_k |\psi_k|^2 e^{-E_k/\mathbf{k}T} \tag{16.70}$$

$$= N!\Lambda^{3N} Q f(\mathbf{r},\mathbf{r}) \tag{16.71}$$

[S] is known as the "Slater sum."

17. Entropy and Irreversibility in Thermodynamics

The second law of thermodynamics states that for any thermodynamic process[1] occurring in a closed, isolated system, ΔS or $dS \geqslant 0$, where the equality sign holds if the process is reversible and the inequality sign holds if

[1] That is, a change from one thermodynamic (equilibrium) state to another, whether passing through equilibrium states only (reversible process) or not (irreversible process).

the process is irreversible. In other words, the entropy increases during spontaneous processes taking place in a closed, isolated system until it reaches a maximum value at the final equilibrium state. As an example, consider a gas in equilibrium, confined by a partition to one-half of a closed, isolated container. The partition is suddenly removed and the spontaneous process of flow of the gas into the entire container takes place, culminating in a final equilibrium state of maximum entropy.

The question naturally arises, can this increasing property of the entropy be deduced from molecular theory? Unfortunately, we cannot discuss this aspect of the entropy in detail, from a molecular point of view, without recourse to the concepts of nonequilibrium thermodynamics and statistical mechanics, which are beyond the scope of this book. We confine ourselves, therefore, to a brief and rather qualitative discussion. We use classical statistical mechanics, and consider a closed, isolated system of N identical monatomic molecules.

We first digress to define a reduced distribution function $f^{(1)}(p_i,q_i;t)$ for the ith molecule in terms of the ensemble distribution function $f(p,q;t)$, which we now denote by $f^{(N)}(p,q;t)$, to be specific. The definition is

$$f^{(1)}(p_i,q_i;t) = \int f^{(N)}(p,q;t) \prod_{\substack{j=1 \\ j \neq i}}^{N} dp_j\, dq_j \tag{17.1}$$

where dp_j includes the three components of momentum of the jth particle, and similarly for dq_j. The integration is over all p's and q's except those belonging to the ith molecule. From Eq. (3.1),

$$\int f^{(1)}(p_i,q_i;t)\, dp_i\, dq_i = 1 \tag{17.2}$$

The quantity $f^{(1)}(p_i,q_i;t)\, dp_i\, dq_i$ is the probability that the momenta and coordinates of the ith particle[1] are within a range $dp_i\, dq_i$ about p_i,q_i. The effects of interactions between the ith molecule and all other molecules are of course taken into account in the definition of $f^{(1)}$, Eq. (17.1).

Boltzmann defined a quantity $\mathfrak{H}$ by

$$\mathfrak{H}(t) = \int f^{(1)}(p_i,q_i;t) \ln f^{(1)}(p_i,q_i;t)\, dp_i\, dq_i \tag{17.3}$$

and was able to show by collision theory,[2] for gases sufficiently dilute that only binary collisions need be considered, that

$$\frac{d\mathfrak{H}}{dt} \leqslant 0 \tag{17.4}$$

This result, the Boltzmann $\mathfrak{H}$ theorem, may also be obtained from the

[1] The six-dimensional space with the p_i and q_i as coordinates was called the "μ space" by Boltzmann. $f^{(1)}$ is a probability density in μ space.

[2] See Tolman (Gen. Ref.).

Maxwell-Boltzmann integrodifferential transport equation.[1] Qualitatively, it is clear from Eq. (17.3) that a decrease in $\mathfrak{H}$ is associated with a "spreading" of $f^{(1)}$ with time (subject to the restraint of constant energy of the system of N molecules), where the "spread" in $f^{(1)}$ is analogous to that discussed in connection with Eq. (14.113).

Although Eqs. (17.3) and (17.4) are strongly reminiscent of Eq. (14.113) and $dS \geqslant 0$, we are not in a position to write down a general connection between $\mathfrak{H}$ and the entropy S for two reasons: (1) Eq. (17.4) has been proved only for a dilute gas; and (2) the entropy has been defined only for equilibrium states, whereas $\mathfrak{H}$ is not restricted to equilibrium states. We make a few remarks below on the attempts that have been made to get over these difficulties.

We need first the classical mechanical expression for S in terms of $f^{(N)}(p,q)$. Let us find this relation for the canonical ensemble, of which the microcanonical ensemble of interest here, is a special case. Consider the function

$$S' = -\mathbf{k}\int f^{(N)}(p,q) \ln f^{(N)}(p,q)\, dp\, dq \tag{17.5}$$

where, according to Eq. (16.63),

$$f^{(N)} = \frac{e^{-H/\mathbf{k}T}}{N!h^{3N}Q} \tag{17.6}$$

Then, if we divide phase space into regions $(\Delta p\, \Delta q)_j$ as in Eq. (16.57),

$$S' = -\mathbf{k}\sum_j \frac{e^{-H_j/\mathbf{k}T}}{Q} \ln \frac{e^{-H_j/\mathbf{k}T}}{N!h^{3N}Q} \tag{17.7}$$

From Eqs. (14.82) and (14.83),

$$S' = S + \mathbf{k} \ln (N!h^{3N}) \tag{17.8}$$

Therefore S' is the entropy except for a constant. Equation (17.8) applies to both the canonical and microcanonical ensembles.

Now for a dilute gas in equilibrium,

$$f^{(N)}(p,q) = \prod_{i=1}^{N} f^{(1)}(p_i,q_i) \tag{17.9}$$

since the $f^{(1)}$'s are independent, and

$$S' = -\mathbf{k}\int\Big[\prod_{i=1}^{N} f^{(1)}(p_i,q_i)\Big]\Big[\sum_{j=1}^{N} \ln f^{(1)}(p_j,q_j)\Big]\, dp\, dq \tag{17.10}$$

$$= -\mathbf{k}\sum_{j=1}^{N}\Big\{\int f^{(1)}(p_j,q_j) \ln f^{(1)}(p_j,q_j)\, dp_j\, dq_j \int \cdots \int \prod_{\substack{i=1\\ i\neq j}}^{N} [f^{(1)}(p_i,q_i)\, dp_i\, dq_i]\Big\} \tag{17.11}$$

[1] See Chapman and Cowling (Gen. Ref.).

$$= -\mathbf{k}\sum_{j=1}^{N}\int f^{(1)}(p_j,q_j)\ln f^{(1)}(p_j,q_j)\,dp_j\,dq_j \tag{17.12}$$

$$= -N\mathbf{k}\mathfrak{H} \qquad \text{(equil.)} \tag{17.13}$$

Equation (17.12) follows from Eq. (17.2), and Eq. (17.13) follows from the fact that all N molecules are equivalent. Equation (17.13) states that S' is identical with $-N\mathbf{k}\mathfrak{H}$ for a *dilute gas at equilibrium*. Incidentally, the form of $f^{(1)}$ at equilibrium, for the microcanonical ensemble, is just the Maxwellian distribution in velocity components (i.e., the p_j), and a uniform density distribution (i.e., the q_j) throughout the container.[1]

The above agreement between $\mathfrak{H}$ and S' in the dilute-equilibrium special case suggests that we extend the definition of the entropy, $S' - \mathbf{k}\ln(N!h^{3N})$, to nonequilibrium states [Eq. (17.5) already applies to nondilute systems in equilibrium] by writing

$$S'(t) = -\mathbf{k}\int f^{(N)}(p,q;t)\ln f^{(N)}(p,q;t)\,dp\,dq \tag{17.14}$$

with the expectation of finding the desired general result $dS'/dt \geqslant 0$. Unfortunately, this does not work out, for by Liouville's theorem it is easy to see that $dS'/dt = 0$ in general, and hence S' is constant instead of increasing to a maximum value.

As time passes, the representative points originally in a given element of volume of phase space are strung out over all of phase space between E and $E + \delta E$ in such a way that their "fine-grained" density $f^{(N)}$, within the now much distorted element of volume moving with the phase points, remains constant (Liouville's theorem) but that a "coarse-grained" density[2] gets more and more uniform in the energy shell.[3] It is clear that the result $dS'/dt = 0$ from Eq. (17.14) is a particular property of the fine-grained density $f^{(N)}$. On the other hand, the above qualitative remarks indicate that a coarse-grained density would achieve the desired increasing property of S' by virtue of "spreading" of the coarse-grained density in phase space.

There have been several attempts to obtain a generalized $\mathfrak{H}$ theorem by using a coarse-grained density. However, none of these attempts has met with complete success. Tolman,[4] for example, divides phase space up into

[1] See Chapman and Cowling (Gen. Ref.).

[2] An average over a *fixed* element of volume in phase space—only part of which is occupied by the original element of volume moving with the phase points.

[3] Gibbs used the analogy of shaking up a nonmiscible drop of ink in a fluid. Eventually the drop is dispersed into very many small droplets so that the fluid has an apparent ("coarse-grained") uniform ink density throughout. But on a finer scale, a small element of volume of fluid is seen to contain both clear fluid and droplets of ink, the ink density ("fine-grained") within each droplet being the same as at the start.

[4] See Tolman (Gen. Ref.).

fixed cells of equal volume ω and defines a coarse-grained density F at all points within a cell as

$$F = \frac{1}{\omega} \int_\omega f^{(N)}(p,q;t)\, dp\, dq \tag{17.15}$$

That is, F is the average density over the cell. Tolman then defines

$$\mathfrak{H}(t) = \int F \ln F\, dp\, dq \tag{17.16}$$

$$= \sum_i \omega F_i \ln F_i \tag{17.17}$$

where F_i is the coarse-grained density in the ith cell. As a special case of the argument given by Tolman, suppose our knowledge at $t = 0$ is such that the state of the system is certainly within a given cell, say the first, and that we have a uniform density over this cell. Thus, at $t = 0$, $F_i = 0$ except F_1, and $F_1 = 1/\omega$, so that $\mathfrak{H}(0) = -\ln \omega$. At a later time, the ensemble, though still having everywhere a fine-grained density of either zero or $1/\omega$ (by Liouville's theorem), will be distorted with phase points strung out over a large number of cells. Then, as may easily be demonstrated, $\mathfrak{H}$ [Eq. (17.17)] has decreased with time and the generalized $\mathfrak{H}$ theorem appears to hold.

But the following criticism of the theory may be made. Let us use the same ensemble above but divide phase space into a different set of cells of volume ω, choosing as one of the cells that volume occupied by the strung-out phase points at, say, $t = 1$ minute. Then, although we take the same $f^{(N)}(p,q)$ at $t = 0$ as above, there will be phase points in many of the new set of cells at $t = 0$. At $t = 1$ min, all these points will, however, be in a single cell, though they will again be spread ($\mathfrak{H}$ decreasing) over many cells for $t > 1$ min. Hence $\mathfrak{H}$ increases between $t = 0$ and $t = 1$ min and has a maximum cusp at $t = 1$ min. It is thus seen that $\mathfrak{H}$ depends not only on the ensemble but also in a critical and artificial way on the choice of cells, and that $\mathfrak{H}$ is not necessarily a decreasing function of t.

An alternative type of coarse graining, which avoids this problem by employing time averaging instead of averaging over a cell, has been introduced by Kirkwood.[1]

In quantum statistics the problem should be somewhat simplified since, owing to the uncertainty principle, a coarse-grained density is quite natural. Tolman, however, superimposes a further coarse-grained density into the quantum statistics.

$\mathfrak{H}$ Theorem and Dynamical Reversibility. Superficially, it appears paradoxical that the $\mathfrak{H}$ theorem, Eq. (17.4), is asymmetric in time while the dynamical equations of motion of the system, on which the $\mathfrak{H}$ theorem is based, are symmetric in time. There is really no conflict, however, for the

[1] J. G. Kirkwood, *J. Chem. Phys.*, **14**, 180, 347 (1946).

derivation of the $\mathfrak{H}$ theorem[1] involves, in addition to particle dynamics, probability concepts associated with only partial knowledge of the state of the system, while dynamical reversibility has meaning only in connection with the behavior of a system whose dynamical state is exactly known. The situation can be summarized as follows. Irreversibility (asymmetry in time) is associated (1) with a macroscopic state of an experimental system, (2) with theoretical representation of our partial knowledge of the state of the system by a probability distribution over microscopic states, and (3) with prediction of the most probable behavior[2] of the experimental system, for example, in the illustration at the beginning of this section, the prediction that, because $\mathfrak{H}$ would decrease, the gas will diffuse into the remainder of the container if the partition is removed. Dynamical reversibility (symmetry in time) is associated with precise knowledge of microscopic state and *certain* prediction of future behavior of the system.

The above remarks, then, dispose of the difficulty that if the motion of each particle is exactly reversed at a certain time t', $\mathfrak{H}$ will increase with time for $t > t'$ instead of decrease, for this proposed exact reversal is not consistent with our very incomplete knowledge of the dynamical state of the system at t'.

GENERAL REFERENCES

Chapman, S., and T. G. Cowling, "The Mathematical Theory of Non-Uniform Gases" (Cambridge, London, 1939).

Fowler, R. H., "Statistical Mechanics" (Cambridge, London, 1936).

Fowler, R. H., and E. A. Guggenheim, "Statistical Thermodynamics" (Cambridge, London, 1939).

ter Haar, D., "Elements of Statistical Mechanics" (Rinehart, New York, 1954).

Mayer, J. E., and M. G. Mayer, "Statistical Mechanics" (Wiley, New York, 1940).

Rushbrooke, G. S., "Introduction to Statistical Mechanics" (Oxford, London, 1949).

Schrödinger, E., "Statistical Thermodynamics" (Cambridge, London, 1948).

Tolman, R. C., "The Principles of Statistical Mechanics" (Oxford, London, 1938).

[1] See Tolman (Gen. Ref.).

[2] There is, though, a very small chance that the "most probable" behavior will not actually be followed by the experimental system, if the system just happens to be—though we have no way of knowing it—in one of certain rare microscopic states (consistent with the observed macroscopic state and carrying virtually no weight in statistical averaging).

CHAPTER 4

FLUCTUATIONS

18. Introduction

For the most part, fluctuations in thermodynamic functions are insignificant in magnitude and, consequently, of little interest or importance in thermodynamics and statistical mechanics. There are, however, several reasons why fluctuations must be considered in order to complete our brief summary of the principles of statistical mechanics.

In the first place, by showing for the various ensembles that deviations from mean values are ordinarily completely negligible, we justify the use of any ensemble to calculate thermodynamic functions in a particular problem, irrespective of the actual experimental arrangement of the system of interest (closed, open, heat-bath, constant-pressure, etc.). Thus, we shall find that, in a canonical ensemble, the energy probability distribution is peaked so high about the mean energy that this ensemble is a microcanonical ensemble for all practical purposes; again, in a grand canonical ensemble, the probability distribution will turn out to have such a sharp maximum at the mean energy and mean composition that this ensemble is virtually a canonical or a microcanonical ensemble; etc. Because of this "mean value" or *thermodynamic* equivalence of the different ensembles, one is free to choose a particular ensemble to work with, when calculating thermodynamic functions, on grounds of computational convenience. The canonical ensemble has, in fact, proved to be the most useful of the various ensembles from this point of view, though there are exceptions (see Sec. 20), and the microcanonical ensemble the least useful.

Second, certain nonthermodynamic properties can be related to fluctuations, such as, for example, the dependence of light scattering on composition fluctuations (see Sec. 21).

Third, in the neighborhood of a critical point or two-phase (or multiphase) region fluctuations in some quantities are no longer of a thermodynamically negligible order of magnitude.

Finally, suppose we represent a system which contains a relatively *small* number of molecules, and which is in equilibrium with its surroundings (if any), by a statistical ensemble.[1] Then, in this small system, fluctuations will be relatively important. It should be emphasized here that our statistical mechanical results, including the treatment of fluctuations in the present chapter, are applicable to (weakly interacting[1]) systems containing a small number of particles. However, it is *not* permissible in this case to make all the associations between statistical mechanical quantities and thermodynamic functions, for many of the thermodynamic functions will not be defined.[2] In particular: (1) In a closed, isolated small system (microcanonical ensemble) none of the associations with thermodynamic properties can be made, since none of these properties is a well-defined macroscopic experimental quantity. Statistical mechanical average values and fluctuations can be computed by assigning equal weight ($1/\Omega$) to each basic state, but these averages have no *thermodynamic* significance. (2) In a closed, isothermal small system (canonical ensemble), the temperature of the small system is the temperature of the (large) heat bath in which it is immersed. We can therefore use[3] Eq. (14.83) for p_k to compute average values, fluctuations, etc. But these average values cannot be associated with "thermodynamic properties" of the small system. (3) Similarly, in a small closed system in thermal and mechanical equilibrium with its surroundings (isothermal-isobaric ensemble), T and $\mathbf{X}$ are defined and Eq. (14.62) can be used, but further associations with thermodynamics are meaningless. (4) For the grand canonical ensemble, Eq. (14.43) holds and only T and $\boldsymbol{\mu}$ are defined thermodynamically for the small system. (5) For the generalized ensemble, Eq. (14.16) is valid and only T, $\boldsymbol{\mu}$, and $\mathbf{X}$ are defined thermodynamically for the small system [T, $\boldsymbol{\mu}$, and $\mathbf{X}$ are interrelated by $f(T,\mathbf{X},\boldsymbol{\mu}) = 0$].

Because of the possible application to small systems, we shall make a point in Sec. 19 of writing first in each case the general statistical mechanical fluctuation equation *before* associating statistical averages with thermodynamic properties which are defined only for large systems.

The analysis of fluctuations to be given below is restricted in several ways. (1) We consider only those fluctuations consistent with equilibrium—that

[1] The ensemble must be the "proper" one in this case, depending on the experimental conditions, even for the calculation of mean values. The condition of "weak interaction" with surroundings is quite restrictive for small systems, it should be noted. Of course, for a study of fluctuations, the "proper" ensemble must be used whether N is large or small.

[2] That is, fluctuations about the mean value are relatively large so that the mean loses its usual significance as the (virtually) only possible result of a measurement. Indeed, the "thermodynamic measurement" itself cannot be made on a very small system.

[3] To justify this, the small system and the heat bath, combined, are used to establish associations with thermodynamics, including $1/\mathbf{k}T \leftrightarrow \beta$. Equation (14.83) can then be deduced for the small system alone. See Schrödinger (Gen. Ref.), pp. 9–10.

is, fluctuations associated with the possibility of observing different values of certain quantities as predicted by an equilibrium ensemble. Fluctuations accompanying departures of a system from equilibrium are beyond the scope of equilibrium statistical mechanics and must be omitted here. (2) It will be recalled that, in deriving other ensembles from the equilibrium microcanonical ensemble (Sec. 13), fluctuations from the most probable distribution were made vanishingly small by the use, in effect, of infinitely large surroundings for the system of interest. Contributions to fluctuations in various properties resulting from deviations from the most probable distribution[1] are therefore not included here—we shall be considering fluctuations "within" the most probable distribution itself. (3) The fluctuation equations we derive are strictly valid for a system in "weak" interaction with infinite surroundings (omitting the microcanonical case from the present discussion). If these results are applied, then, to a particular region (say a volume V') of an actual system—considering the region to be the "system" and the rest of the actual system to be the surroundings—incorrect results are obtained in limiting cases (critical points and two phase regions) where fluctuations are very large. This is because[2] the finite size of the surroundings and the actual interaction or coupling between fluctuations in neighboring regions become significant. See also Appendix 9.

Since we are restricting ourselves to equilibrium and to the most probable distribution for any given ensemble, fluctuations (as well as ensemble averages) are not defined for nonmechanical properties (see Sec. 14) such as S, T, $\boldsymbol{\mu}$, etc. As indicated in Eqs. (14.114) to (14.119), we are left, then, with the consideration of fluctuations in $\mathbf{X}$ in the microcanonical ensemble, in $\mathbf{X}$ and E in the canonical ensemble, in $\mathbf{x}$ and E in the isothermal-isobaric ensemble, in $\mathbf{X}$, $\mathbf{N}$, and E in the grand canonical ensemble, and in $\mathbf{x}$, $\mathbf{N}$, and E in the generalized ensemble.

In Sec. 19 we derive expressions for the mean square relative deviation from the mean,

$$\frac{\overline{(G - \bar{G})^2}}{(\bar{G})^2}$$

for each property G mentioned above, but restrict the discussion to one-component systems with one external parameter (the volume V). It is very easy to obtain, using the same methods,[3] similar equations for any number of components and external parameters, for higher-order fluctuations, $\overline{(G - \bar{G})^n}$, $n > 2$, and for cross fluctuations, $\overline{(G - \bar{G})^n(M - \bar{M})^m}$ (where M is a second property), but we omit these complications in Sec. 19. However, as a

[1] These are discussed, for example, by Fowler (Gen. Ref.).

[2] See M. J. Klein and L. Tisza, *Phys. Rev.*, **76**, 1861 (1949).

[3] See Gibbs (Gen. Ref.) and Fowler (Gen. Ref.).

particularly important illustration of a more complicated case, composition fluctuations in a multicomponent system are considered in Sec. 21, using the grand canonical ensemble. Section 20 is devoted to a discussion of some of the implications of Sec. 19.

In ordinary large thermodynamic systems, fluctuations are very small indeed, as we shall see, so that it is quite legitimate to represent the probability of any given deviation from the mean by a Gaussian distribution with standard deviation determined by the second-order fluctuation, $\overline{(G - \bar{G})^2}$. Of course, when there are large fluctuations, this second-order quantity does not tell the whole story, and higher moments (or orders) must also be used to characterize the probability distribution satisfactorily. But returning to the important small fluctuation case, if we write

$$P = c \exp \left[\frac{-a(G - \bar{G})^2}{(\bar{G})^2} \right] \tag{18.1}$$

or

$$P(x) = ce^{-ax^2} \qquad x = \frac{(G - \bar{G})}{\bar{G}} \tag{18.2}$$

where $P(x)\,dx$ is the probability of a relative deviation from the mean between x and $x + dx$, then normalization gives $c = (a/\pi)^{1/2}$. Also, $\overline{x^2} = 1/2a$. That is,

$$P(x) = \left(\frac{1}{2\pi\overline{x^2}} \right)^{1/2} e^{-x^2/2\overline{x^2}} \tag{18.3}$$

where the standard relative deviation σ is $(\overline{x^2})^{1/2}$. We shall find in Sec. 19 that $\overline{x^2}$ is ordinarily very small so that $P(x)$ is virtually a Dirac δ function.

19. Fluctuations According to the Various Ensembles

We shall obtain below expressions for $\overline{x^2}$ in terms of various ensemble averages. In most cases the ensemble averages involved are associated with thermodynamic functions so that $\overline{x^2}$ can be written in terms of purely thermodynamic quantities, provided that N is very large so that the thermodynamic quantities are defined. If N is not large, the relation between $\overline{x^2}$ and ensemble averages is still valid, but we cannot proceed beyond this relation in this case.

As already mentioned, in this section we restrict ourselves to a one-component system with the single external parameter V.

Canonical and Microcanonical Ensembles. We consider first the canonical ensemble. We differentiate

$$\bar{E}Q = \sum_k E_k e^{-E_k/\mathbf{k}T} \tag{19.1}$$

with respect to temperature, and obtain

$$\left(\frac{\partial \bar{E}}{\partial T}\right)_{N,V} Q + \frac{\bar{E}}{\mathbf{k}T^2} \sum_k E_k e^{-E_k/\mathbf{k}T} = \frac{1}{\mathbf{k}T^2} \sum_k E_k{}^2 e^{-E_k/\mathbf{k}T}$$

or

$$\overline{E^2} - (\bar{E})^2 = \overline{(E - \bar{E})^2} = \mathbf{k}T^2 \left(\frac{\partial \bar{E}}{\partial T}\right)_{N,V} \tag{19.2}$$

Making the association with thermodynamics (N large), $E \leftrightarrow \bar{E}$,

$$\frac{\overline{(E - \bar{E})^2}}{(\bar{E})^2} = \frac{\mathbf{k}T^2 C_v}{E^2} \tag{19.3}$$

where[1] C_v is the heat capacity at constant volume (and N). Now experimentally $C_v = O(N\mathbf{k})$ and $E = O(N\mathbf{k}T)$; therefore

$$\frac{\overline{(E - \bar{E})^2}}{(\bar{E})^2} = O\left(\frac{1}{N}\right) \tag{19.4}$$

The standard relative deviation from the mean is then $O(1/N^{1/2})$. If $N = 10^{22}$, this relative deviation is $O(10^{-11})$ or a percentage deviation of $O(10^{-9}$ per cent). If $N = 10^{10}$ (i.e., about 10^{-14} of a mole), the standard percentage deviation from the mean is still very small, $O(10^{-3}$ per cent). We therefore find that fluctuations in energy are extremely small (i.e., the probability distribution in E is extremely sharp), owing directly to the astronomical number of molecules in an ordinary thermodynamic system.

Next, let us differentiate

$$\bar{p}Q = \sum_k \left(-\frac{\partial E_k}{\partial V}\right) e^{-E_k/\mathbf{k}T} \tag{19.5}$$

with respect to V, where p is the pressure in Eq. (19.5). Then

$$\left(\frac{\partial \bar{p}}{\partial V}\right)_{N,T} Q + \frac{\bar{p}}{\mathbf{k}T} \sum_k \left(-\frac{\partial E_k}{\partial V}\right) e^{-E_k/\mathbf{k}T}$$

$$= \sum_k \left(-\frac{\partial^2 E_k}{\partial V^2}\right) e^{-E_k/\mathbf{k}T} + \frac{1}{\mathbf{k}T} \sum_k \left(\frac{\partial E_k}{\partial V}\right)^2 e^{-E_k/\mathbf{k}T} \tag{19.6}$$

or

$$\overline{p^2} - (\bar{p})^2 = \mathbf{k}T\left[\left(\frac{\partial \bar{p}}{\partial V}\right)_{N,T} + \overline{\frac{\partial^2 E}{\partial V^2}}\right] \tag{19.7}$$

$$= \mathbf{k}T\left[\left(\frac{\partial \bar{p}}{\partial V}\right)_{N,T} - \overline{\frac{\partial p}{\partial V}}\right] \tag{19.8}$$

[1] We use C_v instead of C_V to avoid confusion with the constant C_V in Sec. 14 and Appendixes 2 and 3.

With the thermodynamic association $p \leftrightarrow \bar{p}$, we have

$$\frac{\overline{(p-\bar{p})^2}}{(\bar{p})^2} = \frac{\mathbf{k}T}{p^2}\left[\left(\frac{\partial p}{\partial V}\right)_{N,T} - \overline{\frac{\partial p}{\partial V}}\right] \tag{19.9}$$

Let us use the equation of state $pV = N\mathbf{k}T$ to obtain an idea of the order of magnitude of the first term on the right-hand side of Eq. (19.9). We find $-(1/N)$, which is of the same order of magnitude as in Eq. (19.4). The second term, involving $\overline{\partial p/\partial V}$, must be positive and larger in magnitude than the first term since the left-hand side is necessarily positive. The meaning of $\overline{\partial p/\partial V}$, or $\overline{\partial^2 E/\partial V^2}$, becomes clear only on examining in detail the law of force between the molecules and the walls of the container. Fowler (Gen. Ref.) works out an example and concludes that this second term, though larger in absolute value than the first, is also completely negligible.

In the microcanonical ensemble there are only fluctuations in p to discuss. The desired result is just a limiting case of Eq. (19.8). If we consider a canonical ensemble in which we gradually limit the accessibility of energy states until only those in a very small range in energy δE at E remain, Eq. (19.8) still applies but it is understood that the averages are now over only those states in the range δE, and T is defined, if one likes, by Eq. (14.94).

Isothermal-Isobaric Ensemble. First we examine fluctuations in V. If we write

$$\bar{V}\Delta = \sum_{k,V} V e^{-(pV+E_k)/\mathbf{k}T} \tag{19.10}$$

and differentiate with respect to p, we obtain immediately

$$\overline{V^2} - (\bar{V})^2 = -\mathbf{k}T\left(\frac{\partial \bar{V}}{\partial p}\right)_{N,T} \tag{19.11}$$

or, with $V \leftrightarrow \bar{V}$,

$$\frac{\overline{(V-\bar{V})^2}}{(\bar{V})^2} = -\frac{\mathbf{k}T}{V^2}\left(\frac{\partial V}{\partial p}\right)_{N,T} = \frac{\mathbf{k}T\kappa}{V} \tag{19.12}$$

where κ is the compressibility. If $pV = N\mathbf{k}T$, the right-hand side of Eq. (19.12) becomes $1/N$. This will also be the order of magnitude for any slope $(\partial p/\partial V)_{N,T}$ in the pressure-volume curve of a real fluid *except* in a two-phase or critical region, where the fluctuations become very large. In a two-phase region, for example, V will almost always (see Appendix 9) be either Nv_1 or Nv_2 where v_1 is the volume per molecule in phase 1 and v_2 is the volume per molecule in phase 2. Thus $\overline{(V-\bar{V})^2}$ will be of the order of magnitude of $(\bar{V})^2$ and the right-hand side of Eq. (19.12) will be of order unity. We then conclude from Eq. (19.12) that, while ordinarily $-(\partial p/\partial V)_{N,T} = O(N\mathbf{k}T/V^2)$, in a two-phase region we have $-(\partial p/\partial V)_{N,T} = O(\mathbf{k}T/V^2)$, which is essentially zero, relative to $O(N\mathbf{k}T/V^2)$, as required for thermodynamic purposes.

To obtain an expression for the fluctuation in density, N/V, we differentiate both

$$\overline{V^{-1}}\,\Delta = \sum_{k,V} V^{-1} e^{-(pV+E_k)/\mathbf{k}T} \tag{19.13}$$

and

$$\overline{V^{-2}}\,\Delta = \sum_{k,V} V^{-2} e^{-(pV+E_k)/\mathbf{k}T} \tag{19.14}$$

with respect to p, giving

$$\left(\frac{\partial \overline{V^{-1}}}{\partial p}\right)_{N,T} - \frac{1}{\mathbf{k}T}\,\overline{V\,\overline{V^{-1}}} = -\frac{1}{\mathbf{k}T} \tag{19.15}$$

and

$$\left(\frac{\partial \overline{V^{-2}}}{\partial p}\right)_{N,T} - \frac{1}{\mathbf{k}T}\,\overline{V}\,\overline{(V^{-1})^2} = -\frac{1}{\mathbf{k}T}\,\overline{V^{-1}} \tag{19.16}$$

The left-hand side of Eq. (19.15) is then substituted for $-1/\mathbf{k}T$ in the right-hand side of Eq. (19.16), leading to

$$\overline{(V^{-1})^2} - (\overline{V^{-1}})^2 = \frac{\mathbf{k}T}{\overline{V}}\left[\left(\frac{\partial \overline{V^{-2}}}{\partial p}\right)_{N,T} - \overline{V^{-1}}\left(\frac{\partial \overline{V^{-1}}}{\partial p}\right)_{N,T}\right] \tag{19.17}$$

Now N is constant so that $\bar{\rho} = N\overline{V^{-1}}$ and $\overline{\rho^2} = N^2\overline{(V^{-1})^2}$. Also, $V^{-1} \leftrightarrow \overline{V^{-1}}$ and $V^{-2} \leftrightarrow \overline{V^{-2}}$. Thus finally,

$$\frac{\overline{(V^{-1} - \overline{V^{-1}})^2}}{(\overline{V^{-1}})^2} = \frac{\overline{(\rho - \bar{\rho})^2}}{(\bar{\rho})^2} = -\frac{\mathbf{k}T}{V^2}\left(\frac{\partial V}{\partial p}\right)_{N,T} = \frac{\mathbf{k}T\kappa}{V} \tag{19.18}$$

which is the same relative fluctuation as in Eq. (19.12). This equivalence between the relative fluctuation in G and G^{-1} holds in general only in the "thermodynamic" limit[1] (N large).

Before treating fluctuations in E, let us discuss fluctuations in $H = E + pV$. We differentiate

$$\bar{H}\Delta = \sum_{k,V} (E_k + pV) e^{-(pV+E_k)/\mathbf{k}T} \tag{19.19}$$

$$\bar{H} = \bar{E} + p\bar{V}$$

with respect to T and get

$$\overline{H^2} - (\bar{H})^2 = \mathbf{k}T^2\left(\frac{\partial \bar{H}}{\partial T}\right)_{N,p} \tag{19.20}$$

or, in thermodynamic terms,

$$\frac{\overline{(H - \bar{H})^2}}{(\bar{H})^2} = \frac{\mathbf{k}T^2}{H^2}\,C_p \tag{19.21}$$

[1] The equivalence can be deduced from the Gaussian distribution, Eq. (18.1).

where C_p is the heat capacity at constant pressure (and N). The analogy between Eq. (19.21) and Eq. (19.3) should be noted. The fluctuations in H are "normal"[1] except when C_p becomes very large—as, for example, in a two-phase region where C_p is essentially infinite, since heat can be absorbed (converting some of one phase into the other) at constant pressure *and* temperature.[2]

The expression for the fluctuation in E is more complicated, as might be expected, since H rather than E is the natural "heat" function for the variables T, p, and N, as is well known. From $H = E + pV$, we have

$$(\bar{H})^2 = (\bar{E})^2 + 2p\bar{E}\bar{V} + p^2(\bar{V})^2 \tag{19.22}$$

$$\overline{H^2} = \overline{E^2} + 2p\overline{EV} + p^2\overline{V^2} \tag{19.23}$$

and $$\overline{E^2} - (\bar{E})^2 = [\overline{H^2} - (\bar{H})^2] - p^2[\overline{V^2} - (\bar{V})^2] - 2p(\overline{EV} - \bar{E}\bar{V}) \tag{19.24}$$

To find $\overline{EV}$, we differentiate Eq. (19.10) with respect to T, and obtain

$$\overline{EV} = \mathbf{k}T^2 \left(\frac{\partial \bar{V}}{\partial T}\right)_{N,p} + \bar{V}\bar{E} - p[\overline{V^2} - (\bar{V})^2] \tag{19.25}$$

Using Eqs. (19.11), (19.20), and (19.25), Eq. (19.24) becomes

$$\frac{\overline{E^2} - (\bar{E})^2}{\mathbf{k}T^2} = \left(\frac{\partial \bar{H}}{\partial T}\right)_{N,p} - 2p\left(\frac{\partial \bar{V}}{\partial T}\right)_{N,p} - \frac{p^2}{T}\left(\frac{\partial \bar{V}}{\partial p}\right)_{N,T} \tag{19.26}$$

In thermodynamic language, employing

$$C_p - C_v = -\frac{T(\partial V/\partial T)^2_{N,p}}{(\partial V/\partial p)_{N,T}} \tag{19.27}$$

Equation (19.26) may be written

$$\frac{\overline{E^2} - (\bar{E})^2}{\mathbf{k}T^2} = C_v - \frac{[p(\partial V/\partial p)_{N,T} + T(\partial V/\partial T)_{N,p}]^2}{T(\partial V/\partial p)_{N,T}} \tag{19.28}$$

This result should be compared with Eq. (19.2) for the fluctuation in E in the canonical ensemble. The second term on the right-hand side of Eq. (19.28) is of the same order of magnitude as C_v, is necessarily positive, and represents the additional contribution to the fluctuation in E owing to the lack of constancy of V in this case. An equation very similar to Eq. (19.28) may be derived, relating the fluctuation in H to C_p in the canonical ensemble.

[1] That is, the mean-square relative deviation is $O(1/N)$.

[2] Incidentally, C_v is not infinite under these conditions, for when a two-phase system absorbs heat at constant volume both p and T change. As in Eq. (19.12), the right-hand side of Eq. (19.21) is of order unity in a two-phase region and $C_p = O(N^2\mathbf{k})$, which is essentially infinite relative to the usual $C_p = O(N\mathbf{k})$.

Grand Canonical Ensemble. For the fluctuation in p, we start with

$$\bar{p}\Xi = \sum_{k,N} \left(-\frac{\partial E_k}{\partial V}\right) e^{(N\mu - E_k)/\mathbf{k}T} \tag{19.29}$$

Differentiating with respect to V, we find

$$\overline{p^2} - (\bar{p})^2 = \mathbf{k}T \left[\left(\frac{\partial \bar{p}}{\partial V}\right)_{\mu,T} - \overline{\frac{\partial p}{\partial V}}\right] \tag{19.30}$$

which reduces to [compare Eq. (19.9)]

$$\frac{\overline{(p-\bar{p})^2}}{(\bar{p})^2} = -\frac{\mathbf{k}T}{p^2}\overline{\frac{\partial p}{\partial V}} \tag{19.31}$$

on putting $p \leftrightarrow \bar{p}$, since thermodynamically p is a function of μ and T only.

The fluctuation in N is found by differentiating

$$\bar{N}\Xi = \sum_{k,N} N e^{(N\mu - E_k)/\mathbf{k}T} \tag{19.32}$$

with respect to μ. We deduce

$$\overline{N^2} - (\bar{N})^2 = \mathbf{k}T \left(\frac{\partial \bar{N}}{\partial \mu}\right)_{V,T} \tag{19.33}$$

The thermodynamic result is then

$$\frac{\overline{(N-\bar{N})^2}}{(\bar{N})^2} = \frac{\mathbf{k}T}{N^2}\left(\frac{\partial N}{\partial \mu}\right)_{V,T} \tag{19.34}$$

$$= \frac{\overline{(\rho-\bar{\rho})^2}}{(\bar{\rho})^2} \tag{19.35}$$

where Eq. (19.35) follows because V is constant in $\rho = N/V$. Now

$$d\mu = v\,dp \qquad T \text{ const} \tag{19.36}$$

where $v = V/N$; therefore

$$\left(\frac{\partial \mu}{\partial v}\right)_T = v\left(\frac{\partial p}{\partial v}\right)_T \tag{19.37}$$

and hence

$$-\frac{N^2}{V}\left(\frac{\partial \mu}{\partial N}\right)_{V,T} = V\left(\frac{\partial p}{\partial V}\right)_{N,T} \tag{19.38}$$

Therefore, Eq. (19.35) may also be written

$$\frac{\overline{(\rho-\bar{\rho})^2}}{(\bar{\rho})^2} = -\frac{\mathbf{k}T}{V^2}\left(\frac{\partial V}{\partial p}\right)_{N,T} = \frac{\mathbf{k}T\kappa}{V} \tag{19.39}$$

which is the same as Eq. (19.18). This is hardly surprising since in a thermodynamic system constant p and T are equivalent in a sense to constant μ

and T, for μ is determined by p and T only. As pointed out in connection with Eq. (19.12), density fluctuations are large in a critical or two-phase region (see Appendix 9 and Sec. 39). Otherwise, however, the number of molecules in an open system is virtually always just the mean number N, as the mean-square relative fluctuation in N is of order $1/\bar{N}$.

Just as $E + pV$ is the natural heat function in the isothermal-isobaric ensemble, $E - \mu N = TS - pV$ plays this role here.[1] This is clear from the thermodynamic equation

$$dE = T\,dS - p\,dV + \mu\,dN \tag{19.40}$$

for, at constant μ and V,

$$T\,dS = d(E - \mu N) \tag{19.41}$$

Let $J = E - \mu N$. Then we differentiate

$$J\Xi = \sum_{k,N} (E_k - \mu N)e^{(N\mu - E_k)/\mathbf{k}T} \tag{19.42}$$

with respect to T, and arrive at

$$\overline{J^2} - (\bar{J})^2 = \mathbf{k}T^2 \left(\frac{\partial J}{\partial T}\right)_{\mu,V} \tag{19.43}$$

or the thermodynamic equation

$$\frac{\overline{(J - \bar{J})^2}}{(\bar{J})^2} = \frac{\mathbf{k}T^2}{J^2} C_{\mu,V} \tag{19.44}$$

where $C_{\mu,V}$ is the "heat capacity" at constant μ and V. This equation is completely analogous to Eqs. (19.3) and (19.21).

The fluctuation in E is found by essentially the same argument as in Eqs. (19.22) to (19.28). Here J replaces H and $\overline{NE}$ is found by differentiating Eq. (19.32) with respect to T. We obtain in this way

$$\frac{\overline{E^2} - (\bar{E})^2}{\mathbf{k}T^2} = \left(\frac{\partial J}{\partial T}\right)_{\mu,V} + 2\mu \left(\frac{\partial \bar{N}}{\partial T}\right)_{\mu,V} + \frac{\mu^2}{T}\left(\frac{\partial \bar{N}}{\partial \mu}\right)_{T,V} \tag{19.45}$$

Introducing thermodynamic associations,

$$\left(\frac{\partial J}{\partial T}\right)_{\mu,V} = C_{\mu,V} = \left(\frac{\partial E}{\partial T}\right)_{\mu,V} - \mu \left(\frac{\partial N}{\partial T}\right)_{\mu,V} \tag{19.46}$$

$$= C_v + \left(\frac{\partial N}{\partial T}\right)_{\mu,V} \left[\left(\frac{\partial E}{\partial N}\right)_{T,V} - \mu\right] \tag{19.47}$$

[1] This heat function is in general (see Table 1) $TS +$ fundamental function. In the order of Table 1, we have (1) $TS + 0 = TS$, (2) $TS - pV = E - \mu N$, (3) $TS + \mu N = H$, (4) $TS + A = E$, and (5) $TS - TS = 0$.

But
$$\left(\frac{\partial E}{\partial N}\right)_{T,V} = \mu + T\left(\frac{\partial S}{\partial N}\right)_{T,V} = \mu + T\left[\frac{(\partial N/\partial T)_{\mu,V}}{(\partial N/\partial \mu)_{T,V}}\right] \tag{19.48}$$

so that finally

$$\frac{\overline{E^2} - (\bar{E})^2}{\mathbf{k}T^2} = C_v + \frac{[\mu(\partial N/\partial\mu)_{T,V} + T(\partial N/\partial T)_{\mu,V}]^2}{T(\partial N/\partial \mu)_{T,V}} \tag{19.49}$$

This equation is the analogue of Eq. (19.28). It can, in fact, be rewritten, by standard thermodynamic manipulations, as

$$\frac{\overline{E^2} - (\bar{E})^2}{\mathbf{k}T^2} = C_v - \frac{\left[\left(p + \frac{E}{V}\right)\left(\frac{\partial V}{\partial p}\right)_{N,T} + T\left(\frac{\partial V}{\partial T}\right)_{N,p}\right]^2}{T\left(\frac{\partial V}{\partial p}\right)_{N,T}} \tag{19.50}$$

The second term on the right-hand side of Eq. (19.49) is associated with the nonconstancy of N.

Generalized Ensemble. The fluctuations in E, N, and V are essentially infinite in this ensemble; thus a detailed treatment is hardly necessary for these variables. The following argument will, however, probably be worthwhile. If in Υ (with p, T, and μ self-consistent) we sum first over N and k, we find [Eq. (14.107)] that each V in the range $0 \leqslant V \leqslant 2\bar{V}$ has the same probability. Then $\overline{V^2} = 4(\bar{V})^2/3$ and

$$\frac{\overline{(V - \bar{V})^2}}{(\bar{V})^2} = \frac{1}{3} \tag{19.51}$$

Similarly, on summing first over V and k, we find [see Eq. (14.108)] that each N has the same probability, and hence

$$\frac{\overline{(N - \bar{N})^2}}{(\bar{N})^2} = \frac{1}{3} \tag{19.52}$$

These mean square relative fluctuations are thus very large—of order unity rather than the normal $O(1/\bar{N})$.

We shall see below that the fluctuations in E/N are normal so that the fluctuations in $E = N \cdot (E/N)$ are essentially determined by the fluctuations in N (that is, $E/N \simeq$ constant). Hence, the fluctuations in E are also very large.

Incidentally, there will also be large fluctuations in the composite extensive ("heat") function $E - \mu N + pV$ though fluctuations in TS are not defined (i.e., *all* states N, V, k are "included" in S to begin with so that there is no question of S taking on different values in different states). However, TS is equal to the *average value*, $\bar{E} - \mu\bar{N} + p\bar{V}$, of the above function, to within terms of thermodynamic significance.[1]

[1] See footnote 2 on p. 76.

Although there are large fluctuations in the extent of the system and therefore in many extensive properties, fluctuations in intensive properties should, of course, be normal. To illustrate this, we investigate fluctuations in N/V and E/N. We use

$$\bar{G} = \frac{\sum\limits_{N,V,k} G e^{(-pV+N\mu-E_k)/\mathbf{k}T}}{\sum\limits_{N,V,k} e^{(-pV+N\mu-E_k)/\mathbf{k}T}} \tag{19.53}$$

where it is understood here that the values of p, μ, and T are related by $f(p,T,\mu) = 0$, or

$$d\mu = -s\,dT + v\,dp \tag{19.54}$$

$$s = \frac{S}{N} \qquad v = \frac{V}{N}$$

$\bar{G}$ may be regarded as a function of $\bar{V}$ and two of p, T, and μ (see Sec. 14).

For the fluctuation in $N/V = \rho$, we differentiate Eq. (19.53), with $G = N/V^2$, with respect to p, holding T and $\bar{V}$ fixed. This gives

$$\left(\frac{\partial\overline{N/V^2}}{\partial p}\right)_{T,\bar{V}} + \frac{\overline{N/V^2}}{\mathbf{k}T}\left[-\bar{V} + \bar{N}\left(\frac{\partial\mu}{\partial p}\right)_T\right] = \frac{1}{\mathbf{k}T}\left[-\overline{N/V} + \overline{N^2/V^2}\left(\frac{\partial\mu}{\partial p}\right)_T\right] \tag{19.55}$$

If we put $(\partial\mu/\partial p)_T = v$ and make thermodynamic associations at this point, Eq. (19.55) becomes

$$\left(\frac{\partial N/V^2}{\partial p}\right)_{T,V} = \frac{1}{\mathbf{k}T}\left(-\rho + \frac{\overline{N^2/V^2}}{\rho}\right)$$

which can be written

$$\frac{\overline{(\rho-\bar{\rho})^2}}{(\bar{\rho})^2} = -\frac{\mathbf{k}T}{V^2}\left(\frac{\partial V}{\partial p}\right)_{N,T} \tag{19.56}$$

This is the same result as Eqs. (19.18) and (19.39). This might have been anticipated in view of the comments following Eq. (19.39).

We now turn to E/N, and take $G = E/N^2$ in Eq. (19.53). We deduce

$$\left(\frac{\partial\overline{E/N^2}}{\partial\mu}\right)_{T,\bar{V}} + \frac{\overline{E/N^2}}{\mathbf{k}T}\left[-\bar{V}\left(\frac{\partial p}{\partial\mu}\right)_T + \bar{N}\right] = \frac{1}{\mathbf{k}T}\left[-\overline{EV/N^2}\left(\frac{\partial p}{\partial\mu}\right)_T + \overline{E/N}\right] \tag{19.57}$$

and $$\left(\frac{\partial \overline{E/N^2}}{\partial T}\right)_{p,\bar{V}} + \frac{\overline{E/N^2}}{\mathbf{k}T^2}\left[\bar{N}T\left(\frac{\partial \mu}{\partial T}\right)_p + p\bar{V} - \bar{N}\mu + \bar{E}\right]$$

$$= \frac{1}{\mathbf{k}T^2}\left[\overline{E/N}T\left(\frac{\partial \mu}{\partial T}\right)_p + p\overline{EV/N^2} - \mu\overline{E/N} + \overline{E^2/N^2}\right] \tag{19.58}$$

Next we introduce thermodynamic associations, use Eq. (19.54), and eliminate $\overline{EV/N^2}$ between Eqs. (19.57) and (19.58). The result is

$$\frac{\overline{E^2/N^2} - (\overline{E/N})^2}{\mathbf{k}T^2} = \left(\frac{\partial \overline{E/N^2}}{\partial T}\right)_{p,V} + \frac{pv}{T}\left(\frac{\partial \overline{E/N^2}}{\partial \mu}\right)_{T,V} \tag{19.59}$$

Now if we carry out the differentiations indicated in Eq. (19.59) and use

$$\left(\frac{\partial E}{\partial T}\right)_{p,V} = C_v + \left(\frac{\partial E}{\partial N}\right)_{T,V}\left(\frac{\partial N}{\partial T}\right)_{p,V} \tag{19.60}$$

$$\left(\frac{\partial N}{\partial T}\right)_{p,V} = \left(\frac{\partial N}{\partial T}\right)_{\mu,V} - s\left(\frac{\partial N}{\partial \mu}\right)_{T,V} \tag{19.61}$$

$$\left(\frac{\partial E}{\partial \mu}\right)_{T,V} = \left(\frac{\partial E}{\partial N}\right)_{T,V}\left(\frac{\partial N}{\partial \mu}\right)_{T,V} \tag{19.62}$$

and Eq. (19.48), we obtain finally

$$\frac{N^2}{\mathbf{k}T^2}[\overline{(E/N)^2} - (\overline{E/N})^2]$$

$$= C_v + \frac{\left[T\left(\frac{\partial N}{\partial T}\right)_{\mu,V} + \left(\mu - \frac{E}{N}\right)\left(\frac{\partial N}{\partial \mu}\right)_{T,V}\right]\left[T\left(\frac{\partial N}{\partial T}\right)_{\mu,V} + \left(\mu - 2\frac{E}{N}\right)\left(\frac{\partial N}{\partial \mu}\right)_{T,V}\right]}{T\left(\frac{\partial N}{\partial \mu}\right)_{T,V}} \tag{19.63}$$

It is easy to show that Eq. (19.63) can also be written

$$\frac{N^2}{\mathbf{k}T^2}[\overline{(E/N)^2} - (\overline{E/N})^2]$$

$$= C_v - \frac{\left[T\left(\frac{\partial V}{\partial T}\right)_{N,p} + p\left(\frac{\partial V}{\partial p}\right)_{N,T}\right]\left[T\left(\frac{\partial V}{\partial T}\right)_{N,p} + \left(p - \frac{E}{V}\right)\left(\frac{\partial V}{\partial p}\right)_{N,T}\right]}{T\left(\frac{\partial V}{\partial p}\right)_{N,T}} \tag{19.64}$$

Equation (19.63) is very similar to Eq. (19.49), while Eq. (19.64) resembles Eq. (19.28). The order of magnitude of the fluctuation is clearly the same as that found in the other ensembles.

20. Thermodynamic Equivalence of Ensembles

We have seen in the preceding section that the normal mean-square relative fluctuation in a thermodynamic property, from its mean value, is of order $1/\bar{N}$. Since $\bar{N}$ is of the order of 10^{23} in thermodynamic systems, these normal fluctuations are extremely small. This conclusion is, of course, consistent with the point of view of pure thermodynamics, which is developed without any mention of fluctuations. Our statistical analysis thus provides an a posteriori theoretical justification of the tacit neglect of fluctuations in the formulation of thermodynamics.

Large fluctuations are predicted by statistical theory in two-phase and critical regions but this conclusion is also consistent with thermodynamics. For example, in a two-phase equilibrium, density fluctuations are normal within each pure phase but in a given element of volume—in the absence of gravitational forces—density fluctuations would be large, with the density ranging between the two extreme values appropriate to the separate phases.[1] At any particular time the density would depend on the proportion of the two phases in the element of volume. Large fluctuations of this type obviously present no problem in thermodynamics and in fact need not be mentioned in a purely thermodynamic discussion.

We see, then, that thermodynamic deductions from statistical mechanics depend in practice only on *mean values*. This fact eliminates the uniqueness of the "proper" ensemble in a given application so long as we confine our interest to mean values and thermodynamic properties. In fact, with such small fluctuations from the mean as we ordinarily encounter, the most probable value is thermodynamically indistinguishable from the mean value and may be used in its place.

As an example, consider a closed system, in thermal equilibrium with its surroundings, characterized by T, V, and N. The "proper" ensemble here is the canonical ensemble with

$$Q(N,V,T) = \sum_{E} \Omega(E,N,V)e^{-E/\mathbf{k}T} \tag{20.1}$$

where $\Omega(E,N,V)$ is the number of quantum states with energy E. That is, we sum here over energy levels rather than quantum states and Ω is the degeneracy of the level with energy[2] E. Let us examine in detail the transition to the microcanonical ensemble which results when we make use of the fact that fluctuations in the canonical ensemble are thermodynamically

[1] The probability of different densities need not agree with Appendix 9, for we refer here to a subsystem which interacts with its surroundings.

[2] In practice, with N large, Ω is so large and successive values of E so close together, summation may be replaced by integration in many cases. Alternatively, we may regard the summation in Eq. (20.1) as being over values of E at constant intervals δE, with Ω the number of states between E and $E + \delta E$.

negligible. The quantity $\Omega e^{-E/\mathbf{k}T}$ is the weight used in forming mean values. In accordance with the discussion above, we are going to ignore fluctuations and consider only most probable values. This is equivalent to saying that all terms in the sum, Eq. (20.1), are to be ignored except the largest. Let E^* be the value of E giving the maximum term. Then we find E^* from

$$\left(\frac{\partial \Omega e^{-E/\mathbf{k}T}}{\partial E}\right)_{N,V,T} = 0 \tag{20.2}$$

or

$$\left(\frac{\partial \ln \Omega}{\partial E}\right)_{N,V} = \frac{1}{\mathbf{k}T} \tag{20.3}$$

which is the same as Eq. (14.94) for the microcanonical ensemble. In Eq. (14.94), N, V, E, and $\Omega(E,N,V)$ are given and these quantities determine T. Here N, V, T, and $\Omega(E,N,V)$ are given, and determine $E = E^*$. From

$$Q(N,V,T) = \Omega(E^*,N,V)e^{-E^*/\mathbf{k}T} \tag{20.4}$$

we also find [Eqs. (14.85) to (14.90)]

$$\frac{p}{\mathbf{k}T} = \left(\frac{\partial \ln Q}{\partial V}\right)_{N,T} = \left(\frac{\partial \ln \Omega}{\partial V}\right)_{N,E=E^*} \tag{20.5}$$

$$-\frac{\mu}{\mathbf{k}T} = \left(\frac{\partial \ln Q}{\partial N}\right)_{V,T} = \left(\frac{\partial \ln \Omega}{\partial N}\right)_{V,E=E^*} \tag{20.6}$$

$$S = \mathbf{k} \ln Q + \mathbf{k}T \left(\frac{\partial \ln Q}{\partial T}\right)_{N,V} = \mathbf{k} \ln \Omega(E^*,N,V) \tag{20.7}$$

Equations (20.5) to (20.7) are the same as the microcanonical ensemble Eqs. (14.95), (14.96), and (14.92), respectively. We thus see the complete thermodynamic equivalence of these two ensembles.

As another example, consider the transition from the grand canonical ensemble to the canonical ensemble. We have

$$\Xi(\mu,V,T) = \sum_N e^{N\mu/\mathbf{k}T} Q(N,V,T) \tag{20.8}$$

The most probable value of N, N^*, is found from

$$\left(\frac{\partial\, Q e^{N\mu/\mathbf{k}T}}{\partial N}\right)_{\mu,V,T} = 0 \tag{20.9}$$

or

$$-\frac{\mu}{\mathbf{k}T} = \left(\frac{\partial \ln Q}{\partial N}\right)_{V,T} \tag{20.10}$$

which is the canonical ensemble Eq. (14.88). Then from

$$\Xi(\mu,V,T) = e^{N^*\mu/\mathbf{k}T} Q(N^*,V,T) \tag{20.11}$$

we deduce [Eqs. (14.48) to (14.50)]

$$\frac{p}{\mathbf{k}T} = \left(\frac{\partial \ln \Xi}{\partial V}\right)_{T,\mu} = \left(\frac{\partial \ln Q}{\partial V}\right)_{T,N=N^*} \tag{20.12}$$

$$\frac{\bar{N}}{\mathbf{k}T} = \left(\frac{\partial \ln \Xi}{\partial \mu}\right)_{T,V} = \frac{N^*}{\mathbf{k}T} \tag{20.13}$$

$$S = \mathbf{k} \ln \Xi + \mathbf{k}T \left(\frac{\partial \ln \Xi}{\partial T}\right)_{\mu,V} = \mathbf{k} \ln Q + \mathbf{k}T \left(\frac{\partial \ln Q}{\partial T}\right)_{V,N=N^*} \tag{20.14}$$

Equations (20.12) and (20.14) are equivalent to Eqs. (14.90) and (14.89), respectively. Also,

$$pV = \mathbf{k}T \ln \Xi = N^*\mu + \mathbf{k}T \ln Q \tag{20.15}$$

or

$$-\mathbf{k}T \ln Q = N^*\mu - pV = A \tag{20.16}$$

in agreement with Eq. (14.85).

There are of course a number of other cases in which we can convert one ensemble into another by picking out the largest term in a sum (or multiple sum). For example, generalized ensemble → grand canonical ensemble ($V = V^*$), grand canonical ensemble → microcanonical ensemble ($N = N^*$, $E = E^*$), etc.

The above discussion has concerned the "reduction" of a more general to a less general ensemble (i.e., fewer summations in the partition function) by picking out a maximum term. The opposite procedure of changing a less general into a more general ensemble (by introducing more summations) is sometimes helpful for analytical purposes, and of course is thermodynamically legitimate for the same reason that "reduction" is legitimate. However, there is of course no point in introducing a more general ensemble if, by using the device of picking out a maximum term with respect to N, V, and/or E, the ensemble is reduced again to the original ensemble. For example, if Q has been set up in a given problem, nothing is to be gained by forming Ξ [Eq. (20.8)] if the procedure of putting $N = N^*$ has to be introduced to handle Ξ. For this procedure simply reduces Ξ back to Q again. But there are many applications in which a definite advantage results from a more general ensemble. In particular Q is almost always easier to handle than Ω since the awkward restriction E = constant in summing over states is removed. Similarly, Ξ is more convenient than Q in some problems[1] because of the elimination of the restraint N = constant, which restraint may make the summation in Q difficult. Also, it may be noted that, if the equation of state with μ, V, and T as independent variables is of particular interest, then it is natural to use[2] Ξ [Eq. (14.46)].

[1] See, for example, Rushbrooke (Gen. Ref.) and Appendix 5.

[2] See Chaps. 5 and 6.

In many problems a partition function, say Q, is developed in such a way that an extra summation over some nonthermodynamic parameter is included. If this summation (or integration) is difficult, the partition function may be reduced to just the maximum term in the sum over the parameter. This is equivalent (using Q) to minimizing the free energy A with respect to the parameter. This step is, of course, proper only if the relative fluctuations about the most probable value of the parameter are negligible.

Magnitude of δE in the Microcanonical Ensemble. We conclude this section by a few remarks concerning the unimportance of the magnitude of δE for thermodynamic properties, in particular for the entropy, in the microcanonical ensemble. We know from thermodynamic measurements that $S/\mathbf{k} = O(N)$ and therefore that $\Omega = \exp[O(N)]$. Now suppose S and Ω correspond to δE, and S' and Ω' to $\alpha\,\delta E$ where δE and $\alpha\,\delta E$ are both very small compared to E but the constant α may be much greater or much less than unity. Then $\Omega' = \alpha\Omega$ and

$$\frac{S'}{\mathbf{k}} = \ln \alpha + \left(\frac{S}{\mathbf{k}}\right) \tag{20.17}$$

Now $S/\mathbf{k}$ in Eq. (20.17) is of order N; therefore even if α is very far from unity, say $O(N)$ or $O(1/N)$, $\ln \alpha$ is just $O(\ln N)$ which is completely negligible compared to $S/\mathbf{k}$. Thus, because of the tremendous number of quantum states Ω, the size of δE is quite unimportant. We see in the Appendixes that $\Omega = \exp[O(N)]$ is verified by statistical mechanics in particular examples.[1]

21. Composition Fluctuations in Multicomponent Systems

Because of applications to light-scattering theory[2] and to solution theory[3] we include here a discussion of fluctuations in composition in systems with more than one component. We use the grand canonical ensemble in order to include fluctuations in *relative* composition. Alternatively, instead of the grand canonical ensemble, one can employ the ensemble of Eq. (14.97) for the light-scattering problem, as shown by Stockmayer.[4] The isothermal-isobaric ensemble cannot be used because the relative composition is constant, though $(N_1 + \cdots + N_r)/V$ fluctuates.

[1] See Mayer and Mayer (Gen. Ref.) for a detailed discussion of this whole question, including the minimum value of δE consistent with the uncertainty principle.

[2] F. Zernike, "L'Opalescence critique" (Dissertation, Amsterdam, 1915); H. C. Brinkman and J. J. Hermans, *J. Chem. Phys.*, **17**, 574 (1949); J. G. Kirkwood and R. J. Goldberg, *J. Chem. Phys.*, **18**, 54 (1950); W. H. Stockmayer, *J. Chem. Phys.*, **18**, 58 (1950).

[3] J. G. Kirkwood and F. P. Buff, *J. Chem. Phys.*, **19**, 774 (1951).

[4] W. H. Stockmayer, *loc. cit.*

We give first an analysis which treats the various components symmetrically. This point of view is useful in general solution theory. The quantity of interest is

$$\overline{(N_i - \bar{N}_i)(N_j - \bar{N}_j)} = \overline{N_i N_j} - \bar{N}_i \bar{N}_j \tag{21.1}$$

Now

$$\bar{N}_i \Xi = \sum_{k,\mathbf{N}} N_i \exp\left(\frac{\mathbf{N}\cdot\boldsymbol{\mu} - E_k}{\mathbf{k}T}\right) \tag{21.2}$$

and, on differentiating Eq. (21.2) with respect to μ_j,

$$\overline{N_i N_j} - \bar{N}_i \bar{N}_j = \mathbf{k}T\left(\frac{\partial \bar{N}_i}{\partial \mu_j}\right)_{V,T,\mu_k} = \mathbf{k}T\left(\frac{\partial \bar{N}_j}{\partial \mu_i}\right)_{V,T,\mu_k} \tag{21.3}$$

where the subscript μ_k means all μ's except the one indicated in the differentiation. The last form of Eq. (21.3) follows by symmetry. The completely analogous equation which can be deduced in the same way from Eq. (14.97) (taking $i = 1$) is

$$\overline{N_k N_j} - \bar{N}_k \bar{N}_j = \mathbf{k}T\left(\frac{\partial \bar{N}_k}{\partial \mu_j}\right)_{p,T,\mu,N_1} = \mathbf{k}T\left(\frac{\partial \bar{N}_j}{\partial \mu_k}\right)_{p,T,\mu,N_1} \tag{21.4}$$

where the subscript μ means all μ's except μ_1 and the one indicated in the differentiation.

Returning to the grand canonical ensemble, Eq. (21.3) is the required result except that it is not in a useful form for thermodynamic applications, as the thermodynamic quantities $\partial N_i/\partial \mu_j$ are not very familiar. The remaining purely thermodynamic problem, then, is to replace the $\partial N_i/\partial \mu_j$ by the better-known derivatives $(\partial \mu_i/\partial N_j)_{p,T,N_k}$. As an intermediate step, however, we introduce the quantities $(\partial \mu_i/\partial N_j)_{V,T,N_k}$, as follows. At constant V and T, μ_α is a function of $N_1, \ldots, N_r$:

$$d\mu_\alpha = \sum_{i=1}^{r}\left(\frac{\partial \mu_\alpha}{\partial N_i}\right)_{V,T,N_k} dN_i \qquad \alpha = 1, \ldots, r \tag{21.5}$$

If all μ's except μ_j are held constant in addition to V and T, Eqs. (21.5) become

$$\delta_{\alpha j}\, d\mu_\alpha = \sum_{i=1}^{r}\left(\frac{\partial \mu_\alpha}{\partial N_i}\right)_{V,T,N_k} dN_i \qquad \alpha = 1, \ldots, r \tag{21.6}$$

These are r linear nonhomogeneous equations in the unknowns $dN_1, \ldots, dN_r$ which we can solve for, say, dN_i, giving

$$\left(\frac{\partial N_i}{\partial \mu_j}\right)_{V,T,\mu_k} = \frac{|A|_{ij}}{|A|} \tag{21.7}$$

where $|A|$ is the determinant with elements

$$A_{lm} = \left(\frac{\partial \mu_l}{\partial N_m}\right)_{V,T,N_k} \tag{21.8}$$

and $|A|_{ij}$ is the cofactor of the element A_{ij}. That is,

$$|A|_{ij} = \frac{\partial |A|}{\partial A_{ij}} \tag{21.9}$$

A_{lm} is also given by

$$A_{lm} = \left(\frac{\partial \mu_l}{\partial N_m}\right)_{p,T,N_k} + \left(\frac{\partial \mu_l}{\partial p}\right)_{T,\mathbf{N}} \left(\frac{\partial p}{\partial N_m}\right)_{V,T,N_k} \tag{21.10}$$

But, in Eq. (21.10),

$$\left(\frac{\partial \mu_l}{\partial p}\right)_{T,\mathbf{N}} = \bar{v}_l \tag{21.11}$$

where $\bar{v}_l$ is a partial molar volume, and

$$\left(\frac{\partial p}{\partial N_m}\right)_{V,T,N_k} = -\frac{(\partial V/\partial N_m)_{p,T,N_k}}{(\partial V/\partial p)_{T,\mathbf{N}}} \tag{21.12}$$

$$= \frac{\bar{v}_m}{V\kappa} \tag{21.13}$$

where κ, the compressibility, is defined as

$$\kappa = -\frac{1}{V}\left(\frac{\partial V}{\partial p}\right)_{T,\mathbf{N}} \tag{21.14}$$

Our final result is then

$$\overline{N_i N_j} - \bar{N}_i \bar{N}_j = \frac{\mathbf{k}T|A|_{ij}}{|A|} \tag{21.15}$$

where

$$A_{lm} = \left(\frac{\partial \mu_l}{\partial N_m}\right)_{p,T,N_k} + \frac{\bar{v}_l \bar{v}_m}{V\kappa} \tag{21.16}$$

In a one-component system,

$$A_{11} = |A| = \frac{V}{N^2 \kappa} \qquad |A|_{11} = 1 \tag{21.17}$$

and

$$\frac{\overline{(N - \bar{N})^2}}{(\bar{N})^2} = \frac{\mathbf{k}T\kappa}{V} \tag{21.18}$$

in agreement with Eq. (19.39).

In light-scattering theory it is helpful to have separated from each other fluctuations in relative composition (or in concentrations of solutes) and fluctuations in the total number of molecules in the system. To accomplish this[1] we shall call component 1 the solvent and components 2, . . . , r the solutes, and express the relative composition by the concentration variables $c_2 = N_2/N_1, \ldots, c_r = N_r/N_1$. These quantities are directly proportional to molalities, as ordinarily used in chemistry.

We recall that

$$p_k(\mathbf{N}) = \frac{\exp[(\mathbf{N}\cdot\boldsymbol{\mu} - E_k)/\mathbf{k}T]}{\Xi} \tag{21.19}$$

is the probability of observing k and $\mathbf{N}$. The probability of observing $\mathbf{N}$ irrespective of k is

$$p(\mathbf{N}) = \frac{\exp(\mathbf{N}\cdot\boldsymbol{\mu}/\mathbf{k}T)Q(\mathbf{N})}{\Xi} \tag{21.20}$$

$$= \frac{C_{\mathbf{N},V}\exp\{[\mathbf{N}\cdot\boldsymbol{\mu} - A(\mathbf{N})]/\mathbf{k}T\}}{\Xi} \tag{21.21}$$

where $A = -\mathbf{k}T\ln Q$ is the Helmholtz free energy in the canonical ensemble for $\mathbf{N}$, V, and T, and $C_{\mathbf{N},V}$ is essentially a constant [see Eqs. (14.105) and (14.106) and Appendix 9].

We shall confine ourselves here to "normal" fluctuations, that is, we exclude critical and two-phase regions. Then we may anticipate that fluctuations are so small that the mean composition is also the most probable composition and a Gaussian probability distribution about the most probable composition can be employed without error. We therefore expand $\ln p$ in Eq. (21.21) about $\bar{N}_1, \ldots, \bar{N}_r$ and obtain (linear terms drop out because $\ln p$ has a maximum at $\bar{N}_1, \ldots, \bar{N}_r$)

$$p(\mathbf{N}) = c\exp\left[-\frac{1}{2\mathbf{k}T}\sum_{i,j=1}^{r}\left(\frac{\partial^2 A}{\partial N_i\partial N_j}\right)_{T,V,N_k}\Delta N_i\,\Delta N_j\right] \tag{21.22}$$

where c is a constant, $\Delta N_i = N_i - \bar{N}_i$, and it is to be understood throughout that the derivative is to be evaluated at $\bar{N}_1, \ldots, \bar{N}_r$.

From thermodynamics and Eq. (21.16),

$$\left(\frac{\partial^2 A}{\partial N_i\partial N_j}\right)_{T,V,N_k} = \left(\frac{\partial\mu_i}{\partial N_j}\right)_{T,V,N_k} = \left(\frac{\partial\mu_j}{\partial N_i}\right)_{T,V,N_k} = \left(\frac{\partial\mu_i}{\partial N_j}\right)_{p,T,N_k} + \frac{\bar{v}_i\bar{v}_j}{V\kappa} \tag{21.23}$$

Let us introduce Eq. (21.23) into Eq. (21.22) and at the same time replace ΔN_i $(i = 2, \ldots, r)$ by a concentration fluctuation variable. That is, from

$$\ln c_i = \ln N_i - \ln N_1$$

[1] Equation (21.4), used by Stockmayer, introduces component 1 as the solvent at the outset.

we note that

$$\frac{\Delta c_i}{\bar{c}_i} = \frac{\Delta N_i}{\bar{N}_i} - \frac{\Delta N_1}{\bar{N}_1} \equiv \xi_i \qquad i = 2, \ldots, r \tag{21.24}$$

The new variable ξ_i thus gives the relative deviation in the concentration c_i of the ith solute from its most probable or mean value, $\bar{c}_i$. The sum, Σ, in Eq. (21.22) then becomes

$$\Sigma = \sum_{i,j=1}^{r} \frac{\bar{v}_i \bar{v}_j}{V\kappa} \Delta N_i \, \Delta N_j + \sum_{i,j=1}^{r} \left(\frac{\partial \mu_i}{\partial N_j}\right)_{p,T,N_k} \times \bar{N}_i \bar{N}_j \left[\xi_i \xi_j + (\xi_i + \xi_j)\frac{\Delta N_1}{\bar{N}_1} + \left(\frac{\Delta N_1}{\bar{N}_1}\right)^2\right] \tag{21.25}$$

where Eq. (21.25) has been condensed by introducing the symbol ξ_1, where $\xi_1 \equiv 0$. If we make use of the Gibbs-Duhem equation,

$$\sum_{i=1}^{r} \bar{N}_i \left(\frac{\partial \mu_i}{\partial N_j}\right)_{p,T,N_k} = 0 \tag{21.26}$$

we see that only the term $\xi_i \xi_j$ in the square bracket of Eq. (21.25) contributes to the second sum. Eq. (21.25) then simplifies to

$$\Sigma = \frac{\xi^2 V}{\kappa} + \sum_{i,j=2}^{r} \left(\frac{\partial \mu_i}{\partial N_j}\right)_{p,T,N_k} \bar{N}_i \bar{N}_j \xi_i \xi_j \tag{21.27}$$

where

$$\xi \equiv \sum_{i=1}^{r} \frac{\bar{v}_i \, \Delta N_i}{V} \tag{21.28}$$

The physical significance of the new variable ξ is seen from

$$\Delta V = \sum_{i=1}^{r} \left(\frac{\partial V}{\partial N_i}\right)_{p,T,N_k} \Delta N_i \tag{21.29}$$

to be $\xi = \Delta V/V$, where ΔV is the change in volume of the system at constant T and p which would accompany the variations $\Delta N_1, \ldots, \Delta N_r$ if the volume were *not* held fixed. ξ is closely related to the relative variation in the total number of molecules, $\Delta N/\bar{N}$, where $N = N_1 + \cdots + N_r$, though ξ is in general not equal to $\Delta N/\bar{N}$ since

$$\frac{\Delta N}{\bar{N}} = \frac{v \sum_{i=1}^{r} \Delta N_i}{V} \tag{21.30}$$

where $v = V/\bar{N}$. Further discussion of the relation between ξ and $\Delta N/\bar{N}$ in a special case is given below.

The independent variables $N_1, \ldots, N_r$ or $\Delta N_1, \ldots, \Delta N_r$ have now been replaced by $\xi_2, \ldots, \xi_r$ and ξ. Let P be the corresponding new probability distribution function. Then

$$P(\xi_2, \ldots, \xi_r, \xi) = c \exp\left(-\frac{\xi^2 V}{2\mathbf{k}T\kappa} - \sum_{i,j=2}^{r} \beta_{ij}\xi_i\xi_j\right) \tag{21.31}$$

where

$$\beta_{ij} = \frac{\bar{N}_i\bar{N}_j}{2\mathbf{k}T}\left(\frac{\partial\mu_i}{\partial N_j}\right)_{p,T,N_k} \tag{21.32}$$

$$= \frac{\bar{N}_1\bar{c}_i\bar{c}_j}{2\mathbf{k}T}\left(\frac{\partial\mu_i}{\partial c_j}\right)_{p,T,c_k} \tag{21.33}$$

In order to calculate average values such as $\overline{\xi^2}$, etc., we must first evaluate c by normalization. To do this, we make a linear transformation

$$\xi_i = \sum_{k=2}^{r} a_{ik}\eta_k \qquad i = 2, \ldots, r \tag{21.34}$$

to new variables η_k such that

$$\sum_{i,j=2}^{r} \beta_{ij}\xi_i\xi_j = \sum_{k=2}^{r} \eta_k{}^2 \tag{21.35}$$

It is then easy to integrate Eq. (21.31), provided we know the Jacobian of the transformation from $\xi_2, \ldots, \xi_r$ to $\eta_2, \ldots, \eta_r$. The Jacobian is found as follows, by a further consideration of the transformation, Eq. (21.34): If we substitute Eq. (21.34) into Eq. (21.35), we obtain

$$\sum_{i,j} \beta_{ij}\xi_i\xi_j = \sum_{i,j,k,l} \beta_{ij}a_{ik}\eta_k a_{jl}\eta_l \tag{21.36}$$

In order for this result to simplify to

$$\sum_k \eta_k{}^2 = \sum_{k,l} \delta_{kl}\eta_k\eta_l \tag{21.37}$$

as required in Eq. (21.35), we must have

$$\sum_{i,j} \beta_{ij}a_{ik}a_{jl} = \delta_{kl} \tag{21.38}$$

That is, by the rules of matrix multiplication,

$$\begin{aligned}\delta_{kl} &= \sum_i a_{ik}(\beta a)_{il} \\ &= \sum_i \tilde{a}_{ki}(\beta a)_{il} \\ &= (\tilde{a}\beta a)_{kl}\end{aligned} \tag{21.39}$$

where $||\tilde{a}||$ is the transpose of $||a||$; that is, $\tilde{a}_{ki} = a_{ik}$. Now

$$d\xi_2 \cdots d\xi_r = \frac{\partial(\xi_2, \ldots, \xi_r)}{\partial(\eta_2, \ldots, \eta_r)}\, d\eta_2 \cdots d\eta_r \tag{21.40}$$

and, from Eq. (21.34),

$$\frac{\partial(\xi_2, \ldots, \xi_r)}{\partial(\eta_2, \ldots, \eta_r)} = |a| \tag{21.41}$$

But according to Eq. (21.39),

$$1 = |\tilde{a}||\beta||a| \tag{21.42}$$

and, since $|\tilde{a}| = |a|$, we have finally for the Jacobian

$$|a| = \frac{1}{|\beta|^{1/2}} \tag{21.43}$$

where β_{ij} in $|\beta|$ is given by Eq. (21.33).

Returning to Eq. (21.31),

$$\begin{aligned} 1 &= c\int_{-\infty}^{+\infty} \exp\left(-\frac{\xi^2 V}{2\mathbf{k}T\kappa} - \sum_{i,j=2}^{r} \beta_{ij}\xi_i\xi_j\right) d\xi_2 \cdots d\xi_r\, d\xi \\ &= \frac{c}{|\beta|^{1/2}} \int_{-\infty}^{+\infty} \exp\left(-\frac{\xi^2 V}{2\mathbf{k}T\kappa} - \sum_{k=2}^{r} \eta_k{}^2\right) d\eta_2 \cdots d\eta_r\, d\xi \end{aligned}$$

so that we find, on carrying out the integrations,

$$c = \left(\frac{V|\beta|}{2\mathbf{k}T\kappa\pi^r}\right)^{1/2} \tag{21.44}$$

It is, of course, legitimate to use infinite limits in the above integrations because only very small fluctuations contribute appreciably to the integral. Equation (21.31) then becomes, finally,

$$P(\xi_2, \ldots, \xi_r, \xi) = \left(\frac{V|\beta|}{2\mathbf{k}T\kappa\pi^r}\right)^{1/2} \exp\left(-\frac{\xi^2 V}{2\mathbf{k}T\kappa} - \sum_{i,j=2}^{r} \beta_{ij}\xi_i\xi_j\right) \tag{21.45}$$

We see from Eq. (21.45) that the concentration fluctuations ξ_i are orthogonal to (i.e., independent of) the "volume" fluctuations ξ (there are no cross terms $\xi\xi_i$). Analytically, since $\xi \exp(-a\xi^2)$ is an odd function of ξ, Eq. (21.45) yields

$$\overline{\xi\xi_i} = 0 \qquad i = 2, \ldots, r \tag{21.46}$$

A further consequence of the orthogonality between ξ and the ξ_i is that the probabilities of various fluctuations in ξ are the same under some arbitrary restrictions in the ξ_i, say ξ_i = constant ($i = 2, \ldots, r$), as in the absence of any such restrictions. In particular, let us take the $\xi_i = 0$ (constant relative composition). Then, from Eq. (21.28),

$$\xi = \frac{\sum_{i=1}^{r} \bar{v}_i n_i\, \Delta N}{\sum_{i=1}^{r} \bar{v}_i n_i \bar{N}} = \frac{\Delta N}{\bar{N}} \tag{21.47}$$

where the n_i are mole fractions. We may therefore conclude that $\overline{\xi^2}$ *in general* has the same value as $\overline{(\Delta N)^2}/(\bar{N})^2$ under conditions of *constant relative composition.* But it is rather obvious[1] and may easily be confirmed by explicit analysis that, according to the grand canonical ensemble, $\overline{(\Delta N)^2}/(\bar{N})^2$ for a multicomponent system with relative composition held fixed is the same as $\overline{(\Delta N)^2}/(\bar{N})^2$ in a one-component system[2] [Eqs. (19.34) and (19.39)]. This identity of $\overline{\xi^2}$ with the result in Eq. (19.39) is readily verified by the direct computation of $\overline{\xi^2}$ from Eq. (21.45), giving

$$\overline{\xi^2} = \frac{\mathbf{k}T\kappa}{V} \tag{21.48}$$

Since $\overline{(\Delta N)^2}/(\bar{N})^2$ is also the density fluctuation $\overline{(\Delta\rho)^2}/(\bar{\rho})^2$, where $\rho = N/V$, in a grand canonical ensemble (V constant), $\overline{\xi^2}$ is usually referred to as the density fluctuation at constant relative composition. As will be obvious from the above discussion, however, this terminology has its origin in the orthogonality of ξ and the ξ_i and is *not* supposed to imply that all the fluctuations ΔN_i $(i = 2, \ldots, r)$ included in Eq. (21.28) in general are consistent with a fixed relative composition.

Fluctuations in the concentrations of the ith and jth solutes interact (or are coupled) with each other, unless $\beta_{ij} = 0$, according to Eq. (21.45). To express this quantitatively, we observe that

$$\frac{\partial}{\partial\beta_{kl}} \int_{-\infty}^{+\infty} \exp\left(-\frac{\xi^2 V}{2\mathbf{k}T\kappa} - \sum_{i,j=2}^{r} \beta_{ij}\xi_i\xi_j\right) d\xi_2 \cdots d\xi_r\, d\xi$$
$$= -\int_{-\infty}^{+\infty} \xi_k\xi_l \exp\left(-\frac{\xi^2 V}{2\mathbf{k}T\kappa} - \sum_{i,j=2}^{r} \beta_{ij}\xi_i\xi_j\right) d\xi_2 \cdots d\xi_r\, d\xi \tag{21.49}$$

or

$$\frac{\partial}{\partial\beta_{kl}}\left(\frac{1}{c}\right) = -\frac{1}{c}\,\overline{\xi_k\xi_l} \tag{21.50}$$

Then, from Eqs. (21.9) and (21.44),

$$\overline{\xi_k\xi_l} = \frac{1}{2}\frac{|\beta|_{kl}}{|\beta|} \tag{21.51}$$

which is rather analogous to Eq. (21.15).

[1] This is especially easy to see in the isothermal-isobaric ensemble. Actually, all the equations given above for this ensemble hold for a multicomponent system. Fluctuations in volume in this *closed* system are also fluctuations in density, N/V, *at constant relative composition.*

[2] In applying Eqs. (19.34) and (19.39) to a multicomponent system with fixed relative composition, $N = N_1 + \cdots + N_r$ and $\mu = \mathbf{n} \cdot \boldsymbol{\mu}$ where the n_i are mole fractions.

If $r = 1$ (one-component system), only the term in ξ^2 remains in Eq. (21.45), and $\xi = \Delta N/\bar{N}$; thus Eq. (21.48) is in this case completely identical with Eq. (19.39). If $r = 2$ (two-component system), concentration fluctuations occur:

$$\overline{\xi_2{}^2} = \frac{\overline{(\Delta c_2)^2}}{(\bar{c}_2)^2} = \frac{1}{2}\frac{1}{\beta_{22}} = \left[\frac{\bar{N}_1(\bar{c}_2)^2}{\mathbf{k}T}\left(\frac{\partial \mu_2}{\partial c_2}\right)_{p,T}\right]^{-1} \tag{21.52}$$

To verify that these concentration fluctuations are "normal," put $\partial\mu_2/\partial c_2 = O(\mathbf{k}T/c_2)$, whence $\overline{\xi_2{}^2} = O(1/\bar{N}_2)$ as usual.

GENERAL REFERENCES

Fowler, R. H., "Statistical Mechanics" (Cambridge, London, 1936).
Gibbs, J. W., "The Collected Works of J. Willard Gibbs" (Yale, New Haven, 1948), Vol. II.
Mayer, J. E., and M. G. Mayer, "Statistical Mechanics" (Wiley, New York, 1940).
Rushbrooke, G. S., "Introduction to Statistical Mechanics" (Oxford, London, 1949).
Schrödinger, E., "Statistical Thermodynamics" (Cambridge, London, 1948).
Tolman, R. C., "The Principles of Statistical Mechanics" (Oxford, London, 1938).

CHAPTER 5

THEORY OF IMPERFECT GASES AND CONDENSATION

In this chapter we attempt to give a rigorous discussion of a physically simple system in which intermolecular forces play a dominant role, namely, a one-component, classical, monatomic gas with a total intermolecular potential energy given by the sum of pair interactions. As is well known, the treatment of this system can be generalized considerably, and we point out some of these generalizations briefly in the course of this chapter (and also in Sec. 40, Chap. 6). Even with the restrictions indicated, very little progress can be made in the way of calculating thermodynamic functions numerically. The analysis is therefore rather formal, but in spite of this it offers considerable insight into the nature of gas imperfections and condensation. A rigorous treatment of an even simpler physical model with intermolecular forces is discussed in Chap. 7.

Although very important contributions have been made by others, the major advances in this field are due to J. E. Mayer and his collaborators.

At the present time there is considerable controversy[1] over the rather subtle mathematical points involved in the condensation and critical-point theory due[2] to Mayer and collaborators, Born and Fuchs, and Kahn and Uhlenbeck. Also it is generally agreed that this theory is not applicable in its present form to the liquid phase. For these reasons we shall use the Mayer theory (with alternative derivations in part) in Secs. 22 to 25 in considering the non-controversial imperfect-gas region only (no condensed phase present), and then adopt in Sec. 28 the recent approach of Yang and Lee to include condensation. The alternative formalism of Yang and Lee makes possible a

[1] See, for example, S. Katsura and H. Fujita, *J. Chem. Phys.*, **19**, 795 (1951); *Progr. Theoret. Phys. (Japan)*, **6**, 498 (1951); C. N. Yang and T. D. Lee, *Phys. Rev.*, **87**, 404, 410 (1952); B. H. Zimm, *J. Chem. Phys.*, **19**, 1019 (1951); J. E. Mayer, *J. Chem. Phys.*, **19**, 1024 (1951); H. N. V. Temperley, *Proc. Phys. Soc. (London)*, **67A**, 233 (1954).

[2] Mayer and Mayer (Gen Ref.); M. Born and K. Fuchs, *Proc. Roy. Soc. (London)*, **A166**, 391 (1938); B. Kahn and G. E. Uhlenbeck, *Physica*, **5**, 399 (1938).

simple mathematical description of condensation phenomena which avoids the disputed points of the Mayer theory and includes the liquid phase, but which has the disadvantage of being less detailed and physically rather obscure, at least at the present time.

Section 26 is a digression in which we give alternative derivations, which may be found instructive by some readers, of the results of Sec. 23. In addition, a brief discussion of the approximate cluster theory of Frenkel and Band is included in this section. Section 27 contains an exact treatment of the equilibrium between physical clusters in an imperfect gas.

22. The Partition Function and Cluster Integrals

The partition function Q (canonical ensemble), for this problem, is

$$Q = \frac{1}{h^{3N}N!} \int \cdot \cdot \cdot \int e^{-H/\mathbf{k}T} \, d\mathbf{p}_1 \cdot \cdot \cdot d\mathbf{p}_N \, d\mathbf{r}_1 \cdot \cdot \cdot d\mathbf{r}_N \tag{22.1}$$

where $$d\mathbf{p}_i = dp_{xi}\, dp_{yi}\, dp_{zi} \qquad d\mathbf{r}_i = dx_i\, dy_i\, dz_i$$

and $$H = \sum_{i=1}^{N} \frac{1}{2m} (p_{xi}^2 + p_{yi}^2 + p_{zi}^2) + U(\mathbf{r}_1, \ldots, \mathbf{r}_N) \tag{22.2}$$

U is the total potential energy of interaction of all $N(N - 1)/2$ pairs of molecules in the gas, when the molecules are in the configuration $\mathbf{r}_1, \ldots, \mathbf{r}_N$ (the wall forces are treated as due to an infinitely steep and high potential barrier; hence there is no contribution from this source to U). The $\mathbf{r}_i$ integrations are taken over the volume V of the container, so that Q is a function of N, V, and T. Of course, in order for Q to have thermodynamic significance, N must be very large. In fact, for purely thermodynamic purposes (fluctuations of no interest), it is often convenient to consider a system of infinite extent. The limit

$$\lim_{N\to\infty} \frac{1}{N} \ln Q = -\frac{A}{N\mathbf{k}T} \qquad T, \frac{V}{N} \text{ const} \tag{22.3}$$

is then the quantity of primary interest (see Sec. 28).

We substitute Eq. (22.2) into Eq. (22.1) and carry out the momentum integrations, with the result

$$Q = \frac{Z}{N!\Lambda^{3N}} \tag{22.4}$$

where $$\Lambda = \frac{h}{(2\pi m\mathbf{k}T)^{1/2}}$$

and $$Z = \int_V \cdot \cdot \cdot \int e^{-U/\mathbf{k}T} \, d\mathbf{r}_1 \cdot \cdot \cdot d\mathbf{r}_N \tag{22.5}$$

$Z(N,V,T)$ is the so-called "configuration integral." If this function $Z(N,V,T)$ can be found, all thermodynamic properties of the gas follow from Q in Eq. (22.4). Unfortunately, Eq. (22.5) involves a very complicated $3N$-fold integration which in general cannot be carried out. It is still

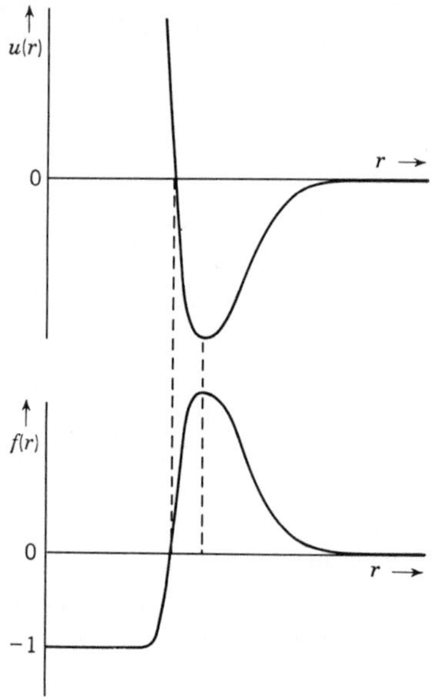

FIG. 4. The form of the functions $u(r)$ and $f(r)$ in a typical case (schematic).

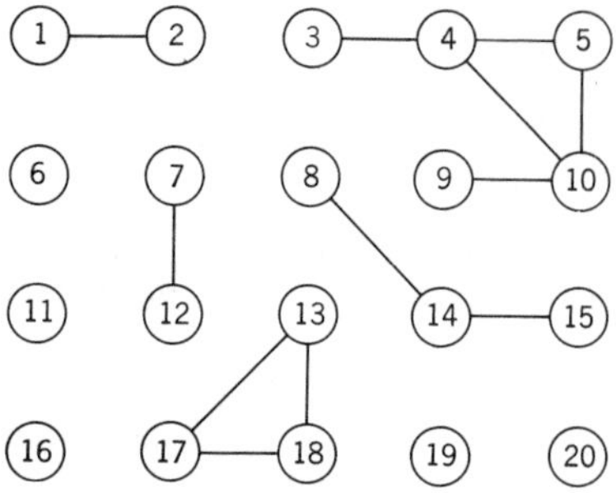

FIG. 5. Diagram representing a typical term in the expansion of exp $(-U/\mathbf{k}T)$, as in Eq. (22.9), for $N = 20$.

illuminating, however, to carry through the analysis in a formal way, and we now proceed to do this.

According to the assumption already mentioned, the potential energy U is merely the sum of potentials due to molecular pairs. That is,

$$U(\mathbf{r}_1, \ldots, \mathbf{r}_N) = \sum_{1 \leqslant i < j \leqslant N} u(r_{ij}) \tag{22.6}$$

where $u(r_{ij})$ is the potential of intermolecular force between molecules i and j as a function of the distance r_{ij} between these molecules. We now introduce the function f_{ij} defined by

$$f_{ij} = e^{-u(r_{ij})/\mathbf{k}T} - 1 \tag{22.7}$$

Then

$$e^{-U/\mathbf{k}T} = e^{-\Sigma u(r_{ij})/\mathbf{k}T} = \prod e^{-u(r_{ij})/\mathbf{k}T}$$
$$= \prod_{1 \leq i < j \leq N} (1 + f_{ij}) \tag{22.8}$$

If this product is multiplied out, we have

$$e^{-U/\mathbf{k}T} = 1 + \Sigma f_{ij} + \Sigma f_{ij} f_{kl} + \cdots \tag{22.9}$$

By introducing the f functions, the effects of intermolecular forces are more clearly exposed. For in the absence of intermolecular forces (perfect gas), all the f's are zero and only the leading term on the right-hand side of Eq. (22.9) remains. In this case $Z = V^N$. In each configuration $\mathbf{r}_1, \ldots, \mathbf{r}_N$ of the N molecules, the f's represent in Eq. (22.9) the contribution to $e^{-U/\mathbf{k}T}$ of nonzero intermolecular forces. In the usual case $u(r)$ and $f(r)$ have

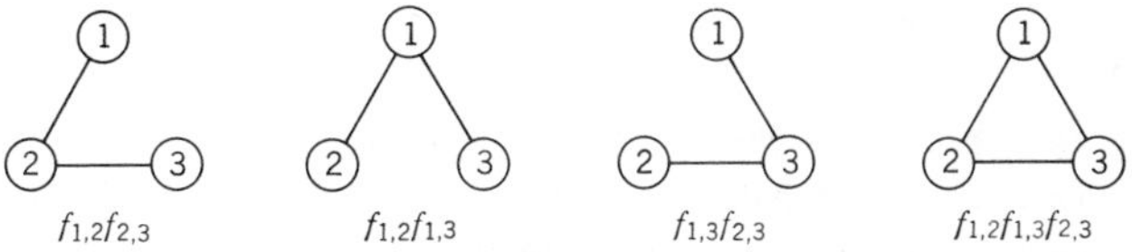

FIG. 6. Clusters of three molecules.

the form shown in Fig. 4. It is seen that $f(r_{ij})$ is zero except when r_{ij} is small—of the order of several molecular diameters or less.

In Eq. (22.9) there is a term corresponding to each possible combination of from zero to all the $N(N-1)/2$ f's. Any such term may be represented by a diagram consisting of circles for the molecules and a line segment between the ith and jth molecule for each factor f_{ij} which occurs in the term. Thus, if $N = 20$, the diagram for the term

$$f_{1,2} f_{3,4} f_{4,5} f_{4,10} f_{5,10} f_{7,12} f_{8,14} f_{9,10} f_{13,17} f_{13,18} f_{14,15} f_{17,18}$$
$$= (f_{1,2})(f_{7,12})(f_{8,14} f_{14,15})(f_{13,17} f_{13,18} f_{17,18})(f_{3,4} f_{4,5} f_{4,10} f_{5,10} f_{9,10}) \tag{22.10}$$

is shown in Fig. 5. For each of the terms in Eq. (22.9) there is one such diagram and for each diagram there is one term. For any particular diagram, molecules which are connected together directly or indirectly by lines are said to form a "cluster." In the example above, there are five clusters of one molecule each, two clusters of two molecules each, two clusters of three molecules each, and one cluster of five molecules.

There are in general many ways in which any given set of molecules may be connected together into a cluster. For example, a set of three molecules may be connected into a cluster in four ways, as indicated in Fig. 6.

We introduce next the cluster sum $S_{i,j,k,\ldots}$ as the sum of all terms which connect in a cluster each of the molecules $i, j, k, \ldots$, no other molecules being connected to this cluster. For example, from Fig. 6,

$$S_{1,2,3} = f_{1,2}f_{2,3} + f_{1,2}f_{1,3} + f_{1,3}f_{2,3} + f_{1,2}f_{1,3}f_{2,3} \tag{22.11}$$

Also,

$$S_{1,2} = f_{1,2} \tag{22.12}$$

and we define for the unit cluster

$$S_i = 1 \tag{22.13}$$

Consider, now, a product of S's such that in the product each molecule occurs once and only once as a subscript. A typical example (compare Fig. 5) for $N = 20$ is

$$S_6S_{11}S_{16}S_{19}S_{20}S_{1,2}S_{7,12}S_{8,14,15}S_{13,17,18}S_{3,4,5,9,10} \tag{22.14}$$

If the defining sums for the S's are substituted into (22.14) and the product expanded, the resulting sum will contain all terms of Eq. (22.9) consistent with the restriction that:

Molecules 6, 11, 16, 19, 20 are each in unit clusters
Molecules 1, 2 are connected in a cluster
Molecules 7, 12 are connected in a cluster
Molecules 8, 14, 15 are connected in a cluster
Molecules 13, 17, 18 are connected in a cluster
Molecules 3, 4, 5, 9, 10 are connected in a cluster

Among these terms would be the one mapped in Fig. 5.

Thus, to include every term in Eq. (22.9), one must take the sum of all possible different products of the S's, each product containing every molecule once and only once as a subscript. Each such product corresponds to a different possible division of the N molecules into groups or clusters.

We now define the cluster integrals b_j by

$$b_j(V,T) = \frac{1}{j!V}\int\cdots\int_V S_{1,2,\ldots,j}\,d\mathbf{r}_1\cdots d\mathbf{r}_j \tag{22.15}$$

In particular,

$$b_1 = \frac{1}{V}\int_V S_1\,d\mathbf{r}_1 = 1 \tag{22.16}$$

Clearly, in the definition of b_j, it is immaterial which j molecules occur as subscripts on the S of the integrand; the same function of V and T is obtained in any case.

Let us consider the multiple integral of (22.14),

$$I = \int\cdots\int_V S_6S_{11}\cdots S_{3,4,5,9,10}\,d\mathbf{r}_1\cdots d\mathbf{r}_{20} \tag{22.17}$$

Since by its definition $S_{i,j,k,\ldots}$ depends only on the coordinates of the molecules appearing as subscripts, the multiple integral over all coordinates of a product of these S's is merely the product of the integrals of the several S's. Hence, from Eqs. (22.15) and (22.17),

$$I = (1!Vb_1)^5(2!Vb_2)^2(3!Vb_3)^2(5!Vb_5) \tag{22.18}$$

In general, a product of the S's which represents m_1 unit clusters, m_2 clusters of two molecules, . . . , m_j clusters of j molecules, . . . , yields an integral over $d\mathbf{r}_1 \cdots d\mathbf{r}_N$ of

$$I = (1!Vb_1)^{m_1}(2!Vb_2)^{m_2} \cdots (j!Vb_j)^{m_j} \cdots \tag{22.19}$$

The question now arises: how many S products out of all possible S products yield the result in Eq. (22.19)? That is, in how many different ways can N molecules be divided into groups or clusters so that there are m_1 unit clusters, m_2 clusters of two molecules, . . . , m_j clusters of j molecules, . . . ? The N molecules are to be regarded as distinguishable (that is, labeled) here since the actual indistinguishability of the molecules has already been taken care of by the factor $(N!)^{-1}$ in Eq. (22.4). The number of ways of dividing N distinguishable objects among labeled boxes so that there is one object in each of the first m_1 boxes, two objects in each of the next m_2 boxes, etc., is

$$\frac{N!}{(1!)^{m_1}(2!)^{m_2} \cdots (j!)^{m_j} \cdots} \tag{22.20}$$

In the cluster problem we do not want to count separately arrangements which differ only in the exchange of all the molecules (objects) in one cluster (box) with all the molecules in another cluster of the same size. The number of such exchanges among clusters with j molecules is $m_j!$; therefore we have to divide (22.20) by $m_1!m_2! \ldots m_j! \ldots$. The final result in the cluster problem is therefore

$$\frac{N!}{\prod_{j=1}^{N} (j!)^{m_j} m_j!} \tag{22.21}$$

The product of (22.19) and (22.21) is then the contribution to the integral of the right-hand side of Eq. (22.9) [that is, to Z in Eq. (22.5)] owing to all terms in Eq. (22.9) associated with those products of S's which represent the cluster partition $m_1, m_2, \ldots, m_j, \ldots$. This contribution to Z is

$$N! \prod_{j=1}^{N} \frac{(Vb_j)^{m_j}}{m_j!} \tag{22.22}$$

There will be a contribution of this form to Z for each different set of positive integers (or zeros) $m_1, m_2, \ldots, m_j, \ldots$ consistent with the condition $\sum_{j=1}^{N} jm_j = N$. Hence we can write

$$Z = N! \sum_{\substack{\mathbf{m} \\ (\sum_{j=1}^{N} jm_j = N)}} \left[\prod_{j=1}^{N} \frac{(Vb_j)^{m_j}}{m_j!} \right] \tag{22.23}$$

and

$$Q = \frac{1}{\Lambda^{3N}} \sum_{\substack{\mathbf{m} \\ (\sum_{j=1}^{N} jm_j = N)}} \left[\prod_{j=1}^{N} \frac{(Vb_j)^{m_j}}{m_j!} \right] \tag{22.24}$$

where the sums are over all sets $\mathbf{m} = m_1, m_2, \ldots$ consistent with the restriction stated. The number of terms in the sum in Eqs. (22.23) and (22.24) is the number of ways in which N may be written as a sum of positive integers, order being immaterial. This number is known as the "partitio numerorum" of N.

Although we have actually merely rearranged terms so far, the essential step being the introduction of cluster sums as in Eq. (22.11), Eq. (22.24) is in a form which is much more convenient to work with than a Q written directly in terms of the disorganized sum in Eq. (22.9). More important, we shall see in the next section that the b_j's turn out to be just the coefficients in the expansion of $p/\mathbf{k}T$ in powers of the activity.

Cluster Integrals. We conclude this section with a few further remarks concerning clusters and cluster integrals. Let us imagine that in Eq. (22.15), for j small compared to V/r_0^3, where r_0 is of the order of a molecular diameter, we keep the jth molecule fixed at some point $\mathbf{r}_j$ and carry out the integrations over $\mathbf{r}_1, \mathbf{r}_2, \ldots, \mathbf{r}_{j-1}$. Since each f_{ij} is nonzero only when r_{ij} is less than a few molecular diameters, each product of f's in $S_{1,2,\ldots,j}$ [for example, Eq. (22.11)] is nonzero only when *all* r_{ij}'s corresponding to the f's in the product are small. Since the j molecules are "connected" to each other directly or indirectly (see Fig. 6) in every product of f's in $S_{1,2,\ldots,j}$, $S_{1,2,\ldots,j}$ is nonzero only in those regions of configuration space where molecules $1, 2, \ldots, j$ are all rather close together. This is the origin of the term "cluster." With the jth molecule fixed at $\mathbf{r}_j$, integration over the coordinates of the other $j - 1$ molecules will give nonzero contributions only in regions of configuration space where all the r_{ij} ($i = 1,2, \ldots, j - 1$) are not too large, that is, when all the $j - 1$ molecules are near the point $\mathbf{r}_j$. Unless $\mathbf{r}_j$ happens to be within a few molecular diameters of the wall, which is a negligible possibility for macroscopic V and j small compared to V/r_0^3, the result of the integration over $\mathbf{r}_1, \ldots, \mathbf{r}_{j-1}$ will be independent of the location of the point $\mathbf{r}_j$ and of the volume V of the container. That is,

$$\int \cdots \int_V S_{1,2,\ldots,j}\, d\mathbf{r}_1 \cdots d\mathbf{r}_{j-1} = F(T)$$

and
$$b_j = \frac{1}{j!V}\int_V F(T)\,d\mathbf{r}_j = \frac{F(T)}{j!} \tag{22.25}$$

Hence b_j is independent of volume for macroscopic V, and j small compared to V/r_0^3. Obviously, for a given V and sufficiently large values of j, the above argument breaks down and b_j will depend on V as well as T.

As a special case, we note that the explicit expression for b_2 is easily obtained from Eqs. (22.12) and (22.25):

$$b_2(T) = \frac{1}{2!V}\iint_V [e^{-u(r)/\mathbf{k}T} - 1]\,d\mathbf{r}_1\,d\mathbf{r}_2$$

$$= \frac{1}{2}\int_0^\infty [e^{-u(r)/\mathbf{k}T} - 1]4\pi r^2\,dr \tag{22.26}$$

where $r = r_{12}$.

We see in Fig. 4 that $f(r)$ is negative for r small. For a "hard-sphere" model, $f(r)$ is never positive, being zero or -1 for all r. Thus negative values of $S_{1,2,\ldots,j}$ and of b_j are easily possible. At high temperatures, especially, when the region of positive $f(r)$ in Fig. 4 is largely "washed out," negative b_j's will be common.

Although it is natural and convenient to discuss the clusters we have introduced above in language appropriate to actual physical clusters or aggregates of molecules, strictly speaking the grouping of terms leading to "clusters" was a purely mathematical step and we shall see in the next section that the physical analogy should not be pushed too far.

23. Pressure of the Gas Expressed as a Power Series in the Activity

We use the grand partition function here to obtain the expansion of $p/\mathbf{k}T$ in powers of the activity. This method[1] is very simple and is also rigorous so long as only the gas phase is present. We have

$$\Xi(T,V,\lambda) = \sum_{N\geqslant 0} Q(N,V,T)\lambda^N \tag{23.1}$$

$$= \sum_{N\geqslant 0} \frac{Z(N,V,T)z^N}{N!} \tag{23.2}$$

where
$$\lambda = e^{\mu/\mathbf{k}T} \tag{23.3}$$

$$z = \frac{\lambda}{\Lambda^3} \tag{23.4}$$

[1] Due to Fowler and Guggenheim, and DeBoer.

λ is the "absolute activity" and z the "activity." Also, in Eq. (23.1), $Q(N = 0)$ is defined as unity.

Let us digress briefly to discuss z. For an infinitely dilute (perfect) gas (Appendix 2), we have

$$\frac{\mu}{\mathbf{k}T} = \ln \Lambda^3 + \ln \frac{\bar{N}}{V} \qquad \frac{\bar{N}}{V} \to 0 \tag{23.5}$$

If we now *define* an "active" number density or activity z which bears the same relation to μ at any density that $\bar{N}/V$ does as $\bar{N}/V \to 0$, then

$$\frac{\mu}{\mathbf{k}T} = \ln \Lambda^3 + \ln z \tag{23.6}$$

$$\lim_{\bar{N}/V \to 0} z = \frac{\bar{N}}{V} \tag{23.7}$$

This is seen to be the same z as in Eq. (23.4). It should be noted that the fugacity f of thermodynamics is defined by

$$\frac{\mu}{\mathbf{k}T} = \ln \frac{\Lambda^3}{\mathbf{k}T} + \ln p \qquad p \to 0 \tag{23.8}$$

$$\frac{\mu}{\mathbf{k}T} = \ln \frac{\Lambda^3}{\mathbf{k}T} + \ln f \qquad \text{(general)} \tag{23.9}$$

$$\lim_{p \to 0} f = p \tag{23.10}$$

Therefore

$$\frac{f}{\mathbf{k}T} = z \tag{23.11}$$

Now consider the infinite series

$$X = \sum_{j \geqslant 1} b_j(V,T) z^j \tag{23.12}$$

Then

$$e^{VX} = \prod_{j \geqslant 1} e^{V b_j z^j}$$

$$= \prod_{j \geqslant 1} \left[\sum_{m_j \geqslant 0} \frac{1}{m_j!} (V b_j)^{m_j} z^{j m_j} \right] \tag{23.13}$$

The coefficient of z^N in Eq. (23.13) (written as a single power series in z after multiplication) is seen to be

$$\sum_{\substack{\mathbf{m} \\ (\sum_{j=1}^{N} j m_j = N)}} \left[\prod_{j=1}^{N} \frac{(V b_j)^{m_j}}{m_j!} \right] \tag{23.14}$$

since necessarily all $m_j = 0$ for $j > N$ in order to satisfy $\sum_j jm_j = N$. But (23.14) is also just $Z(N)/N!$, according to Eq. (22.23); therefore

$$e^{VX} = \sum_{N \geqslant 0} \frac{Z(N)}{N!} z^N = \Xi \tag{23.15}$$

and

$$\frac{1}{V} \ln \Xi = \sum_{j \geqslant 1} b_j(V,T) z^j \tag{23.16}$$

The thermodynamic pressure is then given by [compare Eq. (22.3) and see Sec. 28]

$$\frac{p}{\mathbf{k}T} = \lim_{V \to \infty} \sum_{j \geqslant 1} b_j(V,T) z^j \tag{23.17}$$

Actually, as we have seen in the preceding section, for macroscopic but finite V,

$$b_j(V,T) = b_j(T) \tag{23.18}$$

for the relatively small clusters of importance in a gas. Hence,

$$\frac{p}{\mathbf{k}T} = \sum_{j \geqslant 1} b_j(T) z^j \tag{23.19}$$

In an imperfect gas, only a finite number of terms contribute appreciably to $p/\mathbf{k}T$. That is, $X(z)$ converges. In fact, the convergence of this series is required to justify Eqs. (23.13) and (23.14) so that the treatment given is valid only when $X(z)$ converges.

The above approach (or its equivalent) to the theory of an imperfect gas, excluding condensation, is generally accepted as rigorous. Difficulties in connection with the double limiting process of Eq. (23.17), convergence of Eq. (23.12), volume dependence of the cluster integrals, etc., arise only when one attempts to include condensation, the critical point, and the liquid phase (see Sec. 28).

Equation (23.19) gives $p/\mathbf{k}T$ as a power series in the activity, with coefficients $b_j(T)$. However, in applications we are usually interested in $p/\mathbf{k}T$ as a function of $\bar{N}/V$ or $V/\bar{N}$ instead of z. To introduce $\bar{N}/V = \rho$, we use[1]

$$\bar{N} = \mathbf{k}T \left(\frac{\partial \ln \Xi}{\partial \mu}\right)_{T,V} = z \left(\frac{\partial \ln \Xi}{\partial z}\right)_{T,V} \tag{23.20}$$

From

$$\ln \Xi = V \sum_{j \geqslant 1} b_j(T) z^j \tag{23.21}$$

we then obtain

$$\rho = \sum_{j \geqslant 1} j b_j(T) z^j \tag{23.22}$$

Equation (23.22) gives ρ as a function of T and z, or by inversion, z as a

[1] For simplicity of notation we denote $\bar{N}/V$ here by ρ instead of by $\bar{\rho}$.

function of ρ and T to be inserted for z in Eq. (23.19). The details of this procedure, leading to the usual virial expansion of $pV/\bar{N}\mathbf{k}T$ in powers of ρ, will be given in Sec. 25. But we remark here that for ρ and z sufficiently small, Eqs. (23.22) and (23.19) give

$$\rho = z \tag{23.23}$$

and

$$\frac{p}{\mathbf{k}T} = z = \rho \tag{23.24}$$

which is the perfect-gas law, as expected.

The Helmholtz free energy of the gas is given by

$$\begin{aligned}\frac{A}{\mathbf{k}T} &= \frac{\bar{N}\mu}{\mathbf{k}T} - \frac{pV}{\mathbf{k}T} \\ &= \bar{N}[\ln \Lambda^3 + \ln z - v \sum_{j\geq 1} b_j(T) z^j]\end{aligned} \tag{23.25}$$

where $v = V/\bar{N}$ and z is determined as a function of v and T by Eq. (23.22).

An alternative rather interesting simple expression for $pV/\mathbf{k}T$ will now be derived. Let us *define* a number m_k' by the following equations:

$$m_k''(N) = \frac{\displaystyle\sum_{\substack{\mathbf{m} \\ (\Sigma j m_j = N)}} m_k \prod_j \frac{(Vb_j)^{m_j}}{m_j!}}{\displaystyle\sum_{\substack{\mathbf{m} \\ (\Sigma j m_j = N)}} \prod_j \frac{(Vb_j)^{m_j}}{m_j!}} \tag{23.26}$$

$$m_k' = \frac{\displaystyle\sum_N \left\{ m_k''(N) \left[z^N \sum_{\substack{\mathbf{m} \\ (\Sigma j m_j = N)}} \prod_j \frac{(Vb_j)^{m_j}}{m_j!} \right] \right\}}{\displaystyle\sum_N \left[z^N \sum_{\substack{\mathbf{m} \\ (\Sigma j m_j = N)}} \prod_j \frac{(Vb_j)^{m_j}}{m_j!} \right]} \tag{23.27}$$

$$= \frac{\displaystyle\sum_N \left[z^N \sum_{\substack{\mathbf{m} \\ (\Sigma j m_j = N)}} m_k \prod_j \frac{(Vb_j)^{m_j}}{m_j!} \right]}{\displaystyle\sum_N \left[z^N \sum_{\substack{\mathbf{m} \\ (\Sigma j m_j = N)}} \prod_j \frac{(Vb_j)^{m_j}}{m_j!} \right]} \tag{23.28}$$

The denominator in Eq. (23.26) is $Q\Lambda^{3N}$ and the denominator in Eqs. (23.27) and (23.28) is Ξ. We see from Eq. (23.26) that $m_k''(N)$ has the formal appearance of an "average" over the different values of m_k occurring in the various sets $\mathbf{m}$ consistent with $\Sigma j m_j = N$, using the "weight" $\Pi[(Vb_j)^{m_j}/m_j!]$ for each set $\mathbf{m}$ in computing the average. In Eq. (23.27), $m_k''(N)$ is then averaged over the grand canonical ensemble in the usual way to give m_k'. While

Eq. (23.27) is a true average, the "averaging" within the canonical ensemble in Eq. (23.26) is *not*, since the quantity

$$P(\mathbf{m}) = \frac{\prod_j \frac{(Vb_j)^{m_j}}{m_j!}}{\sum_{\substack{\mathbf{m} \\ (\Sigma jm_j = N)}} \prod_j \frac{(Vb_j)^{m_j}}{m_j!}} \tag{23.29}$$

for a given set $\mathbf{m}$ does *not* have the significance of a probability. In fact, $P(\mathbf{m})$ will often be negative[1] since the b_j can be negative (see Sec. 22). That $P(\mathbf{m})$ is not a probability can be seen from

$$Z = \int_V \cdots \int (1 + \Sigma f_{ij} + \Sigma f_{ij} f_{kl} + \cdots)\, d\mathbf{r}_1 \cdots d\mathbf{r}_N \tag{23.30}$$

Z is given here essentially by a double sum: first, the series is summed to give $e^{-U/\mathbf{k}T}$ for a given configuration, and then a sum (integral) over all possible configurations is carried out. With the sums performed in this order we know from earlier considerations that the result of the first sum, $e^{-U/\mathbf{k}T}$, has the significance of a probability (actually, a probability density). That is,

$$\frac{e^{-U/\mathbf{k}T}\, d\mathbf{r}_1 \cdots d\mathbf{r}_N}{Z}$$

is the probability of the system having a configuration in the element of volume in configuration space $d\mathbf{r}_1 \cdots d\mathbf{r}_N$. On the other hand, if we collect terms in the series into convenient groups and sum (integrate) *first* over configuration space and *second* over the series, which is just the procedure followed in deriving Eq. (22.23), there is no reason why the result of the first sum in this case should be a probability, and in fact it cannot be since the $P(\mathbf{m})$ are not necessarily all positive.

We now proceed with the argument, defining m'_k by Eq. (23.28), but we emphasize that this does not attribute to m'_k the significance of an average of the m_k. If we write Ξ in the form

$$\Xi = \sum_N z^N \sum_{\substack{\mathbf{m} \\ (\Sigma jm_j = N)}} \prod_j \frac{(Vb_j)^{m_j}}{m_j!} \tag{23.31}$$

and differentiate with respect to Vb_k, we find, on comparing the result with Eq. (23.28), that

$$\frac{Vb_k}{\Xi} \frac{\partial \Xi}{\partial Vb_k} = m'_k \tag{23.32}$$

[1] The denominator in Eq. (23.29) will always be positive [see Eqs. (22.4) and (22.5)], but not the numerator.

But from Eq. (23.21) we also have

$$Vb_k \frac{\partial \ln \Xi}{\partial Vb_k} = Vb_k z^k \tag{23.33}$$

so that

$$m'_k = Vb_k(T)z^k \tag{23.34}$$

Equations (23.19) and (23.22) become, then,

$$\frac{pV}{\mathbf{k}T} = \sum_{j\geqslant 1} m'_j \tag{23.35}$$

and

$$\bar{N} = \sum_{j\geqslant 1} jm'_j \tag{23.36}$$

These remarkably simple results account for our interest in the m'_k. Equation (23.34) shows that m'_k is negative whenever b_k is negative, while all the m_j, including m_k, are by definition always zero or positive integers. Experimentally $pV/\mathbf{k}T$ is often greater than $\bar{N}$ and this of course is possible only if some of the m'_k are negative, in view of Eq. (23.36).[1] In a *strictly formal* way, Eq. (23.35) states that the pressure in an imperfect gas is given by a perfect-gas equation with $\bar{N}$ replaced by the "average number of clusters" of all sizes. That these clusters do not have a direct[2] significance as physical clusters or aggregates is clear not only from the possibility of negative values of some of the m'_k but also from the fact that Eq. (23.35) is a perfect-gas equation, instead of being in a form which would take into account the excluded volume effect necessarily associated with physical clusters (see Secs. 26 and 27).

General Imperfect Gas. In deriving Eq. (23.19), which expresses $p/\mathbf{k}T$ as a power series in z, we made use of the definition of the b_j given in Sec. 22. This definition referred specifically to a monatomic, classical gas with a total potential energy which is a sum of pair potentials. We give here an alternative and simpler derivation[3] of Eq. (23.19) which is also more general. The final equation reduces to the special case referred to above, but it is valid as well for a polyatomic, quantum-mechanical gas with a total potential energy not a sum of pair potentials.

In the general case just described, we can write

$$\Xi = \sum_{N\geqslant 0} Q_N \lambda^N = \sum_{N\geqslant 0} \frac{Z_N z^N}{N!} \tag{23.37}$$

$$= 1 + Q_1\lambda + \cdots = 1 + Z_1 z + \cdots \tag{23.38}$$

[1] As a simple illustration, take $\bar{N} = 3$, $m'_1 = 6$, $m'_2 = 0$, and $m'_3 = -1$. Then $\Sigma jm'_j = 3 = \bar{N}$ and $\Sigma m'_j = 5 > \bar{N}$.

[2] An indirect physical significance has been emphasized by Mayer: the m'_k can be considered as excesses (defined in a certain way) over random expectations.

[3] S. Ono, *J. Chem. Phys.*, **19**, 504 (1951); J. E. Kilpatrick, *J. Chem. Phys.*, **21**, 274 (1953).

where
$$Q_N(V,T) \equiv Q(N,V,T)$$
$$\lambda = e^{\mu/\mathbf{k}T} \qquad z = \frac{Q_1}{Z_1}\lambda$$

Aside from the proportionality to λ, we want z defined in such a way that $z \to \rho$ when $\rho \to 0$. From [see Eq. (23.20)]

$$\frac{pV}{\mathbf{k}T} = \ln \Xi = Z_1 z + \cdots$$
$$= \bar{N} + \cdots$$

it is clear that we have, then, to define Z_1 by $Z_1 \equiv V$. Thus,

$$z = \frac{Q_1}{V}\lambda \tag{23.39}$$
$$z \to \rho \qquad \text{as} \qquad \rho \to 0$$

The definition of Z_N is therefore, from Eq. (23.37),

$$Z_N = \left(\frac{V}{Q_1}\right)^N N!Q_N \tag{23.40}$$

The point in adopting the formal definitions of Eqs. (23.39) and (23.40) is that we do not have to commit ourselves to a classical system, monatomic gas, etc. It is simply understood that the appropriate Q_N are used in any given problem.

Now

$$\frac{pV}{\mathbf{k}T} = \ln \Xi = \ln \sum_{N \geqslant 0} \frac{Z_N z^N}{N!} \tag{23.41}$$

But the right-hand side of Eq. (23.41) can be expanded to give a power series in z. Denote the coefficient of z^j in this series by Vb_j. Then

$$\frac{p}{\mathbf{k}T} = \sum_{j \geqslant 1} b_j z^j \tag{23.42}$$

where we can easily verify that

$$\begin{aligned} 1!Vb_1 &= Z_1 = V \\ 2!Vb_2 &= Z_2 - Z_1^2 \\ 3!Vb_3 &= Z_3 - 3Z_1Z_2 + 2Z_1^3 \\ 4!Vb_4 &= Z_4 - 4Z_1Z_3 - 3Z_2^2 + 12Z_1^2Z_2 - 6Z_1^4 \end{aligned} \tag{23.43}$$

The expressions for $j!Vb_j$ are well known in the theory of statistics as the semi-invariants of Thiele. The general relation is

$$j!Vb_j = j!\sum_{\mathbf{n}}(-1)^{\sum_i n_i - 1}\left(\sum_i n_i - 1\right)! \prod_i \left[\frac{(Z_i/i!)^{n_i}}{n_i!}\right] \tag{23.44}$$

where the first sum is over all sets of positive integers or zero, $\mathbf{n} = n_1, n_2, \ldots,$ such that $\sum_i in_i = j$.

To find the inverse of Eq. (23.44), we follow the argument of Eqs. (23.12) to (23.14) and obtain from Eq. (23.42),

$$e^{pV/\mathbf{k}T} = \sum_{N \geqslant 0} C_N z^N$$

where C_N is given by (23.14) except that the b_j's are defined by Eq. (23.44).

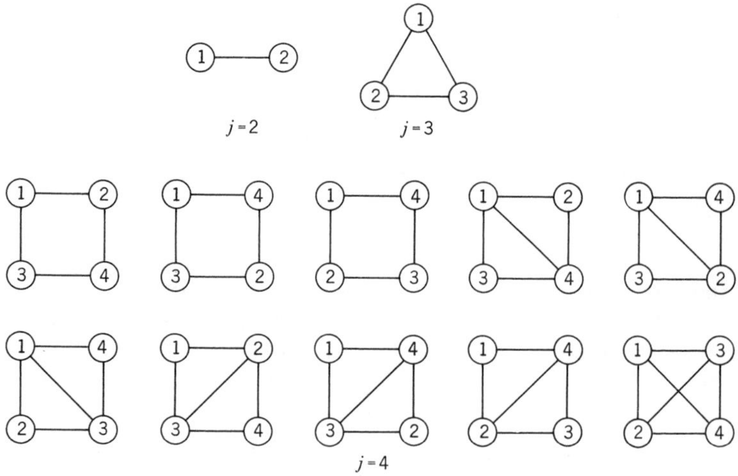

FIG. 7. Irreducible clusters for $j = 2, 3, 4$.

C_N is also given by $Z_N/N!$ in Eq. (23.41). Therefore the required inverse of Eq. (23.44) is

$$Z_N = N! \sum_{\substack{\mathbf{m} \\ (\Sigma j m_j = N)}} \left[\prod_{j=1}^{N} \frac{(Vb_j)^{m_j}}{m_j!} \right] \tag{23.45}$$

This is formally identical with Eq. (22.23), as might have been expected.

The virial coefficients of an imperfect gas follow directly from Eq. (23.42) (see Sec. 25); thus we have here the basis for quite general expressions for the virial coefficients. See also Appendix 10.

24. Irreducible Cluster Integrals

We return now to the special case of Sec. 22. The cluster integral b_j involves an integral (over the configuration space of j molecules) of the sum of all products of f_{ij}'s corresponding to different "singly connected" cluster diagrams with j molecules. If we examine in more detail the integral of any one of these products (corresponding to a single cluster diagram), we find that the integral can be decomposed into what we shall call irreducible cluster integrals, β_k. Further progress then depends on the numerical calculation of the β_k's, but this is, unfortunately, practical only for the simplest β_k's, at least at the present time. But the β_k's have a further special significance which warrants their introduction: they turn out to be (Sec. 25) essentially the coefficients in the virial expansion of $pV/\bar{N}\mathbf{k}T$ in powers of $\bar{N}/V$.

An irreducible cluster is defined as any product of f_{ij}'s associated with a "doubly connected" diagram, except that a cluster of two molecules is also considered "irreducible." In a doubly connected diagram there are at least two entirely independent paths which do not cross at any circle, between each pair of molecules in the diagram. Figure 7 shows the different possible irreducible clusters for $j = 2$, 3, and 4. Thus, the cluster in Fig. 8 is made up of two irreducible clusters with $j = 2$, one irreducible cluster with $j = 3$, and one irreducible cluster with $j = 4$.

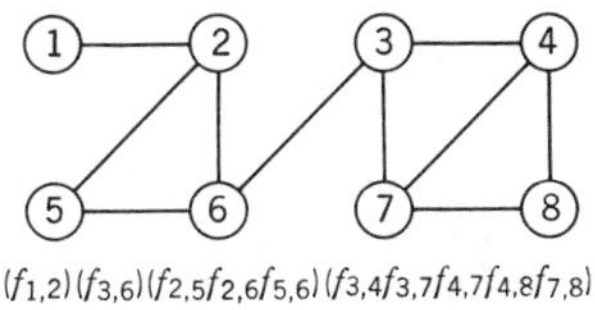

FIG. 8. Cluster of eight molecules containing four irreducible clusters.

We define the irreducible cluster sum $S'_{i,j,k,\ldots}$ as the sum of all different terms (products of f's) that connect molecules $i, j, k, \ldots$ into an irreducible cluster. Then, for example,

$$S'_{1,2} = f_{1,2} \tag{24.1}$$

$$S'_{1,2,3} = f_{1,2}f_{1,3}f_{2,3} \tag{24.2}$$

while $S'_{1,2,3,4}$ is a sum of 10 terms corresponding to the irreducible clusters shown in Fig. 7 for $j = 4$.

The irreducible cluster integral β_k is defined by

$$\beta_k = \frac{1}{k!V} \int \cdot \cdot \cdot \int_V S'_{1,2,\ldots,k+1}\, d\mathbf{r}_1 \cdot \cdot \cdot d\mathbf{r}_{k+1} \tag{24.3}$$

It is obviously immaterial which $k + 1$ molecules appear as subscripts of the S' in the integrand. k is called the index of the irreducible cluster and it should be observed that an irreducible cluster with index k contains $k + 1$ molecules. For k not too large, β_k is independent of volume by the same argument as before.

There is a very close analogy in the argument to be used here (as compared to that in Sec. 22) between Z and b_j, the sum in Eq. (22.9) and $S_{1,2,\ldots,j}$, $S_{1,2,\ldots,j}$ and $S'_{1,2,\ldots,j}$, and b_j and β_k. It will be recalled that the procedure in Sec. 22 consisted essentially of (1) collecting the terms in Eq. (22.9) belonging to all products of S's with m_1 clusters of size one, . . . , m_j clusters of size j, . . . , (2) expressing the integral of the product of S's in terms of the b_j's, and (3) finally finding the number of such products of S's. A completely analogous argument is used here. We (1) collect the terms in $S_{1,2,\ldots,j}$ belonging to all products of S''s with n_1 irreducible clusters of index one, . . . , n_k irreducible clusters of index k, . . . , (2) express the integral of the product of S''s in terms of the β_k's, and (3) finally find the number of such products of the S''s. Our final result will be an expression for b_j in terms of the β_k's.

Consider now, as an example, the product

$$S'_{1,2}S'_{3,6}S'_{2,5,6}S'_{3,4,7,8} \tag{24.4}$$

This product contains all terms in the integrand for b_8 (i.e., in $S_{1,2,\ldots,8}$) satisfying the restriction that:

Molecules 1 and 2 are joined in an irreducible cluster
Molecules 3 and 6 are joined in an irreducible cluster
Molecules 2, 5, and 6 are joined in an irreducible cluster
Molecules 3, 4, 7, and 8 are joined in an irreducible cluster

Among these terms is the one shown in Fig. 8.

The contribution of (24.4) to the cluster integral b_8 is

$$I' = \frac{1}{8!V}\int\cdots\int_V S'_{1,2}S'_{3,6}S'_{2,5,6}S'_{3,4,7,8}\,d\mathbf{r}_1\cdots d\mathbf{r}_8 \tag{24.5}$$

This integral can be decomposed into irreducible integrals β_k by "breaking off" the irreducible clusters one at a time.[1] Let us change to relative coordinates with respect to molecule 2 in order first to "break off" the irreducible cluster of 1 and 2. That is,

$$I' = \frac{1}{8!V}\int\cdots\int_V S'_{1,2}S'_{3,6}S'_{2,5,6}S'_{3,4,7,8}\,d\mathbf{r}_2\,d\mathbf{r}_{12}\,d\mathbf{r}_{23}\cdots d\mathbf{r}_{28} \tag{24.6}$$

For j small compared to $V/r_0{}^3$, which we assume in the remainder of this section, the integral over $\mathbf{r}_2$ gives V. Also, $\mathbf{r}_{12}$ occurs only in $S'_{1,2}$; thus

$$\begin{aligned} 8!I' &= \left(\int_V S'_{1,2}\,d\mathbf{r}_{12}\right)\left(\int\cdots\int_V S'_{3,6}S'_{2,5,6}S'_{3,4,7,8}\,d\mathbf{r}_{23}\cdots d\mathbf{r}_{28}\right) \\ &= \left(\frac{1}{V}\int_V S'_{1,2}\,d\mathbf{r}_1\,d\mathbf{r}_2\right)\left(\frac{1}{V}\int\cdots\int_V S'_{3,6}S'_{2,5,6}S'_{3,4,7,8}\,d\mathbf{r}_2\cdots d\mathbf{r}_8\right) \\ &= \frac{\beta_1}{V}\int\cdots\int_V S'_{3,6}S'_{2,5,6}S'_{3,4,7,8}\,d\mathbf{r}_2\cdots d\mathbf{r}_8 \end{aligned} \tag{24.7}$$

[1] Reference can be made in the following to Fig. 8, but it should be remembered that the product in Fig. 8 is only one term in (24.4).

We again change to relative coordinates, this time with respect to molecule 6, in order to "break off" next the irreducible cluster 2, 5, 6. We find

$$8!I' = \beta_1 \left(\frac{1}{V}\int_V \cdots \int S'_{2,5,6}\, d\mathbf{r}_2\, d\mathbf{r}_5\, d\mathbf{r}_6\right)\left(\frac{1}{V}\int_V \cdots \int S'_{3,6}S'_{3,4,7,8}\, d\mathbf{r}_3\, d\mathbf{r}_4\, d\mathbf{r}_6\, d\mathbf{r}_7\, d\mathbf{r}_8\right)$$

$$= \beta_1(2!\beta_2)\frac{1}{V}\int_V \cdots \int S'_{3,6}S'_{3,4,7,8}\, d\mathbf{r}_3\, d\mathbf{r}_4\, d\mathbf{r}_6\, d\mathbf{r}_7\, d\mathbf{r}_8 \tag{24.8}$$

In the same way, we change to relative coordinates with respect to molecule 3 and obtain finally

$$I' = \frac{1}{8!}(1!\beta_1)^2(2!\beta_2)(3!\beta_3) \tag{24.9}$$

This type of argument can obviously be applied to any product of S''s. In general, the contribution to b_j of a product of the S''s corresponding to n_1 irreducible clusters of index 1, . . . , n_k irreducible clusters of index k, . . . , is

$$\frac{1}{j!}(1!\beta_1)^{n_1}(2!\beta_2)^{n_2}\cdots(k!\beta_k)^{n_k}\cdots \tag{24.10}$$

Finally, we must ask: How many different products of S''s are there which give the contribution to b_j in (24.10)? That is, in how many ways is it possible to distribute j distinguishable molecules into n_1 sets of two molecules, . . . , n_k sets of $k + 1$ doubly connected molecules, . . . , such that every molecule occurs in at least one set; if a molecule occurs in two or more sets it "connects" these sets together; and we require that the sets be singly but not doubly connected *to each other*. This is a difficult combinatorial problem, requiring a lengthy proof, which has been solved by Mayer and Harrison.[1] The result is

$$\frac{j!}{j^2}\prod_{k=1}^{j-1}\left(\frac{j}{k!}\right)^{n_k}\frac{1}{n_k!} \tag{24.11}$$

The product of (24.10) and (24.11),

$$\frac{1}{j^2}\prod_{k=1}^{j-1}\frac{(j\beta_k)^{n_k}}{n_k!} \tag{24.12}$$

is the contribution to b_j of *all* products of S''s corresponding to a given set of numbers $n_1, n_2, \ldots, n_k, \ldots$.

By arranging the S''s in a product of S''s in the order in which we factor them out [Eqs. (24.5) to (24.9)], it is seen that an S' with $k + 1$ indices has k indices which do not occur[2] in succeeding S''s. The single exception is the

[1] For proof, see Mayer and Mayer (Gen. Ref.), and also M. Born and K. Fuchs, *Proc. Roy. Soc.* (*London*), **A166**, 391 (1938).

[2] That is, the sets in the statement of the combinatorial problem above are singly connected to each other.

last S' which has $k + 1$ such indices. Since the number of S''s with $k + 1$ indices is n_k, the total number of different indices (molecules) is

$$j = \sum_{k=1}^{j-1} kn_k + 1 \tag{24.13}$$

Then

$$b_j = \frac{1}{j^2} \sum_{\substack{\mathbf{n} \\ (\sum_{k=1}^{j-1} kn_k = j-1)}} \prod_{k=1}^{j-1} \frac{(j\beta_k)^{n_k}}{n_k!} \tag{24.14}$$

where the summation is over all sets $\mathbf{n}$ consistent with $\Sigma kn_k = j - 1$. This result is very similar in form to Eq. (22.23). One point at which the analogy between Z and b_j breaks down, however, is that the derivation of Eq. (24.14) has made use of the assumption that j is small compared to V/r_0^3. No assumption of this type was used in obtaining Eq. (22.23). However, so long as we avoid condensation and the liquid phase, the above assumption concerning j is legitimate, as we have already seen.

From Eq. (24.14) we find the first few b_j's to be given by

$$\begin{aligned} b_2 &= \frac{\beta_1}{2} \\ b_3 &= \frac{\beta_1^2}{2} + \frac{\beta_2}{3} \\ b_4 &= \frac{2\beta_1^3}{3} + \beta_1\beta_2 + \frac{1}{4}\beta_3 \end{aligned} \tag{24.15}$$

25. The Virial Expansion for the Gas

In this section we obtain the expansion of $pV/\bar{N}\mathbf{k}T$ in powers of $1/v = \rho = \bar{N}/V$. We shall first find the initial terms of the series algebraically and then give a general derivation due to Kahn. The problem is to solve Eq. (23.22) for $z(\rho)$ and substitute this for z in Eq. (23.19).

Algebraic Method. We substitute

$$z = \rho + a_2\rho^2 + a_3\rho^3 + \cdots \tag{25.1}$$

into [Eq. (23.22)]

$$\sum_{j\geqslant 1} jb_jz^j - \rho = 0 \tag{25.2}$$

and equate the coefficient of each power of ρ to zero to find the a_i's. This gives

$$\begin{aligned} a_2 &= -2b_2 \\ a_3 &= -3b_3 - 4a_2b_2 = -3b_3 + 8b_2^2 \\ a_4 &= -4b_4 + 30b_2b_3 - 40b_2^3 \\ &\cdots\cdots\cdots\cdots \end{aligned} \tag{25.3}$$

We then substitute

$$z = \rho - 2b_2\rho^2 + (8b_2^2 - 3b_3)\rho^3 + \cdots \tag{25.4}$$

for z in Eq. (23.19) and obtain

$$\frac{pV}{\bar{N}\mathbf{k}T} = 1 - b_2\rho + (4b_2^2 - 2b_3)\rho^2 + \cdots \tag{25.5}$$

$$= 1 - {}^1\!/_2\beta_1\rho - {}^2\!/_3\beta_2\rho^2 - \cdots \tag{25.6}$$

where Eq. (25.6) follows from Eqs. (24.15) and (25.5). Equation (25.6) suggests that the complete expression is

$$\frac{pV}{\bar{N}\mathbf{k}T} = 1 - \sum_{k\geqslant 1} \frac{k}{k+1}\beta_k\rho^k \tag{25.7}$$

and this will be verified below.

If we write

$$\frac{pV}{\bar{N}\mathbf{k}T} = 1 + B_2\rho + B_3\rho^2 + \cdots \tag{25.8}$$

where B_n is the nth virial coefficient as usually defined, we see that

$$B_n = -\frac{n-1}{n}\beta_{n-1} \tag{25.9}$$

Thus there turns out to be a very simple relationship between the virial coefficients and the irreducible cluster integrals [Eq. (24.3)]. The numerical calculation of the first few virial coefficients is considered in detail by Hirschfelder, Curtiss, and Bird.[1]

Kahn's Derivation.[2] Let us define a function $\varphi(\xi)$ by

$$\varphi(\xi) = \sum_{k\geqslant 1} \beta_k\xi^k \tag{25.10}$$

Proceeding as in Eqs. (23.13) and (23.14), when $\varphi(\xi)$ converges,

$$e^{j\varphi(\xi)} = \prod_{k\geqslant 1} e^{j\beta_k\xi^k} = \prod_{k\geqslant 1}\left[\sum_{n_k\geqslant 0}\frac{1}{n_k!}(j\beta_k)^{n_k}\xi^{kn_k}\right] \tag{25.11}$$

The coefficient of ξ^{j-1} in the expansion of $e^{j\varphi(\xi)}$ is therefore

$$\sum_{\substack{\mathbf{n} \\ (\sum_{k=1}^{j-1} kn_k = j-1)}} \prod_{k=1}^{j-1} \frac{(j\beta_k)^{n_k}}{n_k!} \tag{25.12}$$

But, according to Eq. (24.14), this coefficient is just j^2b_j.

[1] Hirschfelder, Curtiss, and Bird (Gen. Ref.).

[2] B. Kahn (Gen. Ref.).

Next, we expand $e^{j\varphi(\xi)}$ in a Taylor series about $\xi = 0$. The coefficient of ξ^{j-1} in this expansion is

$$\frac{1}{(j-1)!}\left[\frac{d^{j-1}}{d\xi^{j-1}}\,e^{j\varphi(\xi)}\right]_{\xi=0} \tag{25.13}$$

On comparing (25.12) and (25.13), we have

$$jb_j = \frac{1}{j!}\left\{\frac{d^{j-1}}{d\xi^{j-1}}[e^{\varphi(\xi)}]^j\right\}_{\xi=0} \tag{25.14}$$

Then Eq. (25.14) may be substituted for jb_j in

$$\rho = \sum_{j\geqslant 1} jb_j z^j \tag{25.15}$$

giving

$$\rho = \sum_{j\geqslant 1} \frac{z^j}{j!}\left\{\frac{d^{j-1}}{d\xi^{j-1}}[e^{\varphi(\xi)}]^j\right\}_{\xi=0} \tag{25.16}$$

We are now in a position to use a well-known theorem due to Lagrange, on the reversion of series.[1] As applied to our problem, the theorem states that the solution of the equation

$$x = zf(x) \tag{25.17}$$

for $x(z)$ is given by[2]

$$x(z) = \sum_{j\geqslant 1} \frac{z^j}{j!}\left\{\frac{d^{j-1}}{d\xi^{j-1}}[f(\xi)]^j\right\}_{\xi=0} \tag{25.18}$$

and, conversely, the solution of Eq. (25.18) for $z(x)$ is

$$z = \frac{x}{f(x)} \tag{25.19}$$

From Eqs. (25.16) and (25.18) we see that if we put $x = \rho$ and $f(\xi) = e^{\varphi(\xi)}$ in Eq. (25.18), the desired inverse of Eq. (25.15) [or (25.16)] is

$$z = \rho e^{-\varphi(\rho)} \tag{25.20}$$

where

$$\varphi(\rho) = \sum_{k\geqslant 1} \beta_k \rho^k \tag{25.21}$$

To find the virial expansion, we then write [Eq. (23.19)]

$$\begin{aligned} \frac{p}{\mathbf{k}T} &= \int_0^z \Big(\sum_{j\geqslant 1} jb_j z^{j-1}\Big)\,dz \\ &= \int_0^z \frac{\rho(z)}{z}\,dz = \int_0^\rho e^{\varphi(\rho)}d[\rho e^{-\varphi(\rho)}] \end{aligned} \tag{25.22}$$

[1] See E. T. Whittaker and G. N. Watson, "A Course of Modern Analysis" (Cambridge, London, 1935), p. 133.

[2] The first few coefficients can easily be checked by a Taylor expansion of $x(z)$ in Eq. (25.17).

where the last expression follows from Eq. (25.20). Using

$$d[\rho e^{-\varphi(\rho)}] = \left(1 - \rho\frac{d\varphi}{d\rho}\right)e^{-\varphi(\rho)}d\rho$$

and Eq. (25.21) for $\varphi(\rho)$, Eq. (25.22) becomes

$$\frac{p}{\mathbf{k}T} = \rho - \int_0^\rho \Big(\sum_{k\geqslant 1} k\beta_k\rho^k\Big)\, d\rho$$

or

$$\frac{pV}{\bar{N}\mathbf{k}T} = 1 - \sum_{k\geqslant 1}\frac{k}{k+1}\beta_k\rho^k \tag{25.23}$$

which is the required result.

General Imperfect Gas. We continue here the formal discussion of the general imperfect gas which was begun at the end of Sec. 23. Up to this point we have [Eqs. (23.42) and (23.43)] $p/\mathbf{k}T$ as a power series in z with coefficients b_j given in terms of the Z_N or Q_N. We now wish to relate the (virial) coefficients B_n of

$$\frac{p}{\mathbf{k}T} = \rho + \sum_{n\geqslant 2} B_n\rho^n \tag{25.24}$$

to the b_j and thus to the Z_N or Q_N. We cannot use Eq. (24.14) here since it was derived in a special case only, but we can proceed essentially by reversing Kahn's argument.

For convenience, we define quantities β_k in terms of the B_n by Eq. (25.9) so that

$$\frac{p}{\mathbf{k}T} = \rho - \sum_{k\geqslant 1}\frac{k}{k+1}\beta_k\rho^{k+1} \tag{25.25}$$

The β_k are as yet unknown coefficients which we desire to relate to the b_j. We define $\varphi(\rho)$ by Eq. (25.21) and z by Eq. (25.20). Then by reversing the steps in Eqs. (25.22) and (25.23), Eq. (25.25) becomes

$$\frac{p}{\mathbf{k}T} = \int_0^z \frac{\rho}{z}\, dz \tag{25.26}$$

Since $z \to \rho$ as $\rho \to 0$, according to Eq. (25.20), and since Eq. (25.26) is just the thermodynamic relation between p, ρ, and the activity z of Eq. (23.39), the two z's [Eqs. (25.20) and (23.39)] must be identical.

We now want to invert [Eq. (25.20)]

$$\rho = ze^{\varphi(\rho)}$$

to obtain $\rho(z)$. For this purpose we use Eqs. (25.17) and (25.18) with $x = \rho$ and $f(\rho) = e^{\varphi(\rho)}$ and obtain, with the aid of Eq. (25.11),

$$\rho(z) = \sum_{j\geqslant 1}\frac{z^j}{j!}[(j-1)!C_{j-1}] = \sum_{j\geqslant 1}\frac{z^jC_{j-1}}{j} \tag{25.27}$$

where C_{j-1} is given by (25.12). That is, in Eq. (25.18), C_i is the coefficient of ξ^i in the expansion of $[f(\xi)]^j$. But we have also, from Eqs. (23.42) and (23.20),

$$\rho = \sum_{j\geqslant 1} jb_j z^j \tag{25.28}$$

so that

$$jb_j = \frac{C_{j-1}}{j} \qquad \text{or} \qquad j^2 b_j = C_{j-1} \tag{25.29}$$

This is formally the same relation between b_j and the β_k as Eq. (24.14), but Eq. (24.14) was restricted to a special case.

The inverse of Eq. (25.29) is[1]

$$\beta_k = \sum_{\substack{\mathbf{m} \\ (\sum_{j=2}^{k+1}(j-1)m_j = k)}} (-1)^{\sum_j m_j - 1} \frac{(k - 1 + \sum_j m_j)!}{k!} \prod_j \frac{(jb_j)^{m_j}}{m_j!} \tag{25.30}$$

where the first sum is over all sets of positive integers or zero, $\mathbf{m} = m_2, m_3, \ldots$, such that $\sum_j (j-1)m_j = k$. This is the desired result giving β_k (and therefore B_n) in terms of the b_j's. The first few relations are

$$\begin{aligned} \beta_1 &= 2b_2 \\ \beta_2 &= 3b_3 - 6b_2^2 \\ \beta_3 &= 4b_4 - 24b_2b_3 + (80/3)b_2^3 \end{aligned} \tag{25.31}$$

Equations (23.37) to (23.45) and (25.24) to (25.31) show that the nth virial coefficient of a gas can be calculated from $Q_1, Q_2, \ldots, Q_n$ only. For example, to obtain the second virial coefficient B_2, we need consider only Q_1 and Q_2, the partition functions for one and two molecules in the volume V, respectively. In the general case, any one of Eqs. (11.33) to (11.35) can be used for Q_1 and Q_2.

26. Alternative Derivations

This section is, to a large extent, a digression presenting alternative derivations of results already obtained in Sec. 23. However, the methods employed here are useful in many other problems. Also, the approximate cluster theory of Frenkel and Band is discussed briefly in connection with the association equilibrium point of view.

a. Maximum Term Method. Let us apply the familiar method of picking out the largest term in Q and replacing $\ln Q$ by the logarithm of this largest term. In the present case, referring to Eq. (22.24), we wish to find that particular set $\mathbf{m}$, call it $\mathbf{m}^*$, giving the largest term in Eq. (22.24). We are

[1] J. E. Mayer, *J. Chem. Phys.*, **10**, 629 (1942). See also J. E. Kilpatrick, *J. Chem. Phys.*, **21**, 274 (1953).

not justified, it may be recalled, in referring to the set $\mathbf{m}^*$ as the "most probable" set in this case, since the individual terms are not weights or probabilities.

We write

$$t = \prod_{j=1}^{N} \frac{(Vb_j)^{m_j}}{m_j!} \tag{26.1}$$

$$\ln t = \sum_{j=1}^{N} (m_j \ln Vb_j - m_j \ln m_j + m_j) \tag{26.2}$$

and our problem is to maximize $\ln t$ with respect to the m_j, subject to the restraint

$$\sum_{j=1}^{N} jm_j = N \tag{26.3}$$

Using Lagrange's method of undetermined multipliers, we have

$$\frac{\partial(\ln t - \alpha N)}{\partial m_j} = 0 \qquad j = 1, 2, \ldots, N \tag{26.4}$$

or

$$\ln Vb_j - \ln m_j^* - \alpha j = 0$$

or

$$m_j^* = Vb_j s^j \qquad j = 1, 2, \ldots, N \tag{26.5}$$

where

$$s = e^{-\alpha}$$

Equation (26.5) in Eq. (26.3) then gives

$$N = \sum_{j=1}^{N} Vjb_j s^j \tag{26.6}$$

Equation (26.6) determines s as a function of N/V and T.

Now

$$\begin{aligned} -\frac{A}{\mathbf{k}T} &= \ln Q = -N \ln \Lambda^3 + \ln t_{\max} \\ &= -N \ln \Lambda^3 + \sum_{j=1}^{N} (m_j^* \ln Vb_j - m_j^* \ln Vb_j s^j + m_j^*) \\ &= -N \ln \Lambda^3 - N \ln s + \sum_{j=1}^{N} m_j^* \end{aligned} \tag{26.7}$$

and therefore

$$\begin{aligned} -\frac{\mu}{\mathbf{k}T} &= \left(\frac{\partial \ln Q}{\partial N}\right)_{V,T} \\ &= -\ln \Lambda^3 - \ln s - \frac{N}{s}\frac{\partial s}{\partial N} + \sum_{j=1}^{N} Vjb_j s^{j-1} \frac{\partial s}{\partial N} \\ &= -\ln \Lambda^3 - \ln s \end{aligned} \tag{26.8}$$

From Eqs. (23.6) and (26.8) we conclude that s is just what we have called previously the activity z. Since

$$-\frac{A}{\mathbf{k}T} = -\frac{N\mu}{\mathbf{k}T} + \frac{pV}{\mathbf{k}T}$$

Eqs. (26.7) and (26.8) give

$$\frac{pV}{\mathbf{k}T} = \sum_{k=1}^{N} m_j^* = \sum_{k=1}^{N} Vb_j z^j \tag{26.9}$$

or

$$\frac{p}{\mathbf{k}T} = \sum_{k=1}^{N} b_j z^j \tag{26.10}$$

These results are all formally identical with those of Sec. 23 in the limit $N \to \infty$ (T, V/N constant). This is rather remarkable, since the present method breaks down unless all the b_j's are positive. For, according to Eq. (26.5), m_j^* is negative if b_j is negative, but the set of numbers $\mathbf{m}^*$ is one of the sets in Eq. (22.24), and all of the m_j for every set in Eq. (22.24) are by definition zero or positive integers.

b. Association Equilibrium. An alternative, but completely equivalent, point of view to that just considered is the following. In a formal way, consider that the imperfect gas is a perfect-gas mixture of different species (mathematical clusters) in equilibrium with each other. By dissociation and association, clusters may change size but at equilibrium there is a certain mean number m_j^* of clusters of size j. Ignoring fluctuations from these mean values, the partition function of the perfect-gas mixture can be written as [see Eq. (22.24)]

$$Q = \prod_{j=1}^{N} \mathcal{Q}_j \tag{26.11}$$

where $\mathcal{Q}_j = \dfrac{1}{m_j^*!} q_j^{m_j^*} \qquad \ln \mathcal{Q}_j = m_j^* \ln q_j - m_j^* \ln m_j^* + m_j^*$ (26.12)

$$q_j = \frac{Vb_j}{\Lambda^{3j}} \tag{26.13}$$

$\mathcal{Q}_j$ is the partition function for the m_j^* clusters of size j and q_j is the partition function per cluster for clusters of size j.

The condition for association equilibrium between clusters of different sizes is

$$\mu_j = j\mu \qquad j = 2, 3, \ldots, N \tag{26.14}$$

where μ_j is the chemical potential of clusters of size j and μ is the chemical potential of clusters of size one. The chemical potentials are given by

$$-\frac{\mu_j}{\mathbf{k}T} = \left(\frac{\partial \ln \mathcal{Q}_j}{\partial m_j^*}\right)_{V,T} = \ln q_j - \ln m_j^* \qquad j = 1, \ldots, N \tag{26.15}$$

Then
$$m_j^* = q_j e^{\mu_j/\mathbf{k}T}$$
$$= \frac{Vb_j}{\Lambda^{3j}} e^{j\mu/\mathbf{k}T} = Vb_j z^j \tag{26.16}$$

where we have defined z by [see Eq. (23.4)]

$$z = \frac{e^{\mu/\mathbf{k}T}}{\Lambda^3} \tag{26.17}$$

We may note that $m_1^*/V = z$.

The partial pressure p_j due to clusters of size j is

$$p_j = \mathbf{k}T \left(\frac{\partial \ln \mathcal{Q}_j}{\partial V}\right)_{m_j^*,T} = \frac{m_j^* \mathbf{k}T}{V} \tag{26.18}$$

and
$$\frac{pV}{\mathbf{k}T} = \sum_{j=1}^{N} \frac{p_j V}{\mathbf{k}T} = \sum_{j=1}^{N} m_j^* \tag{26.19}$$

or
$$\frac{p}{\mathbf{k}T} = \sum_{j=1}^{N} b_j z^j \tag{26.20}$$

The total number of constituent molecules is

$$N = \sum_{j=1}^{N} j m_j^* = \sum_{j=1}^{N} V j b_j z^j \tag{26.21}$$

The conventional type of equilibrium quotient or mass-action relation between clusters of size j and those of size one is [Eq. (26.14)]

$$\ln q_j - \ln m_j^* = j(\ln q_1 - \ln m_1^*)$$

or
$$\frac{m_1^{*j}}{m_j^*} = \frac{q_1^j}{q_j} \tag{26.22}$$

Thus the correct results are obtained by adopting this formal equilibrium association point of view. The analogy with an actual equilibrium between physical clusters or aggregates is limited, however, by the fact that for negative b_j, m_j^* and q_j are negative.[1]

Frenkel-Band approximate theory. Frenkel and Band,[2] independently, pointed out that an approximate theory very analogous to Mayer's could be developed, based on the equilibrium association of *physical* clusters. We

[1] See also J. E. Kilpatrick, *J. Chem. Phys.*, **21**, 1366 (1953).

[2] J. Frenkel, *J. Chem. Phys.*, **7**, 200 (1939); W. Band, *J. Chem. Phys.*, **7**, 324, 927 (1939). See also R. H. Fowler, "Statistical Mechanics" (Cambridge, London, 1936); J. O. Hirschfelder, F. T. McClure, and I. F. Weeks, *J. Chem. Phys.*, **10**, 201 (1942); S. G. Reed, Jr., *J. Chem. Phys.*, **20**, 208 (1952); H. W. Woolley, *J. Chem. Phys.*, **21**, 236 (1953); W. Weltner, Jr., *J. Chem. Phys.*, **22**, 153 (1954).

confine ourselves here to outlining the general approach without discussing details.

We assume that in an imperfect gas actual clusters of various sizes exist and that there is an association-dissociation equilibrium between these clusters. Attractive forces between clusters are neglected but repulsive forces can be included to the extent of introducing approximate excluded volume corrections. However, for simplicity, in the present brief review, we omit these excluded volume effects.[1]

A cluster of size j has many possible configurations but, as an approximation, only one or a few most probable configurations are taken into account. For example, for j fairly large, say $j > 30$, one might include only a spherical configuration. Let J_j be the partition function of a cluster of size j. A cluster of this size has $3j$ degrees of freedom, which can be divided into translation, rotation, and vibration in the usual way, with the zero of potential energy corresponding to complete separation of the j molecules. The details of writing down J_j depend of course on the kind of approximations one wishes to use.

Then [compare Eqs. (26.11) to (26.22)]

$$Q = \prod_{j=1}^{N} \mathcal{Q}_j \tag{26.23}$$

$$\mathcal{Q}_j = \frac{1}{m_j^*!} J_j^{m_j^*} \tag{26.24}$$

$$J_1 = \frac{V}{\Lambda^3} \tag{26.25}$$

Equation (26.14) is used again, where

$$-\frac{\mu_j}{\mathbf{k}T} = \ln J_j - \ln m_j^* \qquad j = 1, \ldots, N \tag{26.26}$$

so that

$$m_j^* = J_j e^{j\mu/\mathbf{k}T} \tag{26.27}$$

$$= \Lambda^{3j} J_j z^j \tag{26.28}$$

$$\frac{m_1^*}{V} = z \tag{26.29}$$

As before

$$\frac{pV}{\mathbf{k}T} = \sum_{j=1}^{N} m_j^* \tag{26.30}$$

and

$$N = \sum_{j=1}^{N} j m_j^* \tag{26.31}$$

[1] See Band, *loc. cit.*

where m_j^* is given by Eq. (26.28). The above equations are, of course, more complicated if an excluded volume correction is introduced (see also Sec. 27).

The equilibrium constant expression is

$$\frac{m_1^{*j}}{m_j^*} = \frac{J_1{}^j}{J_j} \tag{26.32}$$

The onset of condensation can be understood qualitatively in terms of the following argument[1] which does not pretend to be rigorous. In the limit of very large j, J_j will approach the partition function of j molecules of bulk liquid. Then

$$\ln J_j = -\frac{A_L}{\mathbf{k}T} = -\frac{j\mu_L}{\mathbf{k}T} \tag{26.33}$$

where A_L and μ_L refer to the liquid state and we have ignored the difference between the Helmholtz and Gibbs free energies in the liquid state. Equation (26.33) states that for large enough j, $\ln J_j$ changes linearly with j. Equation (26.27) becomes[2] then

$$m_j^* = x^j \tag{26.34}$$

$$x = e^{(\mu - \mu_L)/\mathbf{k}T} \tag{26.35}$$

For μ appreciably less than μ_L, $x < 1$ and, from Eq. (26.34), m_j^* is (essentially) zero for very large j. That is, very large clusters[3] are thermodynamically unimportant in the gas phase when $\mu < \mu_L$, as expected. On the other hand, when $\mu > \mu_L$ and $x > 1$, m_j^* is large and increases rapidly with j. This corresponds to bulk liquid being more stable than the gas phase, also as expected thermodynamically (since $\mu > \mu_L$). Clearly the transition point—the point of condensation—is at $x = 1$, and, further, this transition from $m_j^* = 0$ to m_j^* very large for large j is extremely sharp as x passes through unity. This is in qualitative agreement with the thermodynamically sharp transition point occurring at $x = 1$, that is, at $\mu = \mu_L$.

For intermediate-sized clusters (say $j = 50$), we will have roughly

$$-\mathbf{k}T \ln J_j = j\mu_L + cj^{2/3} \tag{26.36}$$

where the new term in $j^{2/3}$ is a surface contribution (c is a positive constant proportional to the surface tension). For intermediate values of j, Eq. (26.27) becomes then

$$m_j^* = x^j e^{-cj^{2/3}/\mathbf{k}T} \tag{26.37}$$

[1] See J. Frenkel, "Kinetic Theory of Liquids" (Oxford, London, 1946).

[2] Actually, x^j in Eq. (26.34) should in general be multiplied by a constant $C(j)$, where, for example, $C(j)$ is of order j and $\ln C(j)$ of order $\ln j$. For this corresponds to adding $\ln C(j)$ to the right-hand side of Eq. (26.33), a term which is thermodynamically negligible compared to $-j\mu_L/\mathbf{k}T$ (of order j). The same remarks apply to Eqs. (26.36) and (26.37).

[3] Clusters large enough so that $\ln J_j$ is proportional to j.

When $x < 1$ (gas phase stable), we have the same qualitative results as before, but when $x > 1$ (supersaturated vapor), the first factor (x^j) increases with j while the second decreases with j. The value of j at which m_j^* is a minimum is determined by

$$\frac{d \ln m_j^*}{dj} = 0 \tag{26.38}$$

or

$$\ln x = \frac{2c}{3\mathbf{k}Tj^{1/3}} \tag{26.39}$$

Clusters of this size represent a minimum in stability or a maximum in free energy. These "critical" clusters or embryos of liquid play a crucial role in the kinetic theory of nucleation of the liquid phase from a supersaturated vapor, since they represent a free-energy barrier in the path of growth of large clusters (liquid state) from small clusters.

c. Steepest Descent or Saddle Point Method.[1] This method is used to evaluate contour integrals of a certain type. Our first task, in order to be able to use the method in the present connection, is to express $Z(N)/N!$ in terms of such an integral.

Consider the function $e^{VX(\xi)}$, where

$$X(\xi) = \sum_{j \geqslant 1} b_j(T)\xi^j$$

and ξ is complex now. Within the circle of convergence of $X(\xi)$, the coefficient of ξ^N in the expansion of $e^{VX(\xi)}$ as a power series in ξ is, as we have seen in Eq. (23.13), just $Z/N!$. That is,

$$e^{VX(\xi)} = \sum_{N \geqslant 0} \frac{Z(N)}{N!} \xi^N \tag{26.40}$$

Now define the function $f(\xi)$ as

$$\begin{aligned} f(\xi) &= \frac{e^{VX(\xi)}}{\xi^{N'+1}} \\ &= \sum_{N \geqslant 0} \frac{Z(N)}{N!} \xi^{N-N'-1} \end{aligned} \tag{26.41}$$

The coefficient of ξ^{-1} in the expansion of $f(\xi)$ is, from Eq. (26.41), $Z(N')/N'!$. This coefficient is also, by the residue theorem,

$$\frac{1}{2\pi i} \oint f(\xi)\, d\xi$$

[1] This particular application is due to Born and Fuchs. The method itself is due to Debye.

where the path of integration encloses the origin (the only pole) and lies within the circle of convergence of $X(\xi)$. Dropping the prime on N', we have

$$\frac{Z(N)}{N!} = \frac{1}{2\pi i} \oint \frac{e^{VX(\xi)}}{\xi^{N+1}} \, d\xi \tag{26.42}$$

This is the required contour integral.

We now evaluate the integral in Eq. (26.42) by the saddle-point method. Choosing a circle of radius r as the path of integration, $\xi = re^{i\theta}$, the integral becomes

$$I = i \int_{-\pi}^{+\pi} g(r,\theta) \, d\theta \tag{26.43}$$

where

$$g(r,\theta) = e^{F(r,\theta)} \tag{26.44}$$

$$F(r,\theta) = V \sum_{j \geqslant 1} b_j r^j e^{ji\theta} - N \ln r - Ni\theta \tag{26.45}$$

Now we notice that

$$\frac{\partial g}{\partial \theta} = ig(r,\theta)[V \sum_{j \geqslant 1} j b_j r^j e^{ji\theta} - N] \tag{26.46}$$

and

$$\frac{\partial g}{\partial r} = g(r,\theta) \left[V \sum_{j \geqslant 1} j b_j r^{j-1} e^{ji\theta} - \frac{N}{r} \right] \tag{26.47}$$

According to Eqs. (26.46) and (26.47), both $\partial g/\partial\theta$ and $\partial g/\partial r$ are zero at the point $\theta = 0$ and $r = t$, where t is determined by the equation

$$N = V \sum_{j \geqslant 1} j b_j t^j \tag{26.48}$$

Also, we find easily that[1]

$$\left(\frac{\partial^2 g}{\partial \theta^2}\right)_{\theta=0,r=t} = -\, g(t,0)t \frac{\partial N}{\partial t} < 0 \tag{26.49}$$

and

$$\left(\frac{\partial^2 g}{\partial r^2}\right)_{\theta=0,r=t} = \frac{g(t,0)}{t} \frac{\partial N}{\partial t} > 0 \tag{26.50}$$

The point $\theta = 0$, $r = t$ is therefore a saddle point of the integrand $g(r,\theta)$: along the real axis $g(r,\theta)$ has a minimum at $\theta = 0$, $r = t$, while perpendicular to the real axis $g(r,\theta)$ has a maximum at this point (g is real on the circle $r = t$ in the immediate neighborhood of $\theta = 0$; see below).

We choose as the path of integration the circle with radius $r = t$, passing through the saddle point. We shall see that on this circle the value of $|g|$ falls off *extremely* rapidly on either side of $\theta = 0$, so that the only appreciable

[1] Ideally, if the b_j were known, we would want to prove from Eq. (26.48) that $t > 0$ and $\partial N/\partial t > 0$. But since the b_j are not known (i.e., the integrations cannot be carried out), we have to resort to the statement that t will turn out to be the activity z and, on thermodynamic grounds, we know that z and $\partial N/\partial z$ are positive.

contribution to the integral comes from the immediate neighborhood of $\theta = 0$ (where g is real). To see this, expand $F(t,\theta)$ in a Taylor series about $\theta = 0$:

$$F(t,\theta) = (V \sum_{j\geqslant 1} b_j t^j - N \ln t) - \frac{\theta^2}{2} (V \sum_{j\geqslant 1} b_j j^2 t^j) + \cdots \tag{26.51}$$

so that

$$g(t,\theta) = \frac{e^{VX(t)}}{t^N} \exp \left[-\frac{\theta^2}{2} (V \sum_{j\geqslant 1} b_j j^2 t^j) \right] \tag{26.52}$$

If we let $V \rightarrow \infty$ keeping N/V and T constant, t and the b_j remain constant [Eq. (26.48)] but the coefficient of θ^2 in Eq. (26.52) approaches infinity because of V. Hence $g(t,\theta)$ has a *very* sharp maximum at $\theta = 0$ for large V and higher terms in Eq. (26.51) may be dropped. Also, the limits $\pm \pi$ in Eq. (26.43) can be replaced by $\pm \infty$; thus we have

$$I = \frac{ie^{VX(t)}}{t^N} \int_{-\infty}^{+\infty} \exp \left[-\frac{\theta^2}{2} (V \sum_{j\geqslant 1} b_j j^2 t^j) \right] d\theta \tag{26.53}$$

and

$$\frac{Z(N)}{N!} = Q(N)\Lambda^{3N} = \frac{e^{VX(t)}}{t^N \sqrt{2\pi V \Sigma b_j j^2 t^j}} \tag{26.54}$$

That is,

$$-\frac{A}{\mathbf{k}T} = \ln Q = -N \ln \Lambda^3 - N \ln t + V \sum_{j\geqslant 1} b_j(T) t^j \tag{26 55}$$

where t is determined by Eq. (26.48). In writing Eq. (26.55), we have dropped a negligible term[1] of order $\ln N$, as the other terms are of order N.

We then find $\mu/\mathbf{k}T$ from Eq. (26.55) just as in Eq. (26.8), and therefore conclude that t is the activity z. Also

$$\frac{pV}{\mathbf{k}T} = -\frac{A}{\mathbf{k}T} + \frac{N\mu}{\mathbf{k}T} = V \sum_{j\geqslant 1} b_j z^j \tag{26.56}$$

or

$$\frac{p}{\mathbf{k}T} = \sum_{j\geqslant 1} b_j z^j \tag{26.57}$$

as before.

This method is rigorous for a gas but breaks down in considering condensation.[2] The grand partition function method is preferable in the present problem because it is simpler.

27. Exact Treatment of Physical Clusters

In Sec. 26 it was shown [Eqs. (26.11) to (26.22)] that, in a strictly formal way, an imperfect gas can be considered, without approximation, as an

[1] That is, only the leading term in Eq. (26.51), $F(t,0)$, actually contributes to the final result.

[2] Fowler and Guggenheim (Gen. Ref.).

equilibrium perfect-gas mixture of "mathematical clusters" of various sizes. Also, in the same section, the Frenkel-Band *approximate* theory of *physical* cluster equilibrium was introduced [Eqs. (26.23) to (26.39)]. The purpose of the present section is to give an *exact* treatment of physical clusters.[1]

Formal Relations. We begin by deriving certain relations which must be satisfied in any exact formulation of cluster equilibria in an imperfect gas. We can refer here, without disadvantage, to the general imperfect gas of Eqs. (23.37) to (23.45) and (25.24) to (25.31). The grand partition function of the gas is

$$\Xi = \sum_{N \geqslant 0} Q_N \lambda^N \tag{27.1}$$

If Eq. (27.1) is used as the basis for a study of the properties of an imperfect gas, the Q_N contain all the intermolecular forces of the system. Thus, if physical clusters between pairs, triplets, etc., of molecules are present in the gas because of strong attractive intermolecular forces, their existence and their influence on all thermodynamic properties are automatically taken care of without ever introducing them explicitly. That is, Eq. (27.1) leads to an *implicit* consideration of clusters.

Let us now consider an alternative approach in which clusters appear in an *explicit* way. We regard the gas as made up of many different species, clusters of one, two, three, etc., molecules. Let λ_s be the absolute activity of an s cluster and N_s the number of s clusters. Then we have, in place of Eq. (27.1),

$$\Xi = \sum_{\mathbf{N} \geqslant 0} \left(\prod_s \lambda_s^{N_s}\right) Q_{\mathbf{N}} \tag{27.2}$$

where $Q_{\mathbf{N}}$ is the partition function for a set of clusters $\mathbf{N} = N_1, N_2, \ldots$. However, the clusters are in equilibrium with each other so that

$$\mu_s = s\mu_1 \qquad \lambda_s = \lambda_1^s \qquad s = 1, 2, \ldots \tag{27.3}$$

It should be emphasized at this point that the zero of potential energy for intermolecular interactions between the single molecules of clusters of all sizes is taken at infinite separation of the single molecules. Equation (27.2) becomes, then,

$$\Xi = \sum_{\mathbf{N} \geqslant 0} \lambda_1^{\sum_s s N_s} Q_{\mathbf{N}} \tag{27.4}$$

Now Eqs. (27.1) (clusters implicit) and (27.4) (clusters explicit) refer to the same system and thus must be identical. We establish the detailed correspondence by first noting the limiting forms as λ and λ_1 approach zero:

$$\begin{aligned} \Xi &= 1 + \lambda Q_1 + \cdots \\ &= 1 + \lambda_1 Q_{1000\ldots} + \cdots \end{aligned} \tag{27.5}$$

[1] T. L. Hill, *J. Chem. Phys.*, **23**, 617 (1955).

Q_1 and $Q_{1000\cdots}$ are the same quantity, the partition function of a single molecule in the volume V at T; hence $\lambda = \lambda_1$. Therefore the partition functions in Eqs. (27.1) and (27.4) must be related by

$$Q_N = \frac{Z_N}{N!\Lambda^{3N}} = \sum_{\mathbf{N}} Q_{\mathbf{N}} \tag{27.6}$$

where the sum is over all cluster sets $\mathbf{N}$ such that

$$\sum_{s=1}^{N} sN_s = N$$

and Λ is defined [Eq. (23.40)], for the general gas, by

$$\Lambda^3 = \frac{V}{Q_1} \tag{27.7}$$

For example,

$$\begin{aligned}
Q_1 &= Q_{1000\cdots} = \frac{V}{\Lambda^3} \\
Q_2 &= Q_{2000\cdots} + Q_{0100\cdots} \\
Q_3 &= Q_{3000\cdots} + Q_{1100\cdots} + Q_{00100\cdots} \\
Q_4 &= Q_{4000\cdots} + Q_{2100\cdots} + Q_{0200\cdots} + Q_{10100\cdots} + Q_{00010\cdots}
\end{aligned} \tag{27.8}$$

The equilibrium number of s clusters is obtained by applying

$$\bar{N}_s = \lambda_s \frac{\partial \ln \Xi}{\partial \lambda_s} \tag{27.9}$$

to Eq. (27.2); after the differentiation, we set $\lambda_s = \lambda^s$. This gives

$$\begin{aligned}
\bar{N}_1 &= \lambda Q_{100} + \lambda^2(2Q_{200} - Q_{100}{}^2) + \lambda^3(Q_{110} + 3Q_{300} - Q_{100}Q_{010} \\
&\qquad - 3Q_{100}Q_{200} + Q_{100}{}^3) + \cdots \\
\bar{N}_2 &= \lambda^2 Q_{010} + \lambda^3(Q_{110} - Q_{100}Q_{010}) + \cdots \\
&\cdots\cdots\cdots\cdots \\
\bar{N}_n &= \lambda^n Q_{00\cdots 1} + \lambda^{n+1}(Q_{10\cdots 1} - Q_{100}Q_{00\cdots 1}) + \cdots \\
&\cdots\cdots\cdots\cdots
\end{aligned} \tag{27.10}$$

From the nature of Eqs. (27.2) and (27.9) (all $Q_{\mathbf{N}}$'s positive), the numbers $\bar{N}_1, \bar{N}_2, \ldots$ are necessarily all positive (unlike the numbers of Mayer's "mathematical clusters").

Equations (27.10) lead to the following "equilibrium quotients":

$$\frac{\bar{N}_1{}^n}{\bar{N}_n} = \frac{Q_{100\cdots}{}^n}{Q_{00\cdots 1}}\left\{1 + \lambda\left[\frac{2nQ_{200\cdots}}{Q_{100\cdots}} - (n-1)Q_{100\cdots} - \frac{Q_{100\cdots 1}}{Q_{00\cdots 1}}\right] + \cdots\right\}$$
$$n = 2, 3, \ldots \tag{27.11}$$

The terms other than the leading terms in Eqs. (27.10) and (27.11) correct for the interactions between clusters; these interactions cannot properly be neglected in a discussion of *physical* clusters.

We can replace the variable λ in Eqs. (27.10) by

$$\rho = \frac{\bar{N}}{V} = \frac{\bar{N}_1 + 2\bar{N}_2 + 3\bar{N}_3 + \cdots}{V} \tag{27.12}$$

From

$$z = \frac{\lambda}{\Lambda^3} = \rho \exp\left(-\sum_{k\geqslant 1} \beta_k \rho^k\right)$$

and Eq. (25.29), we have, as in Eq. (25.4),

$$\frac{\lambda}{\Lambda^3} = \rho - 2b_2\rho^2 + (8b_2{}^2 - 3b_3)\rho^3 + \cdots \tag{27.13}$$

where the b_j are given in terms of the Z_N by Eq. (23.44). Equation (27.13) may be introduced in Eq. (27.10) to eliminate λ. The resulting equations simplify somewhat if we use

$$Y_{\mathbf{N}} \equiv N!\,\Lambda^{3N} Q_{\mathbf{N}} \tag{27.14}$$

$$N = N_1 + 2N_2 + 3N_3 + \cdots$$

With this definition, Eqs. (27.8) hold also between the Z_N's and $Y_{\mathbf{N}}$'s:

$$Z_1 = Y_{1000\ldots} = V$$

$$Z_2 = Y_{2000\ldots} + Y_{0100\ldots} \tag{27.15}$$

etc.

We obtain, to terms in ρ^3,

$$\rho_1 = \frac{\bar{N}_1}{V} = \rho - \frac{Y_{010}}{Y_{100}}\rho^2 + \left[2\frac{Y_{010}{}^2}{Y_{100}{}^2} - \frac{Y_{001}}{2Y_{100}} + \left(-\frac{1}{3}\frac{Y_{110}}{Y_{100}} - Y_{010} + 2\frac{Y_{200}Y_{010}}{Y_{100}{}^2}\right)\right]\rho^3 + \cdots$$

$$\rho_2 = \frac{1}{2}\frac{Y_{010}}{Y_{100}}\rho^2 + \left[-\frac{Y_{010}{}^2}{Y_{100}{}^2} + \left(-\frac{Y_{200}Y_{010}}{Y_{100}{}^2} + \frac{1}{2}Y_{010} + \frac{1}{6}\frac{Y_{110}}{Y_{100}}\right)\right]\rho^3 + \cdots \tag{27.16}$$

$$\rho_3 = \frac{1}{6}\frac{Y_{001}}{Y_{100}}\rho^3 + \cdots$$

If, in the procedure just outlined [Eqs. (27.12) to (27.16)], only first terms on the right-hand side of Eqs. (27.10) are used, the terms enclosed in ( ) are missing from Eqs. (27.16).

In conclusion, it may be stressed again that the equations of this subsection are formal but exact.

Definition of a Cluster. It was not necessary above to define just what was meant by an s cluster, but we now turn to this point. First, it is clear that for strictly thermodynamic purposes the definition of a cluster is arbitrary. Equations (27.8) are the crux of the matter. In the absence of approximations, any definition will lead to the same thermodynamic predictions (e.g., virial coefficients), provided only that it specifies: (1) how to divide the phase space (or energy states) of two molecules into two regions (or classes), one region (class) corresponding to $N_1 = 2$, $N_2 = 0$ and the other to $N_1 = 0$, $N_2 = 1$; (2) how to divide the phase space (energy states) of three molecules into three regions (classes), corresponding to $N_1 = 3$, $N_2 = N_3 = 0$, or $N_1 = N_2 = 1, N_3 = 0$, or $N_1 = N_2 = 0, N_3 = 1$; etc. That is, the definition must give the recipe for assigning any particular element of volume in the phase space (energy state) of N molecules to a subphase space (subclass) corresponding to a particular cluster set $\mathbf{N}$. The equations of the preceding section will be applicable to any such definition. Of course the numbers $\bar{N}_1$, $\bar{N}_2$, etc., will in general be different for different definitions, but these are not measurable thermodynamic quantities.

A cluster is, however, a fairly definite physical concept (say, in nucleation theory) so that all physically reasonable definitions should lead to rather similar predictions of $\bar{N}_1$, $\bar{N}_2$, etc. In general, the stronger the forces leading to cluster formation, the greater the tendency of different reasonable definitions to converge on the same predictions.

Cluster of Two Molecules. In the special case of two monatomic molecules in classical mechanics, there is a rather obvious single choice of a recipe for dividing Q_2 into two parts, Q_{20} and Q_{01}. Namely, when the two molecules are in an element of phase space such that the relative kinetic energy of the molecules exceeds the negative of the potential energy of interaction, then the two molecules are not "bound" to each other and the element of phase space belongs to Q_{20}. Otherwise, the molecules are "bound" and the element of phase space belongs to Q_{01}. Q_2, in this case, is

$$Q_2 = \frac{1}{2h^6} \int e^{-H/\mathbf{k}T} \, dx\, dy\, dz\, d\theta\, d\varphi\, dr\, dp_x\, dp_y\, dp_z\, dp_\theta\, dp_\varphi\, dp_r \tag{27.17}$$

$$H = \frac{1}{4m}(p_x^2 + p_y^2 + p_z^2) + \frac{1}{mr^2}\left(p_\theta^2 + \frac{p_\varphi^2}{\sin^2\theta}\right) + \frac{p_r^2}{m} + u(r) \tag{27.18}$$

where x, y, and z refer to the center of mass, θ, φ, and r are the usual relative coordinates, and $u(r)$ is the intermolecular potential. The changes of variables

$$P_\theta = \frac{p_\theta}{(mr^2\mathbf{k}T)^{1/2}} \qquad P_\varphi = \frac{p_\varphi}{(mr^2\mathbf{k}T\sin^2\theta)^{1/2}} \qquad P_r = \frac{p_r}{(m\mathbf{k}T)^{1/2}}$$

followed by integration over x, y, z, φ, θ, p_x, p_y, and p_z give

$$Q_2 = \frac{V}{2\Lambda^6\pi^{3/2}} \int e^{-u(r)/\mathbf{k}T} 4\pi r^2\, dr \int e^{-(P_\theta{}^2+P_\varphi{}^2+P_r{}^2)}\, dP_\theta\, dP_\varphi\, dP_r \quad (27.19)$$

$$= \frac{V}{2\Lambda^6} \int_V e^{-u(r)/\mathbf{k}T} 4\pi r^2\, dr \quad (27.20)$$

According to Eq. (27.18), the condition for a bound pair is

$$P_\theta{}^2 + P_\varphi{}^2 + P_r{}^2 \leqslant -\frac{u(r)}{\mathbf{k}T} \quad (27.21)$$

Thus Q_{01} is obtained by confining the integration over P_θ, P_φ, and P_r in Eq. (27.19), for each r with negative $u(r)$, to the region satisfying (27.21). Contributions to Q_{20} come from values of r such that $u(r) > 0$, or when $u(r) \leqslant 0$ but (27.21) is not satisfied.

Define $F(r)$ by

$$F(r) = \frac{1}{\pi^{3/2}} \int_{(27.21)} e^{-(P_\theta{}^2+P_\varphi{}^2+P_r{}^2)}\, dP_\theta\, dP_\varphi\, dP_r$$

$$= \frac{4}{\pi^{1/2}} \int_0^{[-u(r)/\mathbf{k}T]^{1/2}} e^{-R^2} R^2\, dR$$

$$= \frac{2}{\pi^{1/2}} \int_0^{-u(r)/\mathbf{k}T} e^{-y} y^{1/2}\, dy$$

$$= \frac{\Gamma\left[{}^3/_2, -\dfrac{u(r)}{\mathbf{k}T}\right]}{\Gamma({}^3/_2)} \quad (27.22)$$

where $\Gamma(n,x)$ is the incomplete Γ function. Then we have, finally,

$$Q_{01} = \frac{V}{2\Lambda^6} \int_{u(r)\leqslant 0} e^{-u(r)/\mathbf{k}T} 4\pi r^2 F(r)\, dr \quad (27.23)$$

$$Q_{20} = \frac{V}{2\Lambda^6} \left\{ \int_{u(r)>0} e^{-u(r)/\mathbf{k}T} 4\pi r^2\, dr + \int_{\substack{V \\ u(r)\leqslant 0}} e^{-u(r)/\mathbf{k}T} 4\pi r^2 [1 - F(r)]\, dr \right\} \quad (27.24)$$

Q_{01} is the partition function of a pair of bound molecules, while Q_{20} is the partition function of a pair of molecules which are not bound. It is rather illuminating to rewrite Eqs. (27.23) and (27.24) as complete configuration integrals but using *effective* intermolecular potentials to distinguish the cases where the two molecules are bound or not bound. That is [compare Eq. (27.20)],

$$Q_{01} = \frac{V}{2\Lambda^6} \int_V e^{-u^\ddagger(r)/\mathbf{k}T} 4\pi r^2 \, dr \tag{27.25}$$

$$Q_{20} = \frac{V}{2\Lambda^6} \int_V e^{-u^*(r)/\mathbf{k}T} 4\pi r^2 \, dr \tag{27.26}$$

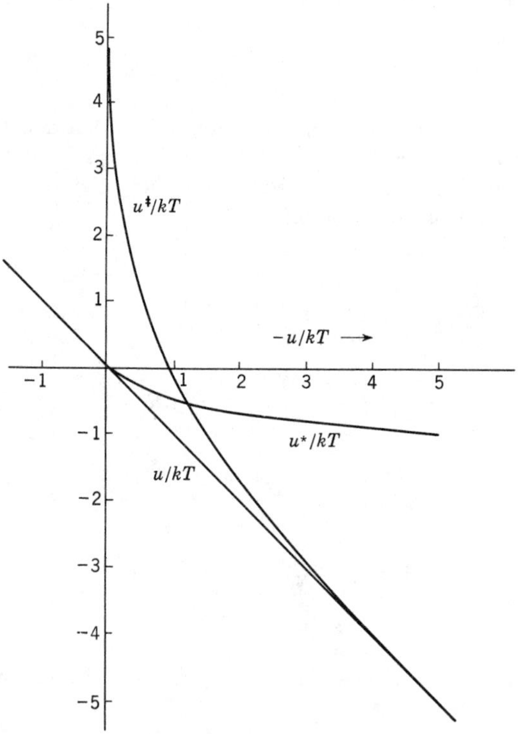

FIG. 9. Effective potentials between bound (‡) and unbound (*) molecular pairs.

where the effective potential between bound molecules is, from Eqs. (27.23) and (27.25),

$$\begin{aligned} u^\ddagger(r) &= +\infty \qquad u(r) > 0 \\ &= u(r) - \mathbf{k}T \ln F(r) \qquad u(r) \leqslant 0 \end{aligned} \tag{27.27}$$

and the effective potential between unbound molecules is

$$\begin{aligned} u^*(r) &= u(r) \qquad u(r) > 0 \\ &= u(r) - \mathbf{k}T \ln [1 - F(r)] \qquad u(r) \leqslant 0 \end{aligned} \tag{27.28}$$

The Lennard-Jones potential

$$u(r) = 4\varepsilon \left[\left(\frac{r_0}{r} \right)^{12} - \left(\frac{r_0}{r} \right)^{6} \right] \tag{27.29}$$

may be used to illustrate Eqs. (27.27) and (27.28). Figure 9 gives $u^{\ddagger}/\mathbf{k}T$ and $u^*/\mathbf{k}T$ as functions of $-u/\mathbf{k}T$, while Figs. 10 and 11 show $u/\mathbf{k}T$, $u^{\ddagger}/\mathbf{k}T$,

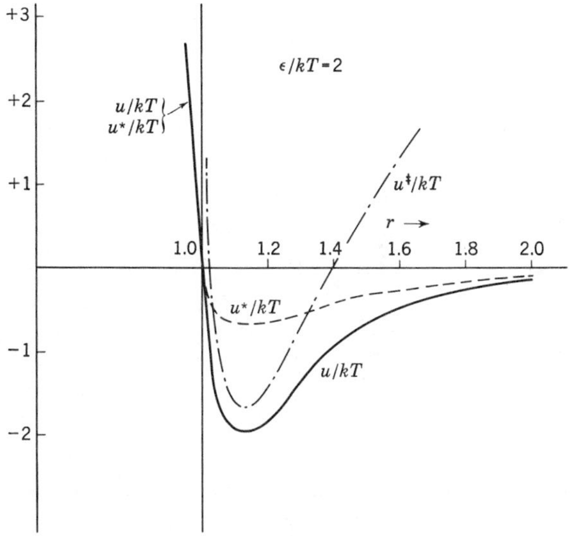

FIG. 10. Effective bound and unbound potentials using the Lennard-Jones potential and $\varepsilon/\mathbf{k}T = 2$.

and $u^*/\mathbf{k}T$ as functions of r for $\varepsilon/\mathbf{k}T = 2$ and 5, using Eq. (27.29). As an aid in interpreting the figures, we note, from the properties of $\Gamma(n,x)$, that the asymptotic behavior of $u^{\ddagger}$ and u^* is

$$\left. \begin{aligned} &\frac{u^{\ddagger}}{\mathbf{k}T} \to -\frac{3}{2} \ln \left(-\frac{u}{\mathbf{k}T} \right) \to +\infty \\ &\frac{u^*}{\mathbf{k}T} \to \frac{u}{\mathbf{k}T} \end{aligned} \right\} \text{as } -\frac{u}{\mathbf{k}T} \to 0+ \tag{27.30}$$

$$\left.\begin{aligned} \frac{u^\ddagger}{\mathbf{k}T} &\to \frac{u}{\mathbf{k}T} \\ \frac{u^*}{\mathbf{k}T} &\to -\frac{1}{2}\ln\left(-\frac{u}{\mathbf{k}T}\right) \end{aligned}\right\} \text{as } -\frac{u}{\mathbf{k}T} \to +\infty \tag{27.31}$$

When ε/kT is very large (strong binding), $u^\ddagger$ is indistinguishable from u in the region of the minimum in u and rises to infinity on both sides of the minimum, while u^* is indistinguishable from u for large r, has a relatively

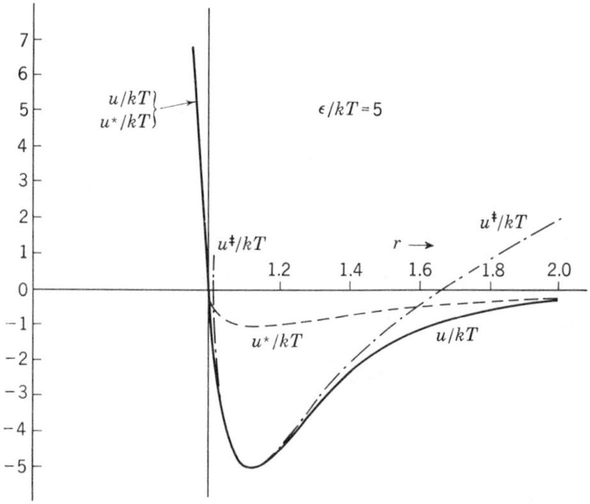

FIG. 11. Effective bound and unbound potentials using the Lennard-Jones potential and $\varepsilon/\mathbf{k}T = 5$.

shallow minimum ($u^*/u^\ddagger \to 0$ as $-u/\mathbf{k}T \to \infty$) where u has a minimum, and is equal to u (rises to $+\infty$) for small r. As a first approximation, in the absence of long-range forces [where the long tail $u^* = u(r)$ for large r cannot be neglected], the interaction u^* between nonbinding molecules could be considered a hard-sphere interaction; the explicit introduction of pairs substantially takes care of the strong attractive interaction. This justifies Band's approximate hard-sphere correction (Sec. 26).

Larger Clusters. Of the various possible definitions of clusters of three and more monatomic molecules in classical mechanics, one stands out as being especially simple and this is the only one we discuss. To introduce this definition, we consider first $N = 2$ (Sec. 22):

$$Z_2 = \int e^{-u_{12}/\mathbf{k}T}\, d\mathbf{r}_1\, d\mathbf{r}_2 \tag{27.32}$$

where $u_{12} = u(r_{12})$. Now we use Eqs. (27.27) and (27.28) to write

$$e^{-u_{12}/\mathbf{k}T} = e^{-u_{12}^{\ddagger}/\mathbf{k}T} + e^{-u_{12}^{*}/\mathbf{k}T} \tag{27.33}$$

and substitute Eq. (27.33) into Eq. (27.32). If this result is put in Eq. (27.6) for Z_2, we obtain the assignment of Q_{01} and Q_{20} already made in Eqs. (27.25) and (27.26). Proceeding now to higher clusters, for $N = 3$ we introduce Eq. (27.33) for each pair interaction, so that

$$\begin{aligned}\frac{Z^3}{3!\Lambda^9} &= \frac{1}{3!\Lambda^9}\int e^{-(u_{12}+u_{13}+u_{23})/\mathbf{k}T}\, d\mathbf{r}_1\, d\mathbf{r}_2\, d\mathbf{r}_3 \\ &= \frac{1}{3!\Lambda^9}\int [e^{-u_{12}^{\ddagger}/\mathbf{k}T} + e^{-u_{12}^{*}/\mathbf{k}T}][e^{-u_{13}^{\ddagger}/\mathbf{k}T} + e^{-u_{13}^{*}/\mathbf{k}T}] \\ &\qquad\qquad \times [e^{-u_{23}^{\ddagger}/\mathbf{k}T} + e^{-u_{23}^{*}/\mathbf{k}T}]\, d\mathbf{r}_1\, d\mathbf{r}_2\, d\mathbf{r}_3 \\ &= Q_{3000\cdots} + Q_{1100\cdots} + Q_{0010\cdots}\end{aligned} \tag{27.34}$$

where

$$Q_{3000\cdots} = \frac{1}{3!\Lambda^9}\int e^{-(u_{12}^{*}+u_{13}^{*}+u_{23}^{*})/\mathbf{k}T}\, d\mathbf{r}_1\, d\mathbf{r}_2\, d\mathbf{r}_3 \tag{27.35}$$

$$Q_{1100\cdots} = \frac{3}{3!\Lambda^9}\int e^{-(u_{12}^{\ddagger}+u_{13}^{*}+u_{23}^{*})/\mathbf{k}T}\, d\mathbf{r}_1\, d\mathbf{r}_2\, d\mathbf{r}_3 \tag{27.36}$$

$$\begin{aligned}Q_{0010\cdots} &= \frac{3}{3!\Lambda^9}\int e^{-(u_{12}^{\ddagger}+u_{13}^{\ddagger}+u_{23}^{*})/\mathbf{k}T}\, d\mathbf{r}_1\, d\mathbf{r}_2\, d\mathbf{r}_3 \\ &\quad + \frac{1}{3!\Lambda^9}\int e^{-(u_{12}^{\ddagger}+u_{13}^{\ddagger}+u_{23}^{\ddagger})/\mathbf{k}T}\, d\mathbf{r}_1\, d\mathbf{r}_2\, d\mathbf{r}_3\end{aligned} \tag{27.37}$$

Equations (27.35) to (27.37) correspond to the diagrams in Fig. 12, where a line between two molecules represents a "bound" interaction (‡) and the absence of a line indicates a "nonbound" interaction (*). Thus, the three molecules form a 3 cluster, by the present definition, if they are all at least singly connected to each other by "bound" pairwise interactions. Similarly, Fig. 13 represents $N = 4$.

This scheme can obviously be extended to arbitrary N and it is clearly analogous to Mayer's (pairwise, mathematical) cluster expansion of Z_N (Sec. 22). Whereas in this section we have put, for each pair,

$$e^{-u_{ij}/\mathbf{k}T} = e^{-u_{ij}^{\ddagger}/\mathbf{k}T} + e^{-u_{ij}^{*}/\mathbf{k}T} \tag{27.38}$$

Mayer uses

$$e^{-u_{ij}/\mathbf{k}T} = (e^{-u_{ij}/\mathbf{k}T} - 1) + 1 \tag{27.39}$$

Mayer's treatment is eventually simplified, however, because of the fact that unity in Eq. (27.39) is independent of r_{ij}, whereas in the present discussion

$e^{-u_{ij}^*/\mathbf{k}T}$ in Eq. (27.38) depends on r_{ij}. On the other hand, the present clusters have some direct physical significance, whereas Mayer's are introduced for mathematical convenience only.

Mayer's procedure amounts formally to defining $u^\ddagger$ and u^* not by Eqs. (27.27) and (27.28) but by

$$u_M^\ddagger = -\mathbf{k}T \ln (e^{-u/\mathbf{k}T} - 1) \tag{27.40}$$

$$u_M^* = 0 \tag{27.41}$$

Since ( ) in Eqs. (27.39) and (27.40) can be negative, we have the possibility of complex values of $u_M^\ddagger$ and negative values of the corresponding Q_N's

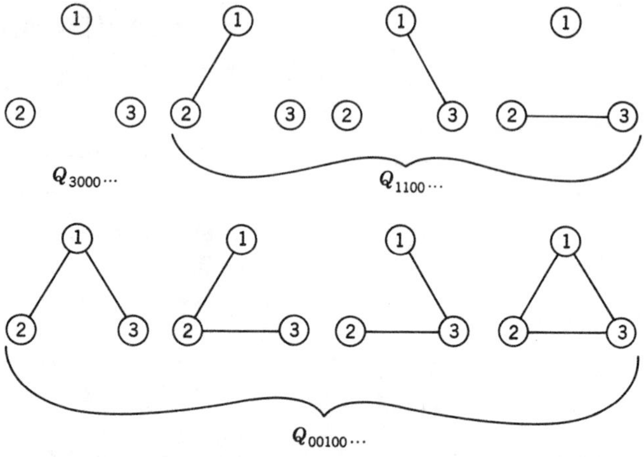

FIG. 12. Partition functions for clusters of three molecules.

[Eqs. (27.8)] and $\bar{N}_i$'s [Eqs. (27.10)]. Also, in view of Eq. (27.41), molecules in different (mathematical) clusters do not interact and the gas can be considered a perfect-gas mixture of (mathematical) clusters, which is not the case with physical clusters or with any pairwise definition of clusters except Mayer's [Eqs. (27.40) and (27.41)]. A consequence of this is that, with Mayer's clusters, all terms but the leading terms on the right-hand side of Eqs. (27.10) drop out. As expected, if Eqs. (27.40) and (27.41) are introduced in Eqs. (27.10) and (27.33) to (27.37), it is easy to verify that the results obtained are identical with those of Eqs. (26.11) to (26.22). The correspondence in notation, in this special case [Eqs. (27.40) and (27.41)], is

$$\bar{N}_n = m_n^* \qquad Q_{\underbrace{00\cdots1}_{n}} = q_n = \frac{Vb_n}{\Lambda^{3n}}$$

As a simple example of an error arising from the treatment of an imperfect gas as a perfect-gas mixture of physical clusters, consider the second virial coefficient B_2. From Sec. 25 and Eq. (27.14), we have

$$B_2 = -b_2 = \frac{Y_{100}{}^2 - Y_{200} - Y_{010}}{2Y_{100}} \tag{27.42}$$

whereas if we use, incorrectly,

$$\frac{p}{\mathbf{k}T} = \rho_1 + \rho_2 + \cdots \tag{27.43}$$

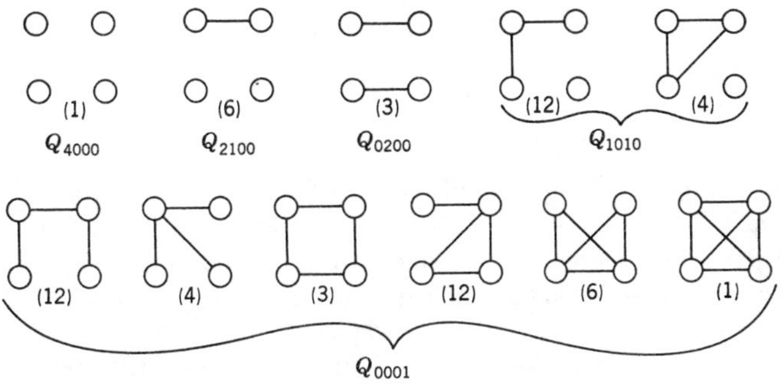

FIG. 13. Partition functions for clusters of four molecules.

we find from Eq. (27.16) that

$$\frac{p}{\mathbf{k}T} = \rho - \frac{1}{2}\frac{Y_{010}}{Y_{100}}\rho^2 + \cdots \tag{27.44}$$

or

$$B_2 = -\frac{1}{2}\frac{Y_{010}}{Y_{100}} \tag{27.45}$$

Equation (27.45) is correct in general only if two unbound molecules do not interact ($Y_{200} = Y_{100}{}^2$), as in Eq. (27.41) (Mayer). To obtain the correct result from Eq. (27.43), one must add the proper term in $\rho_1{}^2$ for an imperfect binary gas mixture.

The above definition [Eqs. (27.27) and (27.28)] of a cluster in terms of pairwise bound interactions is formally simple and is interesting because of the analogy to Mayer's theory. For molecules with very strong attractive interactions [e.g., when $\varepsilon/\mathbf{k}T$ in Eq. (27.29) is large], this definition is physically satisfactory for clusters of any size. However, if the interactions are weak, the present definition would exclude from a cluster a molecule which is not "bound" to any of the other molecules of the cluster taken singly, yet which would ordinarily be considered bound to the entire cluster because of the accumulation of attractive interactions with a number of molecules of the cluster.

In conclusion, it might be mentioned that, if one wishes to use an alternative pairwise definition of physical clusters based simply on a "bound" interaction for $r < r'$ (where r' is some arbitrary but fixed intermolecular distance) and a "nonbound" interaction for $r > r'$, then the appropriate effective potentials are clearly

$$\begin{aligned} u_c^{\ddagger} &= u & r &< r' \\ &= +\infty & r &> r' \\ u_c^{*} &= +\infty & r &< r' \\ &= u & r &> r' \end{aligned} \tag{27.46}$$

28. Theory of Condensation

Our main purpose here is to discuss the Yang-Lee theory of condensation. However, we make a few preliminary remarks on partition functions and grand partition functions in relation to phase changes (see Appendix 9 for further details), and also outline briefly Mayer's original theory of condensation because of its importance in the development of this field.

Partition Functions and Phase Changes. The partition function Q in Eq. (22.4) or (22.24) is exact (for the model adopted) and is a function of N, V, and T. No assumptions have been made at this stage about the dependence of b_j on V, for example. The following properties of a rigorous Q such as this have been rather generally accepted for some time but have been proved recently by van Hove.[1] If we let $N \to \infty$, keeping N/V and T constant, $N^{-1} \ln Q$ approaches a function of N/V (or V/N) and T only. This is the intensive thermodynamic quantity $-A/N\mathbf{k}T$. That is,

$$\lim_{N\to\infty} \frac{1}{N} \ln Q = -\frac{A}{N\mathbf{k}T} = f(v,T) \tag{28.1}$$

Then the thermodynamic pressure p is defined as

$$p(v,T) = -\left(\frac{\partial A}{\partial V}\right)_{N,T} = \mathbf{k}T\left(\frac{\partial f}{\partial v}\right)_T \tag{28.2}$$

The slope $(\partial p/\partial v)_T$ is never positive. At high temperatures it is negative for all $v > 0$ (Fig. 14*a*), but at low enough temperatures there can be a region (or regions) in which $(\partial p/\partial v)_T = 0$ (two-phase region), as in Fig. 14*b*. A loop of the van der Waals type (Fig. 14*c*) is not obtained. The completely flat

[1] L. van Hove, *Physica*, **15**, 951 (1949). See also Appendix 9.

portion in Fig. 14*b*, with mathematical singularities in $p(v)$ at the two ends, is a consequence of the limiting process $N \to \infty$. The absence of a loop is due to the fact that the complete and rigorous configuration integral Z includes *every* possible configuration, including configurations associated with the simultaneous existence of two phases in the volume V.

Approximate evaluations of Z invariably introduce implicitly the restraint of uniform macroscopic density throughout V in enumerating configurations, as well as other approximations. It is not possible under this restraint for the two phases (of different density) to exist together in the container. Mathematically, the result is a loop, as in Fig. 14*c*. The flat portion required

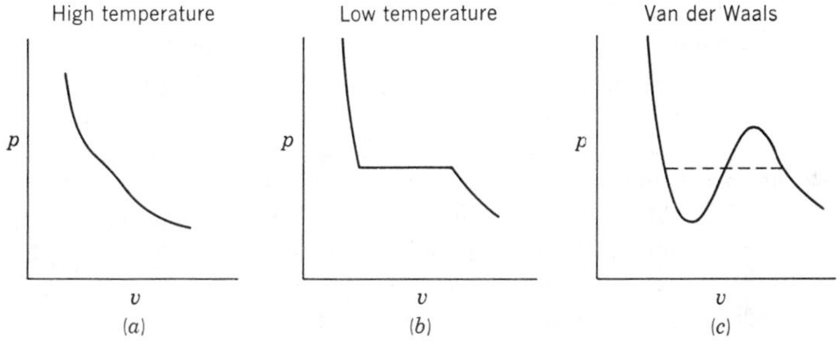

FIG. 14. Pressure-volume isotherms for gas-liquid system.

thermodynamically is then drawn in, using Maxwell's theorem of equal areas.

In principle, one could introduce the restraint of uniform macroscopic density in calculating Z, but no other approximation (see Chaps. 6 and 8). In this case a loop would be obtained, but if the horizontal portion were inserted from the equal-areas theorem, the resulting curve (three analytical parts) would be completely identical with the exact curve (Fig. 14*b*). This procedure, if it could be carried out, would have the advantage of giving exact thermodynamic properties as well as providing the necessary framework for an exact theory of the properties of metastable states.

If N is finite but large and the exact $Q(N,V,T)$ is used, and if we define p' by

$$p' = \mathbf{k}T\left(\frac{\partial \ln Q}{\partial V}\right)_{N,T} \tag{28.3}$$

then p' has no singularities (the corners in Fig. 14*b* are rounded off) and the horizontal region in Fig. 14*b* is not quite flat (see Appendix 9 for the rather complicated details). In this case p' is a function of v and T and also depends, in a thermodynamically unimportant way for large N, on N (or V).

If the grand partition function is used with a rigorous $Q(N,V,T)$ and V is finite but large, and if we define p' by

$$p' = \frac{\mathbf{k}T}{V} \ln \Xi \tag{28.4}$$

then p' will be a decreasing function of v, in Fig. 14*b* the corners will be rounded and the slope of the "flat" region will be slightly negative, and p' will be a function of v and T and will also depend (slightly) on V (or $\bar{N}$). These conclusions are reached from the following argument. First, consider the sum

$$\Xi(V,T,\lambda) = \sum_{N \geqslant 0} Q(N,V,T)\lambda^N \tag{28.5}$$

For real molecules and for given V and T, after N reaches a value in the sum corresponding to tight packing with repulsive forces becoming dominant, succeeding values of N will make the potential energy U in Eq. (22.5) rapidly approach $+\infty$. Thus for large enough N, Q will be zero; therefore, Ξ is, for all practical purposes, a polynomial. Then $\ln \Xi$ is a well-behaved function of λ, V, and T without singularities.[1] No sharp corners or discontinuities are possible in thermodynamic functions derived from Ξ so long as V is finite. Singularities are associated with the limit $V \to \infty$. Second, from Eq. (28.5) we derive easily, as in Chap. 4,

$$\bar{N}(V,T,\lambda) = \lambda \left(\frac{\partial \ln \Xi}{\partial \lambda}\right)_{V,T} \tag{28.6}$$

and

$$\left(\frac{\partial \bar{N}}{\partial \lambda}\right)_{V,T} = \frac{\overline{N^2} - (\bar{N})^2}{\lambda} \tag{28.7}$$

Then

$$\left(\frac{\partial \ln \Xi}{\partial \bar{N}}\right)_{V,T} = \left(\frac{\partial \ln \Xi}{\partial \lambda}\right)_{V,T} \left(\frac{\partial \lambda}{\partial \bar{N}}\right)_{V,T}$$

$$= \frac{\bar{N}}{\overline{N^2} - (\bar{N})^2} \tag{28.8}$$

But, writing $v = V/\bar{N}$ and using the definition of p' in Eq. (28.4),

$$\left(\frac{\partial \ln \Xi}{\partial \bar{N}}\right)_{V,T} = -\frac{V^2}{(\bar{N})^2\mathbf{k}T}\left(\frac{\partial p'}{\partial v}\right)_{V,T} \tag{28.9}$$

Then, finally, on combining Eqs. (28.8) and (28.9), we get

$$\left(\frac{\partial p'}{\partial v}\right)_{V,T} = -\frac{\mathbf{k}T}{V^2}\frac{(\bar{N})^3}{\overline{N^2} - (\bar{N})^2} \tag{28.10}$$

[1] Q and λ are necessarily positive; therefore $\Xi > 0$.

This equation is similar to Eq. (19.39) but a nonthermodynamic derivation is necessary here. Equation (28.10) shows that $\partial p'/\partial v$ cannot be positive.

The thermodynamic pressure is defined as

$$p(v,T) = \mathbf{k}T \lim_{V\to\infty} \frac{1}{V} \ln \Xi \tag{28.11}$$

Equation (28.11) will lead to Fig. 14*b* with a flat horizontal portion, and with two singularities and three analytical parts in $p(v)$.

It should be emphasized that the properties of the grand partition function described in the above paragraphs, based on a rigorous Q, also hold for an approximate Q since the argument makes no use of the precise form of Q. Thus, even if a given approximate Q leads to a loop (Fig. 14*c*) using the canonical ensemble, the same Q in the grand partition function leads to a $p(v)$ curve without a loop[1] (Fig. 14*b*).

Mayer's Theory of Condensation. For our purposes here, let us accept Eqs. (26.5), (26.6), and (26.10), for N large but finite. Since N is finite, the "point" of condensation will not be a singular point; therefore we assume that these equations are valid not only in the gas region but also in at least part of the two-phase region—so long as b_j is independent of V for all j, $1 \leqslant j \leqslant N$.

The argument is very similar to that already given for the Frenkel-Band theory.[2] We want to examine the behavior of m_j^* for large j and thus locate the condensation point. This requires a consideration of b_j for large j.

When j is large, we assume that $\ln b_j$ is equal to the logarithm of the largest term in Eq. (24.14), or the method of steepest descents can be used. Let

$$t' = \frac{1}{j^2} \prod_{k=1}^{j-1} \frac{(j\beta_k)^{n_k}}{n_k!} \tag{28.12}$$

and

$$\ln t' = \sum_{k=1}^{j-1} (n_k \ln j\beta_k - n_k \ln n_k + n_k) - 2 \ln j \tag{28.13}$$

The n_k are restricted by

$$\sum_{k=1}^{j-1} kn_k = j - 1 \tag{28.14}$$

Using the method of undetermined multipliers just as in Eq. (26.4), we find

$$n_k^* = j\beta_k\gamma^k \tag{28.15}$$

where $\gamma = e^{-\alpha}$ (α is the undetermined multiplier) is determined by

$$\sum_{k=1}^{j-1} k\beta_k\gamma^k = 1 \tag{28.16}$$

[1] See, for example, T. L. Hill, *J. Phys. Chem.*, **57**, 324 (1953), and Appendix 9.

[2] Chronologically, Mayer's theory came first.

We have neglected the difference between j and $j - 1$ in writing Eq. (28.16). This gives, from Eq. (28.13),

$$\ln b_j = \ln t'_{\text{max}} = j\left(\sum_{k=1}^{j-1} \beta_k \gamma^k - \ln \gamma\right) \tag{28.17}$$

where the negligible term in $\ln j$ has been dropped. For large j, $\ln b_j$ is therefore

$$\ln b_j = j \ln b_0(T) \tag{28.18}$$

$$\ln b_0(T) = \sum_{k=1}^{\infty} \beta_k \gamma^k - \ln \gamma \tag{28.19}$$

But b_j itself is[1]

$$b_j = C(j,\boldsymbol{\beta}) b_0{}^j \tag{28.20}$$

where C depends on j and the β_k's and $\ln C$ is of negligible order in Eq. (28.18).

Then, again for large j, from Eq. (26.5),

$$m_j^* = VC(b_0 z)^j \tag{28.21}$$

For small enough z so that $b_0 z < 1$, m_j^* is zero since j is very large (i.e., large clusters are not present in thermodynamically important amounts). But, for z such that $b_0 z > 1$, m_j^* is extremely large. Therefore, as z increases and passes through the value $b_0{}^{-1}$, a very sudden formation of large clusters takes place. This is the condensation point. The value of v, $v = v_s$, at which this takes place is [Eq. (26.6)]

$$\frac{1}{v_s} = \sum_{j=1}^{N} j b_j b_0{}^{-j} \tag{28.22}$$

For large j, the general term $jb_j z^j$ in

$$\frac{1}{v} = \sum_{j=1}^{N} j b_j z^j \tag{28.23}$$

can be written $jC(b_0 z)^j$. If z is increased only very slightly above the value $b_0{}^{-1}$, these terms for large j become quite large, making the right-hand side of Eq. (28.23) large and therefore v small. In other words, v is *extremely* sensitive to z for $z > b_0{}^{-1}$; v decreases rapidly while z stays virtually constant,[2] as in Fig. 15. The horizontal portion in Fig. 15 is the two-phase region. The argument breaks down before v gets very small, because the b_j's become volume dependent.

The pressure in Eq. (26.10) is a function of T and z (and depends also on N in a thermodynamically negligible way for N large). Since T is constant

[1] See the footnote to Eq. (26.34).

[2] Since N is finite, z is not quite constant and the corner in Fig. 15 is not quite sharp.

and z virtually so in the two-phase region of Fig. 15, p is also virtually constant in this region. Thus a plot of p versus v has the same qualitative appearance as Fig. 15 (replacing z by p and b_0^{-1} by $\mathbf{k}T \sum_{k=1}^{N} b_j b_0^{-j}$).

The above argument was put on a more rigorous basis by Kahn and Uhlenbeck and Born and Fuchs,[1] using complex variable theory and considering strictly the limit function in Eq. (28.1) and the resulting singularities in the various properties. As already pointed out at the beginning of this chapter, there is some doubt about the validity of the conclusions reached in this way (on condensation) because of the double limiting process involved [see, for example, Eq. (23.17)]. In any case, the liquid state is not included; therefore instead of pursuing this approach further we shall turn below to the alternative method of Yang and Lee.

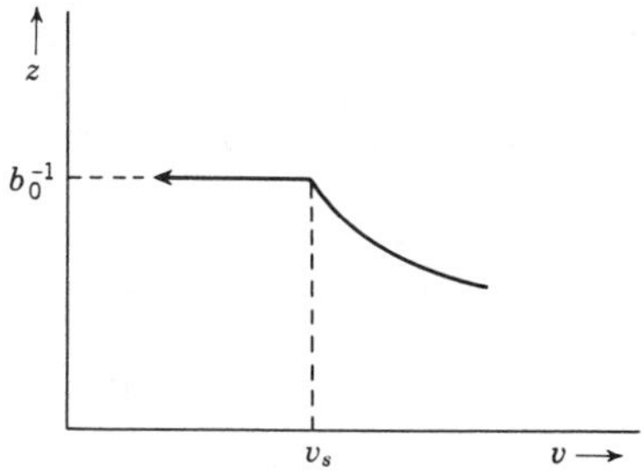

FIG. 15. Activity z versus volume per molecule v in the gas-liquid condensation region.

A further point of controversy and of considerable mathematical complexity, which should be mentioned in this context, concerns the nature of the critical point. There is considerable doubt, both theoretically and experimentally, whether there is one critical point (as in the classical van der Waals picture) or two critical points. Minute amounts of impurities and gravitational effects are sufficient to make the experimental picture somewhat unclear, and the theoretical problem is far from settled.[2]

Yang-Lee Theory of Condensation.[3] The formalism of Ursell and Mayer (Sec. 22) is developed in a straightforward way, with the intermolecular force between pairs of molecules occupying an obvious and important role from the outset. However, this approach has the disadvantage that the theory must be carried through mathematically complicated and sophisticated stages before condensation and the liquid state can be included in a rigorous way. In fact, this part of Mayer's theory has not yet been completed. On the other hand, Yang and Lee have suggested an alternative formalism in terms of which a rigorous mathematical discussion of condensation and the liquid state can be included as easily as a discussion of the gas phase. But the

[1] M. Born and K. Fuchs, *Proc. Roy. Soc. (London)*, **A166**, 391 (1938); B. Kahn and G. E. Uhlenbeck, *Physica*, **5**, 399 (1938).

[2] See, for example, B. H. Zimm, *J. Chem. Phys.*, **19**, 1019 (1951); J. E. Mayer, *J. Chem. Phys.*, **19**, 1024 (1951); W. G. Schneider, *Compt. rend. réunion ann. union intern. phys.* (Paris, 1952), p. 69.

[3] C. N. Yang and T. D. Lee, *Phys. Rev.*, **87**, 404, 410 (1952).

connection with intermolecular forces becomes very remote in this theory. In fact, the theory is sufficiently general that the form of the configuration integral is never made use of.[1] A detailed application of the theory is given in Chap. 7.

We shall introduce at an appropriate point an electrostatic analogy which is very helpful in understanding in physical terms the *mathematical* behavior of the thermodynamic functions, but this (electrostatic) physical analogy is related in no direct way to the actual physical problem under consideration (gas condensation).

The details of the proofs given by Yang and Lee actually depend on the assumptions:[2] (1) that the molecules interact in pairs only [Eq. (22.6)]; (2) that $u(r) = +\infty$ for $r \leqslant a$; and (3) that $u(r) = 0$ for $r \geqslant b$. But assumptions 1 and 3 can be broadened to include many body forces and a weak long tail in $u(r)$ as in van der Waals' forces. Also, assumption 2 is not a real restriction since a can be chosen as small as is necessary to make the thermodynamic results of interest independent of the choice of a.

a. General theory. The grand partition function is [Eq. (23.2)]

$$\Xi(T,V,z) = \sum_{N=0}^{B} \frac{Z(N,V,T)}{N!} z^N \tag{28.24}$$

where B is the maximum possible value of N. B is proportional to V and depends on the choice of a: $V = \alpha(a)B$, where α is a constant. Equation (28.24) is a polynomial of degree B in z, with real and positive coefficients. Therefore none of the zeros of Ξ (that is, roots of $\Xi = 0$) is real and positive, and if z_k is a zero of Ξ and is complex, then the complex conjugate of z_k, z_k^*, is also a zero.

Equation (28.24) can be rewritten in terms of the roots z_k of $\Xi = 0$ as follows:

$$\begin{aligned}\Xi(T,V,z) &= \frac{Z(B)}{B!}(z - z_1)(z - z_2)\cdots(z - z_B)\\ &= \frac{Z(B)}{B!}\left(\frac{z}{z_1} - 1\right)\left(\frac{z}{z_2} - 1\right)\cdots\left(\frac{z}{z_B} - 1\right) z_1 z_2 \cdots z_B\\ &= \prod_{k=1}^{B}\left(1 - \frac{z}{z_k}\right)\end{aligned} \tag{28.25}$$

where $z_k = z_k(V,T)$. Then

$$\ln \Xi = \sum_{k=1}^{B} \ln\left(1 - \frac{z}{z_k}\right) \tag{28.26}$$

The thermodynamic functions are thus expressible in terms of z and the

[1] It applies, for example, to approximate theories as well as rigorous ones.

[2] See also L. Witten, *Phys. Rev.*, **93**, 1131 (1954).

distribution of zeros $z_k(V,T)$ in the complex z plane. Incidentally, this distribution is symmetrical about the real axis, as has already been implied. Strictly, we are interested in the limit $V \to \infty$, and

$$\frac{p(T,z)}{\mathbf{k}T} = \lim_{V\to\infty} \frac{1}{V} \ln \Xi \tag{28.27}$$

$$\rho(T,z) = \lim_{V\to\infty} \frac{1}{V} \left(\frac{\partial \ln \Xi}{\partial \ln z}\right)_{V,T} = \lim_{V\to\infty} \left(\frac{\partial (1/V) \ln \Xi}{\partial \ln z}\right)_{V,T} \tag{28.28}$$

Of course, as $V \to \infty$, the number of zeros of Ξ increases indefinitely so that a density distribution of zeros in the complex plane may be introduced, if convenient, in this limit.

Although only zero and real and positive values of z are of physical interest, in order to discuss the analytical behavior of the thermodynamic functions on the positive real axis it is necessary to consider neighborhoods in the complex z plane of points on the positive real axis. We shall therefore understand that for mathematical purposes z in Eqs. (28.24) to (28.28) may take on complex values.

Since the coefficients in Eq. (28.24) are real and positive, Ξ increases monotonically with z (real and positive) for any V. Then $\ln \Xi$ and $(1/V) \ln \Xi$ also increase monotonically with z (real and positive) for any V. Yang and Lee prove that

$$\lim_{V\to\infty} \frac{1}{V} \ln \Xi \tag{28.29}$$

exists for positive real z and that the limit is independent of the shape of V. It therefore follows that this limit increases monotonically with z (real and positive). Furthermore, the limit is a continuous function of z (real and positive). For with V finite, on differentiating Ξ in Eq. (28.24) with respect to z,

$$\frac{\bar{N}}{V} = \left(\frac{\partial (1/V) \ln \Xi}{\partial \ln z}\right)_{V,T} \tag{28.30}$$

But necessarily $\bar{N} \leqslant B$ and $\bar{N}/V \leqslant B/V = 1/\alpha =$ constant. This implies that as we let V get larger and larger, at no point z (real and positive) does the slope $\partial(1/V) \ln \Xi/\partial z$ increase indefinitely, as it would if there were a discontinuity in $(1/V) \ln \Xi$ at some z in the limit $V \to \infty$. Instead

$$z \left(\frac{\partial (1/V) \ln \Xi}{\partial z}\right)_{V,T} \leqslant \frac{1}{\alpha} \tag{28.31}$$

for all V. We may therefore conclude that $p/\mathbf{k}T$ [Eq. (28.27)] is a continuous, monotonically increasing function of the activity z.

For finite V, the polynomial $\Xi(z)$, Eq. (28.24), is an analytic function of z for all z in the complex plane. The function $(1/V) \ln \Xi$ is also an analytic function of z except at the zeros of Ξ, none of which can be on the positive real axis. Therefore $(1/V) \ln \Xi$ is analytic at every point on the positive real axis. Since the derivative of an analytic function is also analytic,

$$\frac{\partial}{\partial \ln z}\left(\frac{1}{V} \ln \Xi\right), \frac{\partial^2}{\partial \ln z^2}\left(\frac{1}{V} \ln \Xi\right), \text{ etc.} \tag{28.32}$$

are also analytic everywhere on the positive real axis. Thus no singularities in any of these quantities can arise so long as V is finite (z real and positive).

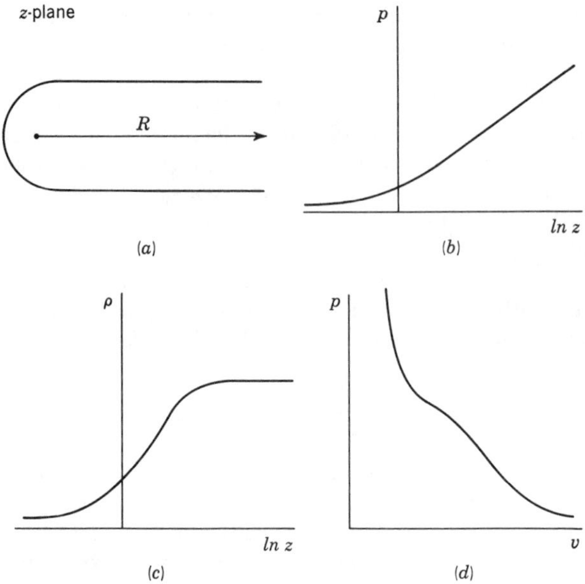

FIG. 16. Relation between pressure, activity, and density above the critical temperature of a gas.

Consider a region R of the complex plane containing a segment of the positive real axis, and such that R remains free of zeros of Ξ as $V \to \infty$. Then, for large V, $(1/V) \ln \Xi$ and the derivatives (28.32) are analytic functions of z in R. Yang and Lee then prove that the limits

$$\lim_{V\to\infty} \frac{1}{V} \ln \Xi, \lim_{V\to\infty} \frac{\partial}{\partial \ln z}\left(\frac{1}{V} \ln \Xi\right), \text{ etc.} \tag{28.33}$$

are also analytic functions of z in R and that the operations $\partial/\partial \ln z$ and $V \to \infty$ commute, so that, for example,

$$\lim_{V\to\infty} \frac{\partial}{\partial \ln z}\left(\frac{1}{V} \ln \Xi\right) = \frac{\partial}{\partial \ln z}\left(\lim_{V\to\infty} \frac{1}{V} \ln \Xi\right) \tag{28.34}$$

We obtain then, from Eqs. (28.28) and (28.34), the thermodynamic equation

$$\rho = \frac{\partial}{\partial \ln z}\left(\frac{p}{\mathbf{k}T}\right) \tag{28.35}$$

for positive real z in R.

Since we have seen that no zero of Ξ can be on the positive real axis (V finite), we might expect offhand that some region R free of zeros of Ξ

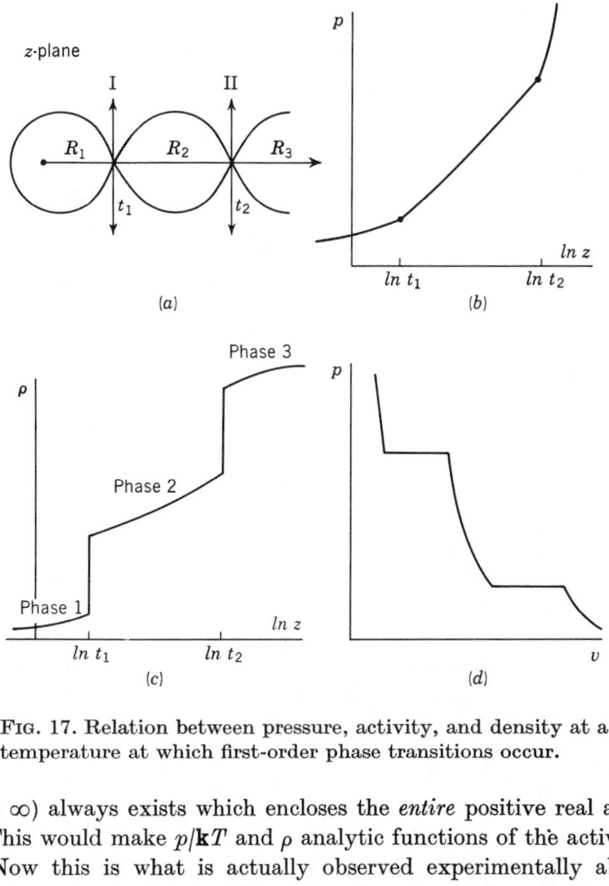

FIG. 17. Relation between pressure, activity, and density at a temperature at which first-order phase transitions occur.

(as $V \to \infty$) always exists which encloses the *entire* positive real axis (Fig. 16*a*). This would make $p/\mathbf{k}T$ and ρ analytic functions of the activity z for all z. Now this is what is actually observed experimentally above the critical temperature of a gas (Fig. 16*b* and *c*), but such analytic functions could not account, for example, for singularities of the type shown in Fig. 17*b* and *c* associated with first-order phase changes. Clearly, the resolution of this difficulty is that although no zero of Ξ can be on the positive real axis for finite V, as $V \to \infty$ zeros of Ξ can "close in" on the positive real axis in such a way that for certain real positive values of z, say $z = t_1$ and t_2 (Fig. 17*a*, *b*, and *c*), every *neighborhood*, however small, of the points t_1 and t_2 contains zeros of Ξ. In this case, $\lim_{V\to\infty} (1/V) \ln \Xi$ is not an analytic function of z at t_1 and t_2. (In Fig. 17*a*, the regions R_1, R_2, and R_3 are supposed free of zeros of Ξ in the limit $V \to \infty$.) We shall see below that the type of density distribution of zeros in the limit $V \to \infty$ which leads to a first-order phase change is a *linear* distribution in the z plane (necessarily symmetric about the real axis) with nonzero (linear) density at the singular points on the positive real axis. For example, *schematically*, we might have linear distributions of zeros on the lines I and II in Fig. 17*a* (as well as possibly having zeros elsewhere in the z plane). Or we shall see in the example of Chap. 7 that the zeros are all on a circle in the complex plane with center at $z = 0$. The intersection of this circle with the positive real axis locates the singular point $z = t_1$.

It should be pointed out that although the Yang-Lee theory shows the type of distribution of zeros of Ξ in the complex plane necessary to lead to a first-order phase change, proof that the partition function in Eq. (22.4) for a system of interacting particles actually leads to such a distribution would be very difficult and has not been achieved as yet. Some information is available, however, about the distribution of zeros in the two-dimensional Ising problem (Chap. 7).

There is in general a different distribution of zeros for each temperature T. If below $T = T_c$ the zeros close in, as $V \to \infty$, on the positive real axis at, say, $z = t_1(T)$, but do not close in above $T = T_c$, then T_c is the critical temperature for the phase change at $z = t_1$. If on the other hand, the points $t_1(T)$ and $t_2(T)$ approach each other as T changes and merge at $T = T_t$, then T_t is the triple-point temperature.

A few further remarks about Figs. 16 and 17 may be worthwhile. Equation (28.7) shows that ρ is an increasing function of z (for any V, including $V \to \infty$). The discontinuities in ρ (Fig. 17*c*) at $z = t_1$ and $z = t_2$ are of course associated with the discontinuities in slope in the $p/\mathbf{k}T$ versus $\ln z$ curve (Fig. 17*b*), as required by Eq. (28.35). From the relation

$$\left(\frac{\partial(p/\mathbf{k}T)}{\partial v}\right)_T = \left(\frac{\partial(p/\mathbf{k}T)}{\partial \ln z}\right)_T \left(\frac{\partial \ln z}{\partial 1/\rho}\right)_T = -\frac{\rho^3}{(\partial\rho/\partial \ln z)_T} \tag{28.36}$$

where $v = 1/\rho$, we see that $p/\mathbf{k}T$ is a decreasing function of v and that this curve is flat (Fig. 17*d*) where the ρ versus $\ln z$ curve is vertical.

The Yang-Lee theory is not restricted to first-order phase changes. Some discussion of the higher-order problem is included in the original papers.

b. Electrostatic Analogue. The manner in which the functional dependence of $p/\mathbf{k}T$ and ρ on the activity z is determined by the distribution of zeros in the complex plane can be understood in physical terms through an electrostatic analogy. Each zero z_k of Ξ is represented in this analogy by a uniform line charge perpendicular to the z plane and passing through it at the point z_k. It can be shown that the pressure and density of the gas at an activity z are simply related to the total (due to all zeros) electric potential and electric field at z on the positive real axis in the electrostatic analogue.

It will be recalled that at a distance r from a uniformly charged line with charge per unit length σ, the magnitude of the electric field is $E = 2\sigma/r$ and the electric potential is given by $\varphi = -2\sigma \ln r$, if the zero of potential is chosen for convenience at $r = 1$. Now suppose a perpendicular line charge

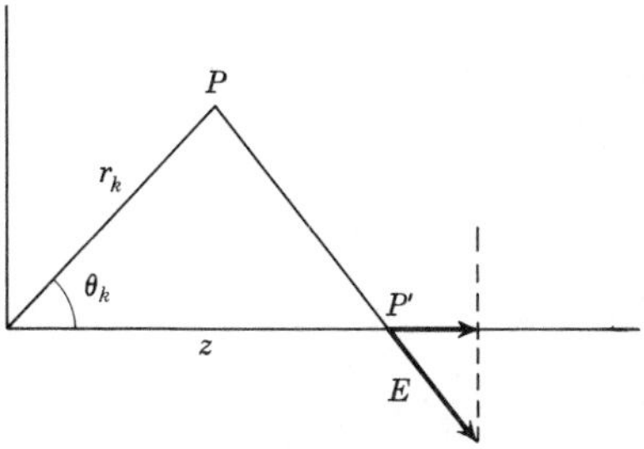

FIG. 18. Electric field at P' due to perpendicular line charge at P.

σ is placed in a plane at a point P (Fig. 18) with coordinates (r_k,θ_k). The potential due to this charge at a point P' with coordinates $(z,0)$ is then

$$\varphi = -2\sigma \ln (z^2 - 2zr_k \cos \theta_k + r_k^2)^{1/2} \tag{28.37}$$

and the component of the electric field at P' along the line $\theta = 0$ is

$$E_{\theta=0} = \frac{2\sigma(z - r_k \cos \theta_k)}{z^2 - 2zr_k \cos \theta_k + r_k^2} \tag{28.38}$$

Let us put a line charge $\sigma = 1/V$ at the location $z_k = r_k e^{i\theta_k}$ of every zero of Ξ in the complex plane (which now becomes the physical plane of Fig. 18). The *total* linear charge density is then $B/V = 1/\alpha =$ constant (independent of V). Because the zeros are distributed symmetrically about the real axis, the total electric field is directed along the real axis. Then according to Eqs. (28.37) and (28.38), the total potential and field at z (real and positive) are given by

$$\varphi(z) = -\frac{2}{V} \sum_{k=1}^{B} \ln (z^2 - 2zr_k \cos \theta_k + r_k^2)^{1/2} \tag{28.39}$$

and

$$E(z) = \frac{2}{V} \sum_{k=1}^{B} \frac{z - r_k \cos \theta_k}{z^2 - 2zr_k \cos \theta_k + r_k^2} \tag{28.40}$$

Also $$\varphi(0) = -\frac{2}{V}\sum_{k=1}^{B} \ln r_k \tag{28.41}$$

Consider any pair of terms in Eq. (28.26) associated with a zero z_k and its complex conjugate z_k^*. The sum of these two terms is

$$\ln\left(1 - \frac{z}{r_k e^{i\theta_k}}\right) + \ln\left(1 - \frac{z}{r_k e^{-i\theta_k}}\right)$$
$$= 2 \ln \frac{(z^2 - 2zr_k \cos\theta_k + r_k^2)^{1/2}}{r_k} \tag{28.42}$$

so that the contribution of each of the two terms to $\ln \Xi$ is, in effect, one half of this result. We can therefore rewrite Eq. (28.26), including any zeros on the negative real axis, in the form

$$\frac{1}{V}\ln \Xi = \frac{1}{V}\sum_{k=1}^{B} \ln \frac{(z^2 - 2zr_k\cos\theta_k + r_k^2)^{1/2}}{r_k} \tag{28.43}$$

Differentiating Eq. (28.43) with respect to z, we have in addition

$$z\frac{\partial(1/V)\ln \Xi}{\partial z} = \frac{z}{V}\sum_{k=1}^{B} \frac{z - r_k\cos\theta_k}{z^2 - 2zr_k\cos\theta_k + r_k^2} \tag{28.44}$$

On comparing Eqs. (28.39) to (28.41) with Eqs. (28.43) and (28.44), we observe that

$$-2\left(\frac{1}{V}\ln \Xi\right) = \varphi(z) - \varphi(0) \tag{28.45}$$

and $$\frac{2}{z}\left(z\frac{\partial(1/V)\ln \Xi}{\partial z}\right) = E(z) \tag{28.46}$$

So finally, in the limit $V \to \infty$ (*total* charge = constant),

$$-\frac{2p}{\mathbf{k}T} = \varphi(z) - \varphi(0) \tag{28.47}$$

$$\frac{2\rho}{z} = E(z) \tag{28.48}$$

With the aid of Eqs. (28.47) and (28.48), an acquaintance with electrostatic problems, and a given limiting distribution of zeros in the complex plane, it is easy to visualize the behavior of $p/\mathbf{k}T$ and ρ on the positive real axis. Chapter 7 will furnish a specific example. But we make one general remark here. Suppose, as $V \to \infty$, the zeros of Ξ close in on the positive real axis in such a way as to give a continuous linear distribution of zeros with non-zero density at, say, $z = t_1$ on the positive real axis (e.g., Fig. 17*a*). Since each zero corresponds to a uniform line charge perpendicular to the complex plane, the limiting charge distribution is a two-dimensional sheet of charge, perpendicular to the complex plane, which intersects the positive real axis at $z = t_1$. Suppose the surface charge density at $z = t_1$ is $\sigma' > 0$. Then, from elementary electrostatic theory, as we move along the positive real axis through the sheet of charge at $z = t_1$, the potential (and therefore $p/\mathbf{k}T$) is continuous but has a discontinuity in slope, and the electric field E (and therefore ρ) has a discontinuity such that

$$E(z = t_1+) - E(z = t_1-) = 4\pi\sigma' \tag{28.49}$$

This is precisely the qualitative behavior associated with a first-order phase change at the activity $z = t_1$ (Fig. 17).

c. Relation to Cluster Integrals. Consider the limiting ($V \to \infty$) distribution of zeros of Ξ in the complex plane. Suppose the nearest of these zeros to the origin is at a distance r_α from the origin. Draw a circle C with radius just less than r_α and with center at the origin. Then, for any V large enough so that all the zeros of Ξ (for this V) lie outside of C, each term of Eq. (28.26) can be expanded as

$$\ln\left(1 - \frac{z}{z_k}\right) = -\left[\frac{z}{z_k} + \frac{1}{2}\left(\frac{z}{z_k}\right)^2 + \frac{1}{3}\left(\frac{z}{z_k}\right)^3 + \cdots\right] \tag{28.50}$$

for z within C. Then Eq. (28.26) becomes

$$\frac{1}{V}\ln \Xi = \sum_{j\geqslant 1} b_j(V,T)z^j \tag{28.51}$$

where

$$b_j(V,T) = -\frac{1}{Vj}\sum_{k=1}^{B}\left[\frac{1}{z_k(V,T)}\right]^j \tag{28.52}$$

From Eqs. (28.27) and (28.51) we have, then,

$$\frac{p}{\mathbf{k}T} = \lim_{V\to\infty}\sum_{j\geqslant 1} b_j(V,T)z^j \tag{28.53}$$

This has the same formal appearance as Eq. (23.17); therefore we may identify the b_j defined in Eq. (28.52) with Mayer's cluster integrals already discussed. Equation (28.52) shows how the cluster integrals are related mathematically to the zeros of Ξ.

Making use of Eq. (28.52), the series in Eq. (28.51) is easily shown to converge uniformly within C so that it is permissible in this region to interchange the limiting processes in Eq. (28.53), giving

$$\frac{p}{\mathbf{k}T} = \sum_{j\geqslant 1} b_j(T)z^j \tag{28.54}$$

where

$$b_j(T) = \lim_{V\to\infty} b_j(V,T) \tag{28.55}$$

Equation (28.54) is the same as Eq. (23.19).

Equation (28.54) is valid within C and gives the pressure for z (real and positive) $< r_\alpha$. If it happens[1] that in the limiting distribution no zero of Ξ is closer to the origin than $z = t_1$ (Fig. 17), that is, if $r_\alpha = t_1$, then Eq. (28.54) is valid for $z < t_1$. If this is not the case, $p/\mathbf{k}T$ will be given for $r_\alpha \leqslant z < t_1$ by Eqs. (28.26) and (28.27) or by the analytic continuation of the right-hand side of Eq. (28.54). If, as seems probable,[2] there is in the limit $V \to \infty$ a continuous and closed ring of singularities of $\ln \Xi$ (zeros of Ξ) passing through $z = t_1$, the right-hand side of Eq. (28.54) cannot be analytically continued, even in principle, around the singularity at $z = t_1$ and back onto the positive real axis where $z > t_1$ (liquid state).

GENERAL REFERENCES

Beattie, J. A., and W. H. Stockmayer, "A Treatise on Physical Chemistry" (Van Nostrand, New York, 1951), edited by H. S. Taylor and S. Glasstone.

DeBoer, J., *Repts. Progr. Phys.*, **12**, 305 (1949).

Fowler, R. H., and E. A. Guggenheim, "Statistical Thermodynamics" (Cambridge, London, 1939).

Green, H. S., "Molecular Theory of Fluids" (North-Holland, Amsterdam, 1952).

Hirschfelder, J. O., C. F. Curtiss, and R. B. Bird, "Molecular Theory of Gases and Liquids" (Wiley, New York, 1954).

Kahn, B., Dissertation (Utrecht, 1938).

Mayer, J. E., and M. G. Mayer, "Statistical Mechanics" (Wiley, New York, 1940).

Rushbrooke, G. S., "Introduction to Statistical Mechanics" (Oxford, London, 1949).

Zimm, B. H., R. A. Oriani, and J. D. Hoffman, *Ann. Rev. Phys. Chem.*, **4**, 207 (1953).

---

[1] As is the case in the two-dimensional Ising problem (Chap. 7).

[2] For example, a distribution of zeros of Ξ around the circle $|z| = t_1$ (as in Chap. 7).

CHAPTER 6

DISTRIBUTION FUNCTIONS AND THE THEORY OF THE LIQUID STATE

In this chapter we again study a one-component, classical, monatomic fluid (gas or liquid) with a total potential energy given by the sum of pair interactions. Our primary object is to develop a third method (in addition to the methods of Mayer and Yang and Lee already considered in Chap. 5) of attacking this problem, the method of distribution functions.

The method of distribution functions can be extended to polyatomic molecules, the solid-state, multicomponent systems, quantum fluids, more general potential energy expressions, etc. A few remarks concerning some of these generalizations[1] will be made during the course of this chapter, and the last section (Sec. 40) will be devoted to a classical, polyatomic, multicomponent system.

The first part (A; Secs. 29 to 36) of this chapter is concerned with the theory of distribution functions in the canonical ensemble, following largely the work of Kirkwood, Yvon, Born, and Green.[2] The second part (B; Secs. 37 to 40) consists of the extension of the theory to the grand canonical ensemble, a development due primarily to Mayer[3] and his collaborators. A discussion of the new integral equations of Kirkwood and Salsburg is included (Sec. 38) in this second part of the chapter.

[1] For details, see, for example, J. G. Kirkwood and J. E. Mayer in "Phase Transformations in Solids" (Wiley, New York, 1951), edited by R. Smoluchowski, J. E. Mayer, and W. A. Weyl; W. G. McMillan and J. E. Mayer, *J. Chem. Phys.*, **13**, 276 (1945); J. E. Mayer, *J. Chem. Phys.*, **10**, 629 (1942); DeBoer (Gen. Ref.).

[2] J. G. Kirkwood, *J. Chem. Phys.*, **3**, 300 (1935); J. G. Kirkwood, *Chem. Rev.*, **19**, 275 (1936); Yvon (Gen. Ref.); Born and Green (Gen. Ref.).

[3] J. E. Mayer and E. Montroll, *J. Chem. Phys.*, **9**, 2 (1941); J. E. Mayer, *J. Chem. Phys.*, **10**, 629 (1942); W. G. McMillan and J. E. Mayer, *J. Chem. Phys.*, **13**, 276 (1945).

A. CANONICAL ENSEMBLE

29. Definition of Distribution and Correlation Functions

We consider a classical system of N monatomic molecules in a volume V at temperature T. In the configuration $\mathbf{r}_1, \ldots, \mathbf{r}_N$ the potential energy of the system is given by Eq. (22.6). Equations (22.4) and (22.5) give the partition function Q and configuration integral Z.

The distribution functions (defined below) express the probabilities of observing different configurations of sets of n molecules out of the total number N. These functions are useful in statistical mechanics primarily because all the thermodynamic functions of the system can be expressed in terms of the relatively simple distribution functions[1] for $n = 1$ and $n = 2$. Unfortunately, we cannot restrict ourselves entirely to the cases $n = 1$ and $n = 2$ since it turns out, as we shall see, that the theoretical calculation of the $n = 2$ distribution functions involves the $n = 3$ functions, etc.

In order to express the free energy of the system in the most convenient way (Sec. 30), it is necessary to consider here a somewhat more general potential energy expression than Eq. (22.6). That is, we write

$$U(\mathbf{r}_1, \ldots, \mathbf{r}_N, \boldsymbol{\xi}) = \sum_{1 \leqslant i < j \leqslant N} \xi_i \xi_j u(r_{ij}) \tag{29.1}$$

where[2] $\boldsymbol{\xi}$ represents $\xi_1, \ldots, \xi_N$ and the ξ's are so-called coupling parameters, each of which ranges from zero to unity. The real fluid has $\xi_i = 1$ for all i (full coupling of the intermolecular forces):

$$U(\mathbf{r}_1, \ldots, \mathbf{r}_N, \mathbf{1}) = U(\mathbf{r}_1, \ldots, \mathbf{r}_N) \tag{29.2}$$

At the other extreme, in the hypothetical state with $\xi_i = 0$ for all i, $U(\mathbf{r}_1, \ldots, \mathbf{r}_N, \mathbf{0}) = 0$. The point of introducing coupling parameters is to permit the calculation of the free energy of the system by a method which is completely analogous to the familiar charging process employed in the derivation of expressions for the free energy of, for example, an electrolyte solution. As Onsager[3] has pointed out, there is no reason why a "charging process" need be restricted to a coulombic potential, and in fact such a process turns out to be very useful in the present theory. The "discharged" state for the ith molecule obviously corresponds to $\xi_i = 0$, while the fully "charged" state corresponds to $\xi_i = 1$.

[1] If the potential energy of the system involved triplet interactions as well as pair interactions, we would need for this purpose the $n = 1$, 2, and 3 distribution functions, etc.

[2] We shall use $\boldsymbol{\xi}$ for the set $\xi_1, \ldots, \xi_N$. Later, we shall use the symbol ξ when $\xi_2 = \cdots = \xi_N = 1$, where $\xi = \xi_1$. Whenever a function $f(x,\boldsymbol{\xi})$ or $f(x,\xi)$ is written simply as $f(x)$, this is understood to mean $\xi_1 = \cdots = \xi_N = 1$.

[3] L. Onsager, *Chem. Rev.*, **13**, 73 (1933).

We are now in a position to define a set of distribution functions. Suppose we observe the configuration of the system of N particles with U given by Eq. (29.1). The probability that particle 1 will be in $d\mathbf{r}_1$ at $\mathbf{r}_1$, 2 in $d\mathbf{r}_2$ at $\mathbf{r}_2, \ldots$, and N in $d\mathbf{r}_N$ at $\mathbf{r}_N$ is[1]

$$\frac{\exp\left[-U(\xi)/\mathbf{k}T\right] d\mathbf{r}_1 \cdots d\mathbf{r}_N}{Z(\xi)} \tag{29.3}$$

where

$$U(\xi) = U(\mathbf{r}_1, \ldots, \mathbf{r}_N, \xi)$$

and

$$Z(\xi) = \int_V \cdots \int \exp\left[-\frac{U(\xi)}{\mathbf{k}T}\right] d\mathbf{r}_1 \cdots d\mathbf{r}_N \tag{29.4}$$

The probability $P^{(n)}(\mathbf{r}_1, \ldots, \mathbf{r}_n, \xi)\, d\mathbf{r}_1 \cdots d\mathbf{r}_n$ that 1 will be observed in $d\mathbf{r}_1$ at $\mathbf{r}_1$, 2 in $d\mathbf{r}_2$ at $\mathbf{r}_2, \ldots$, and n in $d\mathbf{r}_n$ at $\mathbf{r}_n$, irrespective of the configuration of the remaining $N - n$ molecules, is the sum of all probabilities (29.3) consistent with the specified configuration of molecules 1 to n. That is,

$$P^{(n)}(\mathbf{r}_1, \ldots, \mathbf{r}_n, \xi) = \frac{\int_V \cdots \int \exp\left[-U(\xi)/\mathbf{k}T\right] d\mathbf{r}_{n+1} \cdots d\mathbf{r}_N}{Z(\xi)} \tag{29.5}$$

and

$$\int_V \cdots \int P^{(n)}(\mathbf{r}_1, \ldots, \mathbf{r}_n, \xi)\, d\mathbf{r}_1 \cdots d\mathbf{r}_n = 1 \tag{29.6}$$

Of course $P^{(N)}(\mathbf{r}_1, \ldots, \mathbf{r}_N, \xi)\, d\mathbf{r}_1 \cdots d\mathbf{r}_N$ is just the expression (29.3).

Next, we define $\rho^{(n)}(\mathbf{r}_1, \ldots, \mathbf{r}_n, \xi)$ by

$$\rho^{(n)}(\mathbf{r}_1, \ldots, \mathbf{r}_n, \xi) = \frac{N!}{(N-n)!} P^{(n)}(\mathbf{r}_1, \ldots, \mathbf{r}_n, \xi) \tag{29.7}$$

$$= \frac{N!}{(N-n)!} \frac{\int_V \cdots \int \exp\left[-U(\xi)/\mathbf{k}T\right] d\mathbf{r}_{n+1} \cdots d\mathbf{r}_N}{Z(\xi)} \tag{29.8}$$

When $\xi = 1$, $\rho^{(n)}(\mathbf{r}_1, \ldots, \mathbf{r}_n)\, d\mathbf{r}_1 \cdots d\mathbf{r}_n$ is the probability that, if the configuration of the system of N molecules is observed, a molecule (not necessarily molecule 1) will be found in the element of volume $d\mathbf{r}_1$ at $\mathbf{r}_1$, a second in $d\mathbf{r}_2$ at $\mathbf{r}_2, \ldots$, and another in $d\mathbf{r}_n$ at $\mathbf{r}_n$. This follows because there are N choices for the molecule in $d\mathbf{r}_1$, $N - 1$ for $d\mathbf{r}_2, \ldots$, and $N - n + 1$ for $d\mathbf{r}_n$, or a total of

$$N(N-1) \cdots (N-n+1) = \frac{N!}{(N-n)!}$$

[1] In quantum mechanics, the Slater sum [Eqs. (16.69) to (16.71)] replaces $e^{-U/\mathbf{k}T}$ in Eqs. (29.3) and (29.4).

possibilities. When $\boldsymbol{\xi} \neq \mathbf{1}$, $\rho^{(n)}(\mathbf{r}_1, \ldots, \mathbf{r}_n, \boldsymbol{\xi})\, d\mathbf{r}_1 \cdots d\mathbf{r}_n$, as defined by Eq. (29.7), can still be given the above probability interpretation provided we regard ξ_i as being assigned to the *position*[1] $\mathbf{r}_i$ and therefore to whichever molecule happens to be at $\mathbf{r}_i$. This will always be understood below when $\rho^{(n)}(\mathbf{r}_1, \ldots, \mathbf{r}_n, \boldsymbol{\xi} \neq \mathbf{1})$ is referred to in probability terms.

We note that

$$\int_V \cdots \int \rho^{(n)}(\mathbf{r}_1, \ldots, \mathbf{r}_n, \boldsymbol{\xi})\, d\mathbf{r}_1 \cdots d\mathbf{r}_n = \frac{N!}{(N-n)!} \tag{29.9}$$

We shall call the functions $P^{(n)}$ and $\rho^{(n)}$ "distribution functions." The $\rho^{(n)}$ (often referred to as "generic" distribution functions) turn out to be more convenient for most (but not all) purposes than the $P^{(n)}$ (often referred to as "specific" distribution functions) because of the inclusion of the factor $N!/(N-n)!$.

The simplest distribution function $\rho^{(n)}$ is $\rho^{(1)}(\mathbf{r}_1,\boldsymbol{\xi})$. The quantity $\rho^{(1)}(\mathbf{r}_1,\boldsymbol{\xi})\, d\mathbf{r}_1$ is the probability that one of the molecules of the system will be found in the element of volume $d\mathbf{r}_1$ at $\mathbf{r}_1$ if the configuration of the system is observed. In a crystal $\rho^{(1)}(\mathbf{r}_1,\boldsymbol{\xi})$ is a periodic function of $\mathbf{r}_1$ with sharp maxima at lattice points, but in a fluid all points $\mathbf{r}_1$ inside of V are equivalent.[2] That is $\rho^{(1)}(\mathbf{r}_1,\boldsymbol{\xi})$ is independent of $\mathbf{r}_1$ and, from Eq. (29.9),

$$\frac{1}{V}\int_V \rho^{(1)}(\mathbf{r}_1,\boldsymbol{\xi})\, d\mathbf{r}_1 = \rho^{(1)} = \frac{N}{V} = \rho \qquad \text{(fluid)} \tag{29.10}$$

Thus, in a fluid, $\rho^{(1)}$ is just the macroscopic molecular density ρ. Another implication for a fluid is that

$$Z(\boldsymbol{\xi}) = \int_V \left[\int_V \cdots \int \exp\left[-\frac{U(\boldsymbol{\xi})}{\mathbf{k}T}\right] d\mathbf{r}_2 \cdots d\mathbf{r}_N \right] d\mathbf{r}_1 \tag{29.11}$$

$$= V \int_V \cdots \int \exp\left[-\frac{U(\boldsymbol{\xi})}{\mathbf{k}T}\right] d\mathbf{r}_2 \cdots d\mathbf{r}_N \qquad \text{(fluid)} \tag{29.12}$$

since the quantity in brackets in Eq. (29.11) is independent[2] of $\mathbf{r}_1$.

The quantity $\rho^{(2)}(\mathbf{r}_1,\mathbf{r}_2,\boldsymbol{\xi})\, d\mathbf{r}_1\, d\mathbf{r}_2$ is the probability that one molecule of the system will be found in $d\mathbf{r}_1$ at $\mathbf{r}_1$ and another in $d\mathbf{r}_2$ at $\mathbf{r}_2$, if the configuration of the system is observed. In a crystal, $\rho^{(2)}$ obviously depends in general on both $\mathbf{r}_1$ and $\mathbf{r}_2$, even if $r_{12} = |\mathbf{r}_{12}| = |\mathbf{r}_1 - \mathbf{r}_2|$ is very large. In a fluid, $\rho^{(2)}$ can depend only on r_{12} and $\boldsymbol{\xi}$:

$$\rho^{(2)}(\mathbf{r}_1,\mathbf{r}_2,\boldsymbol{\xi}) = \rho^{(2)}(r_{12},\boldsymbol{\xi}) \qquad \text{(fluid)} \tag{29.13}$$

[1] This point of view is particularly natural in an open system (see Sec. 37) since any given molecule does not necessarily stay in the system.

[2] Except for a region of negligible volume near the walls.

$$\iint_V \rho^{(2)}(r_{12},\xi)\, d\mathbf{r}_1\, d\mathbf{r}_2 = V\int_V \rho^{(2)}(r_{12},\xi)\, d\mathbf{r}_{12} = N(N-1) \qquad \text{(fluid)} \quad (29.14)$$

If the distribution of molecules is completely random,[1] the probability that molecule 1 is in $d\mathbf{r}_1$ at $\mathbf{r}_1$, 2 in $d\mathbf{r}_2$ at $\mathbf{r}_2$, . . . , and n in $d\mathbf{r}_n$ at $\mathbf{r}_n$ is

$$\frac{d\mathbf{r}_1}{V}\cdot\frac{d\mathbf{r}_2}{V}\cdots\frac{d\mathbf{r}_n}{V} = P^{(n)}\, d\mathbf{r}_1\cdots d\mathbf{r}_n \qquad (29.15)$$

Then

$$\rho^{(n)} = \frac{1}{V^n}\frac{N!}{(N-n)!} = \rho^n \frac{N!}{N^n(N-n)!} \qquad (29.16)$$

In particular,

$$\rho^{(1)} = \frac{N}{V} = \rho \qquad (29.17)$$

$$\rho^{(2)} = \frac{N(N-1)}{V^2} = \rho^2\left(1-\frac{1}{N}\right) \qquad (29.18)$$

Equation (29.17) holds in a fluid [Eq. (29.10)] whether the distribution is random or not. A random distribution can exist only in the limit $T \to \infty$ or in a hypothetical fluid with no intermolecular forces. In both cases the fluid is of course a perfect gas.

The behavior of $\rho^{(2)}$ in the limit $r_{12} \to \infty$ (fluid) is discussed in Appendix 7. Also, the limit $\rho \to 0$ is considered in Sec. 32 and Appendix 7.

Correlation Functions. If the probability of molecule 1 being in $d\mathbf{r}_1$ at $\mathbf{r}_1$ is independent of the probability of molecule 2 being in $d\mathbf{r}_2$ at $\mathbf{r}_2$, etc., then the probability of observing 1 in $d\mathbf{r}_1$, . . . , and n in $d\mathbf{r}_n$ is

$$P^{(n)}(\mathbf{r}_1, \ldots, \mathbf{r}_n, \xi)\, d\mathbf{r}_1\cdots d\mathbf{r}_n = [P^{(1)}(\mathbf{r}_1,\xi)\, d\mathbf{r}_1]\cdots[P^{(1)}(\mathbf{r}_n,\xi)\, d\mathbf{r}_n] \qquad (29.19)$$

When the n probabilities above are *not* independent, we might introduce a correlation function $C^{(n)}(\mathbf{r}_1, \ldots, \mathbf{r}_n, \xi)$ which would represent the factor by which $P^{(n)}$ deviates from the "independent" value given in Eq. (29.19). That is, we define $C^{(n)}$ by

$$P^{(n)}(\mathbf{r}_1, \ldots, \mathbf{r}_n, \xi) = P^{(1)}(\mathbf{r}_1,\xi)\cdots P^{(1)}(\mathbf{r}_n,\xi)C^{(n)}(\mathbf{r}_1, \ldots, \mathbf{r}_n, \xi) \qquad (29.20)$$

Clearly, $C^{(1)} = 1$.

Alternatively, we can define another correlation function, used extensively by Kirkwood, by the analogous equation

$$\rho^{(n)}(\mathbf{r}_1, \ldots, \mathbf{r}_n, \xi) = \rho^{(1)}(\mathbf{r}_1,\xi)\cdots\rho^{(1)}(\mathbf{r}_n,\xi)g^{(n)}(\mathbf{r}_1, \ldots, \mathbf{r}_n, \xi) \qquad (29.21)$$

[1] By a "random" distribution we shall mean $P^{(n)} = 1/V^n$ and by an "independent" distribution we shall mean

$$P^{(n)}(\mathbf{r}_1, \ldots, \mathbf{r}_n, \xi) = P^{(1)}(\mathbf{r}_1,\xi)\cdots P^{(1)}(\mathbf{r}_n,\xi)$$

A "random" distribution is necessarily "independent," but not vice versa.

from which $g^{(1)} = 1$. The two correlation functions are obviously related by

$$g^{(n)} = \frac{C^{(n)}N!}{N^n(N-n)!} = C^{(n)}\left[1 - \frac{n(n-1)}{2N} + \cdots\right] \tag{29.22}$$

$$= C^{(n)}\left[1 + O\left(\frac{1}{N}\right)\right] \qquad n \text{ small} \tag{29.23}$$

In an "independent" or in a "random" distribution, $C^{(n)} = 1$ but

$$g^{(n)} = \frac{N!}{N^n(N-n)!} = 1 + O\left(\frac{1}{N}\right) \tag{29.24}$$

We shall use $g^{(n)}$ because $g^{(2)}$ is more convenient than $C^{(2)}$ in discussing thermodynamic functions but $C^{(n)}$ may always be found from Eq. (29.22) and $g^{(n)}$.

In a fluid, the $g^{(n)}$ have the following properties: (1) The definition simplifies to

$$\rho^{(n)}(\mathbf{r}_1, \ldots, \mathbf{r}_n, \xi) = \rho^n g^{(n)}(\mathbf{r}_1, \ldots, \mathbf{r}_n, \xi) \tag{29.25}$$

(2) From Eqs. (29.8) and (29.25),

$$g^{(n)}(\mathbf{r}_1, \ldots, \mathbf{r}_n, \xi) = \frac{V^n N!}{N^n(N-n)!} \frac{\int \cdots \int_V \exp\left[-U(\xi)/\mathbf{k}T\right] d\mathbf{r}_{n+1} \cdots d\mathbf{r}_N}{Z(\xi)} \tag{29.26}$$

When the term of order $1/N$ can be ignored, Eq. (29.26) becomes

$$g^{(n)}(\mathbf{r}_1, \ldots, \mathbf{r}_n, \xi) = \frac{V^n \int \cdots \int_V \exp\left[-U(\xi)/\mathbf{k}T\right] d\mathbf{r}_{n+1} \cdots d\mathbf{r}_N}{Z(\xi)} \tag{29.27}$$

(3) The normalization equation for $g^{(n)}$ is, from Eq. (29.26),

$$\frac{1}{V^n} \int \cdots \int_V g^{(n)}(\mathbf{r}_1, \ldots, \mathbf{r}_n, \xi)\, d\mathbf{r}_1 \cdots d\mathbf{r}_n = \frac{N!}{N^n(N-n)!} \tag{29.28}$$

$$= 1 + O\left(\frac{1}{N}\right) \tag{29.29}$$

(4) From Eqs. (29.13), (29.14), and (29.25), the pair correlation function $g^{(2)}$ has the properties

$$g^{(2)}(\mathbf{r}_1, \mathbf{r}_2, \xi) = g^{(2)}(r_{12}, \xi) \tag{29.30}$$

$$\frac{1}{V} \int_V g^{(2)}(r_{12}, \xi)\, d\mathbf{r}_{12} = 1 - \frac{1}{N} \tag{29.31}$$

Furthermore, this function is just (when $\xi = 1$) the familiar radial distribution function $g(r_{12})$ obtained experimentally from X-ray scattering.[1] This can be verified as follows. The probability that a particular molecule is in $d\mathbf{r}_1$ at $\mathbf{r}_1$ while any other is in $d\mathbf{r}_2$ at $\mathbf{r}_2$ is

$$(N-1)P^{(2)}(r_{12},\xi)\, d\mathbf{r}_1\, d\mathbf{r}_2 = \frac{\rho^2}{N} g^{(2)}(r_{12},\xi)\, d\mathbf{r}_1\, d\mathbf{r}_2$$

Then it follows that if a specified molecule is fixed at $\mathbf{r}_1$, the probability of observing a second (unspecified) molecule in $d\mathbf{r}_2$ at $\mathbf{r}_2$ is

$$\text{const} \times g^{(2)}(r_{12},\xi)\, d\mathbf{r}_2$$

If we integrate this expression over V to evaluate the constant, we obtain

$$\text{const} \times \int_V g^{(2)}(r_{12},\xi)\, d\mathbf{r}_{12} = N - 1 \tag{29.32}$$

since all molecules except the molecule fixed at $\mathbf{r}_1$ must be encountered in the integration over V. From Eqs. (29.31) and (29.32), constant $= \rho$. Hence the probability of observing a second molecule in $d\mathbf{r}_{12}$ at $\mathbf{r}_2$ when a specified molecule is at $\mathbf{r}_1$ is

$$\rho g^{(2)}(r_{12},\xi)\, d\mathbf{r}_{12} \tag{29.33}$$

Or, the average number of molecules at a distance between r and $r + dr$ from a specified molecule in a fluid (taking $\xi = 1$) is

$$\rho g^{(2)}(r) \cdot 4\pi r^2\, dr \tag{29.34}$$

and we have also

$$\rho \int_V g^{(2)}(r) \cdot 4\pi r^2\, dr = N - 1 \tag{29.35}$$

The function $g^{(2)}(r)$ in Eqs. (29.34) and (29.35) is seen to have just the properties[2] of the experimental radial distribution function $g(r)$. For this reason, we shall sometimes omit the superscript on $g^{(2)}$.

A typical experimental $g(r)$ curve is shown in Fig. 19 for liquid argon.[3] The peaks in $g(r)$ are the remnants, in the liquid state, of the first-neighbor, second-neighbor, etc., shells surrounding a given molecule at an equilibrium lattice position in the crystal. Of course $g(r) \to 0$ as $r \to 0$ because of the van der Waals repulsion between two molecules at small intermolecular distances. The limiting value of g as $r \to \infty$ is discussed in Appendix 7.

[1] N. S. Gingrich, *Rev. Mod. Phys.*, **15**, 90 (1943); A. Eisenstein and N. S. Gingrich, *Phys. Rev.*, **62**, 261 (1942).

[2] The function $g^{[2]}(r)$ (for a fluid) introduced below is *by definition* identical with $g(r)$. Equation (29.51) states that $g^{[2]}(r) = g^{(2)}(r)$.

[3] Eisenstein and Gingrich, *loc. cit.*

Our principal task in this chapter is to show how $g(r)$ may be calculated theoretically and how the thermodynamic functions of the fluid are related to $g(r)$.

Unsymmetrical Distribution and Correlation Functions. For some purposes[1] it is convenient to use unsymmetrical distribution and correlation functions in which one of the set of n molecules, say the nth, plays a special role. From (29.3), the probability $P^{[n]}(\mathbf{r}_1, \ldots, \mathbf{r}_n, \boldsymbol{\xi})\, d\mathbf{r}_n$ of observing molecule n in

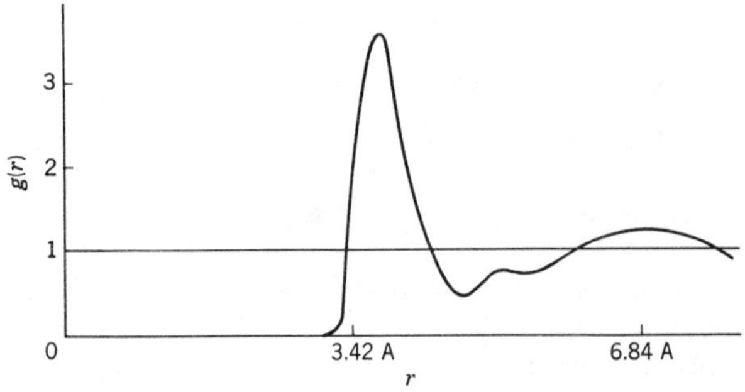

FIG. 19. An experimental radial distribution function for liquid argon.

$d\mathbf{r}_n$ at $\mathbf{r}_n$ if molecule 1 is at $\mathbf{r}_1$, . . . , and molecule $n - 1$ is at $\mathbf{r}_{n-1}$, irrespective of the configuration of the remaining $N - n$ molecules, is

$$P^{[n]}(\mathbf{r}_1, \ldots, \mathbf{r}_n, \boldsymbol{\xi})\, d\mathbf{r}_n = \text{const} \times d\mathbf{r}_n \int \cdots \int_V \exp\left[-\frac{U(\boldsymbol{\xi})}{\mathbf{k}T}\right] d\mathbf{r}_{n+1} \cdots d\mathbf{r}_N \tag{29.36}$$

The "constant" (a function of $\mathbf{r}_1, \ldots, \mathbf{r}_{n-1}$) can be evaluated using

$$\int_V P^{[n]}(\mathbf{r}_1, \ldots, \mathbf{r}_n, \boldsymbol{\xi})\, d\mathbf{r}_n = 1 \tag{29.37}$$

with the result

$$P^{[n]}(\mathbf{r}_1, \ldots, \mathbf{r}_n, \boldsymbol{\xi}) = \frac{\int \cdots \int_V \exp\left[-U(\boldsymbol{\xi})/\mathbf{k}T\right] d\mathbf{r}_{n+1} \cdots d\mathbf{r}_N}{\int \cdots \int_V \exp\left[-U(\boldsymbol{\xi})/\mathbf{k}T\right] d\mathbf{r}_n \cdots d\mathbf{r}_N} \tag{29.38}$$

Now let $\rho^{[n]}(\mathbf{r}_1, \ldots, \mathbf{r}_n, \boldsymbol{\xi})\, d\mathbf{r}_n$ be the probability of finding any molecule in

[1] See below and, for example, J. G. Kirkwood, *J. Chem. Phys.*, **3**, 300 (1935); J. G. Kirkwood and E. M. Boggs, *J. Chem. Phys.*, **10**, 394 (1942); J. G. Kirkwood and E. Monroe, *J. Chem. Phys.*, **9**, 514 (1941).

$d\mathbf{r}_n$ at $\mathbf{r}_n$ (with ξ_n) if molecule 1 is at $\mathbf{r}_1$ (with ξ_1), . . . , and molecule $n-1$ is at $\mathbf{r}_{n-1}$ (with ξ_{n-1}). Then

$$\rho^{[n]}(\mathbf{r}_1, \ldots, \mathbf{r}_n, \boldsymbol{\xi}) = (N - n + 1)P^{[n]}(\mathbf{r}_1, \ldots, \mathbf{r}_n, \boldsymbol{\xi}) \tag{29.39}$$

since any one of the remaining $N - (n - 1)$ molecules can be in $d\mathbf{r}_n$.

The $\rho^{[n]}$ have the following properties, which follow immediately from their definition:

$$\int_V \rho^{[n]}(\mathbf{r}_1, \ldots, \mathbf{r}_n, \boldsymbol{\xi})\, d\mathbf{r}_n = N - n + 1 \tag{29.40}$$

$$\rho^{[1]}(\mathbf{r}_1,\boldsymbol{\xi}) = \rho^{(1)}(\mathbf{r}_1,\boldsymbol{\xi}) \tag{29.41}$$

$$\rho^{[2]}(r_{12},\boldsymbol{\xi}) = \frac{\rho^{(2)}(r_{12},\boldsymbol{\xi})}{\rho} \qquad \text{(fluid)} \tag{29.42}$$

$$\rho^{[n]}(\mathbf{r}_1, \ldots, \mathbf{r}_n, \boldsymbol{\xi}) = \frac{\rho^{(n)}(\mathbf{r}_1, \ldots, \mathbf{r}_n, \boldsymbol{\xi})}{\rho^{(n-1)}(\mathbf{r}_1, \ldots, \mathbf{r}_{n-1}, \boldsymbol{\xi})} \tag{29.43}$$

$$= \frac{N - n + 1}{V} \qquad \text{(random distribution)} \tag{29.44}$$

By analogy with Eq. (29.21) an appropriate correlation function can now be introduced, defined by [note Eq. (29.41)]

$$\rho^{[n]}(\mathbf{r}_1, \ldots, \mathbf{r}_n, \boldsymbol{\xi}) = \rho^{(1)}(\mathbf{r}_n,\boldsymbol{\xi})g^{[n]}(\mathbf{r}_1, \ldots, \mathbf{r}_n, \boldsymbol{\xi}) \tag{29.45}$$

For a fluid,

$$\rho^{[n]}(\mathbf{r}_1, \ldots, \mathbf{r}_n, \boldsymbol{\xi}) = \rho g^{[n]}(\mathbf{r}_1, \ldots, \mathbf{r}_n, \boldsymbol{\xi}) \tag{29.46}$$

and $g^{[n]}(\mathbf{r}_1, \ldots, \mathbf{r}_n, \boldsymbol{\xi})$

$$= \left[1 - \frac{(n-1)}{N}\right] V \frac{\int \cdots \int_V \exp\left[-U(\boldsymbol{\xi})/\mathbf{k}T\right] d\mathbf{r}_{n+1} \cdots d\mathbf{r}_N}{\int \cdots \int_V \exp\left[-U(\boldsymbol{\xi})/\mathbf{k}T\right] d\mathbf{r}_n \cdots d\mathbf{r}_N} \tag{29.47}$$

If we can neglect $(n-1)/N$ compared to unity, the factor $1 - [(n-1)/N]$ in Eq. (29.47) may be omitted.

Other general properties of the $g^{[n]}$ which are easy to establish are

$$g^{[n]}(\mathbf{r}_1, \ldots, \mathbf{r}_n, \boldsymbol{\xi}) = 1 - \frac{n-1}{N} \qquad \text{(random distribution)} \tag{29.48}$$

$$g^{[1]}(\mathbf{r}_1,\boldsymbol{\xi}) = 1 \tag{29.49}$$

$$g^{[n]}(\mathbf{r}_1, \ldots, \mathbf{r}_n, \boldsymbol{\xi}) = \frac{g^{(n)}(\mathbf{r}_1, \ldots, \mathbf{r}_n, \boldsymbol{\xi})}{g^{(n-1)}(\mathbf{r}_1, \ldots, \mathbf{r}_{n-1}, \boldsymbol{\xi})} \tag{29.50}$$

$$g^{[2]}(\mathbf{r}_1,\mathbf{r}_2,\boldsymbol{\xi}) = g^{(2)}(\mathbf{r}_1,\mathbf{r}_2,\boldsymbol{\xi}) \tag{29.51}$$

$$g^{[n]}(\mathbf{r}_1, \ldots, \mathbf{r}_n, \xi) g^{[n-1]}(\mathbf{r}_1, \ldots, \mathbf{r}_{n-1}, \xi) \cdots g^{[2]}(\mathbf{r}_1, \mathbf{r}_2, \xi) = g^{(n)}(\mathbf{r}_1, \ldots, \mathbf{r}_n, \xi) \tag{29.52}$$

$$\frac{1}{V}\int_V g^{[n]}(\mathbf{r}_1, \ldots, \mathbf{r}_n, \xi)\, d\mathbf{r}_n = 1 - \frac{n-1}{N} \qquad \text{(fluid)} \tag{29.53}$$

Finally, it should be noticed that $\rho^{[2]}(\mathbf{r}_1,\mathbf{r}_2)/\rho$ is, by definition, the "radial"[1] distribution function for a liquid or a crystal, that is, the probability of finding a molecule in $d\mathbf{r}_2$ at $\mathbf{r}_2$ if molecule 1 is fixed at $\mathbf{r}_1$, relative to the random expectation ($\rho d\mathbf{r}_2$). In terms of the various distribution functions introduced, the radial distribution function, for a liquid or a crystal, is therefore

$$\frac{\rho^{[2]}(\mathbf{r}_1,\mathbf{r}_2)}{\rho} = \frac{\rho^{(2)}(\mathbf{r}_1,\mathbf{r}_2)}{\rho^{(1)}(\mathbf{r}_1)\rho} = \frac{g^{(2)}(\mathbf{r}_1,\mathbf{r}_2)\rho^{(1)}(\mathbf{r}_2)}{\rho} = \frac{g^{[2]}(\mathbf{r}_1,\mathbf{r}_2)\rho^{(1)}(\mathbf{r}_2)}{\rho} \tag{29.54}$$

Also, if we define $\mathfrak{g}^{(n)}$ by $\rho^{(n)}/\rho^n$ (see Sec. 40), the radial distribution function is

$$\frac{\rho \mathfrak{g}^{(2)}(\mathbf{r}_1,\mathbf{r}_2)}{\rho^{(1)}(\mathbf{r}_1)} \tag{29.55}$$

Of course in a fluid the radial distribution function reduces to

$$g^{(2)}(r_{12}) = g^{[2]}(r_{12}) = \mathfrak{g}^{(2)}(r_{12}) \tag{29.56}$$

30. Thermodynamic Functions of a Fluid and the Radial Distribution Function

In this section we derive the relationships between the thermodynamic functions of a fluid[2] and the radial distribution function. In subsequent sections we shall then consider how the radial distribution function can be calculated theoretically.

Let us first deduce an equation for the internal energy E. We take $\xi = 1$ here. From [see Eqs. (14.85) and (14.89)]

$$E = \mathbf{k}T^2 \left(\frac{\partial \ln Q}{\partial T}\right)_{N,V} \tag{30.1}$$

and [see Eq. (22.4)]

$$Q = \frac{Z}{N!\Lambda^{3N}} \tag{30.2}$$

[1] Quotation marks are used since, for a crystal, $\rho^{[2]}(\mathbf{r}_1,\mathbf{r}_2)/\rho$ does not have a radial ($|\mathbf{r}_2 - \mathbf{r}_1|$) dependence only.

[2] In each case we first derive equations valid also for a crystal, and then specialize to a fluid.

we have

$$E = {}^3/_2 N\mathbf{k}T + \mathbf{k}T^2 \left(\frac{\partial \ln Z}{\partial T}\right)_{N,V} \tag{30.3}$$

$$= {}^3/_2 N\mathbf{k}T + \bar{U} \tag{30.4}$$

where

$$\bar{U} = \frac{\int \cdots \int_V e^{-U/\mathbf{k}T} U \, d\mathbf{r}_1 \cdots d\mathbf{r}_N}{Z} \tag{30.5}$$

In Eq. (30.4), $3N\mathbf{k}T/2$ is the mean kinetic energy of the system and $\bar{U}$ the mean total potential energy. Now U is the sum of $N(N-1)/2$ terms, $u(r_{ij})$, all of which will obviously give the same result on carrying out the integration in the numerator of Eq. (30.5). Therefore

$$\bar{U} = \frac{N(N-1)}{2} \frac{\int \cdots \int_V e^{-U/\mathbf{k}T} u(r_{12}) \, d\mathbf{r}_1 \cdots d\mathbf{r}_N}{Z} \tag{30.6}$$

$$= \frac{N(N-1)}{2} \iint_V u(r_{12}) \left[\frac{\int \cdots \int_V e^{-U/\mathbf{k}T} d\mathbf{r}_3 \cdots d\mathbf{r}_N}{Z} \right] d\mathbf{r}_1 \, d\mathbf{r}_2 \tag{30.7}$$

$$= {}^1/_2 \iint_V u(r_{12}) \rho^{(2)}(\mathbf{r}_1, \mathbf{r}_2) \, d\mathbf{r}_1 \, d\mathbf{r}_2 \tag{30.8}$$

$$= \frac{1}{2} \frac{N^2}{V} \int_0^\infty u(r) g(r) \cdot 4\pi r^2 \, dr \qquad \text{(fluid)} \tag{30.9}$$

The upper limit of infinity is permissible since $u(r) \to 0$ rapidly as $r \to \infty$. Thus,

$$\frac{E}{N\mathbf{k}T} = \frac{3}{2} + \frac{\rho}{2\mathbf{k}T} \int_0^\infty u(r) g(r) \cdot 4\pi r^2 \, dr \qquad \text{(fluid)} \tag{30.10}$$

Equation (30.9) is clearly appropriate if we consider that a given molecule has, on the average, $\rho g(r) 4\pi r^2 \, dr$ neighboring molecules at a distance between r and $r + dr$ from it, and therefore that the average potential energy of interaction of the given molecule with these neighbors is $\rho u(r) g(r) 4\pi r^2 \, dr$. By integrating over r we get the average potential energy of interaction of the given molecule with all other molecules. Since there are N molecules in the fluid, we multiply this result for a single molecule by N to obtain $\bar{U}$, but we must also divide by two to avoid counting each pair interaction twice.

Next, we derive the equation of state using a method due to H. S. Green. For large V we may certainly assume that the pressure is independent of the shape of the container. For convenience, then, we assume that the container is a cube. We have (again with $\xi = 1$), from Eq. (14.90),

$$p = \mathbf{k}T\left(\frac{\partial \ln Q}{\partial V}\right)_{N,T} = \mathbf{k}T\left(\frac{\partial \ln Z}{\partial V}\right)_{N,T} \tag{30.11}$$

where
$$Z = \int_0^{V^{1/3}} \cdots \int_0^{V^{1/3}} e^{-U/\mathbf{k}T} dx_1\, dy_1\, dz_1 \cdots dx_N\, dy_N\, dz_N \tag{30.12}$$

Before differentiating Z with respect to V, we change variables of integration in such a way that the limits of integration become constants and U becomes an explicit function of V. Let the new variables be x_1', y_1', z_1', etc., where

$$\begin{aligned} x_k &= V^{1/3}x_k' \\ y_k &= V^{1/3}y_k' \qquad k = 1, 2, \ldots, N \\ z_k &= V^{1/3}z_k' \end{aligned} \tag{30.13}$$

Then
$$Z = V^N \int_0^1 \cdots \int_0^1 e^{-U/\mathbf{k}T} dx_1' \cdots dz_N' \tag{30.14}$$

in which
$$U = \sum_{1 \leqslant i < j \leqslant N} u(r_{ij}) \tag{30.15}$$

and
$$\begin{aligned} r_{ij} &= [(x_i - x_j)^2 + (y_i - y_j)^2 + (z_i - z_j)^2]^{1/2} \\ &= V^{1/3}[(x_i' - x_j')^2 + (y_i' - y_j')^2 + (z_i' - z_j')^2]^{1/2} \end{aligned} \tag{30.16}$$

From Eqs. (30.14) to (30.16),

$$\begin{aligned} \left(\frac{\partial Z}{\partial V}\right)_{N,T} &= NV^{N-1}\int_0^1 \cdots \int_0^1 e^{-U/\mathbf{k}T} dx_1' \cdots dz_N' \\ &\quad - \frac{V^N}{\mathbf{k}T}\int_0^1 \cdots \int_0^1 e^{-U/\mathbf{k}T} \frac{\partial U}{\partial V} dx_1' \cdots dz_N' \end{aligned} \tag{30.17}$$

where
$$\begin{aligned} \frac{\partial U}{\partial V} &= \sum_{1 \leqslant i < j \leqslant N} \frac{du(r_{ij})}{dr_{ij}} \frac{\partial r_{ij}}{\partial V} \\ &= \sum_{1 \leqslant i < j \leqslant N} \frac{r_{ij}}{3V} \frac{du(r_{ij})}{dr_{ij}} \end{aligned} \tag{30.18}$$

Having carried out the required differentiation with respect to V, we now transform back to the original variables $x_1, \ldots, z_N$. Also, we note that on integrating over the sum in Eq. (30.18), $N(N-1)/2$ identical terms will result. Therefore, Eq. (30.17) becomes

$$\left(\frac{\partial \ln Z}{\partial V}\right)_{N,T} = \frac{N}{V} - \frac{1}{6V\mathbf{k}T}\iint_V r_{12}\frac{du(r_{12})}{dr_{12}} \rho^{(2)}(\mathbf{r}_1,\mathbf{r}_2)\, d\mathbf{r}_1\, d\mathbf{r}_2 \tag{30.19}$$

or
$$\frac{pV}{N\mathbf{k}T} = 1 - \frac{\rho}{6\mathbf{k}T}\int_0^\infty ru'(r)g(r) \cdot 4\pi r^2\, dr \qquad \text{(fluid)} \tag{30.20}$$

This result can be derived in other ways, for example from the virial theorem of Clausius.[1]

There remains the task of relating one more function, such as the entropy, Helmholtz free energy, or chemical potential, to the radial distribution function. All other thermodynamic properties are then expressible in terms of E, pV, and S, A, or μ.

It is clear from Eq. (29.8) that $\rho^{(2)}$ is a function not only of $\mathbf{r}_1$ and $\mathbf{r}_2$, but also of two intensive variables, for example, ρ and T or v and T. With E and pV determined by $\rho^{(2)}(\mathbf{r}_1,\mathbf{r}_2,v,T)$ as in Eqs. (30.4), (30.8), (30.11), and (30.19), a formula for A follows on integration of

$$d\frac{A}{N} = -p\,dv \qquad T \text{ const} \tag{30.21}$$

or

$$d\frac{A}{T} = E\,d\frac{1}{T} \qquad v \text{ const} \tag{30.22}$$

Equation (30.21) is made use of in Secs. 35 and 36. A more convenient procedure for some purposes is to obtain μ by an integration over a coupling parameter ξ, as described below.[2]

From Eq. (30.2) we have

$$-\frac{A}{\mathbf{k}T} = \ln Z(N) - \ln N! - 3N \ln \Lambda \tag{30.23}$$

Now

$$\mu = \left(\frac{\partial A}{\partial N}\right)_{V,T}$$

$$= A(N, V, T) - A(N-1, V, T) \tag{30.24}$$

as N is a very large number. Then, introducing the subscript notation for N (Chap. 5),

$$-\frac{\mu}{\mathbf{k}T} = \ln\frac{Z_N}{Z_{N-1}} - \ln N - \ln \Lambda^3 \tag{30.25}$$

The ratio of configuration integrals in Eq. (30.25) can easily be written in terms of a coupling parameter. To do this, let all the ξ_i's be unity except one, say ξ_1, which we hereafter denote as ξ. When $\xi = 1$, we have the fully coupled real fluid of N molecules, but if $\xi = 0$, molecule 1 does not interact

[1] R. H. Fowler and E. A. Guggenheim, "Statistical Thermodynamics" (Cambridge, London, 1939), p. 271.

[2] Actually, integration with respect to $1/T$ as in Eq. (30.22) is equivalent to integration with respect to a coupling parameter ξ'^2, where all the $\xi_i = \xi'$ in Eq. (29.1). This follows because U occurs only as $U/\mathbf{k}T$ in Z, etc.

at all with any of the remaining $N - 1$ molecules (though the interactions among these $N - 1$ molecules themselves are fully coupled). That is,

$$Z_N(\xi = 1) = Z_N = Z \tag{30.26}$$

$$\begin{aligned} Z_N(\xi = 0) &= \int_V \cdots \int e^{-U_{N-1}/\mathbf{k}T} d\mathbf{r}_1 \cdots d\mathbf{r}_N \\ &= V \int_V \cdots \int e^{-U_{N-1}/\mathbf{k}T} d\mathbf{r}_2 \cdots d\mathbf{r}_N \\ &= V Z_{N-1} \end{aligned} \tag{30.27}$$

where
$$U_{N-1} = \sum_{2 \leqslant i < j \leqslant N} u(r_{ij}) \tag{30.28}$$

We can therefore write, in Eq. (30.25),

$$\begin{aligned} \ln \frac{Z_N}{Z_{N-1}} &= \ln \frac{Z_N(\xi = 1)}{Z_N(\xi = 0)} + \ln V \\ &= \ln V + \int_0^1 \frac{\partial \ln Z(\xi)}{\partial \xi} d\xi \end{aligned} \tag{30.29}$$

where
$$Z(\xi) = \int_V \cdots \int e^{-U(\xi)/\mathbf{k}T} d\mathbf{r}_1 \cdots d\mathbf{r}_N \tag{30.30}$$

$$U(\xi) = \sum_{j=2}^{N} \xi u(r_{1j}) + \sum_{2 \leqslant i < j \leqslant N} u(r_{ij}) \tag{30.31}$$

According to Eqs. (30.30) and (30.31),

$$\frac{\partial Z(\xi)}{\partial \xi} = -\frac{1}{\mathbf{k}T} \int_V \cdots \int e^{-U(\xi)/\mathbf{k}T} [\sum_{j=2}^{N} u(r_{1j})] \, d\mathbf{r}_1 \cdots d\mathbf{r}_N \tag{30.32}$$

All the $N - 1$ terms in Eq. (30.32) have the same value; therefore

$$\frac{\partial \ln Z(\xi)}{\partial \xi} = -\frac{1}{N\mathbf{k}T} \iint_V u(r_{12}) \rho^{(2)}(\mathbf{r}_1, \mathbf{r}_2, \xi) \, d\mathbf{r}_1 \, d\mathbf{r}_2 \tag{30.33}$$

$$= -\frac{\rho}{\mathbf{k}T} \int_0^\infty u(r) g(r, \xi) \cdot 4\pi r^2 \, dr \qquad \text{(fluid)} \tag{30.34}$$

Finally, then, combining Eqs. (30.25), (30.29), and (30.33),

$$\frac{\mu}{\mathbf{k}T} = \ln \rho \Lambda^3 + \frac{1}{N\mathbf{k}T} \int_0^1 \iint_V u(r_{12}) \rho^{(2)}(\mathbf{r}_1, \mathbf{r}_2, \xi) \, d\mathbf{r}_1 \, d\mathbf{r}_2 \, d\xi \tag{30.35}$$

$$= \ln \rho \Lambda^3 + \frac{\rho}{\mathbf{k}T} \int_0^1 \int_0^\infty u(r) g(r, \xi) \cdot 4\pi r^2 \, dr \, d\xi \qquad \text{(fluid)} \tag{30.36}$$

If we define z, as in Eq. (23.6), by

$$\frac{\mu}{\mathbf{k}T} = \ln z + \ln \Lambda^3 \tag{30.37}$$

where z is the activity ($z \to \rho$ as $\rho \to 0$) and introduce an activity coefficient γ defined by $z = \rho\gamma$, then

$$\ln \gamma = \frac{\rho}{\mathbf{k}T} \int_0^1 \int_0^\infty u(r)g(r,\xi) \cdot 4\pi r^2 \, dr \, d\xi \qquad \text{(fluid)} \tag{30.38}$$

Alternatively, if we use [Eq. (23.9)]

$$\frac{\mu}{\mathbf{k}T} = \ln f + \ln \left(\frac{\Lambda^3}{\mathbf{k}T}\right) \tag{30.39}$$

where f is the fugacity ($f \to p$ as $p \to 0$) and introduce an activity coefficient γ_f defined by $f = p\gamma_f$, then

$$\ln \gamma_f = \frac{\rho}{\mathbf{k}T} \int_0^1 \int_0^\infty u(r)g(r,\xi) \cdot 4\pi r^2 \, dr \, d\xi - \ln \frac{pV}{N\mathbf{k}T} \qquad \text{(fluid)} \tag{30.40}$$

Knowledge of the function $g(r)$ for $\xi = 1$ and at, say, $v = v'$ and $T = T'$ suffices, according to Eqs. (30.10) and (30.20), to compute $E/N\mathbf{k}T$ and $pV/N\mathbf{k}T$ at v' and T'. On the other hand, we see from Eqs. (30.21), (30.22), and (30.36) that in order to calculate the complete set of thermodynamic functions at v' and T' we must have available the dependence of g on r over a range of values of v, T, or ξ.

31. Potential of Mean Force and the Superposition Approximation

There are a number of different ways of defining potentials of mean force, both in the canonical and grand canonical ensembles (Sec. 37). We discuss only a few of the possible choices.

Let us first define a quantity $w^{(n)}(\mathbf{r}_1, \ldots, \mathbf{r}_n, \boldsymbol{\xi})$ by

$$\exp\left[-\frac{w^{(n)}(\mathbf{r}_1, \ldots, \mathbf{r}_n, \boldsymbol{\xi})}{\mathbf{k}T}\right] = V^n P^{(n)}(\mathbf{r}_1, \ldots, \mathbf{r}_n, \boldsymbol{\xi}) \tag{31.1}$$

where the multiplication by V^n renders the right-hand side dimensionless. Now suppose we differentiate Eq. (31.1) with respect to the coordinates of one of the n molecules $1, \ldots, n$, say molecule α. Then, from Eq. (29.5), we find

$$\nabla_\alpha w^{(n)} = \overline{\nabla_\alpha U}^{(n)} \tag{31.2}$$

where
$$\overline{\nabla_\alpha U}^{(n)} = \frac{\int \cdots \int_V \exp[-U(\boldsymbol{\xi})/\mathbf{k}T] \, \nabla_\alpha U(\boldsymbol{\xi}) \, d\mathbf{r}_{n+1} \cdots d\mathbf{r}_N}{\int \cdots \int_V \exp[-U(\boldsymbol{\xi})/\mathbf{k}T] \, d\mathbf{r}_{n+1} \cdots d\mathbf{r}_N} \tag{31.3}$$

For any given configuration $\mathbf{r}_1, \ldots, \mathbf{r}_N$ of the N molecules of the system, the force acting on particle α is

$$\mathbf{F}_\alpha = -\nabla_\alpha U(\mathbf{r}_1, \ldots, \mathbf{r}_N, \xi) \tag{31.4}$$

Therefore $-\overline{\nabla_\alpha U^{(n)}}$ in Eq. (31.3) is the *mean* force $\overline{\mathbf{F}}_\alpha^{(n)}$ acting on particle α, averaged over all configurations of molecules $n+1, \ldots, N$, but holding the set of molecules $1, \ldots, n$ fixed at $\mathbf{r}_1, \ldots, \mathbf{r}_n$. That is,

$$\overline{\mathbf{F}}_\alpha^{(n)} = -\overline{\nabla_\alpha U^{(n)}} = -\nabla_\alpha w^{(n)} \qquad \alpha = 1, \ldots, n \tag{31.5}$$

Thus, $w^{(n)}$ is the potential function[1] associated with (that is, whose negative gradient gives) the mean force acting on any one of the molecules $1, \ldots, n$. The potential $w^{(n)}$ is somewhat analogous to the electrostatic potential of a group of n charged particles from which can be derived the force on any one of the particles by taking the appropriate gradient. However, in the present problem there will be not only direct contributions to the potential $w^{(n)}$ owing to the van der Waals interactions between the n fixed molecules themselves, but also an additional averaged contribution from the remaining $N - n$ molecules of the system (see Secs. 32 and 33). Because of the statistical averaging [Eqs. (29.5) and (31.1)] involved in $w^{(n)}$, this potential has the nature of a free energy (see Sec. 32).

An alternative definition of a potential $W^{(n)}$ which leads to a simple relation between the potential and the correlation function $g^{(n)}$ even for a crystal, is the following [see Eq. (29.20)]:

$$\exp\left[-\frac{W^{(n)}(\mathbf{r}_1, \ldots, \mathbf{r}_n, \xi)}{\mathbf{k}T}\right] = \frac{P^{(n)}(\mathbf{r}_1, \ldots, \mathbf{r}_n, \xi)}{P^{(1)}(\mathbf{r}_1,\xi) \cdots P^{(1)}(\mathbf{r}_n,\xi)}$$
$$= C^{(n)}(\mathbf{r}_1, \ldots, \mathbf{r}_n, \xi) \tag{31.6}$$

In the special case of independent probabilities for the n particles, $C^{(n)} = 1$ and $W^{(n)} = 0$. We also find from Eq. (29.22)

$$\exp\left[-\frac{W^{(n)}(\mathbf{r}_1, \ldots, \mathbf{r}_n, \xi)}{\mathbf{k}T}\right] = \frac{(N-n)!N^n}{N!} g^{(n)}(\mathbf{r}_1, \ldots, \mathbf{r}_n, \xi) \tag{31.7}$$

$$= \left[1 + O\left(\frac{1}{N}\right)\right] g^{(n)}(\mathbf{r}_1, \ldots, \mathbf{r}_n, \xi) \tag{31.8}$$

For a fluid, $P^{(1)}(\mathbf{r}_1,\xi) = 1/V$ and $W^{(n)}$ reduces to $w^{(n)}$.

Instead of Eq. (31.2), Eqs. (29.5) and (31.6) lead to

$$\nabla_\alpha W^{(n)} = \overline{\nabla_\alpha U^{(n)}} - \overline{\nabla_\alpha U^{(1)}} \qquad \alpha = 1, \ldots, n \tag{31.9}$$

[1] The function $w^{(n)}$ + constant is also a potential of the mean force $\overline{\mathbf{F}}_\alpha^{(n)}$.

where $\overline{\nabla_\alpha U^{(n)}}$ is given by Eq. (31.3), and

$$\overline{\nabla_\alpha U^{(1)}} = \frac{\int \cdots \int_V \exp\left[-U(\boldsymbol{\xi})/\mathbf{k}T\right] \nabla_\alpha U(\boldsymbol{\xi})\, d\mathbf{r}_1 \cdots d\mathbf{r}_{\alpha-1}\, d\mathbf{r}_{\alpha+1} \cdots d\mathbf{r}_N}{\int \cdots \int_V \exp\left[-U(\boldsymbol{\xi})/\mathbf{k}T\right] d\mathbf{r}_1 \cdots d\mathbf{r}_{\alpha-1}\, d\mathbf{r}_{\alpha+1} \cdots d\mathbf{r}_N} \tag{31.10}$$

$W^{(n)}$ is thus the potential of the mean force on molecule α averaged over the configurations of $N - n$ molecules *relative to* the mean force on molecule α averaged over the configurations of $N - 1$ molecules. Of course $\overline{\nabla_\alpha U^{(1)}}$ will average out to zero in a fluid, by symmetry, but this quantity will not be zero, in general, in a crystal.

Other forms of Eq. (31.5) which are useful follow from the definitions of $w^{(n)}$, $P^{(n)}$, $\rho^{(n)}$, and $g^{(n)}$:

$$-\overline{\nabla_\alpha U^{(n)}} = \mathbf{k}T\nabla_\alpha \ln P^{(n)}(\mathbf{r}_1, \ldots, \mathbf{r}_n, \boldsymbol{\xi}) \tag{31.11}$$

$$= \mathbf{k}T\nabla_\alpha \ln \rho^{(n)}(\mathbf{r}_1, \ldots, \mathbf{r}_n, \boldsymbol{\xi}) \tag{31.12}$$

$$= \mathbf{k}T[\nabla_\alpha \ln \rho^{(1)}(\mathbf{r}_\alpha, \boldsymbol{\xi}) + \nabla_\alpha \ln g^{(n)}(\mathbf{r}_1, \ldots, \mathbf{r}_n, \boldsymbol{\xi})] \tag{31.13}$$

$$= -\overline{\nabla_\alpha U^{(1)}} + \mathbf{k}T\nabla_\alpha \ln g^{(n)}(\mathbf{r}_1, \ldots, \mathbf{r}_n, \boldsymbol{\xi}) \tag{31.14}$$

$$= \mathbf{k}T\nabla_\alpha \ln g^{(n)}(\mathbf{r}_1, \ldots, \mathbf{r}_n, \boldsymbol{\xi}) \qquad \text{(fluid)} \tag{31.15}$$

where $\overline{\nabla_\alpha U^{(n)}}$ is given by Eq. (31.3).

Superposition Approximation. In Secs. 32 and 33 integral equations for $g^{(2)}$ will be derived, the solutions of which require a knowledge of $g^{(3)}$. Similarly, integral equations for $g^{(3)}$ depend on $g^{(4)}$, etc. In order to break this chain of interdependent $g^{(n)}$'s, Kirkwood[1] introduced the superposition approximation which provides an explicit connection between $g^{(3)}$ and $g^{(2)}$. In anticipation of its later use, we discuss this approximation here because of its close relation to the potential of mean force. The approximation consists of assuming that the probability $P^{(3)}(\mathbf{r}_1,\mathbf{r}_2,\mathbf{r}_3,\boldsymbol{\xi})\, d\mathbf{r}_1\, d\mathbf{r}_2\, d\mathbf{r}_3$ of finding the triplet of molecules 1, 2, 3 in the configuration $\mathbf{r}_1,\mathbf{r}_2,\mathbf{r}_3$ within $d\mathbf{r}_1\, d\mathbf{r}_2\, d\mathbf{r}_3$ is proportional to the product of the three pair probabilities:

$$P^{(3)}(\mathbf{r}_1,\mathbf{r}_2,\mathbf{r}_3,\boldsymbol{\xi})\, d\mathbf{r}_1\, d\mathbf{r}_2\, d\mathbf{r}_3 = A'[P^{(2)}(\mathbf{r}_1,\mathbf{r}_2,\boldsymbol{\xi})\, d\mathbf{r}_1\, d\mathbf{r}_2][P^{(2)}(\mathbf{r}_1,\mathbf{r}_3,\boldsymbol{\xi})\, d\mathbf{r}_1\, d\mathbf{r}_3][P^{(2)}(\mathbf{r}_2,\mathbf{r}_3,\boldsymbol{\xi})\, d\mathbf{r}_2\, d\mathbf{r}_3] \tag{31.16}$$

In order for Eq. (31.16) to hold in the limiting case of independent probabilities [Eq. (29.19)] we must have

$$A' = \frac{1}{P^{(1)}(\mathbf{r}_1,\boldsymbol{\xi})P^{(1)}(\mathbf{r}_2,\boldsymbol{\xi})P^{(1)}(\mathbf{r}_3,\boldsymbol{\xi})\, d\mathbf{r}_1\, d\mathbf{r}_2\, d\mathbf{r}_3} \tag{31.17}$$

[1] J. G. Kirkwood, *J. Chem. Phys.*, **3**, 300 (1935).

The superposition approximation can then be expressed in the following equivalent forms (using an obvious shorthand notation):

$$\frac{P^{(3)}(123,\xi)}{P^{(1)}(1,\xi)P^{(1)}(2,\xi)P^{(1)}(3,\xi)} = \frac{P^{(2)}(12,\xi)}{P^{(1)}(1,\xi)P^{(1)}(2,\xi)} \times \frac{P^{(2)}(13,\xi)}{P^{(1)}(1,\xi)P^{(1)}(3,\xi)} \cdot \frac{P^{(2)}(23,\xi)}{P^{(1)}(2,\xi)P^{(1)}(3,\xi)} \tag{31.18}$$

$$W^{(3)}(123,\xi) = W^{(2)}(12,\xi) + W^{(2)}(13,\xi) + W^{(2)}(23,\xi) \tag{31.19}$$

$$\begin{aligned} \nabla_1 W^{(3)}(123,\xi) &= \nabla_1 W^{(2)}(12,\xi) + \nabla_1 W^{(2)}(13,\xi) \\ \nabla_2 W^{(3)}(123,\xi) &= \nabla_2 W^{(2)}(12,\xi) + \nabla_2 W^{(2)}(23,\xi) \\ \nabla_3 W^{(3)}(123,\xi) &= \nabla_3 W^{(2)}(13,\xi) + \nabla_3 W^{(2)}(23,\xi) \end{aligned} \tag{31.20}$$

$$\longrightarrow \rho^{(3)}(123,\xi) = \frac{\rho^{(2)}(12,\xi)\rho^{(2)}(13,\xi)\rho^{(2)}(23,\xi)}{\rho^{(1)}(1,\xi)\rho^{(1)}(2,\xi)\rho^{(1)}(3,\xi)} \tag{31.21}$$

$$\longrightarrow g^{(3)}(123,\xi) = g^{(2)}(12,\xi)g^{(2)}(13,\xi)g^{(2)}(23,\xi) \tag{31.22}$$

where a completely negligible factor

$$\frac{N(N-2)}{(N-1)^2} = 1 + O\left(\frac{1}{N^2}\right) \tag{31.23}$$

has been omitted from the right-hand side of Eqs. (31.21) and (31.22).

The superposition approximation assumes independent *pair* probability densities which is obviously some improvement over the assumption of independent *singlet* probability densities:

$$P^{(3)}(123,\xi) = P^{(1)}(1,\xi)P^{(1)}(2,\xi)P^{(1)}(3,\xi) \tag{31.24}$$

Equation (31.18) is one way of expressing the assumed independence of pair probabilities. Equation (31.19) shows that in introducing the superposition approximation we are assuming the potential $W^{(3)}$ to be pairwise additive [as is U in Eq. (29.1)]. This is equivalent to assuming [Eq. (31.20)] that in a triplet of molecules 1, 2, 3, the relative mean force on molecule 1, say, is just the sum of what it would be if molecules 2 and 3 were held fixed at $\mathbf{r}_2$ and $\mathbf{r}_3$ *separately*. That is, we are neglecting the effect on $W^{(3)}(123,\xi)$ of the perturbation of the $N - 3$ other molecules caused by the actual *simultaneous* presence of molecules 2 and 3.

Unsymmetrical Distribution Functions.[1] By analogy with Eq. (31.6), we define the potential $W^{[n]}$ as

$$\exp\left[-\frac{W^{[n]}(\mathbf{r}_1, \ldots, \mathbf{r}_n, \xi)}{\mathbf{k}T}\right] = \frac{P^{[n]}(\mathbf{r}_1, \ldots, \mathbf{r}_n, \xi)}{P^{(1)}(\mathbf{r}_n,\xi)} \tag{31.25}$$

$$= VP^{[n]}(\mathbf{r}_1, \ldots, \mathbf{r}_n, \xi) \qquad \text{(fluid)} \tag{31.26}$$

[1] J. G. Kirkwood, *J. Chem. Phys.*, **3**, 300 (1935); J. G. Kirkwood and E. M. Boggs, *J. Chem. Phys.*, **10**, 394 (1942).

where $P^{(1)} = P^{[1]}$. If the singlet probability density $P^{[n]}$ for particle n is independent of molecules $1, \ldots, n-1$, $P^{[n]} = P^{[1]}(\mathbf{r}_n,\boldsymbol{\xi})$ and $W^{[n]} = 0$. From Eqs. (29.39) and (29.45),

$$\exp\left[-\frac{W^{[n]}(\mathbf{r}_1, \ldots, \mathbf{r}_n, \boldsymbol{\xi})}{\mathbf{k}T}\right] = \frac{N}{N-n+1} g^{[n]}(\mathbf{r}_1, \ldots, \mathbf{r}_n, \boldsymbol{\xi}) \qquad (31.27)$$

$$= \left[1 + O\left(\frac{1}{N}\right)\right] g^{[n]}(\mathbf{r}_1, \ldots, \mathbf{r}_n, \boldsymbol{\xi}) \qquad (31.28)$$

Taking the gradient of Eq. (31.25) with respect to $\mathbf{r}_n$, using Eq. (29.38), we obtain

$$\nabla_n W^{[n]} = \overline{\nabla_n U^{(n)}} - \overline{\nabla_n U^{(1)}} \qquad (31.29)$$

which is the same as Eq. (31.9) with $\alpha = n$. This equivalence also follows from

$$W^{[n]}(12 \cdots n, \boldsymbol{\xi}) + W^{[n-1]}(12 \cdots n-1, \boldsymbol{\xi}) + \cdots + W^{[2]}(12, \boldsymbol{\xi}) = W^{(n)}(12 \cdots n, \boldsymbol{\xi}) \qquad (31.30)$$

or

$$W^{[n]}(12 \cdots n, \boldsymbol{\xi}) = W^{(n)}(12 \cdots n, \boldsymbol{\xi}) - W^{(n-1)}(12 \cdots n-1, \boldsymbol{\xi}) \qquad (31.31)$$

Equations (31.30) and (31.31) are consequences of Eqs. (29.52), (31.7), and (31.27). Thus, although the potentials $W^{[n]}$ and $W^{(n)}$ differ (by a function of $\mathbf{r}_1, \ldots, \mathbf{r}_{n-1}$), the relative mean force on particle n derivable from the two potentials is the same. From Eq. (31.30), $W^{[2]}(12,\boldsymbol{\xi}) = W^{(2)}(12,\boldsymbol{\xi})$.

Equation (31.31) shows that $W^{[n]}(12 \cdots n, \boldsymbol{\xi})$ is formally analogous to an "interaction" potential of molecule n with molecules $1, \ldots, n-1$. For $W^{(n)}(12 \cdots n, \boldsymbol{\xi})$ includes all "interactions" in the set $1, \ldots, n$ and on subtracting $W^{(n-1)}(12 \cdots n-1, \boldsymbol{\xi})$ we are left with only those "interactions" involving molecule n. In the superposition approximation (or when[1] $\rho \to 0$), this "pairwise" point of view is strictly correct [compare Eqs. (31.19) and (31.38)] though it is not in the general case.

Corresponding to Eqs. (31.11) to (31.15), we have here

$$-\overline{\nabla_n U^{(n)}} = \mathbf{k}T\nabla_n \ln P^{[n]}(\mathbf{r}_1, \ldots, \mathbf{r}_n, \boldsymbol{\xi}) \qquad (31.32)$$

$$= \mathbf{k}T\nabla_n \ln \rho^{[n]}(\mathbf{r}_1, \ldots, \mathbf{r}_n, \boldsymbol{\xi}) \qquad (31.33)$$

$$= \mathbf{k}T[\nabla_n \ln \rho^{(1)}(\mathbf{r}_n, \boldsymbol{\xi}) + \nabla_n \ln g^{[n]}(\mathbf{r}_1, \ldots, \mathbf{r}_n, \boldsymbol{\xi})] \qquad (31.34)$$

$$= -\overline{\nabla_n U^{(1)}} + \mathbf{k}T\nabla_n \ln g^{[n]}(\mathbf{r}_1, \ldots, \mathbf{r}_n, \boldsymbol{\xi}) \qquad (31.35)$$

$$= \mathbf{k}T\nabla_n \ln g^{[n]}(\mathbf{r}_1, \ldots, \mathbf{r}_n, \boldsymbol{\xi}) \quad \text{(fluid)} \qquad (31.36)$$

[1] See Eq. (32.17). This statement of course assumes that Eq. (29.1) is correct.

It is easy to show that the superposition approximation is expressed in "unsymmetrical" form by, for example,

$$\frac{P^{[3]}(123,\boldsymbol{\xi})}{P^{(1)}(3,\boldsymbol{\xi})} = \frac{P^{[2]}(13,\boldsymbol{\xi})}{P^{(1)}(3,\boldsymbol{\xi})} \frac{P^{[2]}(23,\boldsymbol{\xi})}{P^{(1)}(3,\boldsymbol{\xi})} \tag{31.37}$$

$$W^{[3]}(123,\boldsymbol{\xi}) = W^{(2)}(13,\boldsymbol{\xi}) + W^{(2)}(23,\boldsymbol{\xi}) \tag{31.38}$$

$$\nabla_3 W^{[3]}(123,\boldsymbol{\xi}) = \nabla_3 W^{(2)}(13,\boldsymbol{\xi}) + \nabla_3 W^{(2)}(23,\boldsymbol{\xi}) \tag{31.39}$$

$$g^{[3]}(123,\boldsymbol{\xi}) = g^{(2)}(13,\boldsymbol{\xi}) g^{(2)}(23,\boldsymbol{\xi}) \tag{31.40}$$

where the negligible factor in Eq. (31.23) has been omitted in Eq. (31.40).

32. The Kirkwood Integral Equation

The eventual aim of this section is to derive an integral equation for the radial distribution function of a fluid, based on the superposition approximation. We shall, however, start out with a considerably more general treatment.

We begin by differentiating $\rho^{(n)}$ in Eq. (29.8) with respect to one of $\xi_1, \ldots, \xi_n$, say ξ_1, recalling that $U(\boldsymbol{\xi})$ is defined by Eq. (29.1). The result is

$$\begin{aligned}\mathbf{k}T \frac{\partial \ln \rho^{(n)}}{\partial \xi_1} &= \frac{1}{Z(\boldsymbol{\xi})} \left[\sum_{i=2}^{N} \xi_i \int \cdots \int_V \exp\left[-\frac{U(\boldsymbol{\xi})}{\mathbf{k}T}\right] u(r_{1i})\, d\mathbf{r}_1 \cdots d\mathbf{r}_N \right.\\ &\left. - \frac{N!}{(N-n)!\rho^{(n)}} \sum_{i=2}^{N} \xi_i \int \cdots \int_V \exp\left[-\frac{U(\boldsymbol{\xi})}{\mathbf{k}T}\right] u(r_{1i})\, d\mathbf{r}_{n+1} \cdots d\mathbf{r}_N \right]\end{aligned} \tag{32.1}$$

The first integral in Eq. (32.1) can be written as

$$\begin{aligned}&\iint_V u(r_{1i}) \left[\int \cdots \int_V \exp\left[-\frac{U(\boldsymbol{\xi})}{\mathbf{k}T}\right] d\mathbf{r}_2 \cdots d\mathbf{r}_{i-1}\, d\mathbf{r}_{i+1} \cdots d\mathbf{r}_N \right] d\mathbf{r}_1\, d\mathbf{r}_i \\ &\qquad = \frac{Z(\boldsymbol{\xi})}{N(N-1)} \iint_V u(r_{1i}) \rho^{(2)}(\mathbf{r}_1,\mathbf{r}_i,\boldsymbol{\xi})\, d\mathbf{r}_1\, d\mathbf{r}_i\end{aligned}$$

For $i = 2, \ldots, n$, $u(r_{1i})$ can be put outside the integral sign in the second integral of Eq. (32.1). For $i > n$, the second integral becomes

$$\begin{aligned}&\int_V u(r_{1i}) \left[\int \cdots \int_V \exp\left[-\frac{U(\boldsymbol{\xi})}{\mathbf{k}T}\right] d\mathbf{r}_{n+1} \cdots d\mathbf{r}_{i-1}\, d\mathbf{r}_{i+1} \cdots d\mathbf{r}_N \right] d\mathbf{r}_i \\ &\qquad = \frac{Z(\boldsymbol{\xi})(N-n-1)!}{N!} \int_V u(r_{1i}) \rho^{(n+1)}(\mathbf{r}_1, \ldots, \mathbf{r}_n, \mathbf{r}_i, \boldsymbol{\xi})\, d\mathbf{r}_i\end{aligned}$$

With the aid of the above remarks, we have, from Eq. (32.1),

$$\mathbf{k}T\frac{\partial \ln \rho^{(n)}}{\partial \xi_1} = -\sum_{i=2}^{n} \xi_i u(r_{1i}) + \frac{1}{N(N-1)}\sum_{i=2}^{N} \xi_i \int\int_V u(r_{1i})\rho^{(2)}(\mathbf{r}_1,\mathbf{r}_i,\boldsymbol{\xi})\, d\mathbf{r}_1\, d\mathbf{r}_i$$
$$- \frac{1}{(N-n)\rho^{(n)}}\sum_{i=n+1}^{N} \xi_i \int_V u(r_{1i})\rho^{(n+1)}(\mathbf{r}_1, \ldots, \mathbf{r}_n, \mathbf{r}_i, \boldsymbol{\xi})\, d\mathbf{r}_i \quad (32.2)$$

Now if we take $\xi_2 = \xi_3 = \cdots = \xi_N = 1$ and write ξ for ξ_1, Eq. (32.2) simplifies to[1]

$$\mathbf{k}T\frac{\partial \ln \rho^{(n)}}{\partial \xi} = -\sum_{i=2}^{n} u(r_{1i}) + \frac{1}{N}\int\int_V u(r_{12})\rho^{(2)}(\mathbf{r}_1,\mathbf{r}_2,\xi)\, d\mathbf{r}_1\, d\mathbf{r}_2$$
$$-\int_V u(r_{1,n+1})\rho^{[n+1]}(\mathbf{r}_1, \ldots, \mathbf{r}_{n+1}, \xi)\, d\mathbf{r}_{n+1} \quad (32.3)$$

where use has been made of Eq. (29.43) and of the equivalence of the terms in the two sums in Eq. (32.2). Equation (32.3) is valid for a fluid or a crystal. It will be noticed that this differential equation in $\rho^{(n)}$ involves the unknown functions $\rho^{(2)}$ and $\rho^{[n+1]}$ (or $\rho^{(n+1)}$) in the integrands. Since n can take on values from 1 to $N-1$, Eq. (32.3) provides us with a set of $N-1$ interdependent integrodifferential equations the exact solution of which will presumably be just as difficult as the direct evaluation of Z. However, it may turn out that in the distribution function method approximate solutions of the problem can be introduced in a relatively advantageous way.

An alternative form[2] of Eq. (32.3) results on subtracting from this equation the analogous equation with $n-1$ in place of n. We find

$$\mathbf{k}T\frac{\partial \ln \rho^{[n]}}{\partial \xi} = -u(r_{1n}) - \int_V u(r_{1,n+1})\{\rho^{[n+1]}(\mathbf{r}_1, \ldots, \mathbf{r}_{n+1}, \xi)$$
$$- \rho^{[n]}(\mathbf{r}_1, \ldots, \mathbf{r}_{n-1}, \mathbf{r}_{n+1}, \xi)\}\, d\mathbf{r}_{n+1} \quad (32.4)$$

Equations (32.3) and (32.4) can be written in integrated form (that is, as integral equations) by noting that [see Eq. (30.27)]

$$\rho^{(n)}(\mathbf{r}_1, \ldots, \mathbf{r}_n, 0) = \frac{N!}{(N-n)!V}\,\frac{\displaystyle\int\cdots\int_V e^{-U_{N-1}/\mathbf{k}T}d\mathbf{r}_{n+1}\cdots d\mathbf{r}_N}{\displaystyle\int\cdots\int_V e^{-U_{N-1}/\mathbf{k}T}d\mathbf{r}_2\cdots d\mathbf{r}_N}$$
$$= \rho\rho_{N-1}{}^{(n-1)}(\mathbf{r}_2, \ldots, \mathbf{r}_n) \quad (32.5)$$

where the subscript $N-1$ indicates, as before, that the system contains

[1] See Eq. (32.12) below.

[2] J. G. Kirkwood, *J. Chem. Phys.*, **3**, 300 (1935), Eq. (37).

$N - 1$ particles (the absence of a subscript means N, as usual). Also, using Eq. (29.43),

$$\rho^{[n]}(\mathbf{r}_1, \ldots, \mathbf{r}_n, 0) = \rho \rho_{N-1}{}^{[n-1]}(\mathbf{r}_2, \ldots, \mathbf{r}_n) \tag{32.6}$$

Then on integrating Eqs. (32.3) and (32.4) we have

$$\begin{aligned}\mathbf{k}T \ln \rho^{(n)}(\mathbf{r}_1, \ldots, \mathbf{r}_n, \xi) &= \mathbf{k}T \ln \rho + \mathbf{k}T \ln \rho_{N-1}{}^{(n-1)}(\mathbf{r}_2, \ldots, \mathbf{r}_n) \\ &\quad - \xi \sum_{i=2}^{n} u(r_{1i}) + \frac{1}{N} \int_0^\xi \iint_V u(r_{12}) \rho^{(2)}(\mathbf{r}_1, \mathbf{r}_2, \xi)\, d\mathbf{r}_1\, d\mathbf{r}_2\, d\xi \\ &\quad - \int_0^\xi \int_V u(r_{1,n+1}) \rho^{[n+1]}(\mathbf{r}_1, \ldots, \mathbf{r}_{n+1}, \xi)\, d\mathbf{r}_{n+1}\, d\xi \end{aligned} \tag{32.7}$$

and

$$\begin{aligned}\mathbf{k}T \ln \rho^{[n]}(\mathbf{r}_1, \ldots, \mathbf{r}_n, \xi) &= \mathbf{k}T \ln \rho_{N-1}{}^{[n-1]}(\mathbf{r}_2, \ldots, \mathbf{r}_n) - \xi u(r_{1n}) \\ - \int_0^\xi \int_V u(r_{1,n+1}) \{\rho^{[n+1]}(\mathbf{r}_1, \ldots, \mathbf{r}_{n+1}, \xi) &- \rho^{[n]}(\mathbf{r}_1, \ldots, \mathbf{r}_{n-1}, \mathbf{r}_{n+1}, \xi)\}\, d\mathbf{r}_{n+1}\, d\xi \end{aligned} \tag{32.8}$$

The physical significance of the terms in Eq. (32.7) can be seen by rewriting it using the relation

$$\begin{aligned} - \mathbf{k}T \ln \frac{\rho^{(n)}(\mathbf{r}_1, \ldots, \mathbf{r}_n, \xi)}{\rho \rho_{N-1}{}^{(n-1)}(\mathbf{r}_2, \ldots, \mathbf{r}_n)} &= w^{(n)}(\mathbf{r}_1, \ldots, \mathbf{r}_n, \xi) \\ &\quad - w_{N-1}{}^{(n-1)}(\mathbf{r}_2, \ldots, \mathbf{r}_n) \end{aligned} \tag{32.9}$$

which follows from Eqs. (29.7) and (31.1). That is,

$$\begin{aligned} w^{(n)}(\mathbf{r}_1, \ldots, \mathbf{r}_n, \xi) &= w_{N-1}{}^{(n-1)}(\mathbf{r}_2, \ldots, \mathbf{r}_n) + \xi \sum_{i=2}^{n} u(r_{1i}) \\ &\quad + (N - n) \int_0^\xi \int_V u(r_{1,n+1}) P^{[n+1]}(\mathbf{r}_1, \ldots, \mathbf{r}_{n+1}, \xi)\, d\mathbf{r}_{n+1}\, d\xi \\ &\quad - (N - 1) \int_0^\xi \iint_V u(r_{12}) P^{(2)}(\mathbf{r}_1, \mathbf{r}_2, \xi)\, d\mathbf{r}_1\, d\mathbf{r}_2\, d\xi \end{aligned} \tag{32.10}$$

Equation (32.10) states that in a system of N molecules the potential of mean force associated with the set of molecules $1, \ldots, n$ located at $\mathbf{r}_1, \ldots, \mathbf{r}_n$ (with molecule 1 coupled in its interactions with all other molecules to the extent ξ) is made up of two contributions (1) and (2). The first part (1) is the potential of mean force of the set of molecules when $\xi = 0$, while the remaining contribution (2) arises on "charging" molecule 1 up to ξ. When $\xi = 0$, we have in effect a set of molecules $2, \ldots, n$ fixed at $\mathbf{r}_2, \ldots, \mathbf{r}_n$, out of a fully coupled fluid of $N - 1$ molecules, $2, \ldots, N$. The potential of mean force of this set is thus just $w_{N-1}{}^{(n-1)}(\mathbf{r}_2, \ldots, \mathbf{r}_n)$ and this is contribution (1) mentioned above. The second and third terms on the right-hand side of Eq. (32.10) represent the work[1] necessary to "charge" molecule 1 at $\mathbf{r}_1$ from $\xi = 0$ to ξ in the presence of the other $N - 1$ molecules, of which molecules $2, \ldots, n$ are held fixed at $\mathbf{r}_2, \ldots, \mathbf{r}_n$. The second term is the work associated with the interaction between molecule 1 and the fixed molecules $2, \ldots, n$, while the third term is the work arising from the averaged interaction of molecule 1 with the remaining $N - n$ molecules. This is obvious since $P^{[n+1]}\, d\mathbf{r}_{n+1}$ is the probability that any one of these molecules, say molecule $n + 1$, will be in $d\mathbf{r}_{n+1}$ if molecules $1, \ldots, n$ are at $\mathbf{r}_1, \ldots, \mathbf{r}_n$ (with molecule 1 coupled to the extent ξ). The negative of the fourth term on the right-hand side of Eq. (32.10) is clearly the work necessary to "charge" molecule 1 from $\xi = 0$ to ξ in the presence of the other $N - 1$ molecules, averaged over all configurations of molecules 1 to N (no molecule fixed in position). This also follows from the fact that ξ is in effect an external variable and, in view of Eq. (30.33),

$$(N-1)\int_0^{\xi}\iint_V u(r_{12})P^{(2)}(\mathbf{r}_1, \mathbf{r}_2, \xi)\, d\mathbf{r}_1\, d\mathbf{r}_2\, d\xi = A(N, \xi) - A(N, \xi = 0) \tag{32.11}$$

Incidentally, we note in passing that we can write in Eq. (32.3)

$$\frac{1}{N}\iint_V u(r_{12})\rho^{(2)}(\mathbf{r}_1,\mathbf{r}_2,\xi)\, d\mathbf{r}_1\, d\mathbf{r}_2 = \left[\frac{\partial A(N,\xi)}{\partial \xi}\right]_{N,V,T} \tag{32.12}$$

Thus contribution (2) [second, third, and fourth terms on the right-hand side of Eq. (32.10)], above, is the work of "charging" molecule 1 at $\mathbf{r}_1$ in the presence of fixed molecules $2, \ldots, n$ *relative to* the work of "charging" molecule 1, averaged over all configurations of molecules $1, \ldots, N$ (no fixed molecules). This latter work, the fourth term, is independent of $\mathbf{r}_1, \ldots, \mathbf{r}_n$ and thus serves to locate the zero of potential of mean force $w^{(n)}$ as defined in Eq. (31.1). It makes no contribution to the mean force (found by taking the appropriate gradient) on any of the particles of the set $1, \ldots, n$.

For $n = 1$, Eq. (32.7) reads (putting $\xi = 1$)

$$\begin{aligned}\mathbf{k}T \ln \rho^{(1)}(\mathbf{r}_1) = \mathbf{k}T \ln \rho + \frac{1}{N}\int_0^1 \iint_V u(r_{12})\rho^{(2)}(\mathbf{r}_1,\mathbf{r}_2,\xi)\, d\mathbf{r}_1\, d\mathbf{r}_2\, d\xi \\ - \int_0^1 \int_V u(r_{12})\rho^{[2]}(\mathbf{r}_1,\mathbf{r}_2,\xi)\, d\mathbf{r}_2\, d\xi \end{aligned} \tag{32.13}$$

This equation was employed by Kirkwood and Monroe[2] in connection with

[1] This is reversible work done at constant temperature and volume and thus, as already mentioned, $w^{(n)}$ is a free energy.

[2] J. G. Kirkwood and E. Monroe, *J. Chem. Phys.*, **9**, 514 (1941); see also J. G. Kirkwood in "Phase Transformations in Solids" (Wiley, New York, 1951), edited by R. Smoluchowski, J. E. Mayer, and W. A. Weyl.

a theory of fusion of crystals. For a fluid, the two integrals in Eq. (32.13) cancel and we obtain $\rho^{(1)}(\mathbf{r}_1) = \rho$, as expected.

Fluids. The remainder of the discussion in this section is devoted to fluids. Equation (32.7) can be written in this case, using the shorthand notation of Eq. (31.18), as

$$\mathbf{k}T \ln \rho^{(n)}(1 \cdot \cdot \cdot n, \xi) = \mathbf{k}T \ln \rho + \mathbf{k}T \ln \rho_{N-1}{}^{(n-1)}(2 \cdot \cdot \cdot n) - \xi \sum_{i=2}^{n} u(r_{1i})$$
$$- \rho \int_0^\xi \int_V u(r_{1,n+1})[g^{[n+1]}(1 \cdot \cdot \cdot n+1, \xi) - g^{(2)}(1, n+1, \xi)]\, d\mathbf{r}_{n+1}\, d\xi \tag{32.14}$$

From this equation we see that in the limit $\rho \to 0$,

$$- \mathbf{k}T \ln \frac{\rho^{(n)}(1 \cdot \cdot \cdot n, \xi)}{\rho \rho_{N-1}{}^{(n-1)}(2 \cdot \cdot \cdot n)} \to \xi \sum_{i=2}^{n} u(r_{1i}) \tag{32.15}$$

Then using Eq. (32.9), this becomes ($w^{(n)} = W^{(n)}$ for a fluid)

$$W^{(n)}(1 \cdot \cdot \cdot n, \xi) - W_{N-1}{}^{(n-1)}(2 \cdot \cdot \cdot n) \to \xi \sum_{i=2}^{n} u(r_{1i}) \tag{32.16}$$

or

$$W^{(n)}(1 \cdot \cdot \cdot n, \xi) \to \sum_{2 \leqslant i < j \leqslant n} u(r_{ij}) + \xi \sum_{i=2}^{n} u(r_{1i}) \tag{32.17}$$

That is, as $\rho \to 0$, the potential of mean force $W^{(n)}(1 \cdot \cdot \cdot n, \xi)$ approaches just the van der Waals potential energy of the set of molecules $1, \ldots, n$ (taking into account the coupling ξ of molecule 1). This is a consequence of the definition of $W^{(n)}$ and of the fact that the influence of the remaining $N - n$ molecules on the set $1, \ldots, n$ vanishes as $\rho \to 0$. From Eq. (31.8), if we neglect the term of order $1/N$, as $\rho \to 0$ (see Appendix 7),

$$- \mathbf{k}T \ln g^{(n)}(1 \cdot \cdot \cdot n, \xi) \to \sum_{2 \leqslant i < j \leqslant n} u(r_{ij}) + \xi \sum_{i=2}^{n} u(r_{1i}) \tag{32.18}$$

A special case of Eq. (32.18) is

$$g^{(2)}(12,\xi) \to e^{-\xi u(r_{12})/\mathbf{k}T} \tag{32.19}$$

This expression for $g^{(2)}$, taking $\xi = 1$, when substituted in Eq. (30.20), gives a second virial coefficient for a gas in agreement with the results of Chap. 5 [Eq. (22.26) and Sec. 25], as can be seen after an integration by parts.

On introducing correlation functions, the $n = 2$ integral equation is, from Eq. (32.7), (32.8), or (32.14),

$$- \mathbf{k}T \ln g^{(2)}(12,\xi) = \xi u(r_{12}) + \rho \int_0^\xi \int_V u(r_{13})[g^{[3]}(123,\xi) - g^{(2)}(13,\xi)]\, d\mathbf{r}_3\, d\xi \tag{32.20}$$

where a term $\mathbf{k}T \ln [1 - (1/N)]$ has been dropped. If we use Eq. (32.18) as a first approximation to $g^{(2)}$ and $g^{(3)}$, and insert these expressions in the integrand of Eq. (32.20), we obtain as a second approximation to $g^{(2)}$ [valid for higher densities than Eq. (32.19)], with the aid of Eq. (29.50) and an integration with respect to ξ,

$$-\mathbf{k}T \ln g^{(2)}(12,\xi) = \xi u(r_{12}) - \rho \mathbf{k}T \int_V [e^{-\xi u(r_{13})/\mathbf{k}T} - 1][e^{-u(r_{23})/\mathbf{k}T} - 1]\, d\mathbf{r}_3 \tag{32.21}$$

This equation is related to the third virial coefficient of a gas and will be referred to again in Sec. 34.

Superposition Approximation. To obtain an approximate integral equation for $g^{(2)}$ applicable at any density, we resort to the superposition approximation already discussed in Sec. 31. This approximation serves to break the

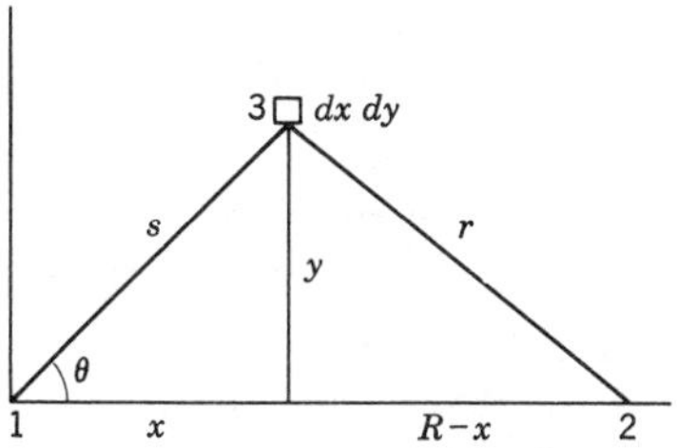

FIG. 20. Bipolar coordinates.

chain of integral equations, Eq. (32.14), at $n = 2$. Since the only coupling parameter is associated with molecule 1, Eq. (31.40) becomes

$$g^{[3]}(123,\xi) = g^{(2)}(13,\xi)g^{(2)}(23) \tag{32.22}$$

Inserting this relation in Eq. (32.20),

$$-\mathbf{k}T \ln g^{(2)}(12,\xi) = \xi u(r_{12}) + \rho \int_0^\xi \int_V u(r_{13}) g^{(2)}(13,\xi)[g^{(2)}(23) - 1]\, d\mathbf{r}_3\, d\xi \tag{32.23}$$

This is the Kirkwood integral equation for $g^{(2)}$.

For computational purposes, it is convenient to introduce bipolar coordinates in Eq. (32.23); see Fig. 20. We put $R = r_{12}$, $s = r_{13}$, and $r = r_{23}$. Rotation of the element of area $dx\,dy$ about the x axis sweeps out $d\mathbf{r}_3$, so that $d\mathbf{r}_3 = 2\pi y\, dx\, dy$. We then transform coordinates x and y to s and r, where

$$s^2 = x^2 + y^2 \qquad r^2 = y^2 + (R - x)^2$$

The Jacobian of this transformation yields

$$dx\,dy = \frac{sr}{yR}\,ds\,dr$$

or

$$d\mathbf{r}_3 = \frac{2\pi sr}{R}\,ds\,dr \tag{32.24}$$

As limits of integration, we use

$$\begin{aligned} 0 \leqslant r \leqslant \infty \\ |R - r| \leqslant s \leqslant R + r \end{aligned} \tag{32.25}$$

Then Eq. (32.23) takes the form

$$-\mathbf{k}T \ln g(R,\xi) = \xi u(R) + \frac{2\pi\rho}{R}\int_0^\infty r[g(r) - 1]\left\{\int_0^\xi \int_{|R-r|}^{R+r} su(s)g(s,\xi)\,ds\,d\xi\right\} dr \tag{32.26}$$

Computations based on this equation will be discussed in Secs. 35 and 36.

33. The Born-Green-Yvon Integral Equation[1]

In Sec. 32, in order to deduce an integrodifferential equation in $\rho^{(n)}(\mathbf{r}_1, \ldots, \mathbf{r}_n, \boldsymbol{\xi})$, we began by differentiating $\rho^{(n)}$ with respect to one of the ξ_i. In this section, an alternative but completely equivalent integrodifferential equation is found by taking the gradient of $\rho^{(n)}$ with respect to one of $\mathbf{r}_1, \ldots, \mathbf{r}_n$ as the initial step. Both of these procedures are special cases of Mayer's general variational method (see Sec. 38).

From Eq. (31.12), taking $\alpha = 1$,

$$-\mathbf{k}T\nabla_1 \ln \rho^{(n)}(\mathbf{r}_1, \ldots, \mathbf{r}_n, \boldsymbol{\xi}) = \overline{\nabla_1 U^{(n)}} \tag{33.1}$$

But from Eq. (29.1),

$$\nabla_1 U = \sum_{i=2}^{N} \xi_1\xi_i\nabla_1 u(r_{1i}) \tag{33.2}$$

so that Eqs. (31.3) and (33.1) become

$$-\mathbf{k}T\nabla_1 \ln \rho^{(n)} = \frac{\xi_1 \sum_{i=2}^{N} \xi_i \int \cdots \int_V \exp\left[-U(\boldsymbol{\xi})/\mathbf{k}T\right] \nabla_1 u(r_{1i})\,d\mathbf{r}_{n+1} \cdots d\mathbf{r}_N}{\int \cdots \int_V \exp\left[-U(\boldsymbol{\xi})/\mathbf{k}T\right] d\mathbf{r}_{n+1} \cdots d\mathbf{r}_N} \tag{33.3}$$

[1] Yvon (Gen. Ref.); Born and Green (Gen. Ref.).

The sum in Eq. (33.3) can be broken up into two parts, $i = 2, \ldots, n$ and $i = n + 1, \ldots, N$. In the first part, the integrals in numerator and denominator cancel on taking $\nabla_1 u(r_{1i})$ outside the integral sign. For $i > n$, we write the integral in the numerator as

$$\int_V \nabla_1 u(r_{1i}) \left\{ \int \cdots \int_V \exp\left[-\frac{U(\boldsymbol{\xi})}{\mathbf{k}T} \right] d\mathbf{r}_{n+1} \cdots d\mathbf{r}_{i-1}\, d\mathbf{r}_{i+1} \cdots d\mathbf{r}_N \right\} d\mathbf{r}_i$$

Then, recalling Eq. (29.38),

$$-\mathbf{k}T\nabla_1 \ln \rho^{(n)} = \xi_1 \sum_{i=2}^{n} \xi_i \nabla_1 u(r_{1i}) + \xi_1 \sum_{i=n+1}^{N} \xi_i \int_V \nabla_1 u(r_{1i}) P^{[n+1]}(\mathbf{r}_1, \ldots, \mathbf{r}_n, \mathbf{r}_i, \boldsymbol{\xi})\, d\mathbf{r}_i \tag{33.4}$$

$$= -\bar{\mathbf{F}}_1^{(n)} \tag{33.5}$$

The physical interpretation of Eq. (33.5) is rather obvious. The mean force on particle 1 at $\mathbf{r}_1$ is the sum of two parts. The first part is the direct force owing to the van der Waals interactions with the fixed particles $2, \ldots, n$. The second part is the averaged force exerted on molecule 1 by molecules $n + 1, \ldots, N$. This is clear since $P^{[n+1]} d\mathbf{r}_i$ is the probability of molecule i being in $d\mathbf{r}_i$ at $\mathbf{r}_i$ if molecules $1, \ldots, n$ are fixed at $\mathbf{r}_1, \ldots, \mathbf{r}_n$.

Let us specialize to the extent of putting all ξ_i's equal to unity except ξ_1 which we call ξ as before. Then

$$-\mathbf{k}T\nabla_1 \ln \rho^{(n)}(\mathbf{r}_1, \ldots, \mathbf{r}_n, \xi) = \xi \sum_{i=2}^{n} \nabla_1 u(r_{1i}) + \xi \int_V \nabla_1 u(r_{1,n+1}) \rho^{[n+1]}(\mathbf{r}_1, \ldots, \mathbf{r}_{n+1}, \xi)\, d\mathbf{r}_{n+1} \tag{33.6}$$

Equations (33.4) and (33.6) again consist of a chain of $N - 1$ interlinked integrodifferential equations, the exact solution of which is presumably as difficult as would be carrying out the integrations in Z in the first place. But again there is the hope of discovering fortunate approximations to use in connection with this alternative procedure. One possibility, the superposition approximation, will be employed below.

The $n = 1$ equation (with $\xi = 1$),

$$-\mathbf{k}T\nabla_1 \ln \rho^{(1)}(\mathbf{r}_1) = \int_V \nabla_1 u(r_{12}) \rho^{[2]}(\mathbf{r}_1, \mathbf{r}_2)\, d\mathbf{r}_2 \tag{33.7}$$

which is equivalent to Eq. (32.13), has been employed in a recent discussion[1]

[1] J. G. Kirkwood in "Phase Transformations in Solids" (Wiley, New York, 1951), edited by R. Smoluchowski, J. E. Mayer, and W. A. Weyl.

of the theory of crystallization. In a fluid, the right-hand side of Eq. (33.7) is zero, by symmetry.

Fluids—Superposition Approximation. We now turn, in the remainder of this section, to the special case of a fluid, utilizing the superposition approximation. Putting $n = 2$ in Eq. (33.6), we have [see Eq. (32.22)]

$$-\mathbf{k}T\nabla_1 \ln g^{(2)}(12,\xi) = \xi\nabla_1 u(r_{12}) + \xi\rho\int_V \nabla_1 u(r_{13})g^{(2)}(13,\xi)g^{(2)}(23)\,d\mathbf{r}_3 \qquad (33.8)$$

In Eq. (33.8), we can write

$$\nabla_1 \ln g^{(2)}(12,\xi) = -\frac{\mathbf{r}_{12}}{r_{12}}\frac{\partial \ln g^{(2)}}{\partial r_{12}}$$

$$\nabla_1 u(r_{12}) = -\frac{\mathbf{r}_{12}}{r_{12}}\frac{du}{dr_{12}}$$

and

$$\nabla_1 u(r_{13}) = -\frac{\mathbf{r}_{13}}{r_{13}}\frac{du}{dr_{13}}$$

Because of the symmetry of the integrand in Eq. (33.8), on carrying out the integration over $d\mathbf{r}_3$ (Fig. 20), only the projection $(\mathbf{r}_{12}/r_{12})\cos\theta$ of the unit vector $\mathbf{r}_{13}/r_{13}$ along the line $1 \rightarrow 2$ will contribute. Then, on canceling $\mathbf{r}_{12}/r_{12}$ in every term and using the notation of Fig. 20, Eq. (33.8) becomes

$$-\frac{\partial}{\partial R}[\mathbf{k}T\ln g(R,\xi) + \xi u(R)]$$
$$= \pi\xi\rho\int_0^\infty u'(s)g(s,\xi)\,ds\int_{|R-s|}^{R+s}\frac{(s^2+R^2-r^2)}{R^2}rg(r)\,dr \qquad (33.9)$$

since

$$\cos\theta = \frac{s^2+R^2-r^2}{2sR}$$

It is convenient to make the choice of limits in Eq. (33.9) opposite to that in (32.25).

Next, we wish to integrate Eq. (33.9) with respect to R to obtain an integral equation. However, we remark first that, in order that the form of the resulting equation may be compared easily with Eq. (32.26), it is legitimate in Eq. (33.8) to replace $g^{(2)}(23)$ by $g^{(2)}(23) - 1$, since for a fluid

$$\int_V \nabla_1 u(r_{13})g^{(2)}(13,\xi)\,d\mathbf{r}_3 = 0 \qquad (33.10)$$

as has already been pointed out [Eq. (33.7)]. Also, in Eq. (33.9), $g(r) - 1$ may be written instead of $g(r)$. Further, in Eq. (33.9), let us extend the range of r to negative values by *defining* $g(-r) = g(r)$. Then we can replace $|R - s|$ by $R - s$ as a limit of integration since the integrand is an odd function of r and the integral over the range $R - s \leqslant r \leqslant s - R$ is zero when $s > R$.

We now integrate over R, between $R = R_1$ and $R = \infty$. At[1] $R = \infty$, $\ln g = 0$ and $u = 0$. Hence we have

$$\mathbf{k}T \ln g(R_1,\xi) = -\xi u(R_1) + \pi\xi\rho \lim_{R_2\to\infty} \int_0^\infty u'(s)g(s,\xi)\,ds \times \int_{R_1}^{R_2} dR \int_{R-s}^{R+s} \frac{s^2 + R^2 - r^2}{R^2} r[g(r) - 1]\,dr \quad (33.11)$$

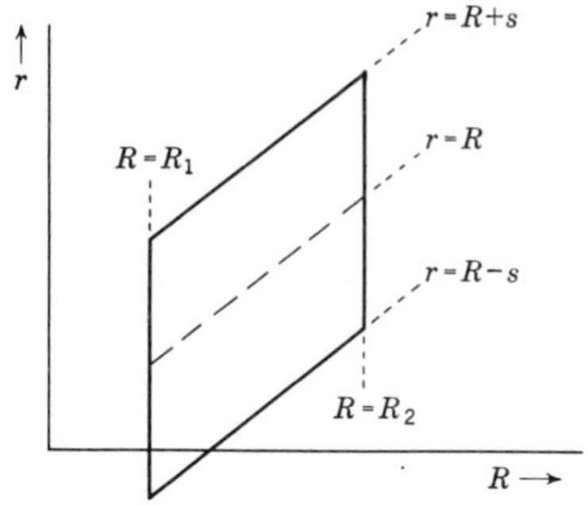

FIG. 21. Limits of integration in Eq. (33.11).

The limits on r and R are indicated in Fig. 21. If we reverse the order of integration over r and R,

$$\lim_{R_2\to\infty} \int_{R_1}^{R_2} dR \int_{R-s}^{R+s} dr = \underbrace{\int_{R_1-s}^{R_1+s} dr \int_{R_1}^{r+s} dR}_{\text{I}} + \underbrace{\lim_{R_2\to\infty} \int_{R_1+s}^{R_2-s} dr \int_{r-s}^{r+s} dR}_{\text{II}} + \underbrace{\lim_{R_2\to\infty} \int_{R_2-s}^{R_2+s} dr \int_{r-s}^{R_2} dR}_{\text{III}}$$

The integration over R,

$$\int_a^b \frac{s^2 + R^2 - r^2}{R^2}\,dR$$

can now be carried out with the limits in I, II, and III. The result in II is zero. In III, it is nonzero, but if we then take the limit $R_2 \to \infty$, integral III vanishes, since[2] $g(r) - 1 \to 0$ as $r \to \infty$. We are left then with integral I, and hence (writing R for R_1)

$$\mathbf{k}T \ln g(R,\xi) = -\xi u(R) + \pi\xi\rho \int_0^\infty u'(s)g(s,\xi)\,ds \int_{R-s}^{R+s} \frac{s^2 - (R - r)^2}{R} \times r[g(r) - 1]\,dr \quad (33.12)$$

[1] Neglecting $1/N$ compared to unity. See Appendix 7.

[2] Neglecting $1/N$ compared to unity. This awkwardness about terms of $O(1/N)$ is eliminated by use of the grand ensemble (see Sec. 37 and Appendix 7).

Figure 22 shows the limits of integration in Eq. (33.12). We reverse the order of integration again here and split the resulting integral into two parts:

$$\int_0^{\infty} ds \int_{R-s}^{R+s} dr = \int_{-\infty}^{+\infty} dr \int_{|R-r|}^{\infty} ds = \int_0^{\infty} dr \int_{|R-r|}^{\infty} ds - \int_0^{-\infty} dr \int_{|R-r|}^{\infty} ds \quad (33.13)$$

In the last integral we then change variables from r to $-r$. The result is

$$\mathbf{k}T \ln g(R,\xi) = -\xi u(R) + \frac{\pi\xi\rho}{R}\int_0^{\infty} r[g(r)-1]\,dr \times \int_{|R-r|}^{\infty} [s^2-(R-r)^2]u'(s)g(s,\xi)\,ds - \frac{\pi\xi\rho}{R}\int_0^{\infty} r[g(r)-1]\,dr \times \int_{R+r}^{\infty} [s^2-(R+r)^2]u'(s)g(s,\xi)\,ds \quad (33.14)$$

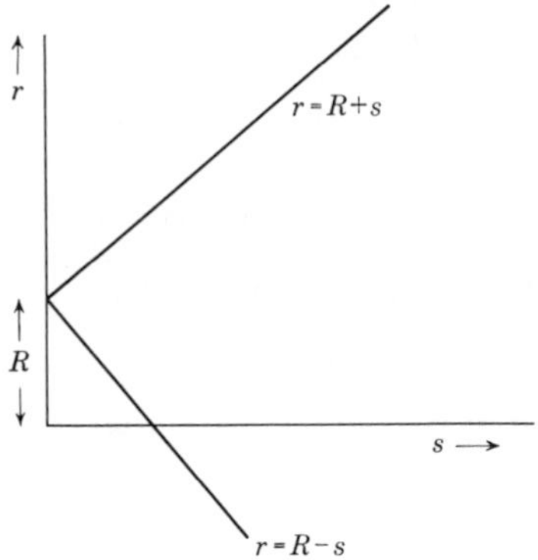

FIG. 22. Limits of integration in Eq. (33.12).

Finally, this equation can be written as

$$\ln g(R,\xi) = -\frac{\xi u(R)}{\mathbf{k}T} + \frac{\pi\rho}{R}\int_0^{\infty} [K(R-r,\,\xi) - K(R+r,\,\xi)]r[g(r)-1]\,dr \quad (33.15)$$

where $$K(t,\xi) = \frac{\xi}{\mathbf{k}T}\int_{|t|}^{\infty} (s^2-t^2)u'(s)g(s,\xi)\,ds \qquad \text{(Born-Green-Yvon)} \quad (33.16)$$

On the other hand, the Kirkwood integral equation for a fluid (employing the superposition approximation), Eq. (32.26), can also be written as in Eq. (33.15), but with the kernel in this case taking the form

$$K(t,\xi) = -\frac{2}{\mathbf{k}T}\int_0^{\xi}\int_{|t|}^{\infty} su(s)g(s,\xi)\,ds\,d\xi \qquad \text{(Kirkwood)} \quad (33.17)$$

The fact that the two kernels are not identical is of course a reflection of the use of the superposition approximation. In the absence of any approximation the two methods (Kirkwood and Born-Green-Yvon) would have to give identical results for $g(R,\xi)$. Equations (33.15) to (33.17) will be discussed further in Secs. 35 and 36.

In conclusion, it should be mentioned that Eq. (33.12) often appears written in a slightly different way. Let us put $y = r - R$. Then

$$\mathbf{k}T \ln g(R,\xi) = -\xi u(R) + \pi\xi\rho \int_0^\infty u'(s) g(s,\xi)\, ds \int_{-s}^{+s} (s^2 - y^2) \times \frac{y + R}{R} [g(y + R) - 1]\, dy \qquad (33.18)$$

34. Radial Distribution Function and Superposition Approximation in Gases

We consider briefly in this section the application of methods discussed above to dilute gases.

Radial Distribution Function. At zero density, according to Eq. (32.19),

$$g(R) = e^{-u(R)/\mathbf{k}T} \qquad (34.1)$$

where a term of order $1/N$ has been dropped (see Appendix 7 for other details). It is therefore convenient to write, for the expansion of $g(R)$ in powers of ρ,

$$g(R,\rho,T) = e^{-u(R)/\mathbf{k}T}[1 + \rho g_1(R,T) + \rho^2 g_2(R,T) + \cdots] \qquad (34.2)$$

If Eq. (34.2) is substituted into Eq. (30.20), the leading term in the series gives the second virial coefficient of the gas [as already mentioned in connection with Eq. (32.19)], the next term gives the third virial coefficient, etc.:

$$B_k = -\frac{1}{6\mathbf{k}T} \int_0^\infty ru'(r) e^{-u(r)/\mathbf{k}T} g_{k-2}(r,T) \cdot 4\pi r^2\, dr \qquad (34.3)$$

where $g_0 \equiv 1$ and B_k is the kth virial coefficient [Eq. (25.8)].

On the other hand, we can write

$$g(R,z,T) = e^{-u(R)/\mathbf{k}T}[1 + zG_1(R,T) + z^2 G_2(R,T) + \cdots] \qquad (34.4)$$

where z is the activity. Since $z \to \rho$ as $\rho \to 0$, $G_1(R,T) = g_1(R,T)$.

DeBoer and Mayer and Montroll[1] give expressions for the G_i in terms of modified cluster integrals.[2] In fact, any $\rho^{(n)}$ can be expanded in powers of z

[1] DeBoer (Gen. Ref.); J. E. Mayer and E. Montroll, *J. Chem. Phys.*, **9**, 2 (1941).

[2] In the development of these modified cluster integrals, following Sec. 22, two molecules are held fixed at $\mathbf{r}_1$ and $\mathbf{r}_2$ where $R = |\mathbf{r}_1 - \mathbf{r}_2|$. Thus a configuration integral over $N - 2$ particles arises, as in Eq. (29.8) for $n = 2$, instead of Z itself. See Appendix 7.

by the methods developed by these authors. Similarly, the g_i can be related to modified irreducible cluster integrals and any $\rho^{(n)}$ can be expanded in powers of ρ. We shall not make use[1] of the general expressions for g_i and G_i (and their generalizations for $\rho^{(n)}$), but confine ourselves to $g_1 = G_1$ as deduced from Eq. (32.21). Equation (32.21) can be written

$$g(r_{12},\rho,T) = e^{-u(r_{12})/\mathbf{k}T} \exp\left\{\rho \int_V [e^{-u(r_{13})/\mathbf{k}T} - 1][e^{-u(r_{23})/\mathbf{k}T} - 1]\, d\mathbf{r}_3\right\}$$

$$= e^{-u(r_{12})/\mathbf{k}T}\left\{1 + \rho \int_V [e^{-u(r_{13})/\mathbf{k}T} - 1][e^{-u(r_{23})/\mathbf{k}T} - 1]\, d\mathbf{r}_3\right\} \quad (34.5)$$

and hence

$$g_1(r_{12},T) = G_1(r_{12},T) = \int_V [e^{-u(r_{13})/\mathbf{k}T} - 1][e^{-u(r_{23})/\mathbf{k}T} - 1]\, d\mathbf{r}_3 \quad (34.6)$$

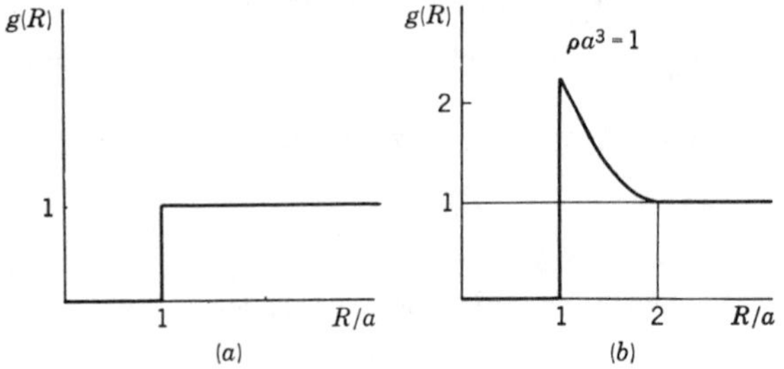

FIG. 23. Radial distribution function for gas of hard spheres according to (*a*) Eq. (34.1) and (*b*) Eq. (34.9) with $\rho a^3 = 1$.

This same result is obtained by the methods of DeBoer and Mayer and Montroll. The third virial coefficient follows from $g_1(r_{12},T)$ and Eq. (34.3), and is in agreement with Eqs. (25.9) and (24.3) of Chap. 5.

Examples. In Fig. 23*a* and *b*, $g(R)$ is plotted for a gas of hard spheres, with

$$\begin{aligned} u(R) &= 0 \qquad & R > a \\ &= +\infty \qquad & R \leqslant a \end{aligned} \quad (34.7)$$

according to Eqs. (34.1) and (34.5) (with $r_{12} = R$), respectively. In Eq. (34.5) the integral is zero except when *both* $r_{13} \leqslant a$ and $r_{23} \leqslant a$. For $a \leqslant R \leqslant 2a$, the only region contributing to the integral is that between molecules 1 and 2 where the center of a third molecule is excluded. Using Eq. (32.24), the integral is in this case

$$2 \cdot \frac{2\pi}{R} \int_{R/2}^{a} r\, dr \int_{R-r}^{r} s\, ds = \frac{4\pi}{3} a^3 \left[1 - \frac{3}{4}\left(\frac{R}{a}\right) + \frac{1}{16}\left(\frac{R}{a}\right)^3\right] \quad (34.8)$$

[1] See Sec. 37 for a discussion of these expansions based on the grand canonical ensemble.

and thus

$$g(R) = 0 \qquad R \leqslant a$$
$$= 1 + \frac{4\pi}{3}(\rho a^3)\left[1 - \frac{3}{4}\left(\frac{R}{a}\right) + \frac{1}{16}\left(\frac{R}{a}\right)^3\right] \qquad a < R \leqslant 2a \qquad (34.9)$$
$$= 1 \qquad R \geqslant 2a$$

When $R > a$, there is no direct force between molecules 1 and 2. Hence the increased value of $g(R)$ over unity, between a and $2a$ in Fig. 23*b*, must be attributed to an effective mean attractive force between molecules 1 and 2

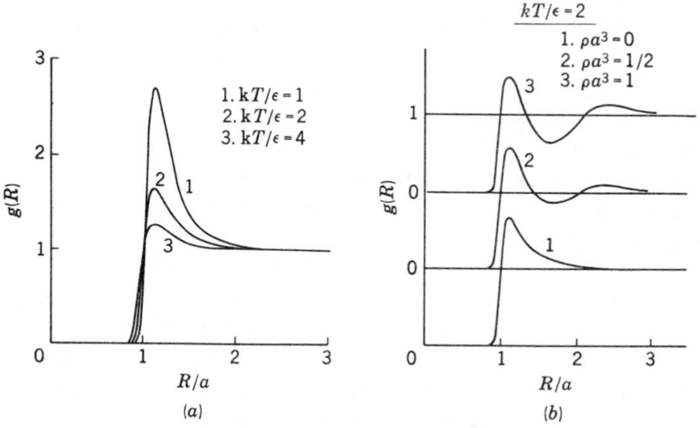

FIG. 24. Radial distribution function for Lennard-Jones gas according to (*a*) Eq. (34.1) and (*b*) Eq. (34.5).

arising from the remaining $N - 2$ molecules [Eq. (33.4)]. The physical explanation of this effective mean force is that molecule 2 shields molecule 1 unsymmetrically from collisions with the other $N - 2$ molecules: molecule 1 is involved more frequently in collisions originating from the side opposite molecule 2 than from the molecule 2 side.

For molecules following the Lennard-Jones potential [Eq. (27.29)]

$$u(R) = 4\varepsilon\left[\left(\frac{a}{R}\right)^{12} - \left(\frac{a}{R}\right)^6\right] \qquad (34.10)$$

g is a function of the reduced variables ρa^3, $\mathbf{k}T/\varepsilon$ and R/a. Figure 24*a* shows the behavior of $g(R/a)$ at zero gas density [Eq. (34.1)] and several temperatures. The single maximum in g occurs of course at that value of R for which $u(R)$ is a minimum, $R/a = 2^{1/6} = 1.123$. With $\rho > 0$ [Eq. (34.5)], the behavior is more complicated, as shown in Fig. 24*b*.[1] A second maximum appears in the $\rho a^3 = {}^1/_2$ case, characteristic of experimental curves for liquids and also for gases at sufficiently high densities.[2] Roughly speaking, this "second-neighbor" shell around a central molecule is just a reflection of the first-neighbor shells around the central molecule's first neighbors. DeBoer (Gen. Ref.) has shown that a detailed analysis of Fig. 24*b* can be given in terms of the potential of mean force. It should be added that a sufficiently dense fluid of hard spheres will also exhibit a second (third, etc.) maximum in $g(R)$ (see Sec. 35).

Superposition Approximation. Although the superposition approximation may be presumed useful at any density, it is not possible to specify just how serious the approximation really is in general because the exact solution of the chain of distribution function integral equations is unknown. However, it *is* possible to test this approximation for gases of low density,[3] since Mayer's theory is available in this region, and is exact.

We start with the expansion of $g(R)$ in Eq. (34.2), and with Eq. (34.3), which expresses the virial coefficients of the gas in terms of the coefficients in the $g(R)$ expansion. If we use coefficients g_k, found by means of the superposition approximation, to deduce virial coefficients B_k from Eq. (34.3), these "superposition" virial coefficients may then be compared with Mayer's exact formulas for the virial coefficients, Eqs. (24.3) and (25.9).

In order to deduce the g_k from the superposition approximation, the expansion in Eq. (34.2) is substituted for $g(R)$ on both sides of the Born-Green-Yvon integral equation [Eq. (33.18) with $\xi = 1$]. It will be recalled that the superposition approximation was resorted to in deriving this integral equation. Coefficients of like powers of ρ on the two sides of the resulting equation are then equated, yielding

$$g_1(R,T) = \frac{\pi}{\mathbf{k}T} \int_0^\infty u'(s) e^{-u(s)/\mathbf{k}T} ds \int_{-s}^{+s} (s^2 - y^2) \frac{y + R}{R} [e^{-u(y+R)/\mathbf{k}T} - 1]\, dy \tag{34.11}$$

and a similar but somewhat more complicated equation for $g_2(R,T)$ involving g_1 in the integrand.

The "superposition" second virial coefficient B_2 agrees of course with the exact second virial coefficient since g_0 in Eq. (34.2) is a constant ($g_0 = 1$) independent of the nature of the approximation. Further, Rushbrooke and Scoins were able to show from g_1 in Eq. (34.11) that the "superposition" B_3 is also identical with Mayer's result[4] but that the "superposition" B_4 is not.

[1] Calculations have been carried out by DeBoer and Michels and Montroll and Mayer.

[2] A. Eisenstein and N. S. Gingrich, *Phys. Rev.*, **62**, 261 (1942).

[3] G. S. Rushbrooke and H. I. Scoins, *Phil. Mag.*, **42**, 582 (1951).

[4] Hence, Eqs. (34.6) and (34.11) are equivalent.

The error in B_4 arises only in connection with the term in $S'_{1,2,3,4}$ [see Eq. (24.3)] corresponding to the bond figure below. That is, the superposition approximation handles the other nine terms (of two types) in $S'_{1,2,3,4}$ correctly (see Fig. 7).

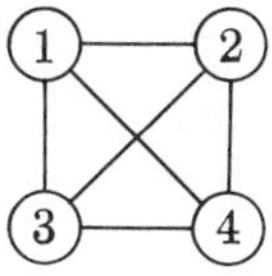

Numerical calculations, to show the extent of this discrepancy, were carried out, independently, for a gas of hard spheres, by[1] Rushbrooke and Scoins, Nijboer and van Hove, and Hart, Wallis, and Pode. The exact coefficients are known[2] to be $B_2 = 2\pi a^3/3 \equiv b$, $B_3 = (5/8)b^2$, and $B_4 = 0.2869b^3$. The "superposition" values, for comparison, are $B_2 = b$, $B_3 = (5/8)b^2$, and $B_4 = 0.2252b^3$.

In summary: The superposition approximation leads to second and third virial coefficients of a gas which are exact, but to a fourth virial coefficient which is not exact. In the case of a gas of hard spheres, the error in the fourth virial coefficient amounts to 21.5 per cent.

35. Fluid of Hard Spheres According to the Superposition Approximation

Kirkwood, Maun, and Alder[3] have carried out numerical computations for a fluid of hard spheres, based on Eqs. (33.15) to (33.17). These results will be summarized in the present section. They not only possess intrinsic interest but also serve as the starting point for the computations with the modified Lennard-Jones potential, described in Sec. 36.

It is convenient to introduce the dimensionless variable $x = R/a$ in Eqs. (33.15) to (33.17), where a is a length characteristic of the molecules. Let us also replace, in Eqs. (33.15) to (33.17), r/a by r, s/a by s, and t/a by t. Then

$$\ln g(x, \xi) = -\frac{\xi u(x)}{\mathbf{k}T} + \frac{\lambda_0}{4x}\int_0^\infty [K(x-r, \xi) - K(x+r, \xi)]r[g(r)-1]\,dr \tag{35.1}$$

where

$$\lambda_0 = 4\pi\rho a^3 \tag{35.2}$$

and $K(t,\xi)$ is given formally by Eqs. (33.16) and (33.17) though with the new definition of s and t, K [Eqs. (33.16) and (33.17)] $= a^2K$ [Eq. (35.1)]. If we replace r by $-r$ in the second integral in Eq. (35.1), involving $K(x + r)$, and define

$$\begin{aligned} g(x) &= g(-x) \\ u(x) &= u(-x) \end{aligned} \tag{35.3}$$

[1] Rushbrooke and Scoins, *loc. cit.*; B. R. A. Nijboer and L. van Hove, *Phys. Rev.*, **85**, 777 (1952); R. W. Hart, R. Wallis, and L. Pode, *J. Chem. Phys.*, **19**, 139 (1951).

[2] See, for example, R. H. Fowler and E. A. Guggenheim, "Statistical Thermodynamics" (Cambridge, London, 1939), p. 289.

[3] J. G. Kirkwood, E. K. Maun, and B. J. Alder, *J. Chem. Phys.*, **18**, 1040 (1950); J. G. Kirkwood and E. M. Boggs, *J. Chem. Phys.*, **10**, 394 (1942).

Eq. (35.1) becomes

$$\ln g(x, \xi) = -\frac{\xi u(x)}{\mathbf{k}T} + \frac{\lambda_0}{4x}\int_{-\infty}^{\infty} K(x - r, \xi) r[g(r) - 1]\, dr \tag{35.4}$$

with $K(x - r, \xi)$ still given (formally) by Eqs. (33.16) and (33.17).

We now restrict ourselves to the case of a fluid of hard spheres of diameter a with

$$\begin{aligned} u(x) &= +\infty \qquad x = \frac{R}{a} \leqslant 1 \\ &= 0 \qquad x > 1 \end{aligned} \tag{35.5}$$

We take $\xi = 1$ and consider the Born-Green-Yvon kernel first:

$$K(t) = \frac{1}{\mathbf{k}T}\int_{|t|}^{\infty} (s^2 - t^2) u'(s) g(s)\, ds \tag{35.6}$$

Now, from Eq. (35.5), $u'(s) = 0$ for $s > 1$; therefore $K(t) = 0$ for $|t| > 1$. To examine the situation with $|t| \leqslant 1$, we note that

$$\begin{aligned} e^{-u(x)/\mathbf{k}T} &= 0 \qquad x \leqslant 1 \\ &= 1 \qquad x > 1 \end{aligned} \tag{35.7}$$

If we differentiate this unit step function,

$$\frac{d}{dx} e^{-u(x)/\mathbf{k}T} = \delta(x - 1 +) = -\frac{u'(x)}{\mathbf{k}T} e^{-u(x)/\mathbf{k}T} \tag{35.8}$$

where $\delta(x - a)$ is the Dirac δ function and $1 +$ means $1 + \varepsilon$, $\varepsilon > 0$, as $\varepsilon \to 0$. Also, from the first term on the right-hand side of Eq. (35.4), $g(s) = 0$ for $s < 1$. Thus, using Eq. (35.8), the only contribution to the integral in Eq. (35.6), for $|t| \leqslant 1$, arises at $s = 1 +$:

$$K(t) = -(1 - t^2) g(1) \tag{35.9}$$

where $g(1) \equiv g(1 +)$. Of course $g(1)$ is a function of ρ or λ_0: $g(1) = g(1,\lambda_0)$. In summary, then, Eq. (35.4) becomes

$$g(x) = 0 \qquad x \leqslant 1 \tag{35.10}$$

$$x \ln g(x) = \frac{\lambda}{4}\int_{-\infty}^{\infty} K_0(x - r) r[g(r) - 1]\, dr \qquad x > 1 \tag{35.11}$$

where

$$K_0(t) = 0 \qquad |t| > 1 \tag{35.12}$$

$$= t^2 - 1 \qquad |t| \leqslant 1 \tag{35.13}$$

and

$$\lambda \equiv \lambda_0 g(1,\lambda_0) \qquad \text{(Born-Green-Yvon)} \tag{35.14}$$

Equation (35.14) determines λ as a function of λ_0 or λ_0 as a function of λ. Solution of Eq. (35.11) will actually provide $g(1)$ as a function of λ so that we may also write

$$\lambda[g(1,\lambda)]^{-1} = \lambda_0 \tag{35.15}$$

We now examine the Kirkwood kernel, Eq. (33.17), for a fluid of hard spheres and $\xi = 1$, following Kirkwood and Boggs.[1] For $|t| > 1$, $K(t) = 0$ because of $u(s)$ in the integrand and Eq. (35.5). For $|t| \leqslant 1$, Eq. (33.17) becomes

$$K(t) = -\frac{2}{\mathbf{k}T}\int_0^1 \int_{|t|}^1 su(s)g(s,\xi)\, ds\, d\xi \tag{35.16}$$

Now we define a function $\chi(s)$ by

$$\chi(s) = \rho\frac{u(s)}{\mathbf{k}T}\int_0^1 g(s,\xi)\, d\xi \tag{35.17}$$

so that

$$K(t) = -\frac{2}{\rho}\int_{|t|}^1 \chi(s)s\, ds \tag{35.18}$$

At this point an additional approximation is introduced into the Kirkwood kernel. This approximation could be avoided at the expense of further computations, but it is particularly advantageous to introduce the approximation since it results (see below) in identical integral equations in the BGY and K cases, except for the definition of λ. The approximation consists in replacing $\chi(s)$ in the interval $0 \leqslant s \leqslant 1$ by its average value $\bar{\chi}$, defined by

$$\bar{\chi} = \frac{\int_0^1 s^2\chi(s)\, ds}{\int_0^1 s^2\, ds} = 3\int_0^1 s^2\chi(s)\, ds \tag{35.19}$$

We therefore put $\bar{\chi}$ for $\chi(s)$ in Eq. (35.18), with the result

$$K(t) = -\frac{\bar{\chi}}{\rho}(1 - t^2) \tag{35.20}$$

Equation (35.4) thus reduces, as in the BGY theory, to Eqs. (35.10) to (35.13), but here

$$\lambda \equiv \frac{\lambda_0\bar{\chi}}{\rho} = 4\pi a^3\bar{\chi} \qquad \text{(Kirkwood)} \tag{35.21}$$

It is desirable to obtain an explicit relation between λ and λ_0 in the Kirkwood formulation, corresponding to Eq. (35.15) for the Born-Green-Yvon formulation. This requires, first, expressions for pV and μ, which we shall

[1] Kirkwood and Boggs, *loc. cit.*

need in any case in the discussion below of the thermodynamic functions of the fluid of hard spheres. The pressure is determined by Eq. (30.20). The derivative $u'(r)$ in the integrand is treated as in Eq. (35.6), and we find

$$\frac{pv}{\mathbf{k}T} = 1 + \frac{2\pi a^3 g(1)}{3v} \qquad \text{(Born-Green-Yvon or Kirkwood)} \qquad (35.22)$$

In Eq. (30.36) for the chemical potential, we write r for r/a, so that

$$\frac{\mu}{\mathbf{k}T} = \ln \rho\Lambda^3 + \frac{\rho a^3}{\mathbf{k}T}\int_0^1\int_0^1 u(r)g(r,\xi)\cdot 4\pi r^2\,dr\,d\xi \qquad (35.23)$$

since the integral over $1 < r \leqslant \infty$ vanishes because of $u(r)$. Then

$$\frac{\mu}{\mathbf{k}T} = \ln \rho\Lambda^3 + 4\pi a^3\int_0^1 \chi(r)r^2\,dr$$

$$= \ln \rho\Lambda^3 + \frac{4\pi a^3\bar{\chi}}{3} = \ln \rho\Lambda^3 + \frac{\lambda}{3} \qquad \text{(Kirkwood)} \qquad (35.24)$$

in view of Eq. (35.21).

The relation between λ and λ_0 in the Kirkwood theory can now be derived from the thermodynamic equation

$$\left[\frac{\partial(\mu/\mathbf{k}T)}{\partial v}\right]_T = v\left[\frac{\partial(p/\mathbf{k}T)}{\partial v}\right]_T \qquad (35.25)$$

On substituting Eqs. (35.22) and (35.24) in Eq. (35.25), we obtain

$$\frac{2\pi a^3}{3}\,v\,\frac{d[g(1)/v^2]}{dv} = \frac{1}{3}\frac{d\lambda}{dv}$$

If we introduce $\gamma = v/g(1)^{1/2}$, this becomes

$$4\pi a^3\frac{d\gamma}{\gamma^2} = -\frac{d\lambda}{g(1)^{1/2}}$$

where we regard $g(1)$ as a function of λ. Then on integration,

$$4\pi a^3\int_\gamma^\infty \frac{d\gamma}{\gamma^2} = \int_0^\lambda \frac{d\lambda}{g(1)^{1/2}}$$

or

$$\frac{1}{[g(1,\lambda)]^{1/2}}\int_0^\lambda \frac{d\lambda}{[g(1,\lambda)]^{1/2}} = 4\pi a^3\rho = \lambda_0 \qquad \text{(Kirkwood)} \qquad (35.26)$$

In summary: To obtain the radial distribution function for a fluid of hard spheres, according to the superposition approximation, the integral equation, Eqs. (35.10) and (35.11), must be solved for various values of the parameter λ, where the kernel K_0 is given in Eqs. (35.12) and (35.13). The connection between the parameter λ and the density of the fluid ρ (or λ_0) is different in the BGY and K cases, the two relations being Eqs. (35.15) and (35.26), where $g(1)$ is obtained as part of the solution of the integral equation for each

TABLE 2. RADIAL DISTRIBUTION FUNCTIONS FOR SEVERAL VALUES OF PARAMETER λ. $x[g(x) - 1]$ AS A FUNCTION OF x

x \ λ	5	10	20	27.4	33
1.00	0.45	0.80	1.36	1.66	1.85
1.08	0.39	0.66	1.08	1.36	1.62
1.16	0.32	0.53	0.83	1.04	1.25
1.24	0.26	0.40	0.59	0.73	0.87
1.32	0.20	0.29	0.37	0.44	0.47
1.40	0.15	0.18	0.18	0.16	0.11
1.48	0.09	0.09	0.01	−0.08	−0.19
1.56	0.05	0.01	−0.12	−0.26	−0.41
1.64	0.01	−0.05	−0.22	−0.39	−0.56
1.72	−0.02	−0.10	−0.29	−0.46	−0.64
1.80	−0.04	−0.13	−0.31	−0.48	−0.63
1.88	−0.05	−0.13	−0.28	−0.41	−0.52
1.96	−0.05	−0.10	−0.18	−0.25	−0.29
2.04	−0.03	−0.05	−0.03	0.02	0.10
2.12	−0.02	−0.01	0.09	0.24	0.44
2.20	0.00	0.02	0.16	0.34	0.63
2.28	0.00	0.04	0.18	0.38	0.65
2.36	0.01	0.04	0.17	0.32	0.52
2.44	0.01	0.04	0.13	0.22	0.30
2.52	0.01	0.03	0.07	0.09	0.06
2.60	0.01	0.02	0.01	−0.03	−0.16
2.68	0	0.01	−0.04	−0.13	−0.32
2.76	....	0.00	−0.07	−0.20	−0.42
2.84	....	−0.01	−0.09	−0.24	−0.45
2.92	....	−0.01	−0.09	−0.21	−0.38
3.00	....	−0.01	−0.07	−0.15	−0.25
3.08	....	−0.01	−0.04	−0.06	−0.03
3.16	....	−0.01	0.00	0.03	0.14
3.24	....	0	0.02	0.09	0.29
3.32	....	....	0.04	0.15	0.41
3.40	....	....	0.05	0.17	0.36

λ. The Kirkwood theory involves the approximation in Eq. (35.20) in addition to the superposition approximation.

The details of the numerical solution of Eqs. (35.10) and (35.11) will be omitted here, as the procedure was necessarily rather involved.[1] The radial

[1] See Kirkwood, Maun, and Alder, *loc. cit.*

distribution function $g(x)$ was calculated for $\lambda = 5$, 10, 20, 27.4, and 33, and checked by direct numerical integration of Eq. (35.11). The results of the calculations are presented (in part) in Table 2 and Fig. 25. All radial distribution functions exhibit their first peak $[g = g(1)]$ at $x = 1$, and then decrease monotonically to the first minimum, which is followed by oscillations of diminishing amplitude resembling those of the experimentally determined radial distribution functions of real liquids.

From the values of $g(1,\lambda)$ obtained from the solutions presented in Table 2, the density corresponding to each λ may be determined by Eqs. (35.15) and

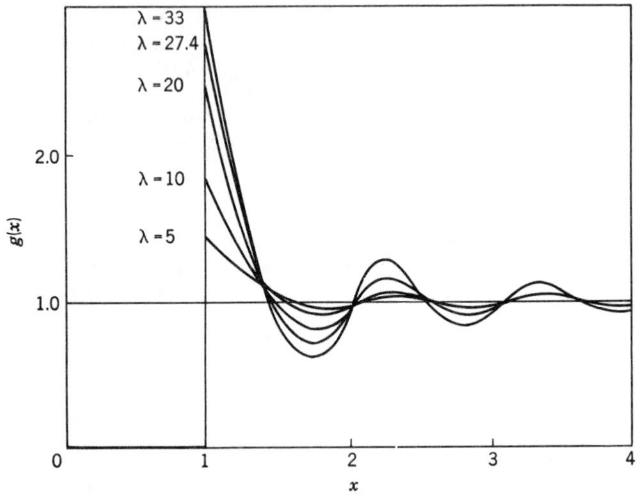

FIG. 25. Radial distribution functions for fluid of hard spheres with several values of λ.

(35.26). These densities are tabulated as a function of λ for the BGY and K theories in Table 3, where v_0 is the minimum (close packing) value of v for hard spheres:

$$v_0 = \frac{a^3}{\sqrt{2}} \tag{35.27}$$

For $\lambda \geqslant 34.8$, no solutions of Eq. (35.11) exist for which $x^2[g(x) - 1]$ [see Eq. (37.20)] is integrable. This value of λ, corresponding to an expansion v/v_0 of 1.24 on the K theory and 1.48 on the BGY theory, probably represents the limit of stability of a fluid phase of hard spheres. For greater densities, a crystalline phase is presumably the stable phase.[1] The transition between fluid and crystalline phases cannot be discussed quantitatively

[1] These conclusions are not definite because of the use of the superposition approximation.

without an investigation of distribution functions in the crystalline phase itself. In the case of hard spheres, it appears likely that the transition may

TABLE 3. FLUID DENSITIES AS FUNCTION OF λ

λ	$g(1,\lambda)$	v/v_0†	v/v_0§
5	1.45	4.74	5.15
10	1.80	2.83	3.20
20	2.36	1.78	2.10
27.4	2.66	1.45	1.73
33	2.85	1.29	1.53
34.8	2.90	1.24	1.48

† From Kirkwood.
§ From Born, Green, and Yvon.

be of second order rather than of first order, although at present this is no more than a surmise.

TABLE 4. EQUATION OF STATE OF THE FLUID OF HARD SPHERES

$\frac{v}{v_0}$	$\frac{pv}{\mathbf{k}T} - 1$†	$\frac{pv}{\mathbf{k}T} - 1$§
8.38	0.44	0.44
4.74	0.91	0.93
3.48	1.39	1.46
2.83	1.89	2.04
2.42	2.40	2.65
2.15	2.91	3.21
1.94	3.43	3.75
1.78	3.93	4.33
1.64	4.44	4.96
1.53	4.95	5.54
1.44	5.46	
1.37	5.99	
1.30	6.50	
1.24	6.93	

† From Kirkwood.
§ From Born, Green, and Yvon.

Turning now to the thermodynamic functions, the equation of state has already been derived [Eq. (35.22)]. We may write it as

$$\frac{pv}{\mathbf{k}T} = 1 + \frac{2\pi\sqrt{2}}{3} \cdot \frac{v_0}{v} \cdot g(1,\lambda) \qquad (35.28)$$

where $g(1,\lambda)$ is given as a function of v/v_0 in Table 3. In Table 4, $(pv/\mathbf{k}T) - 1$ is presented as a function of v/v_0 for the two theories. This function is also plotted in Fig. 26 together, for comparison, with the free-volume expression[1] for hard spheres,

$$\frac{pv}{\mathbf{k}T} - 1 = \frac{1}{(v/v_0)^{1/3} - 1} \tag{35.29}$$

which is frequently used at high densities.

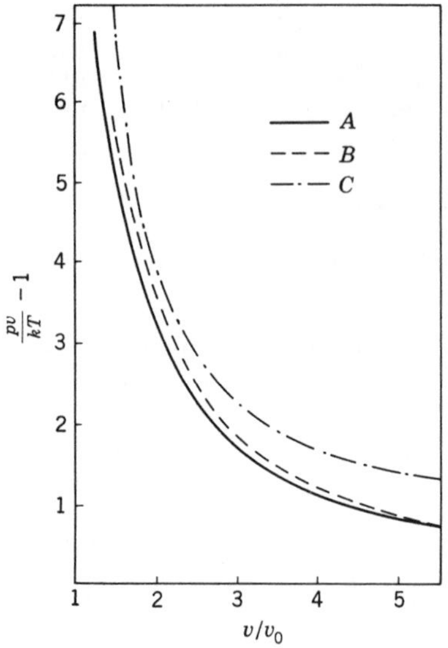

FIG. 26. Equation of state $(pv/\mathbf{k}T) - 1$ as a function of v/v_0, for fluid of hard spheres. *A*, Kirkwood; *B*, Born, Green, and Yvon; *C*, free-volume theory.

We now define certain auxiliary thermodynamic functions. In the limit $\rho \to 0$,

$$\mu = \mathbf{k}T \ln p + \mathbf{k}T \ln \frac{\Lambda^3}{\mathbf{k}T} \tag{35.30}$$

$$\text{E} = {}^3/_2\mathbf{k}T \tag{35.31}$$

$$Ts = \text{E} - \mu + pv \tag{35.32}$$

$$= -\mathbf{k}T \ln p - \mathbf{k}T \ln \frac{\Lambda^3}{\mathbf{k}T} + \frac{5}{2}\mathbf{k}T \tag{35.33}$$

[1] L. Tonks, *Phys. Rev.*, **50**, 955 (1936); H. Eyring and J. O. Hirschfelder, *J. Phys. Chem.*, **41**, 249 (1937).

where
$$\text{E} = \frac{E}{N} \qquad s = \frac{S}{N}$$

We can therefore *define* "excess" quantities μ^{E}, s^{E} and E^{E} at any p by the equations

$$\mu = \mathbf{k}T \ln p + \mathbf{k}T \ln \frac{\Lambda^3}{\mathbf{k}T} + \mu^{\text{E}} \tag{35.34}$$

$$\text{E} = {}^3/_2\mathbf{k}T + \text{E}^{\text{E}} \tag{35.35}$$

$$s = -\mathbf{k} \ln p - \mathbf{k} \ln \frac{\Lambda^3}{\mathbf{k}T} + \frac{5}{2}\mathbf{k} + s^{\text{E}} \tag{35.36}$$

From Eqs. (35.32) to (35.36), we have also

$$\frac{s^{\text{E}}}{\mathbf{k}} = \frac{\text{E}^{\text{E}}}{\mathbf{k}T} - \frac{\mu^{\text{E}}}{\mathbf{k}T} + \left(\frac{pv}{\mathbf{k}T} - 1\right) \tag{35.37}$$

We may also observe from Eq. (30.39) that

$$\mu^{\text{E}} = \mathbf{k}T \ln \gamma_f \tag{35.38}$$

where $\ln \gamma_f$ is given by Eq. (30.40).

For hard spheres, from Eq. (30.10), $\text{E}^{\text{E}} = 0$. Finally, we must consider either s^{E} or μ^{E} [Eq. (35.37)], and we choose s^{E}. In the BGY case, it is convenient to employ Eq. (30.21) and the equation of state. We let a superscript zero refer to a very large volume per molecule, $v^0 \to \infty$. Then

$$\text{A}^0 - \text{A}(v) = -\int_v^{v^0} p\, dv = (\mu^0 - \mathbf{k}T) - (\mu - pv) \tag{35.39}$$

Then, since

$$\mu^0 = -\mathbf{k}T \ln v^0 + \mathbf{k}T \ln \Lambda^3$$

Eq. (35.39) can be written

$$\mu = -\mathbf{k}T \ln v + \mathbf{k}T \ln \Lambda^3 + (pv - \mathbf{k}T) + \int_v^\infty \left(p - \frac{\mathbf{k}T}{v}\right) dv \tag{35.40}$$

If we now equate the right-hand sides of Eqs. (35.34) and (35.40), we obtain an expression for μ^{E}, which, when substituted into Eq. (35.37), leads to (recalling that $\text{E}^{\text{E}} = 0$)

$$\frac{s^{\text{E}}}{\mathbf{k}} = \ln \frac{pv}{\mathbf{k}T} - \int_{v/v_0}^\infty \left(\frac{pv_0}{\mathbf{k}T} - \frac{v_0}{v}\right) d\left(\frac{v}{v_0}\right) \tag{35.41}$$

This equation was used by Kirkwood, Maun, and Alder, in conjunction with

Eq. (35.28) and Tables 3 and 4, to compute $s^E/\mathbf{k}$ for the BGY theory. The results are given in Table 5.

TABLE 5. EXCESS ENTROPY OF THE FLUID OF HARD SPHERES AS A FUNCTION OF DENSITY

v/v_0	$s^E/\mathbf{k}$†	$s^E/\mathbf{k}$§
8.38	−0.03	−0.03
4.74	−0.12	−0.11
3.48	−0.24	−0.23
2.83	−0.39	−0.37
2.42	−0.55	−0.56
2.15	−0.73	−0.76
1.94	−0.92	−1.00
1.78	−1.14	−1.23
1.64	−1.37	−1.49
1.53	−1.60	−1.77
1.44	−1.84	
1.37	−2.07	
1.30	−2.32	
1.24	−2.60	

† From Kirkwood.
§ From Born, Green, and Yvon.

In the Kirkwood theory, we equate the right-hand sides of Eqs. (35.24) and (35.34). This gives μ^E which, when introduced in Eq. (35.37), results in ($\mathrm{E}^E = 0$)

$$\frac{s^E}{\mathbf{k}} = -\frac{\lambda}{3} + \left(\frac{pv}{\mathbf{k}T} - 1\right) + \ln\frac{pv}{\mathbf{k}T} \qquad \text{(Kirkwood)} \tag{35.42}$$

Equations (35.28) and (35.42) and Tables 3 and 4 permit the calculation of $s^E/\mathbf{k}$ with the results shown in Table 5.

It will be observed that the agreement between the K and BGY theories is moderately good both for the equation of state and for the entropy. The discrepancies can be attributed to the superposition approximation and probably also to some extent to the additional approximation, Eq. (35.20), in the Kirkwood theory. In the case of the equation of state, the free volume theory yields a result which does not deviate greatly from either the K or BGY results in the high-density region.

Finally, attention should be called to the recent work done on this problem by the Monte Carlo method using fast computing machines.[1]

[1] N. Metropolis *et al.*, *J. Chem. Phys.*, **21**, 1087 (1953); M. N. Rosenbluth and A. W. Rosenbluth, *J. Chem. Phys.*, **22**, 881 (1954).

36. Fluid with Modified Lennard-Jones Molecular Interaction Potential According to the Superposition Approximation

The Lennard-Jones intermolecular potential [Eq. (27.29)]

$$u_L(R) = 4\varepsilon\left[\left(\frac{a}{R}\right)^{12} - \left(\frac{a}{R}\right)^{6}\right] \tag{36.1}$$

is of course much more realistic than the hard-sphere potential of the previous section. However, in trying to solve Eq. (35.4) for $g(x)$ with the Lennard-Jones potential, very serious difficulties were encountered by Kirkwood, Lewinson, and Alder[1] which were overcome by use of a modified Lennard-Jones potential. The modification consists of using $u_L(R)$ with a hard-sphere cutoff at $R = a$. One might hope that this should have little effect on thermodynamic properties except at high densities, since $u_L(R)$ itself is very steep at $R = a$.

To be more explicit, we write the modified potential $u(x)$ as

$$u(x) = u_0(x) + u_1(x) \qquad x = \frac{R}{a} \tag{36.2}$$

where the hard-sphere part, $u_0(x)$, is

$$\begin{aligned} u_0(x) &= \infty \qquad x \leqslant 1 \\ &= 0 \qquad x > 1 \end{aligned} \tag{36.3}$$

and the Lennard-Jones part, $u_1(x)$, is

$$\begin{aligned} u_1(x) &= 0 \qquad x \leqslant 1 \\ &= u_L(x) \qquad x > 1 \end{aligned} \tag{36.4}$$

We also define for convenience the dimensionless function $\gamma(x)$ by

$$\gamma(x) = \frac{u(x)}{\varepsilon} = \gamma_0(x) + \gamma_1(x) \tag{36.5}$$

where

$$\begin{aligned} \gamma_0(x) &= \infty \qquad x \leqslant 1 \\ &= 0 \qquad x > 1 \end{aligned} \tag{36.6}$$

$$\begin{aligned} \gamma_1(x) &= 0 \qquad x \leqslant 1 \\ &= \gamma_L(x) \qquad x > 1 \end{aligned} \tag{36.7}$$

$$\gamma_L(x) = \frac{u_L(x)}{\varepsilon} = 4(x^{-12} - x^{-6}) \tag{36.8}$$

With the adoption in Eq. (36.2) of a potential function containing two

[1] J. G. Kirkwood, V. A. Lewinson, and B. J. Alder, *J. Chem. Phys.*, **20**, 929 (1952).

parameters a and ε, the system obeys the law of corresponding states[1] and a single set of calculations will be applicable to a number of different nonpolar fluids. All intensive thermodynamic properties can be expressed as functions of, say, $\mathbf{k}T/\varepsilon$ and v/a^3, and the radial distribution function will be a function of x, $\mathbf{k}T/\varepsilon$ and v/a^3.

In the hard-sphere calculations of Sec. 35, g was a function of x and v/a^3. In order to make full use of these hard-sphere results in the present problem, Kirkwood, Lewinson, and Alder followed the procedure of introducing the new variable $\mathbf{k}T/\varepsilon$ in the form of expansions of g and other functions in powers of $\varepsilon/\mathbf{k}T$. To accomplish this, the function $\psi(x)$ is defined by the equation (we take $\xi = 1$)

$$\ln g(x) = -\frac{u_0(x)}{\mathbf{k}T} + \frac{\psi(x)}{x} \tag{36.9}$$

and $\psi(x)$ is expanded as

$$\psi(x) = \psi_0(x) + \frac{\varepsilon}{\mathbf{k}T}\psi_1(x) + \left(\frac{\varepsilon}{\mathbf{k}T}\right)^2 \psi_2(x) + \cdots \tag{36.10}$$

On putting Eq. (36.10) in Eq. (36.9), $g(x)$ can be written as

$$g(x) = g_{(0)}(x)\left\{1 + \frac{\varepsilon}{\mathbf{k}T}\frac{\psi_1(x)}{x} + \left(\frac{\varepsilon}{\mathbf{k}T}\right)^2\left[\frac{\psi_2(x)}{x} + \frac{\psi_1(x)^2}{2x^2}\right]\right\} + \cdots \tag{36.11}$$

where

$$\ln g_{(0)}(x) = -\frac{u_0(x)}{\mathbf{k}T} + \frac{\psi_0(x)}{x} \tag{36.12}$$

Also, the kernel $K(t)$ of Eqs. (33.16), (33.17), and (35.4) is expanded as follows:

$$K(t) = k_0K_0(t) + \frac{\varepsilon}{\mathbf{k}T}K_1(t) + \left(\frac{\varepsilon}{\mathbf{k}T}\right)^2 K_2(t) + \cdots \tag{36.13}$$

where $K_0(t)$ is defined by Eqs. (35.12) and (35.13). Expressions for k_0, $K_1(t)$, $K_2(t)$, etc., in the Kirkwood and Born-Green-Yvon theories may be found by substituting Eqs. (36.5) and (36.11) in the right-hand side of Eqs. (33.16) and (33.17), and comparing coefficients of $\varepsilon/\mathbf{k}T$ with those in Eq. (36.13).

Equations (36.9) and (36.10) are now substituted into the left-hand side of Eq. (35.4) and Eqs. (36.5), (36.11), and (36.13) into the right-hand side. Coefficients of equal powers of $\varepsilon/\mathbf{k}T$ on the two sides of the resulting equation are then equated. In this way a set of integral equations is obtained, the first in $\psi_0(x)$, the second in $\psi_1(x)$, etc. The first equation is in fact just the hard-sphere integral equation already solved (Sec. 35), where $g_{(0)}(x)$ from $\psi_0(x)$ in Eq. (36.12) is the hard-sphere radial distribution function of Sec. 35.

[1] K. S. Pitzer, *J. Chem. Phys.*, **7**, 583 (1939); E. A. Guggenheim, *J. Chem. Phys.*, **13**, 253 (1945). See also E. A. Guggenheim, "Thermodynamics" (North-Holland, Amsterdam, 1949), p. 138.

The equations in $\psi_1(x)$, $\psi_2(x)$, etc., are linear and the solutions depend only on the parameter λ (not on $\varepsilon/\mathbf{k}T$), where λ is related to the density just as in Sec. 35.

Almost all calculations were carried out on the BGY equations, for which

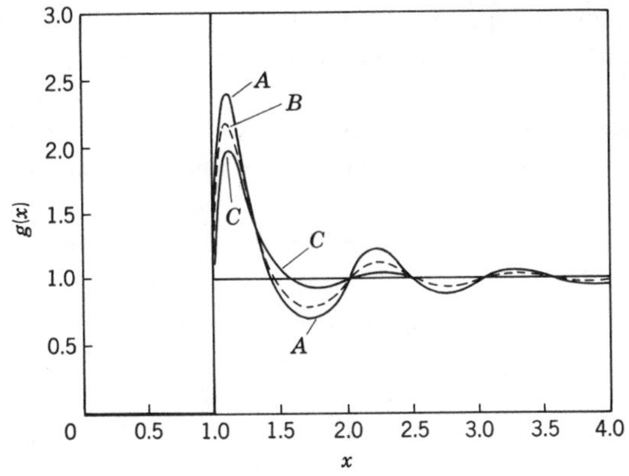

FIG. 27. Theoretical radial distribution functions at $\mathbf{k}T/\varepsilon = 1.25$. A, $\lambda = 27.4$, $v/v_0 = 1.73$; B, $\lambda = 20$, $v/v_0 = 2.10$; C, $\lambda = 5$, $v/v_0 = 5.15$.

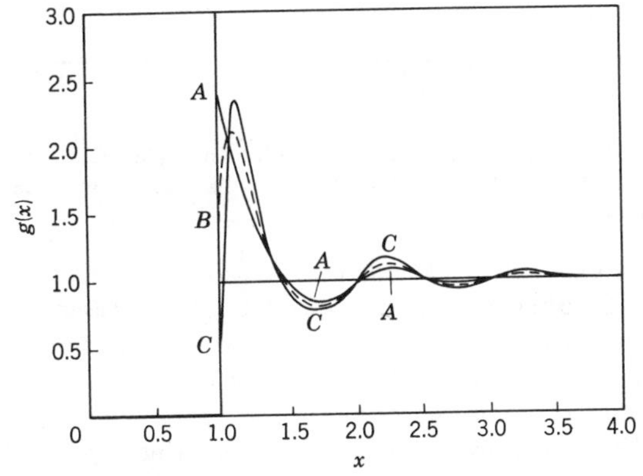

FIG. 28. Theoretical radial distribution functions at $v/v_0 = 2.10$. A, $\mathbf{k}T/\varepsilon = \infty$; B, $\mathbf{k}T/\varepsilon = 1.67$; C, $\mathbf{k}T/\varepsilon = 0.83$.

$\psi_0(x)$ (Sec. 35), $\psi_1(x)$, and $\psi_2(x)$ were computed by a suitable iterative procedure on the appropriate integral equation of the set mentioned above. From the ψ functions tabulated by Kirkwood, Lewinson, and Alder, one can calculate the radial distribution function and the thermodynamic properties of the fluid over a considerable range in density (roughly one-third to three times the density at the critical point) and temperature (from high temperatures down to about two-thirds of the critical temperature).

Figures 27 and 28 exhibit the influence of temperature and of density on $g(x)$, the radial distribution function. It will be seen that increasing density (increasing λ) at constant temperature (Fig. 27), and decreasing temperature (increasing $\varepsilon/\mathbf{k}T$) at constant density (Fig. 28) both tend to pack the molecules into a more orderly array [higher peaks in $g(x)$]. Also, as x, the distance between two molecules, becomes large, the correlation between their positions disappears, and thus $g(x)$ tends to unity. All these facts are qualitatively in

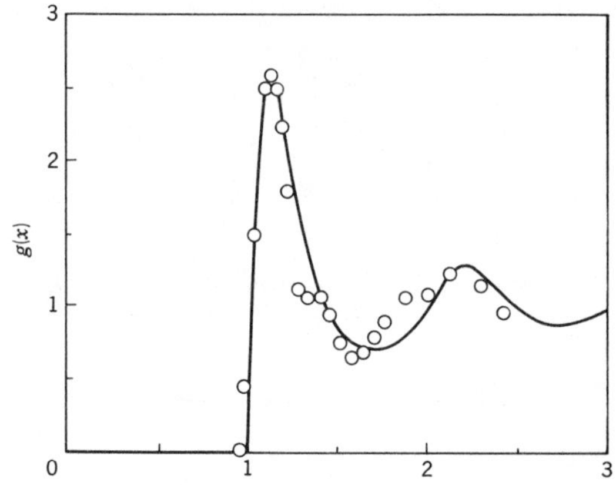

FIG. 29. Radial distribution function. Solid line: theoretical for $v/v_0 = 1.73$, $\mathbf{k}T/\varepsilon = 0.83$. Points: experimental for argon at 91.8°K and $p = 1.8$ atm.

agreement with experiment. A quantitative comparison for argon is presented in Fig. 29, using the experimental data of Eisenstein and Gingrich;[1] here the unit of length a has been chosen so that the experimental and calculated first peaks of $g(x)$ occur at the same value of x. The differences are no larger than the experimental uncertainties.

The theoretical radial distribution function used here, which is correct to terms in $\psi_2(x)$, will be called $g_{(2)}(x)$:

$$\ln g_{(2)}(x) = -\frac{u_0(x)}{\mathbf{k}T} + \frac{1}{x}\left[\psi_0(x) + \frac{\varepsilon}{\mathbf{k}T}\psi_1(x) + \left(\frac{\varepsilon}{\mathbf{k}T}\right)^2 \psi_2(x)\right] \tag{36.14}$$

Since $\psi_2(x)$ turns out to be rather small compared to $\psi_1(x)$, one might hope

[1] A. Eisenstein and N. S. Gingrich, *Phys. Rev.*, **62**, 261 (1942).

that the series, Eq. (36.10), could be cut off after $\psi_2(x)$. However, evidence was obtained which indicated that higher terms are small but not negligible.

The integral equation for $\psi_1(x)$ was also solved for $\lambda = 20$ on the Kirkwood basis, giving $g_{(1)}(x)$ for comparison with the BGY $g_{(1)}(x)$ at $\lambda = 27.4$ (from Table 3 we see that $\lambda^{\mathrm{K}} = 20$ and $\lambda^{\mathrm{BGY}} = 27.4$ correspond to almost the same density, $v/a^3 = 1.26$ and 1.22, respectively). In the important range

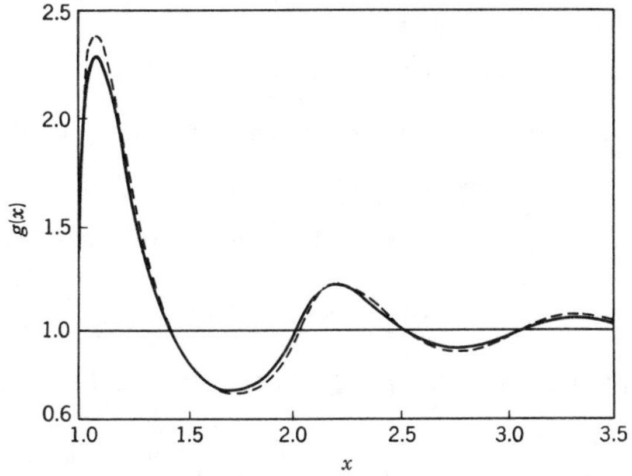

FIG. 30. Comparison of Kirkwood and Born-Green-Yvon $g_{(1)}(x)$. Solid line: Kirkwood, $v/v_0 = 1.78$, $\mathbf{k}T/\varepsilon = 1.67$. Dashed line: Born-Green-Yvon, $v/v_0 = 1.73$, $\mathbf{k}T/\varepsilon = 1.67$.

$0.6 \leqslant \varepsilon/\mathbf{k}T \leqslant 1.0$, $g_{(1)}{}^{\mathrm{K}}(x)$ and $g_{(1)}{}^{\mathrm{BGY}}(x)$ are not greatly different. This is illustrated in Fig. 30.

Thermodynamic properties computed from the radial distribution function may be expressed as functions of reduced variables v^*, p^*, and T^* defined by

$$v^* = \frac{v}{a^3}$$

$$p^* = \frac{pa^3}{\varepsilon} \tag{36.15}$$

$$T^* = \frac{\mathbf{k}T}{\varepsilon}$$

For example, the excess internal energy from Eqs. (30.10), (35.35), and (36.5) may be calculated in dimensionless form from

$$\frac{\mathrm{E}^{\mathrm{E}}}{\varepsilon} = \frac{2\pi}{v^*} \int_0^{\infty} \gamma(x) g(x) x^2 \, dx \tag{36.16}$$

Similarly, the equation of state, from Eqs. (30.20) and (36.5), is given by

$$\frac{p^*v^*}{T^*} = \frac{pv}{\mathbf{k}T} = 1 - \frac{2\pi}{3T^*v^*}\int_0^\infty \gamma'(x)g(x)x^3\,dx \tag{36.17}$$

Because of the simple temperature dependence in Eq. (36.10), the excess entropy is conveniently found from a temperature derivative of the calculated pressure as follows. We rewrite Eq. (35.33) for the entropy in the limit $p \to 0$ or $v \to \infty$ as

$$Ts = \mathbf{k}T \ln v - \mathbf{k}T \ln \Lambda^3 + {}^5\!/_2\mathbf{k}T \tag{36.18}$$

and *define* a new excess entropy per molecule s_v^{E} at any v by the equation

$$s = \mathbf{k} \ln v - \mathbf{k} \ln \Lambda^3 + {}^5\!/_2\mathbf{k} + s_v^{\mathrm{E}} \tag{36.19}$$

To find a suitable expression for s_v^{E} we integrate the thermodynamic equation

$$ds = \left(\frac{\partial p}{\partial T}\right)_v dv \qquad T \text{ const}$$

between v and v^0, where $v^0 \to \infty$. Using Eq. (36.18) for $s(v^0)$, we can show that

$$s = \mathbf{k} \ln v - \mathbf{k} \ln \Lambda^3 + \frac{5}{2}\mathbf{k} - \int_v^\infty \left[\left(\frac{\partial p}{\partial T}\right)_v - \frac{\mathbf{k}}{v}\right] dv \tag{36.20}$$

On comparison with Eq. (36.19),

$$\frac{s_v^{\mathrm{E}}}{\mathbf{k}} = -\int_{v^*}^\infty \left[\left(\frac{\partial p^*}{\partial T^*}\right)_{v^*} - \frac{1}{v^*}\right] dv^* \tag{36.21}$$

Tables 6, 7, and 8 present calculated values of p^*v^*/T^*, $-\mathrm{E}^{\mathrm{E}}/\varepsilon$ and $-s_v^{\mathrm{E}}/\mathbf{k}$ as functions of v^* and T^*. The computations were all made on the BGY equations, using $g_{(2)}(x)$ except for $\lambda = 1$. At $\lambda = 1$, an iterative procedure was employed (the method is applicable only for small λ or small $\varepsilon/\mathbf{k}T$) which avoided the temperature expansion in Eq. (36.10).

TABLE 6. REDUCED EQUATION OF STATE (p^*v^*/T^*) VERSUS v^* AND T^*

λ	v^*	$T^* = 0.833$	1.000	1.250	1.667	2.500	5.000	∞
1	13.82	......	0.629	0.768	0.883	....	....	1.167
5	3.632	−0.594	−0.156	0.264	0.670	1.064	1.456	1.833
10	2.260	−1.445	−0.734	−0.038	0.649	1.326	1.998	2.667
20	1.483	−2.433	−1.268	−0.115	1.018	2.139	3.242	4.333
27.4	1.222	−2.829	−1.382	0.052	1.467	2.856	4.223	5.567

TABLE 7. REDUCED EXCESS INTERNAL ENERGY $-E^E/\varepsilon$ VERSUS v^* AND T^*

v^*	$T^* = 0.833$	1.000	1.250	1.667	2.500	5.000
13.82	....	0.741	0.621	0.537	....	....
3.632	2.280	2.148	2.035	1.939	1.856	1.787
2.260	3.370	3.277	3.192	3.118	3.050	2.990
1.483	5.073	5.024	4.974	4.925	4.873	4.822
1.222	6.313	6.280	6.232	6.181	6.125	6.066

TABLE 8. REDUCED EXCESS ENTROPY
$-s_v^E/\mathbf{k}$ VERSUS v^* AND T^*

v^*	$T^* = 1.000$	1.250	1.667	2.500	5.000
13.82	0.28	0.24	0.21	0.20	0.19
3.632	0.80	0.91	0.99	1.05	1.11
2.260	1.43	1.52	1.56	1.64	1.69
1.483	2.49	2.56	2.57	2.66	2.72
1.222	3.29	3.35	3.35	3.44	3.50

TABLE 9. THEORETICAL AND EXPERIMENTAL EQUATION OF STATE ($pv/\mathbf{k}T$) FOR ARGON

Density d, amagats	0°C isotherm			150°C isotherm		
	Expt.	Theoret.	Error, %	Expt.	Theoret.	Error, %
259.6	0.925	0.982	+6	1.161	1.290	+11
417.1	1.122	1.186	+6	1.486	1.713	+15
635.5	2.138	1.907	−11	2.556	2.776	+9
771.4	3.661	2.569	−30	.....	3.644	

In Tables 9 and 10 we compare the theoretical [based on $g_{(2)}(x)$] and experimental[1] equation of state and excess energy for argon, for which excellent

[1] A. Michels, H. Wijker, and H. Wijker, *Physica*, **15**, 627 (1949); A. Michels, R. J. Lunbeck, and G. J. Wolkers, *Physica*, **15**, 689 (1949).

data are available. The values of a and ε used are those calculated by Michels, Wijker, and Wijker from second virial coefficient data,[1]

$$\varepsilon_L = 1.653 \times 10^{-14} \text{ erg}$$
$$a_L = 3.405 \times 10^{-8} \text{ cm} \tag{36.22}$$

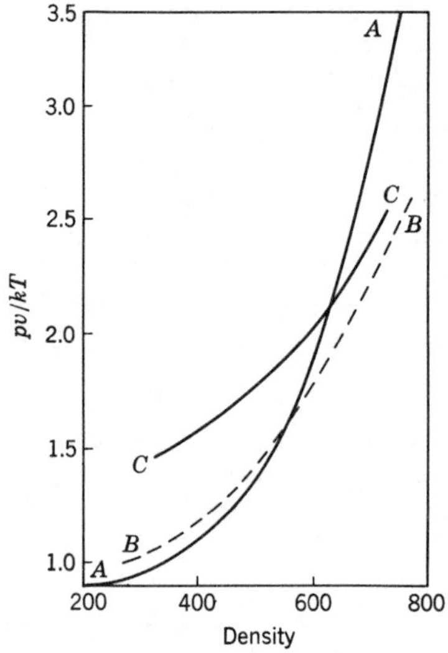

FIG. 31. 0°C isotherm for argon, $pv/\mathbf{k}T$ versus density (amagats). A, experimental; B, Kirkwood-Lewinson-Alder; C, Lennard-Jones and Devonshire (Chap. 8).

TABLE 10. THEORETICAL AND EXPERIMENTAL EXCESS INTERNAL ENERGY ($-E^E$, cal per mole) FOR ARGON

Density d, amagats	0°C isotherm			150°C isotherm		
	Expt.	Theoret.	Error, %	Expt.	Theoret.	Error, %
259.6	399	445	+11	359	431	+20
417.1	625	729	+17	560	717	+28
635.5	918	1162	+27	782	1152	+47

Figure 31 displays the experimental 0°C isotherm, together with the Kirkwood-Lewinson-Alder theoretical curve and the Lennard-Jones and Devon-

[1] This choice of a and ε is slightly arbitrary, for Michels, Wijker, and Wijker used the unmodified Lennard-Jones potential.

shire free-volume theoretical curve.[1] It will be seen that both theoretical isotherms are too flat (too large compressibility), the free-volume one being worse in this respect, and that the Kirkwood-Lewinson-Alder results agree better with experiment at the lower densities, whereas the free-volume theory agrees better at higher densities. Since this comparison is at a fairly high temperature ($T^* > 2$), $g_{(2)}(x)$ should be a good approximation to[2] $g(x)$, the correct solution of Eq. (35.4). The discrepancies with experiment must be

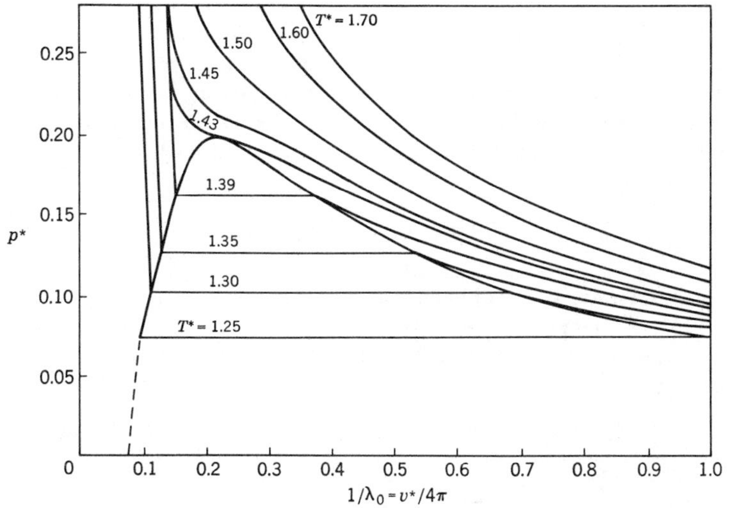

FIG. 32. Theoretical reduced isotherms of fluid showing liquid-vapor transition.

ascribed to the superposition approximation, to inadequacies of the modified $u(x)$, and to the sensitivity of thermodynamic properties to the form of $g(x)$.

The boundary of the two-phase region in the pressure-volume diagram can be established by finding on each isotherm the two volumes v_l^* (liquid) and v_g^* (gas) which have the same pressure p^* and chemical potential μ/ε [as given by Eq. (35.40), rewritten in reduced form]. The two-phase region and a number of pressure-volume isotherms are graphed in Fig. 32. Here the theory reproduces very well the qualitative aspects of experimental fluid isotherms.

[1] R. H. Wentorf, R. J. Buehler, J. O. Hirschfelder, and C. F. Curtiss, *J. Chem. Phys.*, **18**, 1484 (1950); see also Chap. 8.

[2] That is, for small $\varepsilon/\mathbf{k}T$, the powers of $\varepsilon/\mathbf{k}T$ in Eq. (36.10) aid convergence.

Using power series interpolation formulas, the critical constants are found to be

$$p_c^* = 0.199 \qquad v_c^* = 2.585 \qquad T_c^* = 1.433$$

$$\Phi_c = \frac{p_c^* v_c^*}{T_c^*} = \frac{p_c v_c}{\mathbf{k}T_c} = 0.358 \tag{36.23}$$

The critical ratio Φ_c is independent of the parameters a and ε and thus affords an absolute check of the theory. Because a and ε refer to a modified Lennard-Jones potential, it is not legitimate to compare the values in Eq. (36.23) directly with other theories based on the unmodified Lennard-Jones potential. However, since Lennard-Jones parameters are usually evaluated from second virial coefficient data, Kirkwood, Lewinson, and Alder concluded, from a comparison of theoretical expressions for the second virial coefficient based on both the modified and unmodified potentials, that the two sets of parameters are related by

$$\frac{\varepsilon}{\varepsilon_L} = 1.171 \qquad \frac{a_L}{a} = 1.086 \tag{36.24}$$

where ε and a are the parameters of the modified potential, Eq. (36.2), and ε_L and a_L are the parameters of the unmodified potential, Eq. (36.1). If we convert the critical constants in Eq. (36.23) over to the unmodified Lennard-Jones basis, using Eq. (36.24), the values given in the last row of Table 11

TABLE 11. EXPERIMENTAL AND THEORETICAL CRITICAL CONSTANTS

	$\frac{\mathbf{k}T_c}{\varepsilon_L}$	$\frac{p_c a_L^3}{\varepsilon_L}$	$\frac{v_c}{a_L^3}$	$\frac{p_c v_c}{\mathbf{k}T_c}$
Mean value for Ne, N_2, A, CH_4 .	1.277	.121	3.09	.292
Van der Waals	0.296	.0125	8.88	.375
Lennard-Jones and Devonshire.	1.30	.434	1.77	.591
Cernuschi and Eyring	2.74	.469	2.00	.342
Ono	0.75	.128	2.00	.342
Peek and Hill................	1.18	.261	3.25	.719
Mayer and Careri.............	1.42	.414	2.32	.676
KLA–BGY	1.68	.298	2.02	.358

are found. These may be compared with results from experiment, the van der Waals equation, and five free-volume theories discussed in Chap. 8.

In summary we may say that the present theory, based on distribution functions, the superposition approximation, a modified Lennard-Jones potential, and an approximate (i.e., up to ψ_2) temperature expansion of $\psi(x)$, is qualitatively completely satisfactory but only moderately successful quantitatively.

It should be added, in conclusion, that the quantitative agreement of the theory with experiment can, however, be improved considerably by an essentially empirical adjustment of the Kirkwood-Lewinson-Alder radial distribution function[1] (including a transition from the modified Lennard-Jones potential to the unmodified potential).

B. GRAND CANONICAL ENSEMBLE

In many ways the theory of distribution functions is more powerful when developed using the grand canonical ensemble; therefore we now turn to the consideration of open systems. Section 37 is a summary of the fundamental properties of distribution functions in an open, one-component, monatomic, classical system, and is roughly equivalent in content to the canonical ensemble theory of Secs. 29 to 34. Sections 38 and 39 are concerned, for the same system, with the Kirkwood-Salsburg integral equation and with distribution functions at phase transitions, respectively. Finally, in Sec. 40 we discuss a polyatomic-multicomponent system using the McMillan-Mayer theory of solutions.

The coupling parameter ξ is omitted in the following sections except where explicitly required.

37. Distribution Functions in Monatomic, One-component Systems

We denote the distribution functions of Sec. 29 to 36, from this point on, as $\rho_N^{(n)}$, $g_N^{(n)}$, etc. Thus, $\rho_N^{(n)}(\mathbf{r}_1, \ldots, \mathbf{r}_n)\, d\mathbf{r}_1 \cdots d\mathbf{r}_n$ is the probability of observing molecules in $d\mathbf{r}_1, \ldots, d\mathbf{r}_n$ at $\mathbf{r}_1, \ldots, \mathbf{r}_n$ when the volume V contains N molecules. But we recall that the probability that an open system actually does contain exactly N molecules is

$$P_N = \frac{e^{N\mu/\mathbf{k}T} Q(N,V,T)}{\Xi(\mu,V,T)} = \frac{z^N Z_N}{N!\,\Xi} \tag{37.1}$$

where

$$Z_N = Z(N,V,T)$$

$$z = \frac{e^{\mu/\mathbf{k}T}}{\Lambda^3}$$

$$\Xi = \sum_{N \geqslant 0} \frac{z^N Z_N}{N!}$$

Now let $\rho^{(n)}(\mathbf{r}_1, \ldots, \mathbf{r}_n)\, d\mathbf{r}_1 \cdots d\mathbf{r}_n$ be the probability of observing molecules in $d\mathbf{r}_1, \ldots, d\mathbf{r}_n$ at $\mathbf{r}_1, \ldots, \mathbf{r}_n$, irrespective of N (that is, properly averaged over N). Then

$$\rho^{(n)} = \sum_{N \geqslant n} \rho_N^{(n)} P_N \tag{37.2}$$

[1] R. W. Zwanzig, J. G. Kirkwood, K. F. Stripp, and I. Oppenheim, *J. Chem. Phys.*, **21**, 1268 (1953).

This is the generic distribution function for an open system. The probability density $\rho_N^{(n)}$ is necessarily zero for $N < n$. $\rho^{(n)}$ is a function of $\mathbf{r}_1, \ldots, \mathbf{r}_n$ and also of T and μ or z. The dependence of $\rho^{(n)}$ on V is of course negligible for macroscopic V; this dependence vanishes completely in the limit $V \to \infty$.

From Eqs. (29.8) and (37.1),

$$\rho^{(n)}(\mathbf{r}_1, \ldots, \mathbf{r}_n) = \frac{1}{\Xi} \sum_{N \geqslant n} \frac{z^N}{(N-n)!} \int \cdots \int_V e^{-U_N/\mathbf{k}T} d\mathbf{r}_{n+1} \cdots d\mathbf{r}_N \qquad (37.3)$$

where
$$U_N = U(\mathbf{r}_1, \ldots, \mathbf{r}_N)$$

If we put $m = N - n$ in this equation, we can write

$$\left(\frac{1}{z}\right)^n \Xi \, \rho^{(n)}(\mathbf{r}_1, \ldots, \mathbf{r}_n) = \sum_{m \geqslant 0} \frac{z^m}{m!} \int \cdots \int_V e^{-U_{n+m}/\mathbf{k}T} d\mathbf{r}_{n+1} \cdots d\mathbf{r}_{n+m} \qquad (37.4)$$

The normalization equation for $\rho^{(n)}$ is

$$\begin{aligned} \int \cdots \int_V \rho^{(n)}(\mathbf{r}_1, \ldots, \mathbf{r}_n)\, d\mathbf{r}_1 \cdots d\mathbf{r}_n &= \frac{1}{\Xi} \sum_{N \geqslant n} \frac{z^N Z_N}{(N-n)!} = \sum_{N \geqslant n} P_N \frac{N!}{(N-n)!} \\ &= \left\langle \frac{N!}{(N-n)!} \right\rangle_{\text{av}} \qquad (37.5) \\ &= \left\langle N^n - \frac{n(n-1)}{2} N^{n-1} + \cdots \right\rangle_{\text{av}} \\ &= \overline{N^n} \left[1 + O\left(\frac{1}{\bar{N}}\right) \right] \qquad (37.6) \end{aligned}$$

In the limit $V \to \infty$, $\bar{N} \to \infty$ (z and T held fixed). It can then be seen, for example from Eq. (21.22), that in this limit $\overline{N^n}/(\bar{N})^n \to 1$ and hence that

$$\lim_{V \to \infty} \frac{1}{V^n \rho^n} \int \cdots \int_V \rho^{(n)}(\mathbf{r}_1, \ldots, \mathbf{r}_n)\, d\mathbf{r}_1 \cdots d\mathbf{r}_n = 1 \qquad (37.7)$$

where $\rho = \bar{N}/V$. This is the normalization condition used by Mayer[1] (Mayer's function $F_n \equiv \rho^{(n)}/\rho^n$ in any type of phase).

In a fluid,

$$\rho_N^{(1)} = \frac{N}{V}$$

and
$$\rho^{(1)} = \sum_{N \geqslant 1} \frac{N}{V} P_N = \frac{\bar{N}}{V} = \rho \qquad (37.8)$$

[1] J. E. Mayer, *J. Chem. Phys.*, **10**, 629 (1942). See also W. G. McMillan and J. E. Mayer, *J. Chem. Phys.*, **13**, 276 (1945).

Also, since $\rho_N^{(2)}$ depends only on r_{12} instead of on $\mathbf{r}_1$ and $\mathbf{r}_2$ separately, $\rho^{(2)}$ is a function of r_{12} only.

The correlation function $g^{(n)}$ is defined by the equation

$$\rho^{(n)}(\mathbf{r}_1, \ldots, \mathbf{r}_n) = \rho^{(1)}(\mathbf{r}_1) \cdots \rho^{(1)}(\mathbf{r}_n) g^{(n)}(\mathbf{r}_1, \ldots, \mathbf{r}_n) \tag{37.9}$$

It follows that $g^{(1)} = 1$.

In a fluid, (1) Eq. (37.9) reduces to

$$\rho^{(n)} = \left(\frac{\bar{N}}{V}\right)^n g^{(n)}$$

or

$$\rho^{(n)} = \rho^n g^{(n)} \tag{37.10}$$

Thus $g^{(n)}$ is the same as Mayer's F_n in a fluid, but not in a crystal. (2) The normalization equation for $g^{(n)}$ is, from Eq. (37.5),

$$\frac{1}{V^n} \int_V \cdots \int g^{(n)}(\mathbf{r}_1, \ldots, \mathbf{r}_n)\, d\mathbf{r}_1 \cdots d\mathbf{r}_n = \frac{1}{(\bar{N})^n} \left\langle \frac{N!}{(N-n)!} \right\rangle_{\text{av}} \tag{37.11}$$

$$= \frac{\overline{N^n}}{(\bar{N})^n} \left[1 + O\left(\frac{1}{\bar{N}}\right)\right] \to 1 \text{ as } V \to \infty \tag{37.12}$$

(3) If the distribution is random ($T \to \infty$, or in the absence of intermolecular forces), we obtain from Eqs. (29.16), (37.2), and (37.3) (on putting $U_N/\mathbf{k}T = 0$; see also Appendix 2),

$$\rho^{(n)} = \frac{1}{V^n} \left\langle \frac{N!}{(N-n)!} \right\rangle_{\text{av}} = z^n = \rho^n \tag{37.13}$$

and

$$g^{(n)} = 1 \tag{37.14}$$

(4) In the limit $\rho \to 0$, we have $z \to \rho \to 0$ (Sec. 23) and therefore only the term $m = 0$ contributes in Eq. (37.4). Since $Z_N = 1$ for $N = 0$, $\Xi \to 1$ as $z \to 0$. Thus Eq. (37.4) reduces to

$$\rho^{(n)} \to \rho^n e^{-U_n/\mathbf{k}T}$$

or

$$g^{(n)} \to e^{-U_n/\mathbf{k}T} \qquad \rho, z \to 0 \tag{37.15}$$

It should be noted that a factor $1 + O(1/\bar{N})$ is not present in Eq. (37.15) as it is in Eq. (32.18). (5) It is shown in Appendix 7, from an expansion of $\rho^{(n)}$ in powers of z, that

$$g^{(n)} \to 1 \tag{37.16}$$

as all $r_{ij}(1 \leqslant i < j \leqslant n) \to \infty$, whereas this limit for $g_N^{(n)}$ (canonical ensemble) has in addition a term of order $1/N$ [see, for example, Eq. (29.24)].

Thermodynamic Properties of Fluids. For the calculation of thermodynamic properties of macroscopic[1] systems, the radial distribution functions

[1] As in Sec. 30, we derive relations valid for both crystals and fluids before specializing to a fluid.

from the canonical and grand canonical ensembles are of course indistinguishable, except when fluctuations are involved (see below). We shall show this indistinguishability here for the internal energy and equation of state. But we might point out first that our earlier treatment (Chap. 4) of fluctuations in the grand canonical ensemble provides us with a new thermodynamic relation, connecting compressibility and (grand canonical ensemble) distribution functions.

From Eq. (37.5) we have

$$\iint_V \rho^{(2)}(\mathbf{r}_1,\mathbf{r}_2)\, d\mathbf{r}_1\, d\mathbf{r}_2 = \overline{N^2} - \bar{N} \tag{37.17}$$

$$\iint_V \rho^{(1)}(\mathbf{r}_1)\rho^{(1)}(\mathbf{r}_2)\, d\mathbf{r}_1\, d\mathbf{r}_2 = (\bar{N})^2 \tag{37.18}$$

We subtract Eq. (37.18) from Eq. (37.17) and use Eq. (19.39):

$$\frac{1}{\rho V}\iint_V [\rho^{(2)}(\mathbf{r}_1,\mathbf{r}_2) - \rho^{(1)}(\mathbf{r}_1)\rho^{(1)}(\mathbf{r}_2)]\, d\mathbf{r}_1\, d\mathbf{r}_2 = \rho \mathbf{k} T \kappa - 1 \tag{37.19}$$

where
$$\kappa = -\frac{1}{V}\left(\frac{\partial V}{\partial p}\right)_{\bar{N},T}$$

It is interesting that $u(r)$ does not appear in Eq. (37.19), as it does in expressions for other thermodynamic functions.

For a fluid, Eq. (37.19) reduces to

$$\rho \int_0^\infty [g(r) - 1] 4\pi r^2\, dr = \rho \mathbf{k} T \kappa - 1 \qquad \text{(fluid)} \tag{37.20}$$

where $g = g^{(2)}$ as before. The fact that the grand canonical ensemble $g(r)$ in Eq. (37.20) approaches unity exactly (no additional term of order $1/\bar{N}$ as in the canonical ensemble) as $r \to \infty$ is important here, for an extra term of order $1/\bar{N}$, when integrated over the volume V, would give a contribution which could not be neglected. Incidentally, the upper limit in Eq. (37.20) can be taken as infinity, since $g(r) \to 1$ rapidly as $r \to \infty$.

The internal energy follows from Eqs. (14.46) and (14.49):

$$\begin{aligned} \bar{E} &= \bar{N}\mu + TS - \bar{p}V \\ &= \bar{N}\mu + \mathbf{k}T^2 \left(\frac{\partial \ln \Xi}{\partial T}\right)_{V,\mu} \end{aligned} \tag{37.21}$$

where Ξ is given in Eq. (37.1). Hence

$$\left(\frac{\partial \Xi}{\partial T}\right)_{V,\mu} = \frac{1}{\mathbf{k}T^2} \sum_{N \geqslant 2} \frac{z^N}{N!} \int_V \cdots \int e^{-U_N/\mathbf{k}T} U_N\, d\mathbf{r}_1 \cdots d\mathbf{r}_N + \sum_{N \geqslant 0} \frac{z^N}{N!} \left(\frac{3N}{2T} - \frac{N\mu}{\mathbf{k}T^2}\right) Z_N \tag{37.22}$$

The first summation in Eq. (37.22) starts at $N = 2$ since $U_N = 0$ for $N = 0$ and $N = 1$. The second summation arises from $\partial z/\partial T$. Proceeding as in Eqs. (30.5) to (30.8),

$$\mathbf{k}T^2\left(\frac{\partial \ln \Xi}{\partial T}\right)_{V,\mu} = \frac{1}{\Xi}\int\int_V u(r_{12})\left[\sum_{N\geq 2}\frac{z^N}{N!}\frac{N(N-1)}{2}\right.$$

$$\left.\times \int_V\cdots\int e^{-U_N/\mathbf{k}T}\,d\mathbf{r}_3\cdots d\mathbf{r}_N\right] d\mathbf{r}_1\,d\mathbf{r}_2 + \frac{3}{2}\bar{N}\mathbf{k}T - \bar{N}\mu$$

$$= {}^1/_2\int\int_V u(r_{12})\rho^{(2)}(\mathbf{r}_1,\mathbf{r}_2)\,d\mathbf{r}_1\,d\mathbf{r}_2 + {}^3/_2\bar{N}\mathbf{k}T - \bar{N}\mu \tag{37.23}$$

$$= \frac{1}{2}\frac{(\bar{N})^2}{V}\int_0^\infty u(r)g(r)\cdot 4\pi r^2\,dr + \frac{3}{2}\bar{N}\mathbf{k}T - \bar{N}\mu \qquad \text{(fluid)} \tag{37.24}$$

Equations (37.21) and (37.24) then agree with our earlier expression for $\bar{E}/\bar{N}\mathbf{k}T$, Eq. (30.10).

To obtain the equation of state, we use Eq. (14.48) and Green's canonical ensemble method [Eqs. (30.11) to (30.20)]. That is,

$$\frac{\bar{p}}{\mathbf{k}T} = \left(\frac{\partial \ln \Xi}{\partial V}\right)_{z,T} = \frac{1}{\Xi}\sum_{N\geq 0}\frac{z^N}{N!}\frac{\partial Z_N}{\partial V} \tag{37.25}$$

$$= \frac{1}{\Xi}\left\{\sum_{N\geq 0}\frac{z^N}{N!}\frac{N}{V}Z_N - \frac{1}{6V\mathbf{k}T}\int\int_V r_{12}u'(r_{12})\right.$$

$$\left.\times\left[\sum_{N\geq 2}\frac{z^N}{N!}N(N-1)\int_V\cdots\int e^{-U_N/\mathbf{k}T}\,d\mathbf{r}_3\cdots d\mathbf{r}_N\right] d\mathbf{r}_1\,d\mathbf{r}_2\right\}$$

$$= \frac{\bar{N}}{V} - \frac{1}{6V\mathbf{k}T}\int\int_V r_{12}u'(r_{12})\rho^{(2)}(\mathbf{r}_1,\mathbf{r}_2)\,d\mathbf{r}_1\,d\mathbf{r}_2 \tag{37.26}$$

$$= \frac{\bar{N}}{V} - \frac{\rho^2}{6\mathbf{k}T}\int_0^\infty ru'(r)g(r)\cdot 4\pi r^2\,dr \qquad \text{(fluid)} \tag{37.27}$$

This result is the same as Eq. (30.20).

It is convenient to postpone a consideration of the chemical potential until the subsection below on the Kirkwood integral equation.

Potential of Mean Force. We define $W^{(n)}(\mathbf{r}_1, \ldots, \mathbf{r}_n)$ by the equation[1]

$$e^{-W^{(n)}(\mathbf{r}_1,\ldots,\mathbf{r}_n)/\mathbf{k}T} = g^{(n)}(\mathbf{r}_1, \ldots, \mathbf{r}_n) = \frac{\rho^{(n)}(\mathbf{r}_1, \ldots, \mathbf{r}_n)}{\rho^{(1)}(\mathbf{r}_1)\cdots\rho^{(1)}(\mathbf{r}_n)} \tag{37.28}$$

[1] This is analogous to Eq. (31.7) in the canonical ensemble.

from which $W^{(1)} = 0$ since $g^{(1)} = 1$. To study the properties of $W^{(n)}$ as a potential of mean force, we take the logarithm of Eq. (37.28) and then the gradient with respect to, say, $\mathbf{r}_\alpha$, where $\mathbf{r}_\alpha$ is a member of the set of positions $\mathbf{r}_1, \ldots, \mathbf{r}_n$. Then

$$-\frac{1}{\mathbf{k}T}\nabla_\alpha W^{(n)} = \frac{\nabla_\alpha \rho^{(n)}}{\rho^{(n)}} - \frac{\nabla_\alpha \rho^{(1)}(\mathbf{r}_\alpha)}{\rho^{(1)}(\mathbf{r}_\alpha)} \tag{37.29}$$

Now
$$\nabla_\alpha \rho^{(n)} = \sum_{N \geqslant n} P_N \nabla_\alpha \rho_N{}^{(n)}$$

$$= \sum_{N \geqslant n} P_N \rho_N{}^{(n)} \left(- \frac{\overline{\nabla_\alpha U_N}^{(n)}}{\mathbf{k}T} \right) \tag{37.30}$$

where Eq. (37.30) follows from Eq. (31.12) with $\overline{\nabla_\alpha U_N}^{(n)}$ defined by Eq. (31.3). Using Eq. (37.30), Eq. (37.29) becomes

$$\nabla_\alpha W^{(n)} = \overline{\nabla_\alpha U}^{(n)} - \overline{\nabla_\alpha U}^{(1)} \qquad \alpha = 1, \ldots, n \tag{37.31}$$

where
$$\overline{\nabla_\alpha U}^{(n)} = \frac{\sum_{N \geqslant n} P_N \rho_N{}^{(n)} \overline{\nabla_\alpha U_N}^{(n)}}{\sum_{N \geqslant n} P_N \rho_N{}^{(n)}}$$

$$= -\mathbf{F}_\alpha{}^{(n)} \tag{37.32}$$

The quantity $\overline{\nabla_\alpha U}^{(1)}$ is of course also given by Eq. (37.32) on putting $n = 1$, with $\overline{\nabla_\alpha U_N}^{(1)}$ defined by Eq. (31.10). We recall that $-\overline{\nabla_\alpha U_N}^{(n)}$ is the mean force on a particle at $\mathbf{r}_\alpha$ averaged in the canonical ensemble over the configurations of the $N - n$ molecules not in the fixed set $1, \ldots, n$ [Eqs. (31.3) to (31.5)]. In an open system, then, $-\overline{\nabla_\alpha U}^{(n)}$ in Eq. (37.32) is the mean force on a particle at $\mathbf{r}_\alpha$ averaged further with respect to N, using the weight[1] $P_N \rho_N{}^{(n)}$ for each N. We can conclude that $W^{(n)}$ is the potential of the mean force on a molecule at $\mathbf{r}_\alpha$ properly averaged over the configurations of $N - n$ molecules and over the number of molecules N in the system *relative to* the mean force on a molecule at $\mathbf{r}_\alpha$ averaged over the configurations of $N - 1$ molecules and over the number of molecules N. In a fluid, $\overline{\nabla_\alpha U_N}^{(1)}$ and $\overline{\nabla_\alpha U}^{(1)}$ are zero by symmetry and

$$\mathbf{F}_\alpha{}^{(n)} = -\overline{\nabla_\alpha U}^{(n)} = -\nabla_\alpha W^{(n)} \qquad \text{(fluid)} \tag{37.33}$$

[1] This is the weight associated with the actual average force exerted on the particle since $P_N \rho_N{}^{(n)}\, d\mathbf{r}_1 \cdots d\mathbf{r}_n$ is the probability that the system contain N molecules and that also a set of molecules be found in $d\mathbf{r}_1, \ldots, d\mathbf{r}_n$ at $\mathbf{r}_1, \ldots, \mathbf{r}_n$.

Alternative forms of Eq. (37.32) are

$$\overline{\nabla_\alpha U^{(n)}} = -\mathbf{k}T\,\frac{\nabla_\alpha \rho^{(n)}}{\rho^{(n)}}$$

$$= \frac{\displaystyle\sum_{N\geqslant n} \frac{z^N}{(N-n)!} \int_V \cdots \int e^{-U_N/\mathbf{k}T} \nabla_\alpha U_N \, d\mathbf{r}_{n+1} \cdots d\mathbf{r}_N}{\displaystyle\sum_{N\geqslant n} \frac{z^N}{(N-n)!} \int_V \cdots \int e^{-U_N/\mathbf{k}T} \, d\mathbf{r}_{n+1} \cdots d\mathbf{r}_N} \tag{37.34}$$

$$= -\mathbf{k}T[\nabla_\alpha \ln g^{(n)}(\mathbf{r}_1, \ldots, \mathbf{r}_n) + \nabla_\alpha \ln \rho^{(1)}(\mathbf{r}_\alpha)] \tag{37.35}$$

$$= -\mathbf{k}T\nabla_\alpha \ln g^{(n)}(\mathbf{r}_1, \ldots, \mathbf{r}_n) \qquad \text{(fluid)} \tag{37.36}$$

From Eqs. (37.15) and (37.28) we note that

$$W^{(n)}(\mathbf{r}_1, \ldots, \mathbf{r}_n) \to U_n(\mathbf{r}_1, \ldots, \mathbf{r}_n) \text{ as } \rho, z \to 0 \tag{37.37}$$

An alternative definition of a potential of mean force in the grand ensemble[1] is the one used by Mayer:

$$e^{-w^{(n)}(\mathbf{r}_1, \ldots, \mathbf{r}_n)/\mathbf{k}T} = \frac{\rho^{(n)}(\mathbf{r}_1, \ldots, \mathbf{r}_n)}{\rho^n} = \frac{\rho^{(1)}(\mathbf{r}_1) \cdots \rho^{(1)}(\mathbf{r}_n)}{\rho^n} g^{(n)}(\mathbf{r}_1, \ldots, \mathbf{r}_n) \tag{37.38}$$

The relation between $W^{(n)}$ and $w^{(n)}$ is

$$W^{(n)}(\mathbf{r}_1, \ldots, \mathbf{r}_n) = w^{(n)}(\mathbf{r}_1, \ldots, \mathbf{r}_n) - w^{(1)}(\mathbf{r}_1) - \cdots - w^{(1)}(\mathbf{r}_n) \tag{37.39}$$

From Eqs. (37.30) to (37.32) and (37.38),

$$-\frac{1}{\mathbf{k}T}\nabla_\alpha w^{(n)} = \frac{\nabla_\alpha \rho^{(n)}}{\rho^{(n)}} \tag{37.40}$$

or

$$\nabla_\alpha w^{(n)} = \overline{\nabla_\alpha U^{(n)}} = -\overline{\mathbf{F}}_\alpha^{(n)} \tag{37.41}$$

This is simpler than Eq. (37.31), but $w^{(n)}$ is related to $g^{(n)}$ in a more complicated way[2] than is $W^{(n)}$. In a fluid, $w^{(n)} = W^{(n)}$. Hence,

$$w^{(n)} \to U_n \text{ as } \rho, z \to 0 \tag{37.42}$$

Superposition Approximation. Let us introduce the superposition approximation in an open system as

$$W^{(3)}(123) = W^{(2)}(12) + W^{(2)}(13) + W^{(2)}(23) \tag{37.43}$$

This is formally the same as Eq. (31.19) which, in the present notation, should

[1] This is the analogue of Eq. (31.1) in the canonical ensemble.

[2] Of course $w^{(n)}$ is more simply related to Mayer's $F_n = \rho^{(n)}/\rho^n$ than is $W^{(n)}$. Mayer's F_n will be used in Sec. 40, but denoted by $\mathfrak{g}^{(n)}$. See also Eq. (29.55) for a closed system.

have subscript N's inserted. From Eqs. (37.9), (37.28), and (37.43) we see that

$$\nabla_1 W^{(3)}(123) = \nabla_1 W^{(2)}(12) + \nabla_1 W^{(2)}(13), \text{ etc.} \tag{37.44}$$

$$\rho^{(3)}(123) = \frac{\rho^{(2)}(12)\rho^{(2)}(13)\rho^{(2)}(23)}{\rho^{(1)}(1)\rho^{(1)}(2)\rho^{(1)}(3)} \tag{37.45}$$

$$g^{(3)}(123) = g^{(2)}(12)g^{(2)}(13)g^{(2)}(23) \tag{37.46}$$

just as in Eqs. (31.20) to (31.22). The approximation in Eqs. (37.43) to (37.46) destroys exact normalization, but in a completely unimportant way. In fact, one can show from Eq. (21.22) that the missing correction factor is $[1 + O(1/\bar{N}^2)]$, as in Eq. (31.23).

From Eq. (37.39) we may observe that Eq. (37.43) is equivalent to

$$w^{(3)}(123) = w^{(2)}(12) + w^{(2)}(13) + w^{(2)}(23) - w^{(1)}(1) - w^{(1)}(2) - w^{(1)}(3) \tag{37.47}$$

For a fluid, $w^{(1)}(1) = w^{(1)}(2) = w^{(1)}(3) = 0$.

We shall have occasion to make use later of the extended form of Eq. (37.43):

$$W^{(n)}(\mathbf{r}_1, \ldots, \mathbf{r}_n, z, T) = \sum_{1 \leqslant i < j \leqslant n} W^{(2)}(\mathbf{r}_i, \mathbf{r}_j, z, T) \tag{37.48}$$

For a fluid, we shall write this as

$$\begin{aligned} W^{(n)}(\mathbf{r}_1, \ldots, \mathbf{r}_n, z, T) &= w^{(n)}(\mathbf{r}_1, \ldots, \mathbf{r}_n, z, T) \\ &= \sum_{1 \leqslant i < j \leqslant n} w(r_{ij}, z, T) \qquad \text{(fluid)} \end{aligned} \tag{37.49}$$

Equations (37.48) and (37.49) are, of course, formally analogous to

$$U_N(\mathbf{r}_1, \ldots, \mathbf{r}_N) = \sum_{1 \leqslant i < j \leqslant N} u(r_{ij}) \tag{37.50}$$

However, whereas Eq. (37.50) is generally considered an excellent (quantum-mechanical) approximation for many purposes, Eqs. (37.48) and (37.49) represent a much more drastic (statistical mechanical) simplification which one is often forced to make without any particular justification.

Born-Green-Yvon Integral Equation. From Eq. (37.3) we have

$$-\mathbf{k}T\nabla_1\rho^{(n)} = \frac{1}{\Xi}\sum_{N \geqslant n} \frac{z^N}{(N-n)!} \int_V \cdots \int e^{-U_N/\mathbf{k}T}\nabla_1 U_N \, d\mathbf{r}_{n+1} \cdots d\mathbf{r}_N \tag{37.51}$$

Now we put

$$\nabla_1 U_N = \sum_{i=2}^{N} \nabla_1 u(r_{1i}) \tag{37.52}$$

and break up the sum over i into two parts, $2 \leqslant i \leqslant n$ and $n + 1 \leqslant i \leqslant N$. In the first part, $\nabla_1 u(r_{1i})$ can be placed in front of the integral sign for each term, and in the second part each of the $N - n$ terms gives the same result. Equation (37.51) becomes

$$-\mathbf{k}T\nabla_1\rho^{(n)} = \rho^{(n)}\sum_{i=2}^{n}\nabla_1 u(r_{1i}) + \int_V \nabla_1 u(r_{1,n+1})\left[\frac{1}{\Xi}\sum_{N\geqslant n}\frac{z^N}{(N-n)!}\times(N-n)\int\cdots\int_V e^{-U_N/\mathbf{k}T}d\mathbf{r}_{n+2}\cdots d\mathbf{r}_N\right]d\mathbf{r}_{n+1}$$

or

$$-\mathbf{k}T\nabla_1 \ln\rho^{(n)}(\mathbf{r}_1,\ldots,\mathbf{r}_n) = \sum_{i=2}^{n}\nabla_1 u(r_{1i}) + \frac{1}{\rho^{(n)}}\int_V \nabla_1 u(r_{1,n+1})\rho^{(n+1)}(\mathbf{r}_1,\ldots,\mathbf{r}_{n+1})\,d\mathbf{r}_{n+1} \tag{37.53}$$

This is (thermodynamically) identical with Eq. (33.6) when $\xi = 1$. Using the superposition approximation and $n = 2$ for a fluid, Eq. (33.8) is again obtained.

Kirkwood Integral Equation. Let the molecule[1] at $\mathbf{r}_1$ have the coupling parameter ξ, and take all other coupling parameters equal to unity. Then

$$\Xi(\xi)\rho^{(n)}(\mathbf{r}_1,\ldots,\mathbf{r}_n,\xi) = \sum_{N\geqslant n}\frac{z^N}{(N-n)!}\int\cdots\int_V e^{-U_N(\xi)/\mathbf{k}T}\,d\mathbf{r}_{n+1}\cdots d\mathbf{r}_N \tag{37.54}$$

where

$$\Xi(\xi) = \sum_{N\geqslant 0}\frac{z^N}{N!}\int\cdots\int_V e^{-U_N(\xi)/\mathbf{k}T}\,d\mathbf{r}_1\cdots d\mathbf{r}_N \tag{37.55}$$

and

$$U_N(\xi) = \xi\sum_{i=2}^{N}u(r_{1i}) + \sum_{2\leqslant i<j\leqslant N}u(r_{ij}) \tag{37.56}$$

We differentiate Eq. (37.54) with respect to ξ, and collect equivalent terms:

$$\Xi\left(\frac{\partial\rho^{(n)}}{\partial\xi}\right)_{z,V,T} - \frac{\rho^{(n)}}{\mathbf{k}T}\iint_V u(r_{12})\left[\sum_{N\geqslant 2}\frac{z^N}{N!}\times(N-1)\int\cdots\int_V e^{-U_N(\xi)/\mathbf{k}T}\,d\mathbf{r}_3\cdots d\mathbf{r}_N\right]d\mathbf{r}_1\,d\mathbf{r}_2$$

$$= -\frac{1}{\mathbf{k}T}\sum_{N\geqslant n}\frac{z^N}{(N-n)!}\left\{\sum_{i=2}^{n}u(r_{1i})\int\cdots\int_V e^{-U_N(\xi)/\mathbf{k}T}\,d\mathbf{r}_{n+1}\cdots d\mathbf{r}_N + (N-n)\int_V u(r_{1,n+1})\left[\int\cdots\int_V e^{-U_N(\xi)/\mathbf{k}T}\,d\mathbf{r}_{n+2}\cdots d\mathbf{r}_N\right]d\mathbf{r}_{n+1}\right\} \tag{37.57}$$

[1] In an open system we cannot regard the set n as made up of *particular* molecules at $\mathbf{r}_1, \ldots, \mathbf{r}_n$. The coupling parameter ξ is therefore considered to be assigned to whichever molecule happens to be at $\mathbf{r}_1$. See the discussion following Eq. (29.7).

The sum in the second term on the left-hand side requires some manipulation:

$$\sum_{N\geqslant 2} \frac{z^N}{N!}(N-1)\int\cdots\int = \sum_{N\geqslant 2}\frac{1}{N}\frac{z^N}{(N-2)!}\int\cdots\int$$

$$= \int_0^z \left[\sum_{N\geqslant 2}\frac{z'^{N-1}}{(N-2)!}\int\cdots\int\right] dz'$$

$$= \int_0^z \left[\frac{\Xi(z',\xi)\rho^{(2)}(\mathbf{r}_1,\mathbf{r}_2,\xi,z')}{z'}\right] dz' \qquad (37.58)$$

$$= \frac{\rho^{(2)*}(\mathbf{r}_1,\mathbf{r}_2,\xi,z)\Xi(z,\xi)}{\bar{N}} \qquad (37.59)$$

where Eq. (37.59) defines $\rho^{(2)*}$ as a pair distribution function averaged over z as indicated in Eq. (37.58).

Returning now to Eq. (37.57), we divide both sides of the equation by $-\rho^{(n)}\Xi/\mathbf{k}T$ and obtain

$$-\mathbf{k}T\left(\frac{\partial \ln \rho^{(n)}}{\partial \xi}\right)_{z,V,T} + \frac{1}{\bar{N}}\int\int_V u(r_{12})\rho^{(2)*}(\mathbf{r}_1,\mathbf{r}_2,\xi)\,d\mathbf{r}_1\,d\mathbf{r}_2$$

$$= \sum_{i=2}^{n} u(r_{1i}) + \frac{1}{\rho^{(n)}}\int_V u(r_{1,n+1})\rho^{(n+1)}(\mathbf{r}_1,\ldots,\mathbf{r}_{n+1},\xi)\,d\mathbf{r}_{n+1} \qquad (37.60)$$

Equation (37.60) is formally the same as Eq. (32.3) except for the occurrence of $\rho^{(2)*}$ and of the derivative with z, V, T constant instead of $\bar{N}$, V, T. But it is easy to see that the two equations are thermodynamically equivalent. First, we write

$$\left(\frac{\partial \ln \rho^{(n)}}{\partial \xi}\right)_{z,V,T} = \left(\frac{\partial \ln \rho^{(n)}}{\partial \xi}\right)_{\bar{N},V,T} + \frac{(\partial \ln \rho^{(n)}/\partial z)_{\xi,V,T}\,(\partial \bar{N}/\partial \xi)_{z,V,T}}{(\partial \bar{N}/\partial z)_{\xi,V,T}} \qquad (37.61)$$

It is apparent from Eq. (37.60) that $(\partial \ln \rho^{(n)}/\partial \xi)_{z,V,T}$ is of order unity. Hence the last term in Eq. (37.61) is negligible (of order $1/\bar{N}$), for one can see, for example, by explicit differentiation of the expressions defining $\bar{N}$ and $\rho^{(n)}$, that $\partial \bar{N}/\partial \xi$ is of order unity, $\partial \bar{N}/\partial z$ of order $\bar{N}/z$ and $\partial \ln \rho^{(n)}/\partial z$ of order $1/z$. Turning now to $\rho^{(2)*}$, we retain only the term in $N = \bar{N}$ in the two sums in Eq. (37.3) (because $N = \bar{N}$ gives the only important contribution to Ξ for large V) and find

$$\rho^{(n)} = \frac{\bar{N}!}{(\bar{N}-n)!}\,\frac{\int\cdots\int_V \exp(-U_{\bar{N}}/\mathbf{k}T)\,d\mathbf{r}_{n+1}\cdots d\mathbf{r}_{\bar{N}}}{Z_{\bar{N}}} = \rho_{\bar{N}}^{(n)} \qquad (37.62)$$

as we should expect. Also keeping only terms in $N = \bar{N}$,

$$\frac{\rho^{(2)*}}{\bar{N}} = \frac{\sum_{N \geq 2} \frac{z^N}{N!} (N-1) \int \cdots \int_V e^{-U_N/\mathbf{k}T} d\mathbf{r}_3 \cdots d\mathbf{r}_N}{\Xi}$$

$$= \frac{1}{\bar{N}} \bar{N}(\bar{N}-1) \frac{\int \cdots \int_V \exp(-U_{\bar{N}}/\mathbf{k}T)\, d\mathbf{r}_3 \cdots d\mathbf{r}_{\bar{N}}}{Z_{\bar{N}}} = \frac{\rho_{\bar{N}}^{(2)}}{\bar{N}} \tag{37.63}$$

From Eqs. (37.62) and (37.63) we then have the required result

$$\frac{\rho_{\bar{N}}^{(2)}}{\bar{N}} = \frac{\rho^{(2)*}}{\bar{N}} = \frac{\rho^{(2)}}{\bar{N}} \tag{37.64}$$

to terms of thermodynamic significance.

The thermodynamic equivalence of the terms in $\rho_N^{(2)}$ and $\rho^{(2)*}$ in Eqs. (32.3) and (37.60) can also be shown (more precisely) in the following way. We have seen [Eq. (32.12)] that the second term on the right-hand side of Eq. (32.3) can be written

$$\frac{1}{N} \iint_V u(r_{12}) \rho_N^{(2)}(\mathbf{r}_1,\mathbf{r}_2,\xi)\, d\mathbf{r}_1\, d\mathbf{r}_2 = \left(\frac{\partial A_N(\xi)}{\partial \xi}\right)_{N,V,T} \tag{37.65}$$

In the grand canonical ensemble the Helmholtz free energy is, by Eq. (14.46),

$$A(\xi,z,V,T) = \bar{N}(\xi,z,V,T)\mu(z) - \mathbf{k}T \ln \Xi(\xi,z,V,T) \tag{37.66}$$

where ξ is regarded as a second external parameter. Then

$$\left(\frac{\partial A}{\partial \xi}\right)_{z,V,T} = \mu \left(\frac{\partial \bar{N}}{\partial \xi}\right)_{z,V,T} - \mathbf{k}T \left(\frac{\partial \ln \Xi}{\partial \xi}\right)_{z,V,T} \tag{37.67}$$

Also
$$\left(\frac{\partial A}{\partial \xi}\right)_{z,V,T} = \left(\frac{\partial A}{\partial \xi}\right)_{\bar{N},V,T} + \left(\frac{\partial A}{\partial \bar{N}}\right)_{\xi,V,T} \left(\frac{\partial \bar{N}}{\partial \xi}\right)_{z,V,T} \tag{37.68}$$

Combining Eqs. (37.67) and (37.68),

$$\left(\frac{\partial A}{\partial \xi}\right)_{\bar{N},V,T} = -\mathbf{k}T \left(\frac{\partial \ln \Xi}{\partial \xi}\right)_{z,V,T} \tag{37.69}$$

To prove the required thermodynamic equivalence, we have to show, then, that

$$-\mathbf{k}T \left(\frac{\partial \ln \Xi}{\partial \xi}\right)_{z,V,T} = \frac{1}{\bar{N}} \iint_V u(r_{12}) \rho^{(2)*}(\mathbf{r}_1,\mathbf{r}_2,\xi)\, d\mathbf{r}_1\, d\mathbf{r}_2$$

$$= \frac{1}{\Xi(z,\xi)} \iint_V u(r_{12}) \left[\int_0^z \frac{\Xi(z',\xi)\rho^{(2)}(\mathbf{r}_1,\mathbf{r}_2,\xi,z')\, dz'}{z'}\right] d\mathbf{r}_1\, d\mathbf{r}_2 \tag{37.70}$$

But by direct differentiation of $\Xi(z,\xi)$ we find easily

$$\frac{z}{\Xi(z,\xi)}\frac{\partial^2\Xi}{\partial z\,\partial\xi} = -\frac{1}{\mathbf{k}T}\iint\limits_V u(r_{12})\rho^{(2)}(\mathbf{r}_1,\mathbf{r}_2,\xi,z)\,d\mathbf{r}_1\,d\mathbf{r}_2 \tag{37.71}$$

Equation (37.70) follows from Eq. (37.71) on multiplying by Ξ/z and integrating with respect to z between $z = 0$ and z.

It is convenient to derive at this point, as promised in the discussion of thermodynamic properties, the thermodynamic equivalent of Eq. (30.35) for the chemical potential from the grand partition function. Let $\xi = 0$ in $\Xi(\xi)$, and we have

$$\Xi(0) = \sum_{N\geqslant 0}\frac{z^N}{N!}V\int\cdots\int\limits_V e^{-U_{N-1}(\mathbf{r}_2,\ldots,\mathbf{r}_N)/\mathbf{k}T}d\mathbf{r}_2\cdots d\mathbf{r}_N$$

$$= 1 + zV\sum_{N\geqslant 0}\frac{1}{N+1}\frac{z^N}{N!}Z_N \tag{37.72}$$

Then
$$\frac{\partial\Xi(0)}{\partial z} = V\Xi(1)$$

But, from Eq. (14.50),

$$\frac{z}{\Xi(0)}\frac{\partial\Xi(0)}{\partial z} = \overline{N(0)}$$

therefore
$$\frac{\Xi(0)}{\Xi(1)} = \frac{zV}{\overline{N(0)}}$$

For large $\bar{N}$, $\overline{N(0)}$ is essentially[1] $\bar{N} + 1$ (i.e., there is room for approximately one more molecule when $\xi = 0$), or for thermodynamic purposes $\overline{N(0)} = \bar{N}$.

Thus
$$\frac{\Xi(0)}{\Xi(1)} = \frac{z}{\rho} \tag{37.73}$$

Equation (37.73) becomes, on taking logarithms and using Eq. (37.70),

$$\frac{\mu}{\mathbf{k}T} = \ln\rho\Lambda^3 - \int_0^1\left(\frac{\partial\ln\Xi}{\partial\xi}\right)_{z,V,T}d\xi \tag{37.74}$$

$$= \ln\rho\Lambda^3 + \frac{1}{\bar{N}\mathbf{k}T}\int_0^1\iint\limits_V u(r_{12})\rho^{(2)*}(\mathbf{r}_1,\mathbf{r}_2,\xi)\,d\mathbf{r}_1\,d\mathbf{r}_2\,d\xi \tag{37.75}$$

$$= \ln\rho\Lambda^3 + \frac{\rho}{\mathbf{k}T}\int_0^1\int_0^\infty u(r)g^*(r,\xi)\cdot 4\pi r^2\,dr\,d\xi \qquad \text{(fluid)} \tag{37.76}$$

[1] It is easy to prove that, to be exact,

$$\frac{1}{\overline{N(0)}} = \frac{1}{zV\,\Xi} + \left(\overline{\frac{1}{N+1}}\right)$$

where $$g^* \equiv \frac{\rho^{(2)*}}{\rho^2} \qquad \text{(fluid)}$$

Equation (37.75) is equivalent, thermodynamically, to Eq. (30.35).

Expansion of Distribution Functions. The way in which distribution functions may be expanded in powers of z or ρ is much more readily seen from the grand canonical than from the canonical ensemble. We shall, however, merely sketch the procedure here without carrying it very far in detail. From Eq. (37.4), we have

$$\rho^{(n)}(\mathbf{r}_1, \ldots, \mathbf{r}_n) = \frac{z^n}{\Xi}\left(e^{-U_n/\mathbf{k}T} + zI_1^{(n)} + \frac{z^2}{2!}I_2^{(n)} + \cdots\right) \tag{37.77}$$

where $$I_1^{(n)}(\mathbf{r}_1, \ldots, \mathbf{r}_n) = \int_V e^{-U_{n+1}/\mathbf{k}T} d\mathbf{r}_{n+1}$$

$$I_2^{(n)}(\mathbf{r}_1, \ldots, \mathbf{r}_n) = \iint_V e^{-U_{n+2}/\mathbf{k}T} d\mathbf{r}_{n+1}\, d\mathbf{r}_{n+2}$$

. . . .

and $$\Xi = 1 + zZ_1 + \frac{z^2}{2!}Z_2 + \cdots \qquad Z_1 = V \tag{37.78}$$

This expansion for Ξ is now introduced in Eq. (37.77) leading to

$$\rho^{(n)}(\mathbf{r}_1, \ldots, \mathbf{r}_n) = z^n \left\{e^{-U_n/\mathbf{k}T} + z[I_1^{(n)} - Z_1 e^{-U_n/\mathbf{k}T}] + z^2\left[\frac{I_2^{(n)}}{2} - Z_1 I_1^{(n)} + \left(Z_1^2 - \frac{Z_2}{2}\right)e^{-U_n/\mathbf{k}T}\right] + \cdots\right\} \tag{37.79}$$

This expansion of $\rho^{(n)}$ is equivalent to DeBoer's,[1] obtained from the canonical ensemble. One can now easily obtain, from Eq. (37.79), the first few terms of an expansion of $g^{(n)}$ in powers of z by using[2]

$$g^{(n)} = \frac{\rho^{(n)}}{\rho^n}$$

and [from Eq. (23.22)]

$$\left(\frac{\rho}{z}\right)^n = 1 + 2b_2nz + [3b_3n + 2n(n-1)b_2^2]z^2 + \cdots \tag{37.80}$$

where the b_j's are related to the Z_N's by Eq. (23.43).

[1] DeBoer (Gen. Ref.), Eq. (7.15).

[2] The system of interest is a fluid since we are considering the neighborhood of $z = \rho = 0$.

To convert the z expansions mentioned above into ρ expansions one must introduce $z(\rho)$ from Eq. (25.4).

These expansions, with $n = 2$, are of course identical with those of Sec. 34 for the radial distribution function, except for terms of order $1/\bar{N}$ which are discussed in Appendix 7.

Distribution Functions at Different Activities. We derive here certain equations, due to Mayer,[1] which are the one-component equivalent of the fundamental equations of the solution theory of McMillan and Mayer (Sec. 40) and which serve as the starting point for Mayer's integral equation theory mentioned in Sec. 38. We confine ourselves to a rather simplified treatment.

In Eq. (37.4), let

$$f(z) = \Xi(z)\left(\frac{1}{z}\right)^n \rho^{(n)}(\mathbf{r}_1, \ldots, \mathbf{r}_n, z) = \sum_{m\geqslant 0} \frac{z^m}{m!} \int_V \cdots \int e^{-U_{n+m}/\mathbf{k}T} \times d\mathbf{r}_{n+1} \cdots d\mathbf{r}_{n+m} \tag{37.81}$$

We now expand[2] $f(z)$ about $z = z^*$ by Taylor's theorem:

$$f(z) = \sum_{t\geqslant 0} \frac{(z - z^*)^t}{t!} \left(\frac{\partial^t f}{\partial z^t}\right)_{z=z^*} \tag{37.82}$$

where, from Eq. (37.81),

$$\frac{\partial^t f}{\partial z^t} = \sum_{m\geqslant t} \frac{z^{m-t}}{(m-t)!} \int_V \cdots \int e^{-U_{n+m}/\mathbf{k}T} d\mathbf{r}_{n+1} \cdots d\mathbf{r}_{n+m}$$

or, putting $m - t = q$,

$$\frac{\partial^t f}{\partial z^t} = \sum_{q\geqslant 0} \frac{z^q}{q!} \int_V \cdots \int e^{-U_{n+t+q}/\mathbf{k}T} d\mathbf{r}_{n+1} \cdots d\mathbf{r}_{n+t+q} \tag{37.83}$$

Next, in Eq. (37.81), we replace n by $n + t$ and m by q:

$$\Xi(z)\left(\frac{1}{z}\right)^{n+t} \rho^{(n+t)}(\mathbf{r}_1, \ldots, \mathbf{r}_{n+t}, z) = \sum_{q\geqslant 0} \frac{z^q}{q!} \int_V \cdots \int e^{-U_{n+t+q}/\mathbf{k}T} \times d\mathbf{r}_{n+t+1} \cdots d\mathbf{r}_{n+t+q} \tag{37.84}$$

On comparing Eqs. (37.83) and (37.84), we observe that

$$\frac{\partial^t f}{\partial z^t} = \Xi(z)\left(\frac{1}{z}\right)^{n+t} \int_V \cdots \int \rho^{(n+t)}(\mathbf{r}_1, \ldots, \mathbf{r}_{n+t}, z)\, d\mathbf{r}_{n+1} \cdots d\mathbf{r}_{n+t} \tag{37.85}$$

[1] J. E. Mayer, *J. Chem. Phys.*, **10**, 629 (1942).
[2] S. Ono, *Progr. Thoret. Phys.* (*Japan*), **6**, 447 (1951).

This is the expression we wish to use for $(\partial^t f/\partial z^t)_{z=z^*}$ in Eq. (37.82). On introducing Eq. (37.85) in Eq. (37.82), putting m for t, we have

$$\Xi(z)\left(\frac{1}{z}\right)^n \rho^{(n)}(\mathbf{r}_1, \ldots, \mathbf{r}_n, z) = \Xi(z^*) \sum_{m \geqslant 0} \frac{(z - z^*)^m}{m!} \left(\frac{1}{z^*}\right)^{n+m} \times \int_V \cdots \int \rho^{(n+m)}(\mathbf{r}_1, \ldots, \mathbf{r}_{n+m}, z^*)\, d\mathbf{r}_{n+1} \cdots d\mathbf{r}_{n+m} \quad (37.86)$$

Equation (37.86) relates the distribution function $\rho^{(n)}$ at activity z to the distribution functions $\rho^{(n)}, \rho^{(n+1)}, \ldots$ at activity z^*. In Eq. (37.86), one can put, if desired,

$$\Xi(z) = e^{pV/\mathbf{k}T}$$
$$\Xi(z^*) = e^{p^*V/\mathbf{k}T} \quad (37.87)$$

$$\left(\frac{1}{z}\right)^n \rho^{(n)}(z) = \left(\frac{\rho}{z}\right)^n g^{(n)}(z) = \left(\frac{1}{\gamma}\right)^n g^{(n)}(z) \quad \text{(fluid)}$$
$$\left(\frac{1}{z^*}\right)^{n+m} \rho^{(n+m)}(z^*) = \left(\frac{\rho^*}{z^*}\right)^{n+m} g^{(n+m)}(z^*) = \left(\frac{1}{\gamma^*}\right)^{n+m} g^{(n+m)}(z^*) \quad \text{(fluid)} \quad (37.88)$$

where the activity coefficient γ is defined as z/ρ, $p^* = p(z^*)$, and $\rho^* = \rho(z^*)$.

Certain special cases of Eq. (37.86) are of interest:

(1) If we put $z^* = 0$, Eq. (37.86) reduces to Eq. (37.81) since

$$\Xi(z^*) \to 1 \qquad \frac{\rho^*}{z^*} \to 1$$
$$g^{(n+m)}(z^*) \to e^{-U_{n+m}/\mathbf{k}T}$$

Setting $n = 0$ in Eq. (37.81) of course just gives the grand partition function, $\Xi(z)$. (2) If we take $z = 0$ in Eq. (37.86) and replace z^* by z,

$$e^{-U_n(\mathbf{r}_1, \ldots, \mathbf{r}_n)/\mathbf{k}T} = \frac{\Xi(z)}{z^n} \sum_{m \geqslant 0} \frac{(-1)^m}{m!} \int_V \cdots \int \rho^{(n+m)}(\mathbf{r}_1, \ldots, \mathbf{r}_{n+m}, z) \times d\mathbf{r}_{n+1} \cdots d\mathbf{r}_{n+m} \quad (37.89)$$

which may be regarded as the inverse of Eq. (37.81). Incidentally, the right-hand side of Eq. (37.89) must be independent of z since the left-hand side does not depend on z. (3) If we put $n = 0$ in Eq. (37.86), $\rho^{(0)} = 1$ and

$$e^{(p-p^*)V/\mathbf{k}T} = \sum_{m \geqslant 0} \frac{(z - z^*)^m}{m!} \left(\frac{1}{z^*}\right)^m \int_V \cdots \int \rho^{(m)}(\mathbf{r}_1, \ldots, \mathbf{r}_m, z^*)\, d\mathbf{r}_1 \cdots d\mathbf{r}_m \quad (37.90)$$

$$= \sum_{m \geqslant 0} \frac{1}{m!} \left(\frac{z - z^*}{\gamma^*}\right)^m \int_V \cdots \int e^{-w^{(m)}(\mathbf{r}_1, \ldots, \mathbf{r}_m, z^*)/\mathbf{k}T} d\mathbf{r}_1 \cdots d\mathbf{r}_m \quad (37.91)$$

Equation (37.91) is formally very similar to

$$\Xi(z) = e^{pV/\mathbf{k}T} = \sum_{m\geqslant 0} \frac{z^m}{m!} \int \cdots \int_V e^{-U_m(\mathbf{r}_1, \ldots, \mathbf{r}_m)/\mathbf{k}T} d\mathbf{r}_1 \cdots d\mathbf{r}_m \quad (37.92)$$

Thus we can use the method of Eqs. (23.41) to (23.45) to expand $(p - p^*)/\mathbf{k}T$ in powers of $(z - z^*)/\gamma^*$. That is, we write

$$\frac{p - p^*}{\mathbf{k}T} = \sum_{j\geqslant 1} b_j'(z^*, T) \left(\frac{z - z^*}{\gamma^*}\right)^j \quad (37.93)$$

and

$$e^{(p-p^*)V/\mathbf{k}T} = \sum_{m\geqslant 0} \frac{1}{m!} \left(\frac{z - z^*}{\gamma^*}\right)^m Z_m'(z^*) \quad (37.94)$$

where the b_j' are introduced in Eq. (37.93) as the (as yet unknown) coefficients in the expansion indicated and, in Eq. (37.94),

$$Z_m'(z^*) = \int \cdots \int_V e^{-w^{(m)}(z^*)/\mathbf{k}T} d\mathbf{r}_1 \cdots d\mathbf{r}_m \quad (37.95)$$

with $Z_1' = V$, in view of Eqs. (37.5) and (37.38). Then, just as in Eqs. (23.41) to (23.45), we find that the b_j' in Eq. (37.93) are given by

$$\begin{aligned} 1!Vb_1' &= Z_1' = V \\ 2!Vb_2' &= Z_2' - Z_1'^2 \\ 3!Vb_3' &= Z_3' - 3Z_1'Z_2' + 2Z_1'^3 \\ 4!Vb_4' &= Z_4' - 4Z_1'Z_3' - 3Z_2'^2 + 12Z_1'^2Z_2' - 6Z_1'^4 \end{aligned} \quad (37.96)$$

etc.

The general relations between the b_j' and Z_N' are formally the same as Eqs. (23.44) and (23.45). It is to be emphasized that we are not making use here of the pair (superposition) assumption, Eq. (37.48).

We can also use the method[1] of Eqs. (25.1) to (25.9) to obtain a "density" expansion for $(p - p^*)/\mathbf{k}T$. We use [Eq. (14.50)]

$$\frac{\rho}{z} = \left(\frac{\partial p/\mathbf{k}T}{\partial z}\right)_{T,V} \quad (37.97)$$

and Eq. (37.93) (holding p^*, z^* and γ^* constant) to obtain

$$\frac{(z - z^*)\rho}{z} = \frac{z - z^*}{\gamma} = \sum_{j\geqslant 1} b_j'(z^*, T) j \left(\frac{z - z^*}{\gamma^*}\right)^j \quad (37.98)$$

[1] Or the more general argument of Eqs. (25.24) to (25.31).

which is the analogue of Eq. (25.2). Therefore, following Eq. (25.1), we write

$$\frac{z - z^*}{\gamma^*} = \frac{z - z^*}{\gamma} + a_2'(z^*,T)\left(\frac{z - z^*}{\gamma}\right)^2 + a_3'(z^*,T)\left(\frac{z - z^*}{\gamma}\right)^3 + \cdots \tag{37.99}$$

Then, as in Eq. (25.3), we find the a_i' to be

$$\begin{aligned} a_2' &= -2b_2' \\ a_3' &= -3b_3' + 8b_2'^2 \\ a_4' &= -4b_4' + 30b_2'b_3' - 40b_2'^3 \\ &\text{etc.} \end{aligned} \tag{37.100}$$

We now substitute Eq. (37.99) for $(z - z^*)/\gamma^*$ in Eq. (37.93) and obtain

$$\frac{p - p^*}{\mathbf{k}T} = \frac{z - z^*}{\gamma}\left[1 - \sum_{k\geqslant 1} \frac{k}{k+1}\beta_k'(z^*,T)\left(\frac{z - z^*}{\gamma}\right)^k\right] \tag{37.101}$$

where

$$\begin{aligned} \beta_1' &= 2b_2' \\ \beta_2' &= 3b_3' - 6b_2'^2 \\ \beta_3' &= 4b_4' - 24b_2'b_3' + (80/3)b_2'^3 \\ &\text{etc.} \end{aligned} \tag{37.102}$$

with the b_j' given by Eq. (37.96). The general relations between the β_k' and b_j' are formally the same as Eqs. (24.14) and (25.30).

In summary: Eqs. (37.93) and (37.101) are generalizations of the expansion of $p/\mathbf{k}T$ in powers of z and ρ in Chap. 5. The potential of mean force $w^{(m)}(z^*)$ plays the role here that U_m does in Chap. 5. In the limit $z^* \to 0$, Eqs. (37.93) and (37.101) reduce to the equations of Chap. 5, in view of Eq. (37.42) and $\gamma^* \to 1$. Equation (37.101) becomes particularly important, as we shall see, in a multicomponent system (Sec. 40).

Order of a Phase Transition.[1] By the conventional definition,[2] in an nth-order phase transition

$$\left(\frac{\partial^n p/\mathbf{k}T}{\partial z^n}\right)_T \text{ is discontinuous}$$

and

$$\left(\frac{\partial^k p/\mathbf{k}T}{\partial z^k}\right)_T \text{ is continuous} \qquad k = 0, 1, 2, \ldots, n-1 \tag{37.103}$$

[1] J. E. Mayer, *J. Chem. Phys.*, **16**, 665 (1948).

[2] See, for example, J. E. Mayer and S. F. Streeter, *J. Chem. Phys.*, **7**, 1019 (1939).

across the phase boundary, in the limit $V \to \infty$. As z has the same value in both phases, say z', we can also state that the nth derivative above is a discontinuous function of z at $z = z'$, while all lower derivatives are continuous at $z = z'$. For example, in a first-order phase transition $p/\mathbf{k}T$ is continuous at $z = z'$ but $\partial(p/\mathbf{k}T)/\partial z$ and ρ [see Eq. (37.97)] are discontinuous at $z = z'$. Figure 17 (see page 173) illustrates this case.

Consider, now, Eq. (37.93) with z in place of z^* and $z + y$ in place of z:

$$\frac{p(z + y)}{\mathbf{k}T} = \frac{p(z)}{\mathbf{k}T} + \sum_{j\geqslant 1} b_j'(z,T)\left[\frac{y\rho(z)}{z}\right]^j \tag{37.104}$$

Differentiate Eq. (37.104) n times with respect to y, holding z and T constant:

$$\frac{1}{\mathbf{k}T}\frac{\partial^n p(z + y)}{\partial(z + y)^n} = \sum_{j\geqslant n} b_j'(z,T)\left[\frac{\rho(z)}{z}\right]^j \frac{j!}{(j - n)!}y^{j-n}$$

Now let $y \to 0$, and we have

$$\frac{\partial^n p(z)/\mathbf{k}T}{\partial z^n} = b_n'(z,T)\left[\frac{\rho(z)}{z}\right]^n n! \tag{37.105}$$

For a first-order phase change ($n = 1$), both sides of Eq. (37.105) are clearly discontinuous at $z = z'$ (recall that $b_1' = 1$). For a phase change of order $n > 1$, by definition the left-hand side of Eq. (37.105) is discontinuous at $z = z'$ but all lower derivatives and $\rho(z)$ are continuous at $z = z'$. The right-hand side of Eq. (37.105) must of course have the same behavior as the left-hand side; thus it follows that $b_n'(z,T)$ is discontinuous at $z = z'$ but $b_k'(z,T)$ with $k = 1, 2, \ldots, n - 1$ is continuous. Further, from Eq. (37.96) we see that $Z_n'(z,T)$ is discontinuous at $z = z'$ but not $Z_k'(z,T)$ with $k = 1, 2, \ldots, n - 1$. Finally, we conclude from Eqs. (37.38) and (37.95) that in an nth-order phase transition ($n > 1$), $\rho^{(n)}(z)$ and $g^{(n)}(z)$ are discontinuous across the phase boundary (that is, at $z = z'$) while $\rho^{(k)}(z)$ and $g^{(k)}(z)$, $k = 1, 2, \ldots, n - 1$, are continuous.[1]

In a first-order phase transition ρ is discontinuous at $z = z'$; therefore $\rho^{(1)}$ must be also, for

$$\frac{1}{V}\int_V \rho^{(1)}(\mathbf{r}_1)\,d\mathbf{r}_1 = \rho \tag{37.106}$$

We may summarize, then, by stating that *in a phase transition of any order* ($n \geqslant 1$), $\rho^{(n)}(z)$ *is discontinuous across the phase boundary while* $\rho^{(1)}(z)$, $\rho^{(2)}(z)$,

[1] As Mayer points out, this conclusion for $k = 1, 2, \ldots, n - 1$ is almost but not quite certain since the integral in Eq. (37.95) can conceivably be continuous at $z = z'$ but the integrand discontinuous.

. . . , $\rho^{(n-1)}(z)$ *are continuous. Furthermore,* $\rho^{(n+1)}(z)$, $\rho^{(n+2)}(z)$, . . . *are also discontinuous.* This last remark follows from Eq. (37.107) below:

$$\rho^{(n+m)}(\mathbf{r}_1, \ldots, \mathbf{r}_{n+m}) = \sum_{N \geqslant n+m} \rho_N{}^{(n+m)}(\mathbf{r}_1, \ldots, \mathbf{r}_{n+m}) P_N$$

$$\int_V \cdots \int \rho^{(n+m)}\, d\mathbf{r}_{n+1} \cdots d\mathbf{r}_{n+m} = \sum_{N \geqslant n+m} P_N \int_V \cdots \int \rho_N{}^{(n+m)}\, d\mathbf{r}_{n+1} \cdots d\mathbf{r}_{n+m}$$
$$= \sum_{N \geqslant n+m} \frac{(N-n)!}{(N-n-m)!} \rho_N{}^{(n)} P_N$$
$$= \left[\frac{(N-n)!}{(N-n-m)!}\right]_{\text{av}} \rho^{(n)}(\mathbf{r}_1, \ldots, \mathbf{r}_n) \tag{37.107}$$

where
$$\left[\frac{(N-n)!}{(N-n-m)!}\right]_{\text{av}} = \frac{\displaystyle\sum_{N \geqslant n+m} \frac{(N-n)!}{(N-n-m)!} \rho_N{}^{(n)} P_N}{\displaystyle\sum_{N \geqslant n} \rho_N{}^{(n)} P_N} \tag{37.108}$$

That is, Eq. (37.107) shows that if $\rho^{(n)}$ is discontinuous, the integral of $\rho^{(n+m)}$ on the left-hand side is discontinuous, as is therefore $\rho^{(n+m)}$ itself.

38. The Kirkwood-Salsburg Integral Equation

To obtain the Born-Green-Yvon and Kirkwood integral equations, already discussed, one begins by differentiating the equation defining the distribution function $\rho^{(n)}$ with respect to a variable on which $\rho^{(n)}$ depends. In the Born-Green-Yvon case, the variable is the position of a particle in space and in the Kirkwood case the variable is a coupling parameter. Mayer[1] has shown that it is in fact possible to carry through a very much more general argument of this type, in which an arbitrary parameter y, on which the distribution functions in the grand ensemble depend, is used in place of the particular variables mentioned above. This general variational method of Mayer's leads to integral equations which include the equations of Born-Green-Yvon and Kirkwood as special cases.[2] We shall, however, not attempt to discuss Mayer's theory here since the two special cases just mentioned have already been derived and examined in detail and these are the only consequences of the general theory which we shall have space to consider.

Kirkwood and Salsburg[3] have recently derived a new set of integral equations without use of a variational procedure. The integral equations are obtained directly, as will be seen below, instead of requiring integration of integrodifferential equations as in the variational method. An interesting

[1] J. E. Mayer, *J. Chem. Phys.*, **15**, 187 (1947).

[2] See S. Ono, *Progr. Theoret. Phys.* (*Japan*), **5**, 822 (1950).

[3] J. G. Kirkwood and Z. Salsburg, *Discussions Faraday Soc.*, **15**, 28 (1953).

feature of these equations is that a particular molecule,[1] say molecule 1, of the set $1, \ldots, n$ is singled out for special treatment so that a close relationship to cell theories (Chap. 8) may be anticipated. In fact, approximate solution of the Kirkwood-Salsburg integral equations will probably depend on the formulation of relatively simple kernels using cell theory as a guide. This type of approximation, incidentally, will make it possible to avoid use of the superposition approximation. It should be mentioned that the Kirkwood-Salsburg equations resemble to some extent earlier equations due to Mayer[2] and Mayer and Montroll.[3]

We now proceed to outline the Kirkwood-Salsburg derivation, which is based on a partial cluster expansion of the Mayer type (Sec. 22). We write the potential energy U_N in the form

$$U_N(1, \ldots, N) = \sum_{i=2}^{n} u(r_{1i}) + \sum_{i=n+1}^{N} u(r_{1i}) + U_{N-1}(2, \ldots, N) \qquad (38.1)$$

where $(1, \ldots, N)$ means $(\mathbf{r}_1, \ldots, \mathbf{r}_N)$ as before. That is, the interactions of particle 1 with the other members of the set $1, \ldots, n$ and with molecules $n+1, \ldots, N$ are separated out of the total potential energy. Then

$$e^{-U_N/\mathbf{k}T} = e^{-U_n^{(1)}(1,\ldots,n)/\mathbf{k}T} e^{-U_{N-1}(2,\ldots,N)/\mathbf{k}T} \prod_{\sigma=n+1}^{N} [1 + f_{1\sigma}(r_{1\sigma})] \qquad (38.2)$$

where

$$U_n^{(1)}(1, \ldots, n) = \sum_{i=2}^{n} u(r_{1i}) \qquad (38.3)$$

Equation (37.3) for $\rho^{(n)}$ then becomes

$$\rho^{(n)}(1, \ldots, n) = \frac{e^{-U_n^{(1)}(1,\ldots,n)/\mathbf{k}T}}{\Xi} \sum_{N \geqslant n} \frac{z^N}{(N-n)!} \times \int \cdots \int_V e^{-U_{N-1}(2,\ldots,N)/\mathbf{k}T} \prod_{\sigma=n+1}^{N} (1 + f_{1\sigma})\, d\mathbf{r}_\sigma \qquad (38.4)$$

We expand the product in Eq. (38.4):

$$\prod_{\sigma=n+1}^{N} (1 + f_{1\sigma}) = 1 + \sum_{\sigma} f_{1\sigma} + \sum_{\sigma} \sum_{\sigma'} f_{1\sigma} f_{1\sigma'} + \cdots \qquad (38.5)$$

There are

$$\frac{(N-n)!}{(N-n-s)!s!}$$

[1] In the grand ensemble, it will be recalled, this means strictly whichever molecule is at $\mathbf{r}_1$.

[2] J. E. Mayer, *J. Chem. Phys.*, **15**, 187 (1947).

[3] J. E. Mayer and E. Montroll, *J. Chem. Phys.*, **9**, 2 (1941).

terms in the multiple sum involving products of s f's in Eq. (38.5) and each such term gives the same result on carrying out the integrations in Eq. (38.4). Thus we have

$$\rho^{(n)}(1, \ldots, n) = \frac{e^{-U_n^{(1)}(1,\ldots,n)/\mathbf{k}T}}{\Xi} \sum_{N \geqslant n} \sum_{s=0}^{N-n} \frac{z^N}{(N-n-s)!s!} \times \int_V \cdots \int e^{-U_{N-1}(2,\ldots,N)/\mathbf{k}T} \left(\prod_{\sigma=n+1}^{n+s} f_{1\sigma} \right) d\mathbf{r}_{n+1} \cdots d\mathbf{r}_N \quad (38.6)$$

Let us define

$$K_s(1; n+1, \ldots, n+s) = \prod_{\sigma=n+1}^{n+s} f_{1\sigma}(r_{1\sigma}) \quad (38.7)$$

and reverse the order of summation in Eq. (38.6). The limits are then $0 \leqslant s \leqslant \infty$ and $s + n \leqslant N \leqslant \infty$. If the term with $s = 0$ is written separately and z is factored out of every term, Eq. (38.6) becomes

$$\rho^{(n)}(1, \ldots, n) = \frac{z e^{-U_n^{(1)}(1,\ldots,n)/\mathbf{k}T}}{\Xi} \left\{ \sum_{N \geqslant n} \frac{z^{N-1}}{(N-n)!} \times \int_V \cdots \int e^{-U_{N-1}(2,\ldots,N)/\mathbf{k}T} \, d\mathbf{r}_{n+1} \cdots d\mathbf{r}_N + \sum_{s \geqslant 1} \sum_{N \geqslant s+n} \frac{z^{N-1}}{(N-n-s)!s!} \times \int_V \cdots \int e^{-U_{N-1}(2,\ldots,N)/\mathbf{k}T} K_s(1; n+1, \ldots, n+s) \, d\mathbf{r}_{n+1} \cdots d\mathbf{r}_N \right\} \quad (38.8)$$

Finally, if we replace $N - 1$ by N in Eq. (38.8),

$$\rho^{(n)}(1, \ldots, n) = z e^{-U_n^{(1)}(1,\ldots,n)/\mathbf{k}T} \left[\rho^{(n-1)}(2, \ldots, n) + \sum_{s \geqslant 1} \frac{1}{s!} \int_V \cdots \int \times K_s(1; n+1, \ldots, n+s) \rho^{(n+s-1)}(2, \ldots, n+s) \, d\mathbf{r}_{n+1} \cdots d\mathbf{r}_{n+s} \right]$$

$$n = 1, 2, \ldots, \infty \quad (38.9)$$

If we define [by analogy with Eq. (35.38)] an excess chemical potential as

$$\mu_v^{\mathrm{E}} = \mathbf{k}T \ln \gamma \quad (38.10)$$

where $\gamma = z/\rho$, we may replace z in Eq. (38.9) by

$$z = \rho e^{\mu_v^{\mathrm{E}}/\mathbf{k}T} \quad (38.11)$$

as is done by Kirkwood and Salsburg in some of their equations.

Equations (38.9) constitute a determinate system of integral equations for the $\rho^{(n)}$ under the conditions that $\rho^{(n)}$ is a symmetric function of $\mathbf{r}_1, \ldots, \mathbf{r}_n$ and that $\rho^{(n)} \to 0$ as $n \to \infty$ in the volume V (because of intermolecular repulsion at short range between real molecules).

If we put $n = 1$ in Eq. (38.9) and integrate over $\mathbf{r}_1$, using $\rho^{(0)} = 1$ and Eq. (37.5), we find an expression for μ or z:

$$\frac{\rho}{z} = e^{-\mu_v{}^E/\mathbf{k}T} = 1 + \frac{1}{V}\sum_{s\geqslant 1}\frac{1}{s!} \times \int \cdots \int_V K_s(1; 2, \ldots, s+1)\rho^{(s)}(2, \ldots, s+1)\, d\mathbf{r}_1 \cdots d\mathbf{r}_{s+1} \tag{38.12}$$

Equation (38.12) has also been obtained by Mayer,[1] using a different method.

In a fluid, from Eq. (37.10), Eqs. (38.9) and (38.12) can be written as[2]

$$g^{(n)}(1, \ldots, n) = \frac{z}{\rho} e^{-U_n{}^{(1)}(1,\ldots,n)/\mathbf{k}T}\left[g^{(n-1)}(2, \ldots, n) + \sum_{s\geqslant 1}\frac{\rho^s}{s!} \times \int \cdots \int_V K_s(1; n+1, \ldots, n+s) g^{(n+s-1)}(2, \ldots, n+s)\, d\mathbf{r}_{n+1} \cdots d\mathbf{r}_{n+s}\right]$$
$$n = 1, 2, \ldots, \infty \tag{38.13}$$

$$\frac{\rho}{z} = e^{-\mu_v{}^E/\mathbf{k}T} = 1 + \sum_{s\geqslant 1}\frac{\rho^s}{s!} \times \int \cdots \int_V K_s(1; 2, \ldots, s+1) g^{(s)}(2, \ldots, s+1)\, d\mathbf{r}_2 \cdots d\mathbf{r}_{s+1} \tag{38.14}$$

Equation (38.9) rather resembles Eq. (32.7) with $\xi = 1$. The comparison is a little more direct if we use the canonical ensemble version of Eq. (38.9), which we first derive. We substitute Eqs. (38.3) and (38.5) into Eq. (29.8) ($\boldsymbol{\xi} = \mathbf{1}$) and find

$$\rho_N{}^{(n)}(1, \ldots, n) = e^{-U_n{}^{(1)}(1,\ldots,n)/\mathbf{k}T}\frac{Z_{N-1}}{Z_N}\sum_{s=0}^{N-n}\frac{N!}{(N-n-s)!s!Z_{N-1}} \times \int \cdots \int_V e^{-U_{N-1}(2,\ldots,N)/\mathbf{k}T} K_s(1; n+1, \ldots, n+s)\, d\mathbf{r}_{n+1} \cdots d\mathbf{r}_N \tag{38.15}$$

[1] J. E. Mayer, *J. Chem. Phys.*, **15**, 187 (1947), Eq. (53).

[2] It should be noted that Kirkwood and Salsburg define $g^{(n)}$ by $\rho^{(n)} = \rho^n g^{(n)}$ for a fluid *or* a crystal (as for Mayer's F_n or our $\mathfrak{g}^{(n)}$).

Then from Eqs. (29.8) and (30.25),

$$\rho_N^{(n)}(1, \ldots, n) = ze^{-U_n^{(1)}(1,\ldots,n)/\mathbf{k}T}\Bigg[\hat{\rho_{N-1}}^{(n-1)}(2, \ldots, n) + \sum_{s=1}^{N-n}\frac{1}{s!} \times \int_V \cdots \int K_s(1; n+1, \ldots, n+s)\rho_{N-1}^{(n+s-1)}(2, \ldots, n+s) \times d\mathbf{r}_{n+1} \cdots d\mathbf{r}_{n+s}\Bigg] \qquad n = 1, 2, \ldots, N \tag{38.16}$$

This is the equivalent of Eq. (38.9). Now we recall from Eq. (30.35) that the fourth term on the right-hand side of Eq. (32.7) is equal to $\mathbf{k}T \ln (z/\rho)$. Hence, after taking the logarithm of both sides of Eq. (38.16), a term-by-term comparison of Eqs. (32.7) and (38.16) shows that

$$\int_0^1 \int_V u(r_{1,n+1})\rho_N^{[n+1]}(1, \ldots, n+1, \xi)\, d\mathbf{r}_{n+1}\, d\xi = -\mathbf{k}T \ln \Sigma \tag{38.17}$$

where

$$\Sigma = 1 + \sum_{s=1}^{N-n} \frac{(N-n)!}{(N-n-s)!s!} \int_V \cdots \int K_s(1; n+1, \ldots, n+s) \times \left[\frac{(N-n-s)!\rho_{N-1}^{(n+s-1)}(2, \ldots, n+s)}{(N-n)!\rho_{N-1}^{(n-1)}(2, \ldots, n)}\right] d\mathbf{r}_{n+1} \cdots d\mathbf{r}_{n+s} \tag{38.18}$$

In Eq. (38.17), Σ plays the role of a "partition function" associated with[1] a "free energy," for the left-hand side of Eq. (38.17) [see the discussion of Eq. (32.10)] is the contribution to the work of "charging" particle 1 from $\xi = 0$ to $\xi = 1$ arising from the interactions between particle 1 and the $N - n$ molecules not in the fixed set $1, \ldots, n$. The integral in Eq. (38.18) is the average value of the product K_s of "excess" ($\xi = 1$ compared to $\xi = 0$) Boltzmann factors associated with the interaction between molecule 1 and molecules $n + 1, \ldots, n + s$:

$$K_s = \prod_{\sigma=n+1}^{n+s} [e^{-u(r_{1\sigma})/\mathbf{k}T} - e^{-\xi_0 u(r_{1\sigma})/\mathbf{k}T}] \qquad \xi_1 = 1,\ \xi_0 = 0$$

$$= \prod_{\sigma=n+1}^{n+s} [e^{-u(r_{1\sigma})/\mathbf{k}T} - 1] \tag{38.19}$$

[1] That is, (free energy) = $-\mathbf{k}T$ ln (partition function).

The probability density [ ] by which K_s is multiplied in Eq. (38.18) is, from Eq. (29.8),

$$[\ \ ] = \frac{\int \cdot \cdot \cdot \int_V e^{-U_{N-1}(2,\ldots,N)/\mathbf{k}T}\, d\mathbf{r}_{n+s+1} \cdot \cdot \cdot d\mathbf{r}_N}{\int \cdot \cdot \cdot \int_V e^{-U_{N-1}(2,\ldots,N)/\mathbf{k}T}\, d\mathbf{r}_{n+1} \cdot \cdot \cdot d\mathbf{r}_N} \tag{38.20}$$

That is, [ ] $d\mathbf{r}_{n+1} \cdot \cdot \cdot d\mathbf{r}_{n+s}$ is the probability of observing molecules in $d\mathbf{r}_{n+1}$ at $\mathbf{r}_{n+1}, \ldots,$ and $d\mathbf{r}_{n+s}$ at $\mathbf{r}_{n+s}$ if other molecules are fixed at $\mathbf{r}_2, \ldots, \mathbf{r}_n$ [and $\xi = 0$ on molecule 1 at $\mathbf{r}_1$ since U_{N-1} in Eq. (38.20) refers to $\mathbf{r}_2, \ldots, \mathbf{r}_N$]. The factor in front of the integral sign in Eq. (38.18) is the number of ways of selecting s particles (to interact with particle 1) out of the $N - n$ particles not in the set $1, \ldots, n$.

Returning now to Eq. (38.9), we note that for real molecules with short-range van der Waals interactions the cluster sums over s terminate after a small finite number of terms, thus obviating any possible concern over convergence of the infinite series. This termination of the series can be seen by considering an idealized form of $u(r)$ appropriate to two molecules with rigid impenetrable cores of radius b exerting on each other a force of finite range a. Any real van der Waals intermolecular potential $u(r)$ can be made to satisfy these conditions for all practical purposes by suitable choices of a and b. Suppose a and b are such that ν but not $\nu + 1$ spheres of radius b can be packed into a sphere of radius a. Then when $s > \nu$, in carrying out the integration in Eq. (38.9), $\rho^{(n+s-1)}$ vanishes because of infinite intermolecular repulsion whenever all the points $\mathbf{r}_{n+1}, \ldots, \mathbf{r}_{n+s}$ are inside a sphere of radius a about $\mathbf{r}_1$; but K_s vanishes if any one of these points, say $\mathbf{r}_\sigma$, is outside of this sphere, for in this case $f_{1\sigma}$ is zero. Thus the integrand is always zero when $s > \nu$ and the term $s = \nu$ is the last of the series. For hard spheres without attractive forces, $a = 2b$ and $\nu = 12$.

Equation (38.9) can be put in an alternative form. We have

$$\rho^{(n)}(1, \ldots, n) = ze^{-U_n^{(1)}(1,\ldots,n)/\mathbf{k}T}\rho^{(n-1)}(2, \ldots, n) + ze^{-U_n^{(1)}(1,\ldots,n)/\mathbf{k}T}$$
$$\times \sum_{s\geq 1} \frac{1}{s!} \int_V \rho^{(n)}(2, \ldots, n+1)\left[\int \cdot \cdot \cdot \int_V K_s(1; n+1, \ldots, n+s)\right.$$
$$\left. \times \frac{\rho^{(n+s-1)}(2, \ldots, n+s)}{\rho^{(n)}(2, \ldots, n+1)}\, d\mathbf{r}_{n+2} \cdot \cdot \cdot d\mathbf{r}_{n+s}\right] d\mathbf{r}_{n+1}$$

We replace s by $s + 1$ in this equation and obtain

$$\rho^{(n)}(1, \ldots, n) = ze^{-U_n^{(1)}(1,\ldots,n)/\mathbf{k}T}\rho^{(n-1)}(2, \ldots, n)$$
$$+ \int_V \rho^{(n)}(2, \ldots, n+1)K^{(n)}(1, \ldots, n+1)\, d\mathbf{r}_{n+1}$$
$$n = 1, 2, \ldots, \infty \tag{38.21}$$

where the kernel is

$$K^{(n)}(1, \ldots, n+1) = zf_{1,n+1}e^{-U_n^{(1)}(1,\ldots,n)/\mathbf{k}T}$$
$$\times \left\{1 + \sum_{s\geqslant 1} \frac{1}{(s+1)!} \int_V \cdots \int \frac{\rho^{(n+s)}(2, \ldots, n+s+1)}{\rho^{(n)}(2, \ldots, n+1)} \prod_{\sigma=n+2}^{n+s+1} f_{1\sigma}\, d\mathbf{r}_\sigma\right\} \tag{38.22}$$

The factor { } in Eq. (38.22) can be written in a form (canonical ensemble) completely analogous to Eq. (38.18) and given an equivalent interpretation. In practice, the final term in the series in Eq. (38.22) is $s = \nu - 1$.

In a fluid, with $n = 2$, Eqs. (38.21) and (38.22) become

$$g^{(2)}(1,2) = \frac{z}{\rho}\, e^{-u(r_{12})/\mathbf{k}T} + \int_V g^{(2)}(2,3)K^{(2)}(1,2,3)\, d\mathbf{r}_3 \tag{38.23}$$

where

$$K^{(2)}(1,2,3) = zf_{13}e^{-u_{12}/\mathbf{k}T}$$
$$\times \left\{1 + \sum_{s\geqslant 1} \frac{\rho^s}{(s+1)!} \int_V \cdots \int \frac{g^{(s+2)}(2, \ldots, s+3)}{g^{(2)}(2,3)} \prod_{\sigma=4}^{s+3} f_{1\sigma}\, d\mathbf{r}_\sigma\right\} \tag{38.24}$$

Equation (38.23) is the integral equation for the radial distribution function in a fluid.

Equation (34.5) for $g^{(2)}$ (equivalent to including the second and third virial coefficients of a gas) follows from Eqs. (38.23) and (38.24) on putting

$$K^{(2)} \to \rho f_{13}e^{-u_{12}/\mathbf{k}T}$$
$$\frac{z}{\rho} \to 1 - 2b_2\rho \tag{38.25}$$

Calculations based on these new integral equations have not been reported as yet.

39. Distribution Functions at a Phase Transition

First-order Transition. In a closed one-component system at temperature T and volume V (where V is very large, $V \to \infty$), suppose only phase a exists for $N \leqslant N_a$ and only phase b for $N \geqslant N_b$. When $N_a < N < N_b$, the two phases a and b exist in equilibrium. If $\rho_N^{(n)}(\mathbf{r}_1, \ldots, \mathbf{r}_n)$ is the generic distribution function for a set of n particles in this system, let

$$\begin{aligned} \rho_N^{(n)} &= \rho_{(a)N}^{(n)} & N &\leqslant N_a \\ &= \rho_{(b)N}^{(n)} & N &\geqslant N_b \end{aligned} \tag{39.1}$$

We now consider the interval $N_a < N < N_b$. In the absence of a gravitational field, the probability that any small region in V is occupied by phase a, when $N_a < N < N_b$, is just equal to the volume fraction $x = (N_b - N)/(N_b - N_a)$ of phase a in the entire two-phase system of volume V. Let us assume that the points $\mathbf{r}_1, \ldots, \mathbf{r}_n$ are close enough together and the single-phase regions within the two-phase system are large enough[1] in extent so that we can with negligible error consider all the points $\mathbf{r}_1, \ldots, \mathbf{r}_n$ to be within the same single-phase region at any given time. Then it follows, as Mayer[2] has pointed out, that

$$\rho_N^{(n)} = x(N)\rho_{(a)N_a}^{(n)} + [1 - x(N)]\rho_{(b)N_b}^{(n)}] \qquad N_a < N < N_b \quad (39.2)$$

In the two-phase interval above let $p = p'$, $\mu = \mu'$, and $z = z'$ be the pressure, chemical potential, and activity, respectively. Then phase a is the stable phase when $z < z'$ and phase b is the stable phase when $z > z'$. In an open system,

$$\rho^{(n)} = \sum_{N \geqslant n} P_N \rho_N^{(n)} \quad (39.3)$$

When $z < z'$, the only values of N which contribute anything appreciable to the sum in Eq. (39.3) are those with $N \leqslant N_a$; thus $\rho^{(n)}$ is in this case characteristic of pure phase a:

$$\rho^{(n)} = \rho_{(a)}^{(n)} = \rho_{(a)\bar{N}}^{(n)} \qquad z < z' \quad (39.4)$$

where the last form follows in the limit $V \to \infty$ [only the term $N = \bar{N}$ contributes appreciably in Eq. (39.3)], with $\bar{N}(z,V,T) \leqslant N_a$. Similarly,

$$\rho^{(n)} = \rho_{(b)}^{(n)} = \rho_{(b)\bar{N}}^{(n)} \qquad z > z' \quad (39.5)$$

where $\bar{N}(z,V,T) \geqslant N_b$. When $z = z'$, only N_a and N_b contribute appreciably to the sum in Eq. (39.3), as is pointed out in Appendix 9. In Eq. (A9.22) (see also the footnote to this equation), which is valid at the ends of the two-phase region (no interfacial effects), let $C(N_a) = C_a$ and $C(N_b) = C_b$. Then

$$\rho^{(n)} = p_a\rho_{(a)N_a}^{(n)} + p_b\rho_{(b)N_b}^{(n)} \qquad z = z' \quad (39.6)$$

where

$$p_a = \frac{C_a}{C_a + C_b} \qquad p_b = \frac{C_b}{C_a + C_b}$$

Now $\rho_{(a)N_a}^{(n)}$ is the limit of $\rho_{(a)\bar{N}}^{(n)}$ or $\rho_{(a)}^{(n)}(z)$ as $z \to z'-$ and $\bar{N} \to N_a -$ and $\rho_{(b)N_b}^{(n)}$ is the limit of $\rho_{(b)\bar{N}}^{(n)}$ or $\rho_{(b)}^{(n)}(z)$ as $z \to z'+$ and $\bar{N} \to N_b+$. Therefore

$$\rho^{(n)}(z') = p_a\rho_{(a)}^{(n)}(z'-) + p_b\rho_{(b)}^{(n)}(z'+) \quad (39.7)$$

Thus we have found that in the two-phase region of a closed system [Eq. (39.2)], $\rho_N^{(n)}$ is given by a linear combination of the two distribution functions

[1] The most stable two-phase configurations will be those with the smallest interfacial areas. This favors large single-phase regions within the two-phase system.

[2] J. E. Mayer, *J. Chem. Phys.*, **10**, 629 (1942).

$\rho_{(a)N_a}{}^{(n)}$ and $\rho_{(b)N_b}{}^{(n)}$ belonging to the separate phases in equilibrium with each other. The coefficients $x(N)$ and $1 - x(N)$ vary between zero and unity but of course have definite values for any given N (or average density $\rho = N/V$) between N_a and N_b. In an open system [Eq. (39.7)] we also have a linear combination of the two extreme distribution functions, but the coefficients are not variable because an average over the possible values of N has been taken.

The three sets of integrodifferential or integral equations we have discussed are

$$-\mathbf{k}T\nabla_1\rho^{(n)} = \rho^{(n)} \sum_{i=2}^{n} \nabla_1 u(r_{1i}) + \int_V \nabla_1 u(r_{1,n+1})\rho^{(n+1)}(\mathbf{r}_1 \ . \ . \ . \ , \mathbf{r}_{n+1})\, d\mathbf{r}_{n+1} \qquad \text{(Born-Green-Yvon)} \quad (39.8)$$

$$-\mathbf{k}T \frac{\partial \rho^{(n)}}{\partial \xi} + \rho^{(n)} \left(\frac{\partial A}{\partial \xi}\right)_{N,V,T} = \rho^{(n)} \sum_{i=2}^{n} u(r_{1i}) + \int_V u(r_{1,n+1}) \times \rho^{(n+1)}(\mathbf{r}_1, \ . \ . \ . \ , \mathbf{r}_{n+1}, \xi)\, d\mathbf{r}_{n+1} \qquad \text{(Kirkwood)} \quad (39.9)$$

$$\rho^{(n)} = ze^{-U_n{}^{(1)}/\mathbf{k}T}\rho^{(n-1)}(\mathbf{r}_2, \ . \ . \ . \ , \mathbf{r}_n) + \int_V K^{(n)}(\mathbf{r}_1, \ . \ . \ . \ , \mathbf{r}_{n+1}) \times \rho^{(n)}(\mathbf{r}_2, \ . \ . \ . \ , \mathbf{r}_{n+1})\, d\mathbf{r}_{n+1} \qquad \text{(Kirkwood and Salsburg)} \quad (39.10)$$

where the integrand in Eq. (39.10) is

$$K^{(n)}\rho^{(n)} = zf_{1,n+1}e^{-U_n{}^{(1)}/\mathbf{k}T} \left\{\rho^{(n)}(\mathbf{r}_2, \ . \ . \ . \ , \mathbf{r}_{n+1}) + \sum_{s\geqslant 1} \frac{1}{(s+1)!} \times \int \cdots \int_V \rho^{(n+s)}(\mathbf{r}_2, \ . \ . \ . \ , \mathbf{r}_{n+s+1}) \prod_{\sigma=n+2}^{n+s+1} f_{1\sigma}\, d\mathbf{r}_\sigma \right\} \quad (39.11)$$

These equations have been derived for both open and closed systems. In a closed system, when $N = N_a$ let the solution of the equations be $\rho_{(a)N_a}{}^{(n)}$ and when $N = N_b$ let the solution be $\rho_{(b)N_b}{}^{(n)}$. Then for a value of N in the two-phase region ($N_a < N < N_b$) it is easy to see that the linear combination in Eq. (39.2) is the solution, for Eqs. (39.8) to (39.10) are *linear in the distribution functions*, and all other quantities appearing in the equations have the same value in the two phases[1] a and b. In an open system the distribution functions are functions of z and T. Let $\rho_{(a)}{}^{(n)}(z'-)$ be the solution of Eqs. (39.8), (39.9), and (39.10) at $z = z'-$ and let $\rho_{(b)}{}^{(n)}(z'+)$ be the solution at $z = z'+$. Then, again because of linearity in the distribution functions and the equality in the two phases of all other quantities appearing in the

[1] From a thermodynamic point of view, as we have pointed out before, ξ is a second external variable (the first being V). Then $X = -(\partial A/\partial \xi)_{N,V,T}$ is the associated generalized force. In the two-phase equilibrium the two phases must have equal values of μ, p, T, and X.

equations, the linear combination $\rho^{(n)}(z')$ in Eq. (39.7) is the solution of Eqs. (39.8), (39.9), and (39.10) when $z = z'$. Actually, any linear combination of $\rho_{(a)}{}^{(n)}(z'-)$ and $\rho_{(b)}{}^{(n)}(z'+)$ satisfies the equations at $z = z'$, but only the particular combination in Eq. (39.7) has the physical significance of a two-phase distribution function in an open system.

From the above discussion, we have the following formal "theory" of a first-order phase transition in an open system. In a certain interval of z values we imagine the set of Eqs. (39.10) [or (39.8) or (39.9)] solved for $\rho^{(1)}$, $\rho^{(2)}, \ldots, \rho^{(n)}, \ldots$. We notice, though, that in the neighborhood of a particular activity, $z = z'$, the solutions are discontinuous,[1] the set $\rho_{(a)}{}^{(1)}(z'-)$, $\rho_{(a)}{}^{(2)}(z'-), \ldots$ being approached as $z \to z'-$ and the set $\rho_{(b)}{}^{(1)}(z'+)$, $\rho_{(b)}{}^{(2)}(z'+), \ldots$ as $z \to z'+$. We therefore conclude that a first-order phase transition occurs at $z = z'$. Exactly at $z = z'$, the (physically significant) solution of the set of integral equations is Eq. (39.7) with $n = 1, 2, \ldots$.

Finally, let us examine the behavior of the compressibility κ at $z = z'$. Equation (19.39) is valid for a two-phase system:

$$\frac{\overline{N^2} - (\bar{N})^2}{\bar{N}} = \frac{\bar{N}}{V}\mathbf{k}T\kappa = \rho\mathbf{k}T\kappa \tag{39.12}$$

We have here

$$\bar{N} = p_a N_a + p_b N_b$$

$$\overline{N^2} = p_a N_a{}^2 + p_b N_b{}^2$$

and

$$\begin{aligned}\overline{N^2} - (\bar{N})^2 &= (p_a - p_a{}^2)N_a{}^2 - 2p_a p_b N_a N_b + (p_b - p_b{}^2)N_b{}^2 \\ &= O((\bar{N})^2)\end{aligned} \tag{39.13}$$

In a one-phase system $\overline{N^2} - (\bar{N})^2$ is of order $\bar{N}$ and κ of order $1/\rho\mathbf{k}T$. From Eqs. (39.12) and (39.13) we see that in a two-phase system $\overline{N^2} - (\bar{N})^2$ is of order $(\bar{N})^2$ and κ of order $\bar{N}/\rho\mathbf{k}T$, which is larger by a factor of order $\bar{N}$ than the usual (one-phase) compressibility. Of course as $V \to \infty$, $\bar{N} \to \infty$ and $\kappa \to \infty$; this corresponds to the completely flat ($V \to \infty$) two-phase portion in a p-V isotherm. If the two phases happen to have the same density, $N_a = N_b$ and the above equations are inapplicable, for all values of N except N_a and N_b have been neglected.

Second- and Higher-order Phase Transitions. The above arguments have to be modified somewhat here since now the two phases in equilibrium have the same density. First consider a closed system. Let N' be the value of N at which the phase equilibrium exists for given T and V. Then, as before, let

$$\begin{aligned}\rho_N{}^{(n)} &= \rho_{(a)N}{}^{(n)} \qquad N < N' \\ &= \rho_{(b)N}{}^{(n)} \qquad N > N'\end{aligned} \tag{39.14}$$

[1] See the last subsection of Sec. 37.

When $N = N'$ and two phases are present,

$$\rho_{N'}{}^{(n)}(N_a) = \frac{N_a}{N'}\rho_{(a)N'-}{}^{(n)} + \frac{(N' - N_a)}{N'}\rho_{(b)N'+}{}^{(n)} \tag{39.15}$$

where N_a/N' is the volume or mole fraction of phase a. The partition function at $N = N'$ is the sum

$$\sum_{N_a=0}^{N'} Q(N_a, N_b = N' - N_a)$$

where $Q(N_a,N_b)$ refers to N_a molecules of phase a and N_b molecules of phase b in V. Proceeding as in Eq. (A9.21) we find, if we neglect interfacial effects,

$$\begin{aligned} Q(N_a, N' - N_a) &= Ce^{-(N_a\mu' - p'V_a)/\mathbf{k}T}e^{-[(N'-N_a)\mu' - p'(V - V_a)]/\mathbf{k}T} \\ &= Ce^{-(N'\mu' - p'V)/\mathbf{k}T} \\ &\cong \text{const} \end{aligned} \tag{39.16}$$

However, if we include interfacial contributions (see Appendix 9), the value of $Q(N_a,N_b)$ in Eq. (39.16) is reduced to a negligible quantity except for $N_a = N'$ and $N_a = 0$. Therefore the distribution function $\rho_{N'}{}^{(n)}(N_a)$, properly averaged over the canonical ensemble, is

$$\rho_{N'}{}^{(n)} = p_a'\rho_{(a)N'-}{}^{(n)} + p_b'\rho_{(b)N'+}{}^{(n)} \tag{39.17}$$

where
$$p_a' = \frac{C(N_a = N')}{C(N_a = N') + C(N_a = 0)} \qquad p_b' = 1 - p_a'$$

In an open system, for very large V, only the value $N = N'$ contributes appreciably to the sum in Eq. (39.3) when $z = z'$; thus we have

$$\begin{aligned} \rho^{(n)} &= \rho_{(a)}{}^{(n)} & z &< z' \\ &= \rho_{(b)}{}^{(n)} & z &> z' \\ &= p_a'\rho_{(a)}{}^{(n)}(z'-) + p_b'\rho_{(b)}{}^{(n)}(z'+) & z &= z' \end{aligned} \tag{39.18}$$

where $(V \to \infty)$

$$\begin{aligned} \rho_{(a)}{}^{(n)}(z'-) &= \rho_{(a)N'-}{}^{(n)} \\ \rho_{(b)}{}^{(n)}(z'+) &= \rho_{(b)N'+}{}^{(n)} \end{aligned} \tag{39.19}$$

Two different sets of solutions of Eqs. (39.8) to (39.10) are obtained as we approach $N = N'$, $z = z'$ from the two sides (phase a and phase b). At $N = N'$, $z = z'$ the linear combinations in Eqs. (39.17) and (39.18) are the physically significant solutions.

The compressibility, given by Eq. (39.12), is of conventional (one-phase) order of magnitude at $z = z'$ for second- and higher-order phase transitions. It is discontinuous at $z = z'$ for a second-order transition and continuous for higher orders.

40. Distribution Functions in Polyatomic, Multicomponent Systems

The extension to this more general case is rather obvious once the notation is suitably generalized. We therefore give a rather condensed discussion.

We confine ourselves primarily to the solution theory of McMillan and Mayer.[1] It should, however, be mentioned that Kirkwood and Buff[2] have recently published the first part of an alternative solution theory which is necessarily equivalent to the McMillan-Mayer theory since both are formally exact. It would appear that the McMillan-Mayer approach is more convenient to use in applications if one wishes to adopt an *unsymmetrical* point of view with respect to the components (e.g., if the osmotic pressure of a polymer solution[3] is desired: polymer = solute; small molecules = solvent). On the other hand, the Kirkwood-Buff theory is more natural when one desires a *symmetrical* consideration of the different components, and should therefore be especially useful if all components have an essentially equivalent status (e.g., a solution of alcohol and water).

Notation and Definitions. McMillan and Mayer show how their theory can be carried through either completely (all degrees of freedom) classically or completely quantum mechanically. However, for practical purposes in solution theory, we make the useful approximation here, at the outset, of assuming that vibrational degrees of freedom are separable. While this approximation is not at all essential to the argument, it would presumably almost always be made in applications in the field of solution physical chemistry.

We begin by considering a *closed* system containing a set of molecules $\mathbf{N} = N_1, N_2, \ldots, N_r$; that is, N_1 molecules of species 1, etc. Each molecule has the usual three translational and zero, two, or three external rotational degrees of freedom. The remaining degrees of freedom are classed either as vibrational or internal rotational: one of these degrees of freedom which is separable, to a sufficient approximation, is called "vibrational" and may be treated quantum mechanically; one which is not separable is called "internal rotational" and is treated classically. The translational and external rotational degrees of freedom are also treated classically. The internal rotational degrees of freedom include, in particular, angular coordinates associated with rotation about bonds. Internal rotation is, of course, important in large molecules such as polymers and proteins, and also in certain much simpler molecules. Because of internal rotation, molecules

[1] W. G. McMillan and J. E. Mayer, *J. Chem. Phys.*, **13**, 276 (1945).

[2] J. G. Kirkwood and F. P. Buff, *J. Chem. Phys.*, **19**, 774 (1951). See also F. P. Buff and R. Brout, *J. Chem. Phys.*, **23**, 458 (1955).

[3] The McMillan-Mayer theory has been applied to solutions of large molecules by, for example, B. H. Zimm, *J. Chem. Phys.*, **14**, 164 (1946); L. Onsager, *Ann. N.Y. Acad. Sci.*, **51**, 627 (1949); A. Isihara, *J. Chem. Phys.*, **18**, 1446 (1950); **19**, 397 (1951); T. L. Hill, *J. Chem. Phys.*, **23**, 623, 2270 (1955).

can take on different configurations and shapes. These may have different potential energies owing to interactions between groups, charges, etc., on the same molecule, and this is taken into account below.

The partition function in the canonical ensemble is written as [see Eq. (16.66)]

$$Q_{\mathbf{N}} = \frac{\prod_{s=1}^{r} q_{sv}^{N_s}}{\prod_{s=1}^{r} (N_s!h^{f_sN_s})} \int \exp\left(-\frac{H_{\mathbf{N}}}{\mathbf{k}T}\right) d\{\mathbf{N}\}_x d\{\mathbf{P}\}_x d''\{\mathbf{N}\}_\theta d\{\mathbf{P}\}_\theta \tag{40.1}$$

where q_{sv} is the (separable) vibrational partition function of a molecule of species s, f_s is the number of translational and rotational (internal and external) degrees of freedom of an s molecule, $H_{\mathbf{N}}$ is the classical Hamiltonian function for translation and rotation, $\{\mathbf{N}\}_x$ refers to all the translational coordinates of the set $\mathbf{N}$, and $\{\mathbf{P}\}_x$ represents the conjugate momenta, while $\{\mathbf{N}\}_\theta$ and $\{\mathbf{P}\}_\theta$ have similar meanings for the rotational degrees of freedom. The double prime on d in Eq. (40.1) is introduced to distinguish the magnitude of this element of volume from the magnitudes of $d'\{\mathbf{N}\}_\theta$ and $d\{\mathbf{N}\}_\theta$ encountered below.

In q_{sv}, the zero of energy is chosen at the bottom of the vibrational potential well. $H_{\mathbf{N}}$ includes the potential energy $U_{\mathbf{N}}$; the zero of potential energy for *inter*molecular interactions is chosen at infinite separation of all molecules, but the zero for *intra*molecular interactions is left arbitrary for each species.

We now integrate over the momenta in Eq. (40.1), and obtain

$$Q_{\mathbf{N}} = \frac{\prod_{s} q_{sv}^{N_s}}{\prod_{s} (N_s!\Lambda_{sx}^{3N_s}\Lambda_{s\theta}^{N_s})} \int \exp\left(-\frac{U_{\mathbf{N}}}{\mathbf{k}T}\right) d\{\mathbf{N}\}_x d'\{\mathbf{N}\}_\theta \tag{40.2}$$

where

$$\Lambda_{sx} = \left(\frac{h^2}{2\pi m_s \mathbf{k}T}\right)^{1/2} \tag{40.3}$$

The integration over $\{\mathbf{P}\}_\theta$ gives a constant $1/\prod_s \Lambda_{s\theta}^{N_s}$ and also, in general, a function of rotational coordinates which, together with $d''\{\mathbf{N}\}_\theta$, is included in $d'\{\mathbf{N}\}_\theta$. For example, in the familiar case of a single rigid diatomic molecule of species s, we have[1]

$$\Lambda_{s\theta} = \frac{h^2}{2\pi I_s \mathbf{k}T} \qquad d''(i_s)_\theta = d\theta_{i_s}\, d\varphi_{i_s} \qquad d'(i_s)_\theta = \sin\theta_{i_s}\, d\theta_{i_s}\, d\varphi_{i_s}$$

where (i_s) is used to denote the coordinates of molecule i of species s. The symbolic subscript θ is, of course, not to be confused with the angle θ_{i_s}.

When N_s/V is very small for all s, we have a perfect-gas mixture. $U_{\mathbf{N}}$ is

[1] See G. S. Rushbrooke, "Introduction to Statistical Mechanics" (Oxford, London, 1949), p. 87.

then the sum of separate contributions from each molecule; these contributions arise from *intra*molecular interactions. We have in this case, from Eq. (40.2),

$$Q_{\mathbf{N}} = \frac{(\prod_s q_{sv}{}^{N_s})(\prod_s V^{N_s}\delta_{s\theta}{}^{N_s})}{(\prod_s N_s!\Lambda_{sx}{}^{3N_s}\Lambda_{s\theta}{}^{N_s})} \tag{40.4}$$

where

$$\delta_{s\theta} = \int e^{-U_{i_s}((i_s)_\theta)/\mathbf{k}T} d'(i_s)_\theta \tag{40.5}$$

$U_{i_s} \equiv U_{N_s=1}$ is a function of the (internal) rotational coordinates of molecule i_s. For the rigid rotator above, U_{i_s} = constant and

$$\delta_{s\theta} = e^{-U_{i_s}/\mathbf{k}T}\int \sin\theta_{i_s}\, d\theta_{i_s}\, d\varphi_{i_s} = 4\pi e^{-U_{i_s}/\mathbf{k}T}$$

It is convenient to rewrite Eq. (40.2) in the form

$$Q_{\mathbf{N}} = \frac{Z_{\mathbf{N}}}{\prod_s (N_s!\Lambda_s{}^{3N_s})} \tag{40.6}$$

where

$$\Lambda_s{}^3 = \frac{\Lambda_{sx}{}^3\Lambda_{s\theta}}{q_{sv}\delta_{s\theta}} \tag{40.7}$$

and

$$Z_{\mathbf{N}} = \frac{1}{\prod_s \delta_{s\theta}{}^{N_s}} \int \exp\left(-\frac{U_{\mathbf{N}}}{\mathbf{k}T}\right) d\{\mathbf{N}\}_x d'\{\mathbf{N}\}_\theta \tag{40.8}$$

Or if we define $d(i_s)_\theta$ and $d\{\mathbf{N}\}_\theta$ by the equations

$$\begin{aligned} \frac{d'\{\mathbf{N}\}_\theta}{\prod_s \delta_{s\theta}{}^{N_s}} &= \prod_s \frac{d'(1_s)_\theta}{\delta_{s\theta}} \cdot \frac{d'(2_s)_\theta}{\delta_{s\theta}} \cdot \cdot \cdot \frac{d'(N_s)_\theta}{\delta_{s\theta}} \\ &= \prod_s d(1_s)_\theta d(2_s)_\theta \cdot \cdot \cdot d(N_s)_\theta \\ &= d\{\mathbf{N}\}_\theta \end{aligned} \tag{40.9}$$

then

$$Z_{\mathbf{N}} = \int \exp\left(-\frac{U_{\mathbf{N}}}{\mathbf{k}T}\right) d\{\mathbf{N}\} \tag{40.10}$$

where

$$d\{\mathbf{N}\} = d\{\mathbf{N}\}_x d\{\mathbf{N}\}_\theta \tag{40.11}$$

Thus $Z_{\mathbf{N}}$, the configuration integral, is normalized in such a way that

$$Z_{\mathbf{N}} = \prod_s V^{N_s} \tag{40.12}$$

for a perfect-gas mixture.

Distribution Functions. The probability of observing a configuration within the element of volume $d\{\mathbf{N}\}$ at $\{\mathbf{N}\}$ is

$$\frac{\exp\left[-U_{\mathbf{N}}(\{\mathbf{N}\})/\mathbf{k}T\right] d\{\mathbf{N}\}}{Z_{\mathbf{N}}} \tag{40.13}$$

Let the coordinates of a particular set of n_1 molecules of species 1, . . . , n_r of species r be denoted by $\{\mathbf{n}\}$, where $\mathbf{n} = n_1, n_2, \ldots, n_r$. The probability of observing the set $\mathbf{n}$ in $d\{\mathbf{n}\}$ at $\{\mathbf{n}\}$, irrespective of the coordinates of the remaining molecules $\mathbf{N} - \mathbf{n}$ is, from Eq. (40.13),

$$P_{\mathbf{N}}^{(\mathbf{n})}(\{\mathbf{n}\})d\{\mathbf{n}\} = \frac{d\{\mathbf{n}\}\int \exp\left[-U_{\mathbf{N}}(\{\mathbf{N}\})/\mathbf{k}T\right] d\{\mathbf{N} - \mathbf{n}\}}{Z_{\mathbf{N}}} \tag{40.14}$$

where

$$\int P_{\mathbf{N}}^{(\mathbf{n})}(\{\mathbf{n}\})\, d\{\mathbf{n}\} = 1 \tag{40.15}$$

Let $\rho_{\mathbf{N}}^{(\mathbf{n})}(\{\mathbf{n}\})\, d\{\mathbf{n}\}$ be the probability of observing *any* n_1 molecules of species 1 (out of N_1), . . . , *any* n_r molecules of species r (out of N_r) in the configuration $\{\mathbf{n}\}$ within $d\{\mathbf{n}\}$. Then

$$\rho_{\mathbf{N}}^{(\mathbf{n})}(\{\mathbf{n}\}) = \left[\prod_s \frac{N_s!}{(N_s - n_s)!}\right] P_{\mathbf{N}}^{(\mathbf{n})}(\{\mathbf{n}\}) \tag{40.16}$$

and

$$\int \rho_{\mathbf{N}}^{(\mathbf{n})}(\{\mathbf{n}\})d\{\mathbf{n}\} = \prod_s \frac{N_s!}{(N_s - n_s)!} \tag{40.17}$$

In an *open* system, the grand partition function is

$$\Xi = e^{pV/\mathbf{k}T} = \sum_{\mathbf{N}\geqslant 0} e^{(N_1\mu_1 + \cdots + N_r\mu_r)/\mathbf{k}T} Q_{\mathbf{N}} \tag{40.18}$$

$$= \sum_{\mathbf{N}\geqslant 0} \left(\prod_s \frac{z_s^{N_s}}{N_s!}\right) Z_{\mathbf{N}} \tag{40.19}$$

$$= \sum_{\mathbf{N}\geqslant 0} \left(\prod_s \frac{z_s^{N_s}}{N_s!}\right) \int \exp\left(-\frac{U_{\mathbf{N}}}{\mathbf{k}T}\right) d\{\mathbf{N}\} \tag{40.20}$$

where we have used Eq. (40.6) and

$$z_s = \frac{e^{\mu_s/\mathbf{k}T}}{\Lambda_s^3} \tag{40.21}$$

The quantity z_s is the activity of the species s. The probability of the open system containing exactly the numbers of molecules $\mathbf{N}$ is

$$P_{\mathbf{N}} = \frac{\left(\prod_s \frac{z_s^{N_s}}{N_s!}\right) Z_{\mathbf{N}}}{\Xi} \tag{40.22}$$

Therefore the probability, in an open system, of observing any n_1 molecules of species 1, etc., in $d\{\mathbf{n}\}$ at $\{\mathbf{n}\}$ is $\rho^{(\mathbf{n})}(\{\mathbf{n}\})d\{\mathbf{n}\}$ where

$$\rho^{(\mathbf{n})}(\{\mathbf{n}\}) = \sum_{\mathbf{N}\geqslant \mathbf{n}} P_{\mathbf{N}}\rho_{\mathbf{N}}^{(\mathbf{n})}(\{\mathbf{n}\}) \tag{40.23}$$

From Eqs. (40.17) and (40.23),

$$\int \rho^{(\mathbf{n})}(\{\mathbf{n}\})d\{\mathbf{n}\} = \sum_{\mathbf{N}\geqslant\mathbf{n}} P_{\mathbf{N}} \left[\prod_s \frac{N_s!}{(N_s - n_s)!}\right]$$

$$= \left\langle \prod_s \frac{N_s!}{(N_s - n_s)!} \right\rangle_{\text{av}} \tag{40.24}$$

In particular, if $\mathbf{n}$ is a single molecule of species s,

$$\int \rho^{(1)}((i_s))d(i_s)_x d(i_s)_\theta = \bar{N}_s \tag{40.25}$$

In a fluid, $\rho^{(1)}$ is independent of $(i_s)_x$; therefore we have

$$\int \rho^{(1)}((i_s)_\theta)d(i_s)_\theta = \frac{\bar{N}_s}{V} = \rho_s \qquad \text{(fluid)} \tag{40.26}$$

Equations (40.14), (40.16), and (40.22), when substituted in Eq. (40.23), give

$$\rho^{(\mathbf{n})}(\{\mathbf{n}\},\mathbf{z},T) = \frac{1}{\Xi(\mathbf{z},V,T)} \sum_{\mathbf{N}\geqslant\mathbf{n}} \left[\prod_s \frac{z_s^{N_s}}{(N_s - n_s)!}\right] \int \exp\left[-\frac{U_{\mathbf{N}}(\{\mathbf{N}\})}{\mathbf{k}T}\right] d\{\mathbf{N} - \mathbf{n}\} \tag{40.27}$$

We have indicated explicitly here that $\rho^{(\mathbf{n})}$ is also a function of T and the activity set $\mathbf{z} = z_1, \ldots, z_r$. In a macroscopic system (strictly, $V \to \infty$), $\rho^{(\mathbf{n})}$ is independent of V. If we put $\mathbf{m} = \mathbf{N} - \mathbf{n}$ in Eq. (40.27), we have the alternative form

$$\frac{e^{pV/\mathbf{k}T}}{\prod_s z_s^{n_s}} \rho^{(\mathbf{n})}(\mathbf{z}) = \sum_{\mathbf{m}\geqslant 0} \left(\prod_s \frac{z_s^{m_s}}{m_s!}\right) \int \exp\left(-\frac{U_{\mathbf{m}+\mathbf{n}}}{\mathbf{k}T}\right) d\{\mathbf{m}\} \tag{40.28}$$

In the limit $\mathbf{z} \to \mathbf{0}$ (perfect gas), Eq. (40.20) becomes

$$\begin{aligned} e^{pV/\mathbf{k}T} &= \Xi = 1 + V\sum_s z_s + \cdots \\ \frac{pV}{\mathbf{k}T} &= \ln \Xi = V\sum_s z_s + \cdots \end{aligned} \tag{40.29}$$

From the general relation for an open system

$$\bar{N}_s = z_s \left(\frac{\partial \ln \Xi}{\partial z_s}\right)_{V,T,z_{\alpha\neq s}} \tag{40.30}$$

we then find that $\bar{N}_s = Vz_s$, or $\rho_s = z_s$ in a perfect gas. That is, z_s is normalized so that $z_s \to \rho_s$ as $\mathbf{z}$ (and $\boldsymbol{\rho}$) $\to \mathbf{0}$. In the limit $\mathbf{z} \to \mathbf{0}$, Eq. (40.28) reduces to

$$\rho^{(\mathbf{n})}(\{\mathbf{n}\}) = \left(\prod_s \rho_s^{n_s}\right) \exp\left[-\frac{U_{\mathbf{n}}(\{\mathbf{n}\})}{\mathbf{k}T}\right] \tag{40.31}$$

as only the term $\mathbf{m} = \mathbf{0}$ contributes and $\Xi \to 1$.

We define next the distribution function $\mathfrak{g}^{(\mathbf{n})}$ by[1]

$$\mathfrak{g}^{(\mathbf{n})}(\{\mathbf{n}\}) = \frac{\rho^{(\mathbf{n})}(\{\mathbf{n}\})}{\prod_s \rho_s^{n_s}} \tag{40.32}$$

Because of the probability definition of $\rho^{(\mathbf{n})}$ (see also Appendix 7), when all the molecules of the set $\mathbf{n}$ are widely separated in a fluid,

$$\rho^{(\mathbf{n})}(\{\mathbf{n}\}) \to \prod_s \rho^{(1)}((1_s))\rho^{(1)}((2_s)) \cdot \cdot \cdot \rho^{(1)}((n_s)) \qquad \text{(fluid)} \tag{40.33}$$

If we now integrate this relation over $\{\mathbf{n}\}_\theta$ and use Eq. (40.26), we obtain

$$\int \rho^{(\mathbf{n})}(\{\mathbf{n}\})d\{\mathbf{n}\}_\theta \to \prod_s \rho_s^{n_s} \qquad \text{(fluid)}$$

or

$$\int \mathfrak{g}^{(\mathbf{n})}(\{\mathbf{n}\})d\{\mathbf{n}\}_\theta \to 1 \qquad \text{(fluid)} \tag{40.34}$$

This normalization property of $\mathfrak{g}^{(\mathbf{n})}$ is useful.

When a distribution function is integrated over rotational coordinates, the remaining function is called a spatial distribution function (subscript x); for example,

$$\mathfrak{g}_x^{(\mathbf{n})}(\{\mathbf{n}\}_x) = \int \mathfrak{g}^{(\mathbf{n})}(\{\mathbf{n}\})d\{\mathbf{n}\}_\theta \tag{40.35}$$

Thus Eq. (40.34) shows that $\mathfrak{g}_x^{(\mathbf{n})} \to 1$ for widely separated molecules in a fluid.

We also have

$$\rho_x^{(\mathbf{n})}(\{\mathbf{n}\}_x) = \int \rho^{(\mathbf{n})}(\{\mathbf{n}\})d\{\mathbf{n}\}_\theta \tag{40.36}$$

For example, in Eq. (40.26), $\rho_x^{(1)}((i_s)_x) = \rho_s$ for a fluid. In general (fluid or crystal), the "radial" distribution function (see the end of Sec. 29)—that is, the probability that a molecule of species β will be found in $d(1_\beta)_x$ at $(1_\beta)_x$ when a molecule of species α is fixed at $(1_\alpha)_x$, relative to the random expectation $\rho_\beta d(1_\beta)_x$—is, by the rules of conditional probability,

$$\frac{\rho_x^{(2)}((1_\alpha)_x,(1_\beta)_x)}{\rho_x^{(1)}((1_\alpha)_x)\rho_\beta} \tag{40.37}$$

In a fluid, this reduces to $\mathfrak{g}_x^{(2)}$.

If we define $g^{(\mathbf{n})}$ by [see Eq. (37.9)]

$$g^{(\mathbf{n})} = \frac{\rho^{(\mathbf{n})}}{\prod_s \rho^{(1)}((1_s)) \cdot \cdot \cdot \rho^{(1)}((n_s))} \tag{40.38}$$

it should be noted that

$$\begin{aligned} g^{(\mathbf{n})} &\neq \mathfrak{g}^{(\mathbf{n})} \qquad \text{(polyatomic fluid)} \\ g_x^{(\mathbf{n})} &\neq \mathfrak{g}_x^{(\mathbf{n})} \qquad \text{(polyatomic fluid)} \end{aligned} \tag{40.39}$$

[1] This is the function $F_{\mathbf{n}}$ of McMillan and Mayer. We use $\mathfrak{g}^{(\mathbf{n})}$ in order to maintain a consistent type of notation ($\mathbf{n}$ appearing as a superscript in parentheses).

so that $g_x^{(2)}$ is not the radial distribution function in a polyatomic fluid. The inequalities in Eq. (40.39) become equalities in a monatomic fluid.

Kirkwood-Buff Solution Theory.[1] From Eq. (40.24) we have

$$\int \rho^{(1)}((1_\alpha))d(1_\alpha) = \bar{N}_\alpha \tag{40.40}$$

$$\int \rho^{(2)}((1_\alpha),(2_\alpha))d(1_\alpha)d(2_\alpha) = \overline{N_\alpha^2} - \bar{N}_\alpha \tag{40.41}$$

$$\int \rho^{(2)}((1_\alpha),(1_\beta))d(1_\alpha)d(1_\beta) = \overline{N_\alpha N_\beta} \qquad \alpha \neq \beta \tag{40.42}$$

These equations can be combined to give

$$\rho_\alpha^2 \int [\mathfrak{g}_x^{(2)}((1_\alpha)_x, (2_\alpha)_x) - \mathfrak{g}_x^{(1)}((1_\alpha)_x)\mathfrak{g}_x^{(1)}((2_\alpha)_x)]d(1_\alpha)_x d(2_\alpha)_x = \overline{N_\alpha^2} - (\bar{N}_\alpha)^2 - \bar{N}_\alpha \tag{40.43}$$

$$\rho_\alpha \rho_\beta \int [\mathfrak{g}_x^{(2)}((1_\alpha)_x, (1_\beta)_x) - \mathfrak{g}_x^{(1)}((1_\alpha)_x)\mathfrak{g}_x^{(1)}((1_\beta)_x)]d(1_\alpha)_x d(1_\beta)_x = \overline{N_\alpha N_\beta} - \bar{N}_\alpha \bar{N}_\beta \tag{40.44}$$

Equations (40.43) and (40.44) relate the spatial distribution functions $\mathfrak{g}_x^{(1)}$ and $\mathfrak{g}_x^{(2)}$ to composition fluctuations. Kirkwood and Buff have pointed out that if the composition fluctuations are eliminated from these equations by means of Eq. (21.15), one obtains a connection between $\mathfrak{g}_x^{(1)}$, $\mathfrak{g}_x^{(2)}$ and thermodynamic quantities (such as $\partial\mu_i/\partial N_j$) which can be used as the basis of a general solution theory. For fluid solutions, Eqs. (40.43) and (40.44) simplify to

$$\frac{1}{V}\int_0^\infty [\mathfrak{g}_x^{(2)}(r_{\alpha\alpha}) - 1]4\pi r_{\alpha\alpha}^2\,dr_{\alpha\alpha} = \frac{\overline{N_\alpha^2} - (\bar{N}_\alpha)^2}{(\bar{N}_\alpha)^2} - \frac{1}{\bar{N}_\alpha} \qquad \text{(fluid)} \tag{40.45}$$

$$\frac{1}{V}\int_0^\infty [\mathfrak{g}_x^{(2)}(r_{\alpha\beta}) - 1]4\pi r_{\alpha\beta}^2\,dr_{\alpha\beta} = \frac{\overline{N_\alpha N_\beta} - \bar{N}_\alpha \bar{N}_\beta}{\bar{N}_\alpha \bar{N}_\beta} \qquad \text{(fluid)} \tag{40.46}$$

We return now to the McMillan-Mayer theory.

Potential of Mean Force. Let us define a quantity $w^{(\mathbf{n})}$ by [see Eq. (37.38)]

$$\exp\left[-\frac{w^{(\mathbf{n})}(\{\mathbf{n}\},\mathbf{z},T)}{\mathbf{k}T}\right] = \mathfrak{g}^{(\mathbf{n})}(\{\mathbf{n}\},\mathbf{z},T) \tag{40.47}$$

From Eq. (40.31) we see that

$$w^{(\mathbf{n})} \to U_{\mathbf{n}} \text{ as } \mathbf{z} \to \mathbf{0} \tag{40.48}$$

A normalization property that we shall need follows from Eqs (40.25), (40.32), and (40.47):

$$\int e^{-w^{(1)}((i_s))/\mathbf{k}T} d(i_s) = V \tag{40.49}$$

In a fluid,

$$\int e^{-w^{(1)}((i_s)_\theta)/\mathbf{k}T} d'(i_s)_\theta = \delta_{s\theta} \tag{40.50}$$

which includes Eq. (40.5) as a special case [see Eq. (40.48)].

[1] J. G. Kirkwood and F. P. Buff, *loc. cit.* See also B. H. Zimm, *J. Chem. Phys.*, **21**, 934 (1953), and F. P. Buff and R. Brout, *loc. cit.*

$w^{(\mathbf{n})}$ has the physical significance of a potential of average force, as in Sec. 37. Let $q_{\mathbf{n}}$ represent an arbitrary coordinate associated with one of the molecules of the set $\mathbf{n}$. When there are $\mathbf{N}$ molecules in the system in the configuration $\{\mathbf{N}\}$, including the set $\mathbf{n}$ in the configuration $\{\mathbf{n}\}$, the component of force along $q_{\mathbf{n}}$, owing to intra- and intermolecular interactions, is $-\partial U_{\mathbf{N}}/\partial q_{\mathbf{n}}$. Now if we keep the set of molecules $\mathbf{n}$ fixed in the configuration $\{\mathbf{n}\}$ but average this force over all configurations of the remaining $\mathbf{N}-\mathbf{n}$ molecules, we obtain

$$-\left(\overline{\frac{\partial U_{\mathbf{N}}}{\partial q_{\mathbf{n}}}}\right) = \frac{\int \exp(-U_{\mathbf{N}}/\mathbf{k}T)(-\partial U_{\mathbf{N}}/\partial q_{\mathbf{n}})d\{\mathbf{N}-\mathbf{n}\}}{\int \exp(-U_{\mathbf{N}}/\mathbf{k}T)d\{\mathbf{N}-\mathbf{n}\}} \tag{40.51}$$

We note from Eqs. (40.14), (40.16), and (40.51) that

$$\mathbf{k}T\,\frac{\partial \rho_{\mathbf{N}}^{(\mathbf{n})}/\partial q_{\mathbf{n}}}{\rho_{\mathbf{N}}^{(\mathbf{n})}} = -\left(\overline{\frac{\partial U_{\mathbf{N}}}{\partial q_{\mathbf{n}}}}\right) \tag{40.52}$$

Turning now to an open system, the probability that the system contain $\mathbf{N}$ molecules and that also a set of molecules $\mathbf{n}$ be found in $d\{\mathbf{n}\}$ at $\{\mathbf{n}\}$ is $P_{\mathbf{N}}\rho_{\mathbf{N}}^{(\mathbf{n})}d\{\mathbf{n}\}$. Since $-\overline{\partial U_{\mathbf{N}}/\partial q_{\mathbf{n}}}$ is the average component of force along $q_{\mathbf{n}}$ under these conditions, the further average over all values of $\mathbf{N}$ for the open system is

$$-\left(\overline{\frac{\partial U}{\partial q_{\mathbf{n}}}}\right) = \frac{\sum\limits_{\mathbf{N}\geqslant\mathbf{n}} P_{\mathbf{N}}\rho_{\mathbf{N}}^{(\mathbf{n})}(-\overline{\partial U_{\mathbf{N}}/\partial q_{\mathbf{n}}})}{\sum\limits_{\mathbf{N}\geqslant\mathbf{n}} P_{\mathbf{N}}\rho_{\mathbf{N}}^{(\mathbf{n})}} \tag{40.53}$$

where $d\{\mathbf{n}\}$ has been canceled from numerator and denominator. This is the actual (physical) average force in an open system, when the set of molecules $\mathbf{n}$ is held fixed in the configuration $\{\mathbf{n}\}$. From Eqs. (40.52) and (40.53),

$$-\left(\overline{\frac{\partial U}{\partial q_{\mathbf{n}}}}\right) = \mathbf{k}T\,\frac{\sum\limits_{\mathbf{N}\geqslant\mathbf{n}} P_{\mathbf{N}}(\partial\rho_{\mathbf{N}}^{(\mathbf{n})}/\partial q_{\mathbf{n}})}{\sum\limits_{\mathbf{N}\geqslant\mathbf{n}} P_{\mathbf{N}}\rho_{\mathbf{N}}^{(\mathbf{n})}} \tag{40.54}$$

But, on differentiating, with respect to $q_{\mathbf{n}}$, the relation

$$\exp\left[-\frac{w^{(\mathbf{n})}}{\mathbf{k}T}\right] = \frac{\sum\limits_{\mathbf{N}\geqslant\mathbf{n}} P_{\mathbf{N}}\rho_{\mathbf{N}}^{(\mathbf{n})}}{\prod\limits_{s} \rho_s^{n_s}} \tag{40.55}$$

which follows from Eqs. (40.23), (40.32), and (40.47), we find [comparing Eq. (40.54)] that

$$-\frac{\partial w^{(\mathbf{n})}}{\partial q_{\mathbf{n}}} = -\left(\overline{\frac{\partial U}{\partial q_{\mathbf{n}}}}\right) \tag{40.56}$$

Equation (40.56) states that the average force is derivable from a potential function and that this function (to within an arbitrary constant) is $w^{(\mathbf{n})}$. In the limit $\mathbf{z}$ (or $\boldsymbol{\rho}$) $\rightarrow \mathbf{0}$, the influence of the averaging [see Eq. (40.51)] over the configurations of molecules not in the set $\mathbf{n}$ disappears, and we are left with only the forces within the set $\mathbf{n}$ itself. In this limit, then, the potential of average force degenerates into $U_{\mathbf{n}}$ as already stated in Eq. (40.48).

Distribution Functions at Different Activities. We return now to Eq. (40.28). Denote either side of this equation by $f(\mathbf{z})$. We wish to expand this function about $\mathbf{z} = \mathbf{z}^*$, where $\mathbf{z}^*$ is some particular activity set, using Taylor's theorem [see Eq. (37.81) et seq.]:

$$f(\mathbf{z}) = \sum_{\mathbf{l} \geqslant \mathbf{0}} \left[\prod_s \frac{(z_s - z_s^*)^{l_s}}{l_s!} \right] \left[\frac{\partial^{l_1 + \cdots + l_r} f(\mathbf{z})}{\partial z_1^{l_1} \partial z_2^{l_2} \cdots \partial z_r^{l_r}} \right]_{\mathbf{z}=\mathbf{z}^*} \tag{40.57}$$

First, we calculate the derivative explicitly using the right-hand side of Eq. (40.28) for $f(\mathbf{z})$. On replacing $\mathbf{m} - \mathbf{l}$ by $\mathbf{q}$, we find that

$$\frac{\partial^{l_1 + \cdots + l_r} f(\mathbf{z})}{\partial z_1^{l_1} \cdots \partial z_r^{l_r}} = \sum_{\mathbf{q} \geqslant \mathbf{0}} \left(\prod_s \frac{z_s^{q_s}}{q_s!} \right) \int \exp\left(- \frac{U_{\mathbf{n}+\mathbf{l}+\mathbf{q}}}{\mathbf{k}T} \right) d\{\mathbf{l} + \mathbf{q}\} \tag{40.58}$$

Next, in Eq. (40.28), we replace $\mathbf{n}$ by $\mathbf{n} + \mathbf{l}$ and $\mathbf{m}$ by $\mathbf{q}$, and integrate both sides of the equation over $\{\mathbf{l}\}$. On comparing the result with Eq. (40.58), we observe that

$$\frac{\partial^{l_1 + \cdots + l_r} f(\mathbf{z})}{\partial z_1^{l_1} \cdots \partial z_r^{l_r}} = \frac{\Xi(\mathbf{z})}{\prod_s z_s^{n_s + l_s}} \int \rho^{(\mathbf{n}+\mathbf{l})}(\{\mathbf{n} + \mathbf{l}\}, \mathbf{z})\, d\{\mathbf{l}\} \tag{40.59}$$

On substituting Eq. (40.59) in Eq. (40.57), we have the desired result (writing $\mathbf{m}$ instead of $\mathbf{l}$)

$$\frac{\Xi(\mathbf{z})}{\prod_s z_s^{n_s}} \rho^{(\mathbf{n})}(\{\mathbf{n}\}, \mathbf{z}) = \Xi(\mathbf{z}^*) \sum_{\mathbf{m} \geqslant \mathbf{0}} \left[\prod_s \frac{(z_s - z_s^*)^{m_s}}{m_s! z_s^{*n_s + m_s}} \right] \int \rho^{(\mathbf{n}+\mathbf{m})}(\{\mathbf{n} + \mathbf{m}\}, \mathbf{z}^*) d\{\mathbf{m}\} \tag{40.60}$$

This equation shows how the distribution function $\rho^{(\mathbf{n})}$ at $\mathbf{z}$ is related to distribution functions at another activity set $\mathbf{z}^*$.

There are several important special cases of Eq. (40.60). (1) If we put $\mathbf{z}^* = \mathbf{0}$, Eq. (40.60) reduces to Eq. (40.28), in view of Eq. (40.31), and using $\Xi(\mathbf{z}^*) \rightarrow 1$ and $\mathbf{z}^* \rightarrow \boldsymbol{\rho}^*$. (2) If we take $\mathbf{z} = \mathbf{0}$ and replace $\mathbf{z}^*$ by $\mathbf{z}$,

$$\exp\left[- \frac{U_{\mathbf{n}}(\{\mathbf{n}\})}{\mathbf{k}T} \right] = \Xi(\mathbf{z}) \sum_{\mathbf{m} \geqslant \mathbf{0}} \left[\prod_s \frac{(-1)^{m_s}}{m_s! z_s^{n_s}} \right] \int \rho^{(\mathbf{n}+\mathbf{m})}(\{\mathbf{n} + \mathbf{m}\}, \mathbf{z}) d\{\mathbf{m}\} \tag{40.61}$$

This is essentially the inverse of Eq. (40.28). The right-hand side of Eq. (40.61) must be independent of the choice of $\mathbf{z}$ since the left-hand side does not involve $\mathbf{z}$. (3) If we put $\mathbf{n} = \mathbf{0}$, since from its definition $\rho^{(\mathbf{0})} = 1$,

$$\frac{\Xi(\mathbf{z})}{\Xi(\mathbf{z}^*)} = \exp\left\{\frac{[p(\mathbf{z}) - p(\mathbf{z}^*)]V}{\mathbf{k}T}\right\}$$

$$= \sum_{\mathbf{m}\geqslant\mathbf{0}}\left[\prod_s \frac{1}{m_s!}\left(\frac{z_s - z_s^*}{\gamma_s^*}\right)^{m_s}\right]\int \exp\left[-\frac{w^{(\mathbf{m})}(\{\mathbf{m}\},\mathbf{z}^*)}{\mathbf{k}T}\right] d\{\mathbf{m}\} \tag{40.62}$$

where γ_s is an activity coefficient, defined by $\gamma_s = z_s/\rho_s$, so that $\gamma_s^* = z_s^*/\rho_s^*$. This result is formally very similar to Eq. (40.20). In fact, Eq. (40.62) reduces to Eq. (40.20) as a special case on letting $\mathbf{z}^* \to \mathbf{0}$ [see Eq. (40.48)].

Osmotic Pressure. What is perhaps the most important single result of the McMillan-Mayer theory follows from Eq. (40.62) if we introduce "osmotic conditions." That is, suppose we have a membrane permeable to "solvent" species (subscript τ) but not to "solute" species (subscript σ). Let the state $\mathbf{z}^*$ in Eq. (40.62) refer to a solution on one side of the membrane which contains solvent species only. Let the value of each z_τ be denoted by z_τ^*, while each $z_\sigma = \rho_\sigma = 0$. We summarize this by $\mathbf{z}^* = \mathbf{z}_\tau^*,\mathbf{0}_\sigma$. Let the state $\mathbf{z}$ in Eq. (40.62) refer to a solution on the other side of the membrane which contains both solute and solvent species and which is in equilibrium, with respect to each solvent species, with the first solution. We represent this by $\mathbf{z} = \mathbf{z}_\tau^*,\mathbf{z}_\sigma$, since at equilibrium each z_τ must have the same value, namely z_τ^*, in the two solutions. Under these conditions the pressure difference across the membrane is, by definition, just the osmotic pressure Π. With this particular choice of $\mathbf{z}$ and $\mathbf{z}^*$ in Eq. (40.62), the only terms in the sum that contribute are those for which $\mathbf{m}$ represents a set of *solute* molecules only, $\mathbf{m}_\sigma$. We have, then,

$$e^{\Pi V/\mathbf{k}T} = \sum_{\mathbf{m}_\sigma\geqslant\mathbf{0}}\left[\prod_\sigma \frac{(z_\sigma/\gamma_\sigma{}^0)^{m_\sigma}}{m_\sigma!}\right]\int \exp\left[-\frac{w^{(\mathbf{m}_\sigma)}(\{\mathbf{m}_\sigma\},\mathbf{z}_\tau^*,\mathbf{0}_\sigma)}{\mathbf{k}T}\right] d\{\mathbf{m}_\sigma\} \tag{40.63}$$

where $\gamma_\sigma{}^0$ denotes the activity coefficient of the solute species σ in the solution with activity set $\mathbf{z}_\tau^*,\mathbf{0}_\sigma$, that is, in a solution which is infinitely dilute with respect to all solute species but not with respect to solvent species. Similarly, $w^{(\mathbf{m}_\sigma)}$ in Eq. (40.63) is the potential of average force associated with a fixed set of solute molecules $\mathbf{m}_\sigma$ in a solution which is infinitely dilute with respect to solute molecules; in other words, $w^{(\mathbf{m}_\sigma)}$ refers to a fixed set of solute molecules immersed in a medium which contains solvent species only (with activity set $\mathbf{z}_\tau^*$). Incidentally, although each solvent species has the same activity in the two solutions, in general the solvent will have a different composition on the two sides of the membrane.

Since $z_\sigma = \gamma_\sigma\rho_\sigma$, the quotient $z_\sigma/\gamma_\sigma{}^0$ in Eq. (40.63) can be written as $(\gamma_\sigma/\gamma_\sigma{}^0)\rho_\sigma$. Thus $\gamma_\sigma/\gamma_\sigma{}^0$ is a "concentration activity coefficient" and $z_\sigma/\gamma_\sigma{}^0$ is a "concentration activity" such that, as the solution becomes dilute with respect to all solutes (with $\mathbf{z}_\tau^*$ held constant), $\gamma_\sigma/\gamma_\sigma^0 \to 1$ and $z_\sigma/\gamma_\sigma^0 \to \rho_\sigma$.

Equation (40.63) is a remarkable result. It states that the osmotic pressure depends on z_σ/γ_σ^0 for solute species and on the potential of average force on solute molecules in an infinitely dilute solution in a way which is formally identical with the dependence of the pressure of a gas [Eq. (40.20)] on z_s for all species and on the potential energy (which is the potential of average force when *all* species are infinitely dilute). The analogy is enhanced by noting that, since in the former (osmotic) case $z_\sigma/\gamma_\sigma^0 \to \rho_\sigma$ as $\mathbf{z}_\sigma \to \mathbf{0}$, while in the latter (gas) case $z_s \to \rho_s$ as $\mathbf{z} \to \mathbf{0}$, we have [see Eq. (40.49)]

$$\frac{\Pi}{\mathbf{k}T} \to \sum_\sigma \rho_\sigma \text{ as } \mathbf{z}_\sigma \to \mathbf{0} \tag{40.64}$$

and [see Eq. (40.29)]

$$\frac{p}{\mathbf{k}T} \to \sum_s \rho_s \text{ as } \mathbf{z} \to \mathbf{0} \tag{40.65}$$

Solvent species do not enter explicitly into the determination of Π according to Eq. (40.63). However, the solvent must be important and in fact plays its role implicitly through γ_σ^0 and $w^{(\mathbf{m}_\sigma)}(\mathbf{z}_\tau^*,\mathbf{0}_\sigma)$.

We now obtain the distribution function $\rho^{(\mathbf{n}_\sigma)}$ for a set of solute molecules at $\mathbf{z}_\tau^*,\mathbf{z}_\sigma$ in terms of the osmotic reference state $\mathbf{z}_\tau^*,\mathbf{0}_\sigma$. In Eq. (40.60) we take $\mathbf{n} = \mathbf{n}_\sigma, \mathbf{z} = \mathbf{z}_\tau^*,\mathbf{z}_\sigma$, and $\mathbf{z}^* = \mathbf{z}_\tau^*,\mathbf{0}_\sigma$ and find

$$\frac{e^{\Pi V/\mathbf{k}T}}{\prod_\sigma (z_\sigma/\gamma_\sigma^0)^{n_\sigma}} \rho^{(\mathbf{n}_\sigma)}(\{\mathbf{n}_\sigma\},\mathbf{z}_\tau^*,\mathbf{z}_\sigma)$$
$$= \sum_{\mathbf{m}_\sigma \geqslant \mathbf{0}} \left[\prod_\sigma \frac{(z_\sigma/\gamma_\sigma^0)^{m_\sigma}}{m_\sigma!}\right] \int \exp\left[-\frac{w^{(\mathbf{n}_\sigma+\mathbf{m}_\sigma)}(\{\mathbf{n}_\sigma+\mathbf{m}_\sigma\},\mathbf{z}_\tau^*,\mathbf{0}_\sigma)}{\mathbf{k}T}\right] d\{\mathbf{m}_\sigma\} \tag{40.66}$$

Here we see precisely the same analogy between $\rho^{(\mathbf{n}_\sigma)}$ for a set of solute molecules and $\rho^{(\mathbf{n})}$ for a gas [Eq. (40.28)], as was pointed out above between Π and p.

Next, we ask: What is the probability $\mathfrak{P}_{\mathbf{N}_\sigma}$ of finding the numbers of solute molecules $\mathbf{N}_\sigma$ in the (open) system at the activity set $\mathbf{z} = \mathbf{z}_\tau^*,\mathbf{z}_\sigma$, irrespective of the population of solvent molecules? If $P_{\mathbf{N}_\tau+\mathbf{N}_\sigma}$ is the probability of the system containing $\mathbf{N}_\tau$ solvent molecules and $\mathbf{N}_\sigma$ solute molecules, then of course

$$\mathfrak{P}_{\mathbf{N}_\sigma} = \sum_{\mathbf{N}_\tau \geqslant \mathbf{0}} P_{\mathbf{N}_\tau+\mathbf{N}_\sigma} \tag{40.67}$$

To calculate this probability, we use Eq. (40.22) for $P_{\mathbf{N}_\tau+\mathbf{N}_\sigma}$. We substitute in Eq. (40.22): $\mathbf{N}_\tau + \mathbf{N}_\sigma$ for $\mathbf{N}$; $\mathbf{z}_\tau^*,\mathbf{z}_\sigma$ for $\mathbf{z}$; and employ Eq. (40.61) for $\exp(-U_\mathbf{N}/\mathbf{k}T)$ in $Z_\mathbf{N}$. To be more specific, in Eq. (40.61), before substitution in Eq. (40.22), we put $\mathbf{n} = \mathbf{N}_\tau + \mathbf{N}_\sigma$ and choose $\mathbf{z} = \mathbf{z}_\tau^*,\mathbf{0}_\sigma$. Equation (40.67) becomes

$$\mathfrak{P}_{\mathbf{N}_\sigma} = \frac{1}{e^{\Pi V/\mathbf{k}T}}\left[\prod_\sigma \frac{(z_\sigma/\gamma_\sigma{}^0)^{N_\sigma}}{N_\sigma!}\right]\sum_{\mathbf{N}_\tau \geqslant \mathbf{0}}\sum_{\mathbf{m}_\tau \geqslant \mathbf{0}}\left\{\left(\prod_\tau \frac{\rho_\tau^{*N_\tau}}{N_\tau!}\right)\right.$$
$$\left.\times\left[\prod_\tau \frac{(-\rho_\tau^*)^{m_\tau}}{m_\tau!}\right]\int \exp\left[-\frac{w^{(\mathbf{N}_\tau+\mathbf{N}_\sigma+\mathbf{m}_\tau)}(\mathbf{z}_\tau^*,\mathbf{0}_\sigma)}{\mathbf{k}T}\right] d\{\mathbf{N}_\tau + \mathbf{N}_\sigma + \mathbf{m}_\tau\}\right\} \tag{40.68}$$

In the double sum in Eq. (40.68), let us collect all the terms for which the sum $\mathbf{N}_\tau + \mathbf{m}_\tau$ is the same, say $\mathbf{N}_\tau + \mathbf{m}_\tau = \mathbf{M}_\tau$. The integral has the same value for all these terms and can be factored out. The sum of these terms can then be written (omitting the integral)

$$\frac{1}{\prod_\tau M_\tau!}\prod_\tau\left\{\sum_{\substack{N_\tau,m_\tau=0\\(N_\tau+m_\tau=M_\tau)}}^{M_\tau}\left[M_\tau!\frac{\rho_\tau^{*N_\tau}}{N_\tau!}\cdot\frac{(-\rho_\tau^*)^{m_\tau}}{m_\tau!}\right]\right\}$$
$$= \frac{1}{\prod_\tau M_\tau!}\prod_\tau (\rho_\tau^* - \rho_\tau^*)^{M_\tau} = 0 \qquad \text{unless } \mathbf{M}_\tau = \mathbf{0}$$

Thus the only nonvanishing contribution to the double sum in Eq. (40.68) arises when $\mathbf{M}_\tau = \mathbf{0}$, or $\mathbf{N}_\tau = \mathbf{m}_\tau = \mathbf{0}$. The final result is, therefore,

$$\mathfrak{P}_{\mathbf{N}_\sigma} = \frac{1}{e^{\Pi V/\mathbf{k}T}}\left[\prod_\sigma \frac{(z_\sigma/\gamma_\sigma{}^0)^{N_\sigma}}{N_\sigma!}\right]\int \exp\left[-\frac{w^{(\mathbf{N}_\sigma)}(\{\mathbf{N}_\sigma\},\mathbf{z}_\tau^*,\mathbf{0}_\sigma)}{\mathbf{k}T}\right] d\{\mathbf{N}_\sigma\} \tag{40.69}$$

Here, as another example, we have the same analogy between $\mathfrak{P}_{\mathbf{N}_\sigma}$ for solute molecules in a solution and $P_\mathbf{N}$ for gas molecules [Eq. (40.22)] as was pointed out above between Π and p.

Aside from the elegance of these results, for practical purposes the above analogies make it possible to take over without modification the formal procedures in the theory of the gaseous state (virial expansions, integral equations for distribution functions, etc.) and apply them to solutes in solution. To make real progress, however, the potential of average force and the activity coefficients for solute molecules at infinite dilution must be specified. Although this can usually be done only in an approximate way, we discuss next the exact equations for these quantities. These equations give at least some general insight into the physical significance of $\gamma_\sigma{}^0$ and $w^{(\mathbf{n}_\sigma)}(\mathbf{z}_\tau^*,\mathbf{0}_\sigma)$, and also aid in the choice of reasonable approximations.

Solute Molecules at Infinite Dilution. Equation (40.27) can be written

$$\rho^{(\mathbf{n})} = \frac{\sum_{\mathbf{N}\geqslant\mathbf{n}}\left[\prod_s \frac{z_s{}^{N_s}}{(N_s - n_s)!}\right]\int \exp\left(-\frac{U_\mathbf{N}}{\mathbf{k}T}\right) d\{\mathbf{N}-\mathbf{n}\}}{\sum_{\mathbf{N}\geqslant\mathbf{0}}\left(\prod_s \frac{z_s{}^{N_s}}{N_s!}\right)\int \exp\left(-\frac{U_\mathbf{N}}{\mathbf{k}T}\right) d\{\mathbf{N}\}} \tag{40.70}$$

The largest term in the denominator occurs when $\mathbf{N} = \bar{\mathbf{N}}$, since the most probable composition [Eq. (40.22)] is also the mean composition in a macroscopic open system. Now we observe that if we retain only the term $\mathbf{N} = \bar{\mathbf{N}}$ in both numerator and denominator of Eq. (40.70) [see Eq. (37.62)]

$$\rho^{(\mathbf{n})} = \left[\prod_s \frac{\bar{N}_s!}{(\bar{N}_s - n_s)!}\right] \frac{\int \exp\,(-\,U_{\bar{\mathbf{N}}}/\mathbf{k}T)\, d\{\bar{\mathbf{N}} - \mathbf{n}\}}{\int \exp\,(-\,U_{\bar{\mathbf{N}}}/\mathbf{k}T)\, d\{\bar{\mathbf{N}}\}} \tag{40.71}$$

which is the correct expression for $\rho_{\bar{\mathbf{N}}}^{(\mathbf{n})}$ in a closed system of composition $\bar{\mathbf{N}}$, according to Eq. (40.16). This result is of course to be expected since, except for terms of a negligible order of magnitude, the equilibrium properties of a macroscopic system are independent of the statistical ensemble used. We now wish to apply the above procedure in a special case.

Let $\mathbf{n} = \mathbf{n}_\sigma$ and $\mathbf{N} = \mathbf{N}_\tau + \mathbf{N}_\sigma$ in Eq. (40.70):

$$\frac{\rho^{(\mathbf{n}_\sigma)}}{\prod_\sigma z_\sigma^{\,n_\sigma}} = \frac{\displaystyle\sum_{\substack{\mathbf{N}_\sigma \geqslant \mathbf{n}_\sigma \\ \mathbf{N}_\tau \geqslant \mathbf{0}}} \left[\prod_\tau \frac{z_\tau^{\,N_\tau}}{N_\tau!}\right]\left[\prod_\sigma \frac{z_\sigma^{\,N_\sigma - n_\sigma}}{(N_\sigma - n_\sigma)!}\right] \int \exp\,(-\,U_{\mathbf{N}_\tau + \mathbf{N}_\sigma}/\mathbf{k}T)\, d\{\mathbf{N}_\tau + \mathbf{N}_\sigma - \mathbf{n}_\sigma\}}{\displaystyle\sum_{\substack{\mathbf{N}_\sigma \geqslant \mathbf{0} \\ \mathbf{N}_\tau \geqslant \mathbf{0}}} \left[\prod_\tau \frac{z_\tau^{\,N_\tau}}{N_\tau!}\right]\left[\prod_\sigma \frac{z_\sigma^{\,N_\sigma}}{N_\sigma!}\right] \int \exp\,(-\,U_{\mathbf{N}_\tau + \mathbf{N}_\sigma}/\mathbf{k}T)\, d\{\mathbf{N}_\tau + \mathbf{N}_\sigma\}} \tag{40.72}$$

Now let $\mathbf{z}_\tau \to \mathbf{z}_\tau^*$ and $\mathbf{z}_\sigma \to \mathbf{0}$:

$$\frac{\rho^{(\mathbf{n}_\sigma)}(\mathbf{z}_\tau^*,\mathbf{0}_\sigma)}{\prod_\sigma \gamma_\sigma^{\,0 n_\sigma} \rho_\sigma^{\,0 n_\sigma}} = \frac{\displaystyle\sum_{\mathbf{N}_\tau \geqslant \mathbf{0}} \left(\prod_\tau \frac{z_\tau^{*N_\tau}}{N_\tau!}\right) \int \exp\,(-\,U_{\mathbf{N}_\tau + \mathbf{n}_\sigma}/\mathbf{k}T)\, d\{\mathbf{N}_\tau\}}{\displaystyle\sum_{\mathbf{N}_\tau \geqslant \mathbf{0}} \left(\prod_\tau \frac{z_\tau^{*N_\tau}}{N_\tau!}\right) \int \exp\,(-\,U_{\mathbf{N}_\tau}/\mathbf{k}T)\, d\{\mathbf{N}_\tau\}} \tag{40.73}$$

Finally, if we retain only the term $\mathbf{N}_\tau = \bar{\mathbf{N}}_\tau$ in numerator and denominator, where $\mathbf{N}_\tau = \bar{\mathbf{N}}_\tau$ corresponds to the maximum term in the denominator, then [see Eqs. (40.32) and (40.47)]

$$\frac{\exp\,[-\,w^{(\mathbf{n}_\sigma)}(\mathbf{z}_\tau^*,\mathbf{0}_\sigma)/\mathbf{k}T]}{\prod_\sigma \gamma_\sigma^{\,0 n_\sigma}} = \exp\left[-\,\frac{\mathfrak{W}^{(\mathbf{n}_\sigma)}(\mathbf{z}_\tau^*,\mathbf{0}_\sigma)}{\mathbf{k}T}\right] \tag{40.74}$$

$$= \frac{\int \exp\,(-\,U_{\bar{\mathbf{N}}_\tau + \mathbf{n}_\sigma}/\mathbf{k}T)\, d\{\bar{\mathbf{N}}_\tau\}}{\int \exp\,(-\,U_{\bar{\mathbf{N}}_\tau}/\mathbf{k}T)\, d\{\bar{\mathbf{N}}_\tau\}} \tag{40.75}$$

Equation (40.74) introduces $\mathfrak{W}^{(\mathbf{n}_\sigma)}$, which is a potential of average force defined only at infinite dilution and differing from $w^{(\mathbf{n}_\sigma)}(\mathbf{z}_\tau^*,\mathbf{0}_\sigma)$ only in the choice of the zero of energy:

$$\mathfrak{W}^{(\mathbf{n}_\sigma)}(\{\mathbf{n}_\sigma\},\mathbf{z}_\tau^*,\mathbf{0}_\sigma) - w^{(\mathbf{n}_\sigma)}(\{\mathbf{n}_\sigma\},\mathbf{z}_\tau^*,\mathbf{0}_\sigma) = \mathbf{k}T\sum_\sigma n_\sigma \ln \gamma_\sigma^{\,0} \tag{40.76}$$

where $\gamma_\sigma{}^0$ is a function of $\mathbf{z}_\tau^*$ and T only. $\bar{\mathbf{N}}_\tau$ is the (mean) composition of solvent on the side of the membrane without solutes, that is, for $\mathbf{z} = \mathbf{z}_\tau^*, \mathbf{0}_\sigma$, the infinitely dilute solution.

The meaning of $\mathfrak{W}^{(\mathbf{n}_\sigma)}$ is clear from Eq. (40.75). $\mathfrak{W}^{(\mathbf{n}_\sigma)}$ is an average (free) energy (averaged, that is, over all possible configurations of the solvent molecules) which includes: (1) intramolecular solute interactions; (2) solute-solute interactions between molecules in the set $\mathbf{n}_\sigma$; (3) solute-solvent interactions; and (4) the "perturbation" of solvent-solvent interactions owing to the presence of the solute set [except for this "perturbation," by which we mean a change in weighting of different solvent configurations, solvent-solvent interactions cancel in the two integrals of Eq. (40.75)].

To understand the significance of $\gamma_\sigma{}^0$, we let $\mathbf{n}_\sigma$ be a single molecule, i_σ, of solute species σ, and integrate Eq. (40.75) with respect to (i_σ). Using Eq. (40.49), we find

$$\frac{1}{\gamma_\sigma{}^0} = \frac{\int \exp\,(-\,U_{\bar{\mathbf{N}}_\tau + i_\sigma}/\mathbf{k}T)\, d\{\bar{\mathbf{N}}_\tau\}\, d(i_\sigma)}{V \int \exp\,(-\,U_{\bar{\mathbf{N}}_\tau}/\mathbf{k}T)\, d\{\bar{\mathbf{N}}_\tau\}} \tag{40.77}$$

$$= \frac{[\int \exp\,(-\,U_{\bar{\mathbf{N}}_\tau + i_\sigma}/\mathbf{k}T)\, d\{\bar{\mathbf{N}}_\tau\}\, d'(i_\sigma)_\theta]/[\int \exp\,(-\,U_{\bar{\mathbf{N}}_\tau}/\mathbf{k}T)\, d\{\bar{\mathbf{N}}_\tau\}]}{\int e^{-U_{i_\sigma}/\mathbf{k}T} d'(i_\sigma)_\theta} \quad \text{(fluid)} \tag{40.78}$$

$$= \frac{\rho_\sigma{}^0 \text{ (in solvent)}}{\rho_\sigma \text{ (in vacuum)}} \tag{40.79}$$

Equation (40.79) follows from $1/\gamma_\sigma{}^0 = \rho_\sigma{}^0/z_\sigma{}^0$, where all these quantities refer to the solution which is infinitely dilute with respect to all solute species. If solute species σ in this solution is in equilibrium with solute species σ in a perfect-gas phase, then $z_\sigma{}^0$ is also the activity of σ in the gas phase and $z_\sigma{}^0 = \rho_\sigma(g)$, where $\rho_\sigma(g)$ is the density of σ in the gas phase. Thus $1/\gamma_\sigma{}^0$ is the "equilibrium constant" $K = \rho_\sigma{}^0/\rho_\sigma(g)$ for the process

$$\sigma \text{ in perfect gas} \rightleftarrows \sigma \text{ in infinitely dilute solution} \tag{40.80}$$

The numerator in Eq. (40.78) is in effect the partition function for σ in the infinitely dilute solution, and the denominator is the same quantity for σ in the perfect gas. From Eq. (40.78) we see that $\mathbf{k}T \ln \gamma_\sigma{}^0$ is an average (free) energy (averaged over all solvent configurations and over the rotational configurations of the solute molecule) which includes: (1) solvent-isolated solute interactions; (2) perturbation of intramolecular solute interactions by the solvent; and (3) perturbation of solvent-solvent interactions owing to the presence of the isolated solute molecule. These contributions (averaged over rotational coordinates) for isolated solute molecules are subtracted from $\mathfrak{W}^{(\mathbf{n}_\sigma)}$ to give $w^{(\mathbf{n}_\sigma)}(\mathbf{z}_\tau^*, \mathbf{0}_\sigma)$ [Eq. (40.76)].

The above remarks can be supplemented by a consideration of the corresponding spatial quantities. We define a spatial potential of average force at any $\mathbf{z}$, a function of $\{\mathbf{n}\}_x$, by the equation

$$\exp\left[-\frac{w_x^{(\mathbf{n})}}{\mathbf{k}T}\right] = \int \exp\left[-\frac{w^{(\mathbf{n})}}{\mathbf{k}T}\right] d\{\mathbf{n}\}_\theta \tag{40.81}$$

Then, on taking the gradient with respect to molecule α of the set $\mathbf{n}$, we find

$$-\nabla_\alpha w_x^{(\mathbf{n})} = \frac{\int \exp\left[-w^{(\mathbf{n})}/\mathbf{k}T\right](-\nabla_\alpha w^{(\mathbf{n})})\, d\{\mathbf{n}\}_\theta}{\int \exp\left[-w^{(\mathbf{n})}/\mathbf{k}T\right] d\{\mathbf{n}\}_\theta} \tag{40.82}$$

Since the probability of observing a given rotational configuration of the set $\mathbf{n}$ is proportional to $\rho^{(\mathbf{n})}$ or to $\exp\left[-w^{(\mathbf{n})}/\mathbf{k}T\right]$, we see from this equation that $w_x^{(\mathbf{n})}$ is a potential of the average force acting on molecules of the set $\mathbf{n}$ including now additional averaging of the force over all rotational configurations of the set $\mathbf{n}$. From Eqs. (40.35) and (40.47),

$$\mathfrak{g}_x^{\rho(\mathbf{n})} = \exp\left[-\frac{w_x^{(\mathbf{n})}}{\mathbf{k}T}\right] \tag{40.83}$$

and, from Eq. (40.34),

$$w_x^{(\mathbf{n})} \to 0 \qquad \text{(fluid)} \tag{40.84}$$

if the molecules of the set $\mathbf{n}$ are widely separated.

Now on integrating Eq. (40.74) over $\{\mathbf{n}_\sigma\}_\theta$, we obtain

$$\mathfrak{W}_x^{(\mathbf{n}_\sigma)}(\{\mathbf{n}_\sigma\}_x,\mathbf{z}_\tau^*,\mathbf{0}_\sigma) - w_x^{(\mathbf{n}_\sigma)}(\{\mathbf{n}_\sigma\},\mathbf{z}_\tau^*,\mathbf{0}_\sigma) = \mathbf{k}T\sum_\sigma n_\sigma \ln \gamma_\sigma^0 \tag{40.85}$$

where $\mathfrak{W}_x^{(\mathbf{n}_\sigma)}$ is defined by an equation analogous to Eq. (40.81). As the molecules of the solute set $\mathbf{n}_\sigma$ in the infinitely dilute solution are moved far apart,

$$w_x^{(\mathbf{n}_\sigma)} \to 0 \qquad \text{(fluid)} \tag{40.86}$$

but

$$\mathfrak{W}_x^{(\mathbf{n}_\sigma)} \to \mathbf{k}T\sum_\sigma n_\sigma \ln \gamma_\sigma^0 \qquad \text{(fluid)} \tag{40.87}$$

Equations (40.85) and (40.86) show that the potential of average force $w_x^{(\mathbf{n}_\sigma)}(\mathbf{z}_\tau^*,\mathbf{0}_\sigma)$ includes in a fluid: (1) solute-solute interactions; and (2) perturbation of solvent-solvent, solute-solvent, and intramolecular solute interactions owing to molecules of the set $\mathbf{n}_\sigma$ being near each other. That is, both (1) and (2) have to vanish as the solute molecules are separated. The perturbation of solvent-solvent interactions in (2) is a second-order perturbation—a perturbation of a perturbation [see the discussions of Eqs. (40.75) and (40.78)]. As a simple example, if two monatomic solute molecules are charged and are situated in a nonelectrolyte solvent so that (1) above is ε^2/r, then an approximation to (1) + (2) would be ε^2/Dr, where D is the bulk dielectric constant of the solvent.

Incidentally, a potential of average force which vanishes without integration over rotational coordinates when the solute molecules $\mathbf{n}_\sigma$ are widely separated in a fluid, is, from Eq. (40.33),

$$W^{(\mathbf{n}_\sigma)} = w^{(\mathbf{n}_\sigma)} - \sum_\sigma [w^{(1)}((1_\sigma)) + w^{(1)}((2_\sigma)) + \cdots + w^{(1)}((n_\sigma))] \quad (40.88)$$

where $W^{(\mathbf{n}_\sigma)}$ is defined by [see Eqs. (37.28), (37.39) and (40.38)]

$$\exp\left[-\frac{W^{(\mathbf{n}_\sigma)}}{\mathbf{k}T}\right] = g^{(\mathbf{n}_\sigma)} \quad (40.89)$$

Virial Expansion of Osmotic Pressure. As already indicated, the formal methods used in the theory of the gaseous state[1] may be applied without alteration to the present problem. We consider here the virial expansion for a single[2] solute (corresponding to the virial expansion of a one-component gas). The solvent may consist of any number of different species. Denote the integral in Eq. (40.63) by (we can drop the subscript σ here)

$$Z_m^* = \int \exp\left[-\frac{w^{(m)}(\{m\},\mathbf{z}_\tau^*,0)}{\mathbf{k}T}\right] d\{m\} \quad (40.90)$$

Then

$$e^{\Pi V/\mathbf{k}T} = \sum_{m\geqslant 0} \frac{(z/\gamma^0)^m}{m!} Z_m^* \quad (40.91)$$

We desire, first, an expansion of $\Pi/\mathbf{k}T$ in powers of z/γ^0,

$$\frac{\Pi}{\mathbf{k}T} = \sum_{j\geqslant 1} b_j^* \left(\frac{z}{\gamma^0}\right)^j \quad (40.92)$$

where the coefficients b_j^* are functions of $\mathbf{z}_\tau^*$ and T. By taking the logarithm of both sides of Eq. (40.91) we find that the b_j^* are given by

$$\begin{aligned} 1!Vb_1^* &= Z_1^* = V \\ 2!Vb_2^* &= Z_2^* - Z_1^{*2} \\ 3!Vb_3^* &= Z_3^* - 3Z_1^*Z_2^* + 2Z_1^{*3} \\ 4!Vb_4^* &= Z_4^* - 4Z_1^*Z_3^* - 3Z_2^{*2} + 12Z_1^{*2}Z_2^* - 6Z_1^{*4} \\ &\text{etc.} \end{aligned} \quad (40.93)$$

The first of Eqs. (40.93) follows from Eq. (40.49). The average value of m is

$$\overline{m} = \sum_{m\geqslant 0} \mathfrak{P}_m m$$

[1] See Eqs. (23.41) to (23.45), (25.1) to (25.9), (25.24) to (25.31), and (37.93) to (37.102).

[2] McMillan and Mayer consider any number of solutes.

or, from Eqs. (40.69), (40.91), and (40.92),

$$\rho = \frac{\overline{m}}{V} = \frac{z}{\gamma^0}\left[\frac{\partial\Pi/\mathbf{k}T}{\partial(z/\gamma^0)}\right]_{\mathbf{z}_T^*,T}$$

$$= \sum_{j\geqslant 1} jb_j^* \left(\frac{z}{\gamma^0}\right)^j \tag{40.94}$$

The coefficients $a_2, a_3, \ldots$ in the inverted series

$$\frac{z}{\gamma^0} = \rho + a_2\rho^2 + a_3\rho^3 + \cdots$$

are found by substituting it in Eq. (40.94) for z/γ^0. Having found the coefficients, the inverted series is then introduced in Eq. (40.92) with the result

$$\frac{\Pi}{\mathbf{k}T} = \rho\left[1 - \sum_{k\geqslant 1}\frac{k}{k+1}\beta_k^*\rho^k\right] \tag{40.95}$$

where

$$\begin{aligned} \beta_1^* &= 2b_2^* \\ \beta_2^* &= 3b_3^* - 6b_2^{*2} \\ \beta_3^* &= 4b_4^* - 24b_2^*b_3^* + {}^{80}/_3 b_2^{*3} \end{aligned} \tag{40.96}$$

etc.

If B_n^* is the nth virial coefficient, that is, if

$$\frac{\Pi}{\mathbf{k}T} = \rho[1 + B_2^*\rho + B_3^*\rho^2 + \cdots] \tag{40.97}$$

then

$$B_n^* = -\frac{n-1}{n}\beta_{n-1}^* \tag{40.98}$$

For example, the second virial coefficient for a fluid solvent is [see Eq. (22.26)]

$$B_2^* = -b_2^* = -2\pi\int_0^\infty\left\{\exp\left[-\frac{w_x^{(2)}(r,\mathbf{z}^*,0)}{\mathbf{k}T}\right] - 1\right\} r^2\,dr \tag{40.99}$$

$$= -\frac{2\pi}{3\mathbf{k}T}\int_0^\infty r^3\frac{\partial w_x^{(2)}}{\partial r}e^{-w_x^{(2)}/\mathbf{k}T}dr \tag{40.100}$$

where r is the distance between the centers of mass of two solute molecules. Equation (40.100) follows from Eq. (40.99) on integrating by parts.

Superposition Approximation. It is convenient here to number the members of the set $\mathbf{m}$ from 1 to $m = m_1 + m_2 + \cdots + m_r$, and omit the boldface notation. The superposition approximation can be defined [see Eqs.

(37.39) and (37.48)] by stating that the probability $\rho^{(m)}d\{m\}$ is assumed expressible as a product of independent pair probabilities,

$$\rho^{(m)}\,d\{m\} = C[\rho^{(2)}((1),(2))d(1)d(2)][\rho^{(2)}((1),(3))d(1)d(3)]\cdots [\rho^{(2)}((m-1),(m))d(m-1)d(m)]$$

C is evaluated by using Eq. (40.33) in the limiting case of independent singlet probabilities. We find

$$\frac{\rho^{(m)}}{\rho^{(1)}((1))\cdots\rho^{(1)}((m))} = \frac{\rho^{(2)}((1),(2))}{\rho^{(1)}((1))\rho^{(1)}((2))}\cdot\frac{\rho^{(2)}((1),(3))}{\rho^{(1)}((1))\rho^{(1)}((3))}\cdots \frac{\rho^{(2)}((m-1),(m))}{\rho^{(1)}((m-1))\rho^{(1)}((m))} \tag{40.101}$$

or
$$w^{(m)} - \sum_{\alpha=1}^{m} w^{(1)}((\alpha)) = \sum_{1\leqslant\beta<\gamma\leqslant m} [w^{(2)}((\beta),(\gamma)) - w^{(1)}((\beta)) - w^{(1)}((\gamma))] \tag{40.102}$$

$$= \sum_{1\leqslant\beta<\gamma\leqslant m} w((\beta),(\gamma)) \tag{40.103}$$

where Eq. (40.103) defines w. In a fluid, w is a function of the intermolecular distance r, $(\beta)_\theta$ and $(\gamma)_\theta$; also, $w \to 0$ as $r \to \infty$ [Eq. (40.88)]. The superposition approximation is generally accepted as quite accurate in the potential energy (that is, when $\mathbf{z} = \mathbf{0}$), but it is certainly a much more serious approximation in the potential of average force ($\mathbf{z} \neq \mathbf{0}$).

A condensed form of the osmotic equation of state can be written if the superposition approximation is made in the potential of average force for solute molecules at infinite dilution. But we begin with a more general situation.[1] In Eq. (40.62) we note that $[p(\mathbf{z}) - p(\mathbf{z}^*)]/\mathbf{k}T$ is independent of $V(V \to \infty)$, and hence that $\ln \Sigma$ is proportional to V, and that

$$\left(\frac{\partial \ln \Sigma}{\partial V}\right)_{\mathbf{z},\mathbf{z}^*,T} = \frac{\ln \Sigma}{V} \tag{40.104}$$

where Σ is the right-hand side of Eq. (40.62),

$$\Sigma = \sum_{\mathbf{m}\geqslant\mathbf{0}} \left[\prod_s \frac{(z_s - z_s^*)^{m_s}}{m_s!}\left(\frac{\rho_s^*}{z_s^*}\right)^{m_s}\right]\int \exp\left[-\frac{w^{(\mathbf{m})}(\{\mathbf{m}\},\mathbf{z}^*)}{\mathbf{k}T}\right] d\{\mathbf{m}\} \tag{40.105}$$

Then
$$\frac{p(\mathbf{z}) - p(\mathbf{z}^*)}{\mathbf{k}T} = \frac{1}{\Sigma}\left(\frac{\partial\Sigma}{\partial V}\right)_{\mathbf{z},\mathbf{z}^*,T}$$

$$= \frac{1}{\Sigma}\sum_{\mathbf{m}\geqslant\mathbf{0}} \left[\prod_s \frac{(z_s - z_s^*)^{m_s}}{m_s!}\left(\frac{\rho_s^*}{z_s^*}\right)^{m_s}\right]\frac{\partial}{\partial V}\int \exp\left[-\frac{w^{(\mathbf{m})}(\{\mathbf{m}\},\mathbf{z}^*)}{\mathbf{k}T}\right] d\{\mathbf{m}\} \tag{40.106}$$

[1] See also S. Ono, *Progr. Theoret. Physics (Japan)*, **6**, 447 (1951).

The derivative in Eq. (40.106) becomes, on using Eq. (40.103) (for a fluid $\mathbf{z}^*$ state) and Green's method (Sec. 30) extended in a straightforward way to a multicomponent system,

$$\frac{\partial}{\partial V}\int \exp\left[-\frac{w^{(\mathbf{m})}}{\mathbf{k}T}\right] d\{\mathbf{m}\} = \frac{(m_1 + \cdots + m_r)}{V}\int \exp\left[-\frac{w^{(\mathbf{m})}}{\mathbf{k}T}\right] d\{\mathbf{m}\}$$
$$- \frac{1}{6V\mathbf{k}T}\sum_{\alpha=1}^{r} m_\alpha(m_\alpha - 1)\int r w'_{\alpha\alpha}(\mathbf{z}^*)\left\{\int \exp\left[-\frac{w^{(\mathbf{m})}}{\mathbf{k}T}\right] d\{\mathbf{m} - \alpha\alpha\}\right\} d(1_\alpha)\, d(2_\alpha)$$
$$- \frac{1}{3V\mathbf{k}T}\sum_{1\leqslant\alpha<\beta\leqslant r} m_\alpha m_\beta \int r w'_{\alpha\beta}(\mathbf{z}^*)\left\{\int \exp\left[-\frac{w^{(\mathbf{m})}}{\mathbf{k}T}\right] d\{\mathbf{m} - \alpha\beta\}\right\} d(1_\alpha)\, d(1_\beta) \tag{40.107}$$

where $w_{\alpha\alpha}(\mathbf{z}^*)$ is shorthand for $w(r,(1_\alpha)_\theta,(2_\alpha)_\theta,\mathbf{z}^*)$ and $\{\mathbf{m} - \alpha\alpha\}$ denotes the configuration of the set $\mathbf{m}$ reduced by two molecules of the species α. The symbols $w_{\alpha\beta}$ and $\{\mathbf{m} - \alpha\beta\}$ have a similar meaning for two molecules of different species, α and β.

Consider first the contribution to $[p(\mathbf{z}) - p(\mathbf{z}^*)]/\mathbf{k}T$ [Eq. (40.106)] arising from the term involving $m_1 + \cdots + m_r$ in Eq. (40.107). If we differentiate both sides of Eq. (40.62) with respect to z_α we find

$$\frac{\rho_\alpha}{z_\alpha}(z_\alpha - z_\alpha^*) = \frac{1}{\Sigma}\sum_{\mathbf{m}\geqslant 0}\frac{m_\alpha}{V}\left[\prod_s \frac{(z_s - z_s^*)^{m_s}}{m_s!}\left(\frac{\rho_s^*}{z_s^*}\right)^{m_s}\right]\int \exp\left[-\frac{w^{(\mathbf{m})}}{\mathbf{k}T}\right] d\{\mathbf{m}\} \tag{40.108}$$

Hence the desired contribution to $[p(\mathbf{z}) - p(\mathbf{z}^*)]/\mathbf{k}T$ is seen to be

$$\sum_{s=1}^{r}\frac{\rho_s}{z_s}(z_s - z_s^*)$$

Now we consider the contribution to $[p(\mathbf{z}) - p(\mathbf{z}^*)]/\mathbf{k}T$ associated with the general $\alpha\alpha$ term in Eq. (40.107). In Eq. (40.60) put $\mathbf{m} - \alpha\alpha$ for $\mathbf{m}$ (that is, $m_\alpha - 2$ for m_α) and take $\mathbf{n} = \alpha\alpha$ (that is, $n_\alpha = 2$, $n_s = 0$ if $s \neq \alpha$):

$$\frac{\Xi(\mathbf{z})}{z_\alpha^{\,2}}\rho_{\alpha\alpha}^{(2)}(\mathbf{z}) = \Xi(\mathbf{z}^*)\sum_{\mathbf{m}\geqslant\alpha\alpha}\left[\prod_s \frac{(z_s - z_s^*)^{m_s}}{(m_s - n_s)!}\left(\frac{\rho_s^*}{z_s^*}\right)^{m_s}\right]$$
$$\times \frac{1}{(z_\alpha - z_\alpha^*)^2}\int \exp\left[-\frac{w^{(\mathbf{m})}(\{\mathbf{m}\},\mathbf{z}^*)}{\mathbf{k}T}\right] d\{\mathbf{m} - \alpha\alpha\} \tag{40.109}$$

The general $\alpha\alpha$ term in Eq. (40.107) gives, in Eq. (40.106),

$$-\frac{1}{6V\mathbf{k}T\Sigma}\int r w'_{\alpha\alpha}(\mathbf{z}^*)\left\{\sum_{\mathbf{m}\geqslant\alpha\alpha}\left[\prod_s \frac{(z_s - z_s^*)^{m_s}}{(m_s - n_s)!}\left(\frac{\rho_s^*}{z_s^*}\right)^{m_s}\right]\right.$$
$$\left.\times \int \exp\left[-\frac{w^{(\mathbf{m})}}{\mathbf{k}T}\right] d\{\mathbf{m} - \alpha\alpha\}\right\} d(1_\alpha)\, d(2_\alpha)$$

or, with the aid of Eq. (40.109),

$$-\frac{1}{6V\mathbf{k}T}\left(\frac{z_\alpha - z_\alpha^*}{z_\alpha}\right)^2 \int r\hat{w}'_{\alpha\alpha}(\mathbf{z}^*)\rho_{\alpha\alpha}{}^{(2)}(\mathbf{z})d(1_\alpha)\,d(2_\alpha)$$

since $\Sigma = \Xi(\mathbf{z})/\Xi(\mathbf{z}^*)$.

A similar argument gives the contribution of the general $\alpha\beta$ term in Eq. (40.107) to $[p(\mathbf{z}) - p(\mathbf{z}^*)]/\mathbf{k}T$ as

$$-\frac{1}{3V\mathbf{k}T}\frac{(z_\alpha - z_\alpha^*)(z_\beta - z_\beta^*)}{z_\alpha z_\beta}\int r w'_{\alpha\beta}(\mathbf{z}^*)\rho_{\alpha\beta}{}^{(2)}(\mathbf{z})d(1_\alpha)\,d(1_\beta)$$

Finally, then

$$\frac{p(\mathbf{z}) - p(\mathbf{z}^*)}{\mathbf{k}T} = \sum_{s=1}^{r}\frac{\rho_s(z_s - z_s^*)}{z_s} - \frac{1}{6V\mathbf{k}T}\sum_{s,s'=1}^{r}\frac{(z_s - z_s^*)(z_{s'} - z_{s'}^*)}{z_s z_{s'}}$$
$$\times \int r w'_{ss'}(\mathbf{z}^*)\rho_{ss'}{}^{(2)}(\mathbf{z})d(1_s)\,d(1_{s'}) \quad (40.110)$$

In the special (osmotic) case $\mathbf{z}^* = \mathbf{z}_\tau^*,\mathbf{0}_\sigma$ and $\mathbf{z} = \mathbf{z}_\tau^*,\mathbf{z}_\sigma$, Eq. (40.110) simplifies to

$$\frac{\Pi}{\mathbf{k}T} = \sum_\sigma \rho_\sigma - \frac{1}{6V\mathbf{k}T}\sum_{\sigma,\sigma'}\int r w'_{\sigma\sigma'}(\mathbf{z}_\tau^*,\mathbf{0}_\sigma)\rho_{\sigma\sigma'}{}^{(2)}(\mathbf{z}_\tau^*,\mathbf{z}_\sigma)d(1_\sigma)\,d(1_{\sigma'}) \quad (40.111)$$

If all solutes are monatomic,

$$\frac{\Pi}{\mathbf{k}T} = \sum_\sigma \rho_\sigma - \frac{1}{6\mathbf{k}T}\sum_{\sigma,\sigma'}\rho_\sigma\rho_{\sigma'}\int_0^\infty r w'_{\sigma\sigma'}(r,\mathbf{z}_\tau^*,\mathbf{0}_\sigma)\mathfrak{g}_{\sigma\sigma'}{}^{(2)}(r,\mathbf{z}_\tau^*,\mathbf{z}_\sigma)\cdot 4\pi r^2\,dr$$
$$(\text{fluid at } \mathbf{z}_\tau^*,\mathbf{z}_\sigma) \quad (40.112)$$

In the special case $\mathbf{z}^* = \mathbf{0}$, we get, from Eq. (40.110),

$$\frac{p(\mathbf{z})}{\mathbf{k}T} = \sum_{s=1}^{r}\rho_s - \frac{1}{6V\mathbf{k}T}\sum_{s,s'=1}^{r}\int r u'_{ss'}\rho_{ss'}{}^{(2)}(\mathbf{z})d(1_s)\,d(1_{s'}) \quad (40.113)$$

which is the generalization of Eq. (37.26) to a polyatomic system of r components.

In an osmotic system with a single solute (we can drop the subscript σ here), Eq. (40.111) reads

$$\frac{\Pi}{\mathbf{k}T} = \rho - \frac{1}{6V\mathbf{k}T}\int r_{12}w'_{12}(\mathbf{z}_\tau^*,0)\rho^{(2)}(\mathbf{z}_\tau^*,z)d(1)d(2) \quad (40.114)$$

In the limit z and $\rho \to 0$, it is easy to show that Eq. (40.114) agrees with Eq. (40.97) up to the second virial coefficient, as expected, since superposition does not introduce any approximation if only pair interactions are involved.

Equation (40.114) can also be written, for a fluid (at z as well as at $z = 0$), as

$$\frac{\Pi}{\mathbf{k}T} = \rho - \frac{\rho^2}{6\mathbf{k}T}\int_0^\infty r_{12}\overline{w'_{12}}g_x{}^{(2)}(r_{12},\mathbf{z}_\tau^*,z)\cdot 4\pi r_{12}{}^2\,dr_{12} \tag{40.115}$$

where

$$\overline{w'_{12}} = \frac{\int g^{(2)}(\mathbf{z}_\tau^*,z)w'_{12}(\mathbf{z}_\tau^*,0)d(1)_\theta d(2)_\theta}{\int g^{(2)}(\mathbf{z}_\tau^*,z)d(1)_\theta d(2)_\theta} \tag{40.116}$$

The quantity $\overline{w'_{12}}$, the negative of an average force, is a function of r_{12}, $\mathbf{z}_\tau^*$, and z. The averaging in Eq. (40.116) is similar to that in Eq. (40.82) but is "mixed" with respect to z.

Debye-Hückel Theory. This theory furnishes a simple illustration of Eq. (40.112). Before turning to this, however, we make a few general remarks concerning the application of the McMillan-Mayer theory to ionic solutions. In applying the osmotic equations above to ionic solutions, the mean composition in the (open) systems on both sides of the semipermeable membrane will correspond to electroneutrality except for thermodynamically negligible deviations which, however, are responsible for the electrical potential difference across the membrane when it exists.[1] The chemical potentials μ_s of Eq. (40.18) for ionic species are so-called "electrochemical" potentials.[1]

No complications arise if the configuration integrals encountered above converge. For example, convergence will be ensured if the solvent (that is, the infinitely dilute solution—the solution on the side of the membrane containing no solutes) is an ionic solution, for in this case the potential of mean force between a pair of solute ions in the infinitely dilute solution will fall off with distance approximately as $e^{-\kappa r}/r$, $\kappa > 0$, because of the atmosphere of solvent ions surrounding each of the solute ions.

If the solvent contains no ions, the potential of average force for a pair of solute ions in the infinitely dilute solution will fall off at least approximately as $1/r$. Divergence difficulties arise in configuration integrals involving this potential; for example, the virial expansion has to be abandoned. As an alternative, more complicated procedures[2] must be introduced. However, as will be illustrated below, no divergence problems present themselves in the application of Eq. (40.112) to ionic solutes with a nonionic solvent because $w^{(2)}$ (from $g^{(2)}$) in this equation refers to finite solute concentrations (instead of zero solute concentration as in the potentials of the virial expansion). Solute ions contribute to each other's ion atmospheres at finite solute concentrations and hence $w^{(2)}$ decreases approximately as $e^{-\kappa r}/r$, $\kappa > 0$, again.

[1] See E. A. Guggenheim, "Thermodynamics" (North-Holland, Amsterdam, 1949), pp. 331–334.

[2] H. A. Kramers, *Proc. Roy. Acad. Sci. Amsterdam*, **30**, 145 (1927); J. E. Mayer, *J. Chem. Phys.*, **18**, 1426 (1950); T. H. Berlin and E. W. Montroll, *J. Chem. Phys.*, **20**, 75 (1952).

Consider now an electrolyte solution of monatomic ions, which we treat as point charges. The solvent is nonionic and has a dielectric constant D. We accept here the well-known relations of the Debye-Hückel theory for the potentials of mean force, and merely show that Eq. (40.112) gives the correct osmotic pressure. For $w_{\sigma\sigma'}$ and $\mathfrak{g}_{\sigma\sigma'}{}^{(2)}$ in Eq. (40.112) we have

$$w_{\sigma\sigma'}(\mathbf{z}_\tau^*,\mathbf{0}_\sigma) = \frac{\bar{z}_\sigma \bar{z}_{\sigma'} \varepsilon^2}{Dr} \tag{40.117}$$

and

$$\mathfrak{g}_{\sigma\sigma'}{}^{(2)} = e^{-w_{\sigma\sigma'}{}^{(2)}/\mathbf{k}T}$$

$$w_{\sigma\sigma'}{}^{(2)}(\mathbf{z}_\tau^*,\mathbf{z}_\sigma) = \frac{\bar{z}_\sigma \bar{z}_{\sigma'} \varepsilon^2 e^{-\kappa r}}{Dr} \tag{40.118}$$

where $\varepsilon = |\text{electronic charge}|$, $\bar{z}_\sigma$ = algebraic valence of ion σ, and

$$\kappa^2 = \frac{4\pi\varepsilon^2}{D\mathbf{k}T} \sum_\sigma \rho_\sigma \bar{z}_\sigma{}^2 \tag{40.119}$$

Neutrality requires that

$$\sum_\sigma \rho_\sigma \bar{z}_\sigma = 0 \tag{40.120}$$

If $\exp(-w_{\sigma\sigma'}{}^{(2)}/\mathbf{k}T)$ in Eq. (40.118) is expanded and Eqs. (40.117) and (40.118) substituted into Eq. (40.112),

$$\frac{\Pi}{\mathbf{k}T} = \sum_\sigma \rho_\sigma + \frac{2\pi}{3} \sum_{\sigma,\sigma'} \rho_\sigma \rho_{\sigma'} \int_0^\infty r^3 \left(\frac{\bar{z}_\sigma \bar{z}_{\sigma'} \varepsilon^2}{Dr^2 \mathbf{k}T}\right) \left[1 - \frac{\bar{z}_\sigma \bar{z}_{\sigma'} \varepsilon^2 e^{-\kappa r}}{Dr\,\mathbf{k}T} + \cdots\right] dr \tag{40.121}$$

The leading term in the series gives no contribution because of Eq. (40.120). Of the remaining terms only the linear term in $1/\mathbf{k}T$ is significant here as only linear terms in $1/\mathbf{k}T$ were retained in deriving Eq. (40.118) in the Debye-Hückel theory. From the $1/\mathbf{k}T$ term we find

$$\frac{\Pi}{\mathbf{k}T} = \sum_\sigma \rho_\sigma - \frac{\kappa^3}{24\pi} \tag{40.122}$$

which is the well-known correct result.

Born-Green-Yvon Integral Equation. Let us derive an integrodifferential equation in the distribution function $\rho^{(n)}$ for a single polyatomic solute in a solution, equivalent to Eq. (37.53). From Eq. (40.66) we have (dropping the subscript σ, putting $N = n + m$, and integrating over $\{\mathbf{n}\}_\theta$)

$$\rho_x{}^{(n)}(\{n\}_x,\mathbf{z}_\tau^*,z) = \frac{1}{e^{\Pi V/\mathbf{k}T}} \int d\{n\}_\theta \sum_{N \geqslant n} \frac{(z/\gamma^0)^N}{(N-n)!} \times \int \exp\left[-\frac{w^{(N)}(\{N\},\mathbf{z}_\tau^*,0)}{\mathbf{k}T}\right] d\{N-n\} \tag{40.123}$$

Now for the state $\mathbf{z}_\tau^*,0$ (infinitely dilute solution) we (1) use the superposition approximation in $w^{(N)}$ [Eq. (40.103)] and (2) assume this state is a fluid. Then, just as in Eqs. (37.51) to (37.53), we find

$$-\mathbf{k}T\nabla_1\rho_x{}^{(n)}(\mathbf{z}_\tau^*,z) = \sum_{i=2}^{n} \int \nabla_1 w((1),(i))\rho^{(n)}(\mathbf{z}_\tau^*,z)\, d\{n\}_\theta + \int \nabla_1 w((1),(n+1))\rho^{(n+1)}(\mathbf{z}_\tau^*,z)\, d\{n\}_\theta\, d(n+1) \tag{40.124}$$

or

$$-\mathbf{k}T\nabla_1\rho_x{}^{(n)}(\mathbf{z}_\tau^*,z) = \rho_x{}^{(n)}(\mathbf{z}_\tau^*,z) \sum_{i=2}^{n} \overline{\nabla_1 w((1),(i))} + \int \overline{\nabla_1 w((1),(n+1))}\rho_x{}^{(n+1)}(\mathbf{z}_\tau^*,z)\, d(n+1)_x \tag{40.125}$$

where

$$\overline{\nabla_1 w((1),(i))} = \frac{\int \nabla_1 w((1),(i))\rho^{(n)}(\mathbf{z}_\tau^*,z)\, d\{n\}_\theta}{\int \rho^{(n)}(\mathbf{z}_\tau^*,z)\, d\{n\}_\theta} \tag{40.126}$$

and

$$\overline{\nabla_1 w((1),(n+1))} = \frac{\int \nabla_1 w((1),(n+1))\rho^{(n+1)}(\mathbf{z}_\tau^*,z)\, d\{n+1\}_\theta}{\int \rho^{(n+1)}(\mathbf{z}_\tau^*,z)\, d\{n+1\}_\theta} \tag{40.127}$$

Equation (40.125) has the same formal appearance as the Born-Green-Yvon equation, Eq. (37.53), except that average forces replace forces. Equation (40.125) is of course also the generalization of Eq. (37.53) to a one-component polyatomic system ($\mathbf{z}_\tau^* = \mathbf{0}$, $w = u$). Since w refers to $\mathbf{z}_\tau^*,0$, we see that the averages here are of the same mixed type as in Eq. (40.116).

If the solute in Eq. (40.125) is monatomic, Eqs. (40.125) and Eq. (37.53) are identical except for obvious changes in notation. Therefore, if, in Eq. (40.125), we take $n = 2$ and introduce a further superposition approximation in the state $\mathbf{z}_\tau^*,z$ to reduce $\rho^{(3)}$ to pair distribution functions, we obtain the Born-Green-Yvon equation for the radial distribution function of solute molecules, Eq. (33.8) with $\xi = 1$. For example, the hard-sphere solution of this equation by Kirkwood, Maun, and Alder (Sec. 35) is applicable to solute molecules (e.g., protein molecules) in a solution if we can assume: (1) superposition in the potential of mean force in the infinitely dilute solution ($\mathbf{z}_\tau^*,0$); (2) that this potential has the hard-sphere dependence on r; and (3) also superposition in $\rho^{(3)}$ at $\mathbf{z}_\tau^*,z$.

Equation (40.125), with $n = 1$, can be used to study long-range order (gel formation, precipitation, etc.) of a macromolecular solute in a fluid solvent.[1]

GENERAL REFERENCES

Born, M., and H. S. Green, "A General Kinetic Theory of Liquids" (Cambridge, London, 1949).

DeBoer, J., *Repts. Progr. Phys.*, **12**, 305 (1949).

Green, H. S., "Molecular Theory of Fluids" (North-Holland, Amsterdam, 1952).

---

[1] See, for example, J. G. Kirkwood and J. Mazur, *Compt. rend. réunion ann. union intern. phys.* (*Paris*), 1952, p. 143.

Hildebrand, J. H., and R. L. Scott, "The Solubility of Non-electrolytes" (Reinhold, New York, 1950).
Hirschfelder, J. O., C. F. Curtiss, and R. B. Bird, "Molecular Theory of Gases and Liquids" (Wiley, New York, 1954).
Kimball, G., "A Treatise on Physical Chemistry" (Van Nostrand, New York, 1951), edited by H. S. Taylor and S. Glasstone.
Mayer, J. E., *Ann Rev. Phys. Chem.*, **1**, 175 (1950).
Yvon, J., "Actualités scientifiques et industrielles" (Hermann & Cie, Paris, 1935).

CHAPTER 7

NEAREST-NEIGHBOR LATTICE STATISTICS

The theories of ferromagnetism, antiferromagnetism, localized adsorption or absorption, the "lattice gas," the order-disorder transition in alloys, and binary liquid solutions can all be related, with varying degrees of approximation, to a simple statistical model[1] which we shall discuss in the present chapter. The model consists of a regular lattice of sites in one, two, or three dimensions in which each site can be occupied either in a manner A or a manner B (the meaning of the two kinds of occupation depends on the particular physical problem, as discussed below); there is a potential energy of interaction ε_{AA}, ε_{AB}, or ε_{BB} between nearest-neighbor pairs of sites, depending on the manner of occupation of the pair; but there is no interaction between higher neighbor pairs.[2] The statistical problem is then to compute the thermodynamic properties of the system by an enumeration of the various possible configurations of A's and B's on the lattice sites, giving each configuration its proper weight $\exp(-E_i/\mathbf{k}T)$, where E_i is the total interaction energy of all nearest-neighbor pairs in the ith configuration.

As we shall see, this simple model can lead to phase transitions, and therein lies its great interest. That is, although the model oversimplifies the physical systems it is supposed to represent, it is important to be able to carry out a rigorous mathematical analysis of such a model as an aid to our understanding of the nature of phase transitions.

The summation over discrete configurations here, with weight $\exp(-E_i/\mathbf{k}T)$, is of course the analogue of the (configuration) integral of $\exp(-U/\mathbf{k}T)$ in, say, a gas or liquid (Chaps. 5 and 6). We should expect the present problem to present somewhat less serious mathematical difficulties than those encountered in Chaps. 5 and 6, because of the finite number of configurations to be enumerated, and this turns out to be the case. Specifically, the one-dimensional problem here can be solved completely with ease, the two-dimensional problem has been solved partially (in closed form) but the

[1] Often referred to as the "Ising model."

[2] The higher neighbor case will be mentioned briefly at the end of Sec. 46.

solution requires sophisticated mathematics, and the three-dimensional problem has yielded so far to rigorous analysis only by way of series expansions.

In Sec. 41 we discuss the thermodynamics of the various physical problems mentioned above in relation to the present model, and show the interconnections between the thermodynamic functions arising in the different cases. In Sec. 42 we introduce rigorous approaches which can be used in attacking the problem at hand. Sections 43 to 45 summarize the exact results which have been obtained in one, two, and three dimensions, respectively. In Sec. 46 we give a very brief discussion of the two most familiar approximate methods applicable here. Approximate theories and comparisons with experiment have been treated elsewhere; therefore we devote very little space to these topics.[1]

A glossary of the definitions of a number of interrelated quantities mentioned frequently in the text will be found in Appendix 8.

41. Thermodynamics and Interconnections

For concreteness in the thermodynamic discussions of this section we shall anticipate in a qualitative way some of the macroscopic properties of the present model (for example, phase transitions, critical temperatures, etc.) which follow rigorously from a complete mathematical analysis. The mathematical theory itself is summarized in Secs. 42 to 45.

Ferromagnetism and Antiferromagnetism. The most important theoretical developments in the field of this chapter have been presented in papers using the language of ferromagnetism; therefore we consider this case first. The notation is something of a compromise between Lee and Yang, Newell and Montroll, and Rushbrooke.

We assume that ferromagnetism and antiferromagnetism are caused by interactions between spins of certain electrons in neighboring atoms making up a pure crystal. For simplicity, we assume that each atom can be in either one of two spin states; one state we designate by A, $\downarrow$ or $+1$, and the

[1] For further details on all aspects of this subject, the reader should consult Nix and Shockley (Gen. Ref.); Fowler and Guggenheim (Gen. Ref.); Wannier (Gen. Ref.); L. Onsager, *Phys. Rev.*, **65**, 117 (1944); B. Kaufman and L. Onsager, *Phys. Rev.*, **76**, 1244 (1949); A. J. Wakefield, *Proc. Cambridge Phil. Soc.*, **47**, 419, 799 (1951); G. S. Rushbrooke (Gen. Ref., 1949 and 1952); *Nuovo cimento*, **6** (suppl. 2), 251 (1949); C. Domb, *Proc. Roy. Soc. (London)*, **A196**, 36 (1949); **A199**, 199 (1949); J. E. Brooks and C. Domb, *Proc. Roy. Soc. (London)*, **A207**, 343 (1951); C. Domb, *Compt. rend. réunion ann. union intern. phys.* (Paris, 1952), p. 192; C. Domb and R. B. Potts, *Proc. Roy. Soc. (London)*, **A210**, 125 (1952); Guggenheim (Gen. Ref.); T. D. Lee and C. N. Yang, *Phys. Rev.*, **87**, 410 (1952); Newell and Montroll (Gen. Ref.). In particular, Newell and Montroll include an up-to-date bibliography which should be used to supplement the references given here.

other by B, $\uparrow$ or -1. The total number of spins (or atoms) is[1] $\mathscr{B}$, the number of $+1$ spins is N or N_A and the number of -1 spins is $\mathscr{B} - N$ or $N_B(N_A + N_B = \mathscr{B})$. The intensity of magnetization $\mathscr{I}$ may be conveniently defined as the net number of $\uparrow$ or -1 spins. That is,

$$\mathscr{I} = N_B - N_A = \mathscr{B} - 2N \tag{41.1}$$

The maximum magnetization (all spins $\uparrow$), in these units, is $\mathscr{B}$ and the minimum is $-\mathscr{B}$ (all spins $\downarrow$). The intensity of magnetization per spin, I, is then

$$I = \frac{\mathscr{I}}{\mathscr{B}} = 1 - 2\frac{N}{\mathscr{B}} = 1 - 2\rho \tag{41.2}$$

where $\rho = N/\mathscr{B}$.

We define the magnetic field H by the statement that for an infinitesimal reversible change $d\mathscr{I}$ in $\mathscr{I}$ ($d\mathscr{I}$ positive means conversion of some $\downarrow$ spins to $\uparrow$), the work done on the system is $H\, d\mathscr{I}$. Defined in this way (for convenience), H contains the magnetic moment of a spin and has dimensions of energy. We then have the thermodynamic equation

$$dE_m = T\, dS_m + H\, d\mathscr{I} + \mu_m\, d\mathscr{B} \tag{41.3}$$

where E_m, S_m, and μ_m are the energy, entropy, and chemical potential (Gibbs free energy per spin—see below), respectively. $\mathscr{I}$ is an external parameter (which replaces the volume in ordinary thermodynamics) and an extensive quantity. On integration,

$$\begin{aligned} F_m = \mu_m\mathscr{B} &= E_m - TS_m - H\mathscr{I} \\ &= A_m - H\mathscr{I} \end{aligned} \tag{41.4}$$

where F_m and A_m are the Gibbs and Helmholtz free energies, respectively.

Also

$$dA_m = -S_m\, dT + H\, d\mathscr{I} + \mu_m\, d\mathscr{B} \tag{41.5}$$

$$dF_m = -S_m\, dT - \mathscr{I}\, dH + \mu_m\, d\mathscr{B} \tag{41.6}$$

We have here[2] the thermodynamic stability condition

$$\left(\frac{\partial \mathscr{I}}{\partial H}\right)_{T,\mathscr{B}} > 0 \tag{41.7}$$

so that[3] when $H \to +\infty$, $\mathscr{I} \to \mathscr{B}$ or $I \to +1$ (all spins -1 or $\uparrow$) and when $H \to -\infty$, $\mathscr{I} \to -\mathscr{B}$ or $I \to -1$ (all spins $+1$ or $\downarrow$).

[1] We use $\mathscr{B}$ here instead of B (as elsewhere in the book) in order to avoid confusion with the states A and B.

[2] Instead of $(\partial V/\partial p)_{T,N} < 0$.

[3] See Fig. 33, below.

For given $\mathscr{B}$ and $\mathscr{I}$ (which fix $\mathscr{B}$ and N), the total number of configurations is

$$\Omega = \frac{\mathscr{B}!}{N!(\mathscr{B} - N)!} \tag{41.8}$$

since this is the number of ways of assigning $N \downarrow$ spins on $\mathscr{B}$ sites. Then the partition function (canonical ensemble) is

$$Q_m(\mathscr{B},\mathscr{I},T) = j_m(T)^{\mathscr{B}} \sum_{i=1}^{\Omega} e^{-E_i/\mathbf{k}T} \tag{41.9}$$

where E_i is the sum of nearest-neighbor pair energies for the ith configuration and j_m represents the nonconfigurational partition function of each of the $\mathscr{B}$ atoms of the system (j_m is assumed separable from the configurational partition function). Then, in Eq. (41.5),

$$A_m = -\mathbf{k}T \ln Q_m \tag{41.10}$$

and the usual [Eqs. (14.88) to (14.90)] differential relations between Q_m and μ_m, H, S_m, E_m, and $C_{\mathscr{I}}$ (heat capacity at constant $\mathscr{I}$ and $\mathscr{B}$) follow.

The restriction of fixed N in Eq. (41.9) is a mathematical handicap and most work has actually been done using the isothermal-isobaric partition function Δ [Eqs. 14.64) to (14.70)], since summation over $\mathscr{I}$ at constant $\mathscr{B}$ is the same as summation over N at constant $\mathscr{B}$:

$$\begin{aligned}\Delta(\mathscr{B},H,T) &= \sum_{N=0}^{\mathscr{B}} \exp\left(\frac{H\mathscr{I}}{\mathbf{k}T}\right) Q_m(\mathscr{B},\mathscr{I},T) \\ &= j_m{}^{\mathscr{B}} \exp\left(\frac{H\mathscr{B}}{\mathbf{k}T}\right) \sum_{N=0}^{\mathscr{B}} e^{-2HN/\mathbf{k}T} \sum_{i=1}^{\Omega} e^{-E_i/\mathbf{k}T}\end{aligned} \tag{41.11}$$

The connection with thermodynamics is given by

$$F_m = -\mathbf{k}T \ln \Delta \tag{41.12}$$

Relations such as

$$\mathscr{I} = \mathbf{k}T \left(\frac{\partial \ln \Delta}{\partial H}\right)_{T,\mathscr{B}} (= \mathscr{B} - 2\bar{N}) \tag{41.13}$$

$$E_m - H\mathscr{I} = \mathbf{k}T^2 \left(\frac{\partial \ln \Delta}{\partial T}\right)_{H,\mathscr{B}} \tag{41.14}$$

$$C_H = \left[\frac{\partial}{\partial T}\left(\mathbf{k}T^2 \frac{\partial \ln \Delta}{\partial T}\right)\right]_{H,\mathscr{B}} \tag{41.15}$$

follow from Eq. (41.6). In Eq. (41.15), C_H is the heat capacity at constant H and $\mathscr{B}$.

Equation (41.11) can be written in more explicit form. Suppose that in the ith configuration there are $N_{AB}{}^i$ nearest-neighbor AB (or $\uparrow\downarrow$, $-+$)

pairs, each with energy ε_{AB}, and also $N_{AA}{}^i$ and $N_{BB}{}^i$ nearest-neighbor AA ($\downarrow\downarrow$, $++$) and BB ($\uparrow\uparrow$, $--$) pairs, with energies ε_{AA} and ε_{BB}, respectively. Then

$$E_i = \varepsilon_{AB}N_{AB}{}^i + \varepsilon_{AA}N_{AA}{}^i + \varepsilon_{BB}N_{BB}{}^i \tag{41.16}$$

With assigned values of N and $\mathscr{B}$, only one of $N_{AB}{}^i$, $N_{AA}{}^i$, and $N_{BB}{}^i$ is an independent quantity, for we have the relations

$$\begin{aligned} 2N_{AA}{}^i + N_{AB}{}^i &= cN \\ 2N_{BB}{}^i + N_{AB}{}^i &= c(\mathscr{B} - N) \end{aligned} \tag{41.17}$$

where c is the number of nearest-neighbor sites of any lattice site in the system. Equations (41.17) follow from a consideration of the nearest neighbors to all the A sites or B sites, respectively, in the lattice. Then Eq. (41.16) becomes

$$E_i = wN_{AB}{}^i + \frac{cN\varepsilon_{AA}}{2} + \frac{c(\mathscr{B} - N)\varepsilon_{BB}}{2} \tag{41.18}$$

where

$$w = \varepsilon_{AB} - \frac{\varepsilon_{AA}}{2} - \frac{\varepsilon_{BB}}{2} \tag{41.19}$$

Obviously, $2w$ is the energy change in the process

$$\begin{aligned} AA + BB &\rightarrow 2\,AB \\ \downarrow\downarrow + \uparrow\uparrow &\rightarrow 2\downarrow\uparrow \end{aligned} \tag{41.20}$$

where AA represents an AA interaction, etc. Substitution of Eq. (41.18) in Eq. (41.11) gives

$$\Delta = j_m{}^{\mathscr{B}} \exp\left(\frac{H\mathscr{B}}{\mathbf{k}T}\right) \exp\left(-\frac{c\mathscr{B}\varepsilon_{BB}}{2\mathbf{k}T}\right) \sum_{N=0}^{\mathscr{B}} \exp\left(-\frac{2\mathscr{H}N}{\mathbf{k}T}\right) \sum_{i=1}^{\Omega} e^{-wN_{AB}{}^i/\mathbf{k}T} \tag{41.21}$$

where

$$2\mathscr{H} = 2H + \frac{c}{2}(\varepsilon_{AA} - \varepsilon_{BB}) \tag{41.22}$$

so that H and $\mathscr{H}$ differ only by a constant.

If we define J as $w/2$, and multiply and divide the right-hand side of Eq. (41.21) by $\exp(c\mathscr{B}J/2\mathbf{k}T)$, we have

$$\Delta = j_m{}^{\mathscr{B}} \exp\left(\frac{H\mathscr{B}}{\mathbf{k}T}\right) \exp\left(-\frac{c\mathscr{B}\varepsilon_{BB}}{2\mathbf{k}T}\right) \exp\left(-\frac{c\mathscr{B}J}{2\mathbf{k}T}\right) \times \sum_{N=0}^{\mathscr{B}} \exp\left(-\frac{2\mathscr{H}N}{\mathbf{k}T}\right) \sum_{i=1}^{\Omega} e^{-E_i'/\mathbf{k}T} \tag{41.23}$$

where

$$E_i' = 2JN_{AB}{}^i - \frac{c\mathscr{B}J}{2} \tag{41.24}$$

The double sum in Eq. (41.23) is not restricted to any particular choice of ε's but it now has the same *formal* appearance as the double sum in Eq. (41.11) if we put, in Eq. (41.11), $\varepsilon_{AA} = \varepsilon_{BB} = -J$ and $\varepsilon_{AB} = +J$, which is the usual convention in discussions of ferromagnetism.

If $w > 0$, which means physically that like nearest-neighbor pairs ($\downarrow\downarrow$, $\uparrow\uparrow$) are more stable energetically than unlike pairs ($\uparrow\downarrow$), the system will tend to split into two phases[1] (canonical ensemble) at low enough temperatures, one phase containing a predominance of A or $\downarrow$ and the other a predominance of B or $\uparrow$. This corresponds to *ferromagnetic* behavior, which is discussed further below. If $w < 0$, unlike pairs ($\downarrow\uparrow$) are more stable, and at low temperatures opposite spins will tend to arrange themselves on alternate lattice sites, without a splitting into two phases. This corresponds to *antiferromagnetic* behavior. In both cases ($w > 0$ and $w < 0$), the spins will approach random behavior at high temperatures ($w/\mathbf{k}T \to 0$); a critical temperature T_c separates the low- and high-temperature regions (see below and Sec. 44).

Lattice Gas. A "lattice gas" is a system of N molecules, each of which is restricted to be located in the neighborhood of one of $\mathscr{B}$ lattice sites, there being not more than one molecule on a given site at any one time. If interactions between molecules on nearest-neighbor sites only are taken into account, this system clearly fits into our general classification of nearest-neighbor lattice statistics.

There are several physical problems that correspond to a lattice gas. Localized monolayer adsorption from a gas or from a solution onto a one-dimensional lattice (linear protein or polymer) or a two-dimensional lattice (surface) are the most obvious. When the lattice is three-dimensional, this process would be called absorption rather than adsorption; the localized absorption of hydrogen gas by palladium and other metals is a well-known example.[2] A lattice model for a liquid with holes (vacant sites) has frequently been used in the theory of liquids (see Chap. 8).

In a lattice gas the state A represents, then, an occupied site, there being $N = N_A$ occupied sites, and state B represents a vacant site of which there are $N_B = \mathscr{B} - N$. The volume (or area or length) V of the system is proportional to the total number of sites $\mathscr{B}$, and for convenience we shall use $\mathscr{B}$ as the external parameter (instead of $V = \text{const} \times \mathscr{B}$). The work term is written $-p\,d\mathscr{B}$ (instead of $-p\,dV$), which defines the pressure p with dimensions of energy. In a surface problem (two dimensions), p is a surface or "spreading" pressure. Then

$$dE = T\,dS - p\,d\mathscr{B} + \mu\,dN \tag{41.25}$$

[1] This is not true for a one-dimensional system (see Sec. 43).

[2] See Fowler and Guggenheim (Gen. Ref.).

where the other symbols have their usual significance. We also have

$$dA = -S\,dT - p\,d\mathscr{B} + \mu\,dN \tag{41.26}$$

$$d(p\mathscr{B}) = S\,dT + p\,d\mathscr{B} + N\,d\mu \tag{41.27}$$

In adsorption and absorption, the lattice gas is in equilibrium with another phase (e.g., a gas) in which the chemical potential of the molecules is also μ. The partition function here is

$$Q(N,\mathscr{B},T) = j(T)^N \sum_{i=1}^{\Omega} e^{-E_i/\mathbf{k}T} \tag{41.28}$$

where the sum is the same as the sum in Eq. (41.9), and j is the separable, nonconfigurational partition function of a molecule at a lattice site. In Eq. (41.26),

$$A = -\mathbf{k}T \ln Q \tag{41.29}$$

As in Eq. (41.11), it is again desirable to eliminate the restriction of constant N by summation over N. In this case, however, N is the number of particles rather than being related to the external parameter (as in $\mathscr{I} = \mathscr{B} - 2N$). Hence we encounter the grand partition function here instead of the isothermal-isobaric partition function of Eq. (41.11). That is,

$$\begin{aligned}\Xi(\mu,\mathscr{B},T) &= \sum_{N=0}^{\mathscr{B}} e^{N\mu/\mathbf{k}T} Q(N,\mathscr{B},T) \\ &= \sum_{N=0}^{\mathscr{B}} e^{N\mu/\mathbf{k}T} j^N \sum_{i=1}^{\Omega} e^{-E_i/\mathbf{k}T}\end{aligned} \tag{41.30}$$

and

$$p\mathscr{B} = \mathbf{k}T \ln \Xi \tag{41.31}$$

Other thermodynamic properties follow from Eq. (41.27) [see Eqs. (14.48) to (14.50)].

With the introduction of Eq. (41.18) for E_i, Eq. (41.30) becomes

$$\Xi = \exp\left(-\frac{c\mathscr{B}\varepsilon_{BB}}{2\mathbf{k}T}\right) \sum_{N=0}^{\mathscr{B}} e^{N\mu/\mathbf{k}T} j^N e^{-cN(\varepsilon_{AA}-\varepsilon_{BB})/2\mathbf{k}T} \sum_{i=1}^{\Omega} e^{-wN^i{}_{AB}/\mathbf{k}T} \tag{41.32}$$

where the choice of real interest in this case[1] is $\varepsilon_{AB} = \varepsilon_{BB} = 0$.

When $w > 0$ ($\varepsilon_{AA} < 0$ if $\varepsilon_{AB} = \varepsilon_{BB} = 0$, that is, attraction between the molecules), the splitting into two phases (ρ large and small) at low temperatures (closed system) corresponds to a two-phase liquid-gas system below the critical temperature. When $w < 0$ (repulsion between the molecules), molecules will tend to occupy alternate lattice sites at low temperatures

[1] It is convenient to use the same general notation ($\varepsilon_{AA}, \varepsilon_{BB}, \varepsilon_{AB}$) for all the problems considered in this section. But it should be noted then that, in a lattice gas, as $N \to 0$, $Q \to \exp(-c\mathscr{B}\varepsilon_{BB}/2\mathbf{k}T)$ and $p \to -c\varepsilon_{BB}/2 = \text{constant}$.

without splitting into two phases, but there can still be a critical temperature (see below and Sec. 44).

Binary System. The two physical systems included here are binary liquid and solid solutions (alloys) with the phenomena of particular interest of phase splitting and the order-disorder transition, respectively. A single thermodynamic discussion suffices for both types of solution.

Each of the $\mathscr{B}$ sites of the lattice is occupied either by an A molecule or a B molecule, there being $N = N_A$ A molecules in the system and $\mathscr{B} - N = N_B$ B molecules. The fundamental thermodynamic equation is

$$dE_s = T\,dS_s + \mu_A\,dN_A + \mu_B\,dN_B \tag{41.33}$$

where E_s and S_s refer to the solution and μ_A and μ_B are the chemical potentials of the two components. The pV term is considered negligible (condensed system) and omitted from Eq. (41.33). We have the additional relations

$$dA_s = -S_s\,dT + \mu_A\,dN_A + \mu_B\,dN_B \tag{41.34}$$

$$A_s = E_s - TS_s = \mu_A N_A + \mu_B N_B \tag{41.35}$$

For use below, let us obtain one more thermodynamic equation. Combining Eqs. (41.34) and (41.35),

$$0 = -S_s\,dT - N_A\,d\mu_A - N_B\,d\mu_B \tag{41.36}$$

We now add $d[(N_A + N_B)\mu_B]$ to both sides of Eq. (41.36), and find

$$d(\mathscr{B}\mu_B) = -S_s\,dT + \mu_B\,d\mathscr{B} - N_A\,d(\mu_A - \mu_B) \tag{41.37}$$

The partition function for the binary solution is

$$Q_s(N_A,N_B,T) = j_A(T)^{N_A} j_B(T)^{N_B} \sum_{i=1}^{\Omega} e^{-E_i/\mathbf{k}T} \tag{41.38}$$

where the sum is identical with the sum in Eqs. (41.9) and (41.28), and j_A and j_B are the separable, nonconfigurational partition functions of an A and a B molecule, respectively. In Eqs. (41.34) and (41.35)

$$A_s = -\mathbf{k}T \ln Q_s \tag{41.39}$$

To obtain a partition function strictly analogous to Eqs. (41.11) and (41.30), we wish to carry out a summation over $N = N_A$. This requires that we digress briefly to invent a suitable partition function,[1] Y. Let

$$Y(\alpha,\mathscr{B},T) = \sum_{N=0}^{\mathscr{B}} e^{\alpha N} Q_s(N,\mathscr{B},T) \tag{41.40}$$

[1] This differs from Eq. (15.1) in that $N_A + N_B$ is held constant on summing over N_A in Eq. (41.40) instead of N_B as in Eq. (15.1).

where we now regard Q_s as a function of $N = N_A$, $\mathscr{B} = N_A + N_B$, and T instead of N_A, N_B, and T, and the significance of α is yet to be established. We use the type of argument in Sec. 20. Let

$$t(N) = e^{\alpha N} Q_s(N,\mathscr{B},T)$$

Using only the maximum term in Eq. (41.40),

$$\ln Y = \ln t_{\max} = \alpha N + \ln Q_s(N,\mathscr{B},T) \tag{41.41}$$

where N is determined by

$$\frac{\partial \ln t}{\partial N} = 0 = \alpha + \left(\frac{\partial \ln Q_s}{\partial N}\right)_{\mathscr{B},T} \tag{41.42}$$

Equation (41.42) also relates α to thermodynamic properties:

$$-\alpha = \left(\frac{\partial \ln Q_s}{\partial N}\right)_{\mathscr{B},T} = \left(\frac{\partial \ln Q_s}{\partial N}\right)_{N_B,T} + \left(\frac{\partial \ln Q_s}{\partial N_B}\right)_{N,T} \left(\frac{\partial N_B}{\partial N}\right)_{\mathscr{B}} = -\frac{\mu_A}{\mathbf{k}T} + \frac{\mu_B}{\mathbf{k}T} \tag{41.43}$$

where we have used Eqs. (41.34) and (41.39) Thus we can rewrite Eq. (41.40) as

$$Y(\mu_A - \mu_B, \mathscr{B}, T) = \sum_{N=0}^{\mathscr{B}} e^{N(\mu_A - \mu_B)/\mathbf{k}T} Q_s(N, \mathscr{B}, T) \tag{41.44}$$

On putting Eq. (41.43) in Eq. (41.41), using Eqs. (41.35) and (41.39),

$$-\mathscr{B}\mu_B = \mathbf{k}T \ln Y \tag{41.45}$$

Finally, we note from Eq. (41.37),

$$N_A = \mathbf{k}T \left(\frac{\partial \ln Y}{\partial(\mu_A - \mu_B)}\right)_{T,\mathscr{B}} \tag{41.46}$$

Equation (41.44) can also be written as

$$Y = j_B{}^{\mathscr{B}} \sum_{N=0}^{\mathscr{B}} e^{N(\mu_A - \mu_B)/\mathbf{k}T} j_A{}^N j_B{}^{-N} \sum_{i=1}^{\Omega} e^{-E_i/\mathbf{k}T} \tag{41.47}$$

or

$$Y = j_B{}^{\mathscr{B}} \exp\left(-\frac{c\mathscr{B}\varepsilon_{BB}}{2\mathbf{k}T}\right) \sum_{N=0}^{\mathscr{B}} e^{N(\mu_A - \mu_B)/\mathbf{k}T} j_A{}^N j_B{}^{-N} e^{-cN(\varepsilon_{AA} - \varepsilon_{BB})/2\mathbf{k}T} \times \sum_{i=1}^{\Omega} e^{-wN_{AB}{}^i/\mathbf{k}T} \tag{41.48}$$

When $w > 0$ (AA and BB pairs more stable than AB pairs), the system splits into two phases of different relative composition at low temperatures. When $w < 0$, the disorder $\rightarrow$ order transition in an alloy will occur as the temperature is lowered, with A and B atoms tending to occupy alternate positions on the lattice sites.

Symmetry Properties. Further discussion will be aided if we digress at this point to consider three symmetry properties of the Ising model. For concreteness, we use the language of the magnetic case.

From Eqs. (41.13) and (41.21), we find

$$I = 1 + \frac{1}{\mathscr{B}}\left[\frac{\partial \ln \Sigma}{\partial(\mathscr{H}/\mathbf{k}T)}\right]_{T,\mathscr{B}} \tag{41.49}$$

where
$$\Sigma = \sum_{N=0}^{\mathscr{B}} \exp\left(-\frac{2\mathscr{H}N}{\mathbf{k}T}\right)\sum_{i=1}^{\Omega} e^{-wN_{AB}{}^{i}/\mathbf{k}T} \tag{41.50}$$

$$= \sum_{N=0}^{\mathscr{B}} \exp\left(-\frac{2\mathscr{H}N}{\mathbf{k}T}\right)\sum_{N_{AB}} g(N,N_{AB},\mathscr{B})e^{-wN_{AB}/\mathbf{k}T} \tag{41.51}$$

In Eq. (41.51) we have grouped together the $g(N,N_{AB},\mathscr{B})$ different configurations which have N_{AB} nearest-neighbor AB ($\downarrow\uparrow$) pairs when there are N spins of type A ($\downarrow$) out of the total of $\mathscr{B}$ spins of both types. Of course

$$\sum_{N_{AB}} g(N,N_{AB},\mathscr{B}) = \Omega = \frac{\mathscr{B}!}{N!(\mathscr{B}-N)!} \tag{41.52}$$

On carrying out the differentiation in Eq. (41.49) we obtain

$$I(\mathscr{H}) = 1 - 2\,\frac{\sum_{N=0}^{\mathscr{B}} (N/\mathscr{B}) \exp(-2\mathscr{H}N/\mathbf{k}T)\sum_{N_{AB}} g(N,N_{AB},\mathscr{B})e^{-wN_{AB}/\mathbf{k}T}}{\Sigma} \tag{41.53}$$

Now in each of the $g(N,N_{AB},\mathscr{B})$ configurations with given N, N_{AB}, and $\mathscr{B}$, suppose we change every A to a B and every B to an A. The number of AB pairs N_{AB} will be unaltered but the number of A's changes from N to $\mathscr{B} - N$. Thus we can conclude that

$$g(N, N_{AB}, \mathscr{B}) = g(\mathscr{B} - N, N_{AB}, \mathscr{B}) \tag{41.54}$$

Next, let us find $I(-\mathscr{H})$ from Eq. (41.53):

$$I(-\mathscr{H}) = 1 - 2\,\frac{\sum_{\mathscr{B}-N=0}^{\mathscr{B}}\left(1 - \frac{\mathscr{B}-N}{\mathscr{B}}\right)\exp\left[-\frac{2\mathscr{H}(\mathscr{B}-N)}{\mathbf{k}T}\right]\exp\left[\frac{2\mathscr{H}\mathscr{B}}{\mathbf{k}T}\right] \times \sum_{N_{AB}} g(\mathscr{B}-N, N_{AB}, \mathscr{B})e^{-wN_{AB}/\mathbf{k}T}}{\sum_{\mathscr{B}-N=0}^{\mathscr{B}}\exp\left[-\frac{2\mathscr{H}(\mathscr{B}-N)}{\mathbf{k}T}\right]\exp\left(\frac{2\mathscr{H}\mathscr{B}}{\mathbf{k}T}\right) \times \sum_{N_{AB}} g(\mathscr{B}-N, N_{AB}, \mathscr{B})e^{-wN_{AB}/\mathbf{k}T}} \tag{41.55}$$

where the variable of summation has been changed from N to $\mathscr{B} - N$, and we have written

$$g(N, N_{AB}, \mathscr{B}) = g(\mathscr{B} - N, N_{AB}, \mathscr{B})$$

$$\frac{N}{\mathscr{B}} = 1 - \frac{\mathscr{B} - N}{\mathscr{B}}$$

$$N\mathscr{H} = -(\mathscr{B} - N)\mathscr{H} + \mathscr{B}\mathscr{H}$$

If we now cancel $\exp(2\mathscr{H}\mathscr{B}/\mathbf{k}T)$ in Eq. (41.55) and change variables from $\mathscr{B} - N$ to N, we see on comparing with Eq. (41.53) that the quotient of the

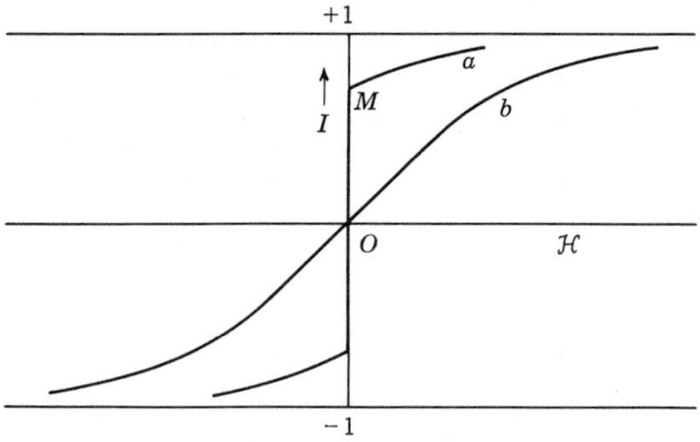

FIG. 33. Magnetization (I) versus magnetic field ($\mathscr{H}$). (a) Ferromagnet below T_c. (b) Ferromagnet above T_c (or antiferromagnet).

two sums in Eq. (41.55) is just {1 − [quotient of sums in Eq. (41.53)]}. That is,

$$I(-\mathscr{H}) = 1 - 2\left\{1 - \left[\frac{1 - I(\mathscr{H})}{2}\right]\right\}$$

or

$$I(\mathscr{H}) + I(-\mathscr{H}) = 0 \qquad I(0) = 0 \tag{41.56}$$

In other words, I is an odd function of $\mathscr{H}$.

Typical schematic $I(\mathscr{H})$ curves are shown in Fig. 33. Curve a represents a ferromagnet ($w > 0$) below the critical (Curie) temperature, with a discontinuity (when $\mathscr{B} \to \infty$, as usual) in I at $\mathscr{H} = 0$. The "spontaneous magnetization" M is defined as

$$M = \lim_{\mathscr{H} \to 0+} I(\mathscr{H}) \tag{41.57}$$

M decreases from $+1$ at $T = 0$ to zero at $T \geqslant T_c$ (see Fig. 34).

Experimentally, the spontaneous magnetization M is the magnetization per spin that a ferromagnet ($T < T_c$) retains if a sample is placed in a magnetic field $\mathscr{H} > 0$ and then the field is removed ($\mathscr{H} \to 0+$).

Curve b of Fig. 33 (schematic) could refer to a ferromagnet with $T > T_c$ or to an antiferromagnet.

One important consequence of Eq. (41.56) is the following. Although in general the two heat capacities C_H and $C_{\mathscr{I}}$ are different, in the special case $\mathscr{H}$ = constant = 0 we also have I = constant = 0 so that

$$C_{\mathscr{H}=0}(T) = C_{I=0}(T) \tag{41.58}$$

$I = 0$ corresponds, of course, to $\rho = \bar{N}/\mathscr{B} = 1/2$.

The second symmetry property is of a more restricted nature. We consider only $\mathscr{H} = 0$ and lattices whose sites can be divided into two groups α and β such that the nearest neighbor sites of an α site are all β, and vice versa.

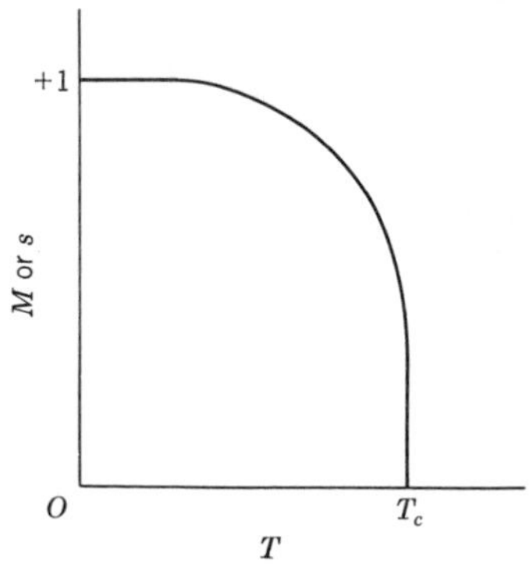

FIG. 34. Spontaneous magnetization (M) of ferromagnet or degree of long-range order (s) of antiferromagnet versus temperature.

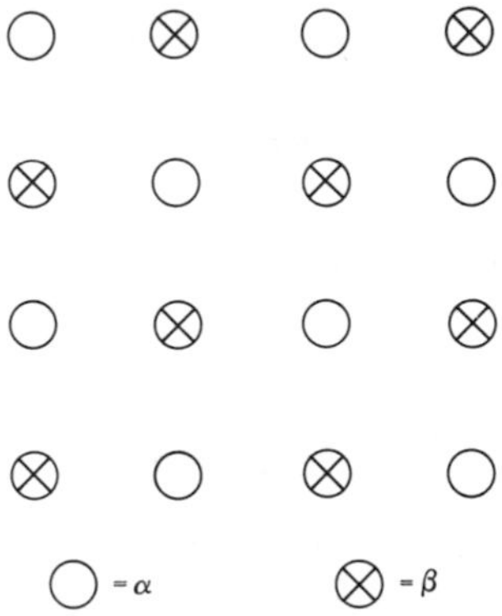

FIG. 35. α and β sites of simple square lattice.

Examples are the simple square ($c = 4$; Fig. 35), simple cubic ($c = 6$), and body-centered cubic ($c = 8$) lattices. In Eq. (41.21), we set $\mathscr{H} = 0$ and obtain

$$\Delta(\mathscr{H} = 0) = j_m^{\mathscr{B}} \exp\left[-\frac{c\mathscr{B}(\varepsilon_{AA} + \varepsilon_{BB})}{4\mathbf{k}T}\right] \sum_{N_{AB}=0}^{c\mathscr{B}/2} G(N_{AB},\mathscr{B}) e^{-wN_{AB}/\mathbf{k}T} \tag{41.59}$$

where $G(N_{AB},\mathscr{B})$ is the number of configurations with N_{AB} nearest-neighbor AB ($\downarrow\uparrow$) pairs, irrespective of the value of N. That is,

$$G(N_{AB},\mathscr{B}) = \sum_N g(N,N_{AB},\mathscr{B}) \tag{41.60}$$

Also, from Eq. (41.52),

$$\sum_{N_{AB}=0}^{c\mathscr{B}/2} G(N_{AB},\mathscr{B}) = 2^{\mathscr{B}} \tag{41.61}$$

Equation (41.61) also follows directly from the fact that when all possible values of N and N_{AB} are included, each of the $\mathscr{B}$ sites can be in one of two states, giving a total of $2^{\mathscr{B}}$ states. Now in each of the $G(N_{AB},\mathscr{B})$ configurations with given N_{AB} and $\mathscr{B}$, suppose we change every A on an α site to B and every B on an α site to A. This will alter every nearest-neighbor pair in the configuration: every AA or BB will become an AB and every AB will become an AA or a BB. Since the total number of nearest-neighbor pairs is $c\mathscr{B}/2$, of which N_{AB} are AB pairs initially, we have

$$G(N_{AB},\mathscr{B}) = G\left(\frac{c\mathscr{B}}{2} - N_{AB},\mathscr{B}\right) \tag{41.62}$$

Let $\Delta(\varepsilon,\mathscr{H} = 0)$ denote the Δ in Eq. (41.59), and consider now $\Delta(-\varepsilon,\mathscr{H} = 0)$, where $-\varepsilon$ indicates that the signs of ε_{AA}, ε_{BB}, and ε_{AB} are reversed (so that $w \to -w$; ferromagnet $\to$ antiferromagnet). Then, from Eq. (41.59),

$$\begin{aligned}\Delta(-\varepsilon,\mathscr{H} = 0) &= j_m{}^{\mathscr{B}} \exp\left[\frac{c\mathscr{B}(\varepsilon_{AA} + \varepsilon_{BB})}{4\mathbf{k}T}\right] \exp\left(\frac{c\mathscr{B}w}{2\mathbf{k}T}\right) \\ &\quad \times \sum_{\frac{c\mathscr{B}}{2}-N_{AB}=0}^{c\mathscr{B}/2} G\left(\frac{c\mathscr{B}}{2} - N_{AB},\mathscr{B}\right) \exp\left\{-\frac{[(c\mathscr{B}/2) - N_{AB}]w}{\mathbf{k}T}\right\} \\ &= \Delta(\varepsilon,\mathscr{H} = 0) \exp\left[\frac{c\mathscr{B}(2\varepsilon_{AB} + \varepsilon_{AA} + \varepsilon_{BB})}{4\mathbf{k}T}\right]\end{aligned} \tag{41.63}$$

The significance of this result is that, at $\mathscr{H} = 0$ (or $\rho = 1/2$), the energy and free energy (as functions of temperature) of a ferromagnet with ε_{AA}, ε_{BB}, and ε_{AB} differ from the energy and free energy of an antiferromagnet with $-\varepsilon_{AA}$, $-\varepsilon_{BB}$, and $-\varepsilon_{AB}$ by at most a constant,[1] and the two heat capacities as a function of temperature are identical. Thus, a singularity in, for example, the heat capacity at $T = T_c$ in the one case will also occur at $T = T_c$ and be of the same form in the other case.

The third symmetry property is closely related to the second and restricted in the same way ($\mathscr{H} = 0$ and the existence of the two groups of sites α and β). We take $w > 0$ throughout. Equation (41.59) can be written

$$\Delta(\mathscr{H} = 0) = j_m{}^{\mathscr{B}} \exp\left[-\frac{c\mathscr{B}(\varepsilon_{AA} + \varepsilon_{BB})}{4\mathbf{k}T}\right] \sum_{N=0}^{\mathscr{B}} \sum_{N_{AB}} g(N,N_{AB},\mathscr{B}) e^{-wN_{AB}/\mathbf{k}T} \tag{41.64}$$

Let $P(N)$ be the probability of observing the value N in this system (at

[1] This constant is zero with the choice $\varepsilon_{AA} = \varepsilon_{BB} = -J$, $\varepsilon_{AB} = J$.

$\mathscr{H} = 0$), recalling that the system is "open" with respect to $\mathscr{I}$ and therefore N [see Eq. (14.62)]. Then

$$P(N) = \frac{\sum_{N_{AB}} g(N,N_{AB},\mathscr{B})e^{-wN_{AB}/\mathbf{k}T}}{\sum_{N=0}^{\mathscr{B}} \sum_{N_{AB}} g(N,N_{AB},\mathscr{B})e^{-wN_{AB}/\mathbf{k}T}} \tag{41.65}$$

$P(N)$ is symmetrical about $N = \mathscr{B}/2$ [see Eq. (41.54)] so that $\bar{N} = \mathscr{B}/2$ at every temperature. Above the critical (Curie) temperature $P(N)$ will have a single extremely sharp (as $\mathscr{B} \to \infty$) peak at $N = \mathscr{B}/2$ of the usual type in statistical mechanics. Below the critical temperature, there are two possible phases (Fig. 33*a*) and $P(N)$ has two peaks at $I = M$ and $-M$ or at $N = N^0 = (\mathscr{B}/2)(1 - M)$ and $N = \mathscr{B} - N^0 = (\mathscr{B}/2)(1 + M)$. This type of behavior follows from Appendix 9. We note in particular for use below that the function $P(N)$ determines M for any $T < T_c$ from the smaller value of N, N^0 ($N^0 \leqslant \mathscr{B}/2$), at which $P(N)$ has a peak:

$$M = 1 - \frac{2N^0}{\mathscr{B}} = 1 - 2\rho^0 \qquad 0 \leqslant N^0 \leqslant \frac{\mathscr{B}}{2} \tag{41.66}$$

We now wish to relate the above discussion to the concept of "long-range order" in the antiferromagnetic case ($T < T_c$). Let an A on an α site and a B on a β site be "right" states, and let an A on a β site and a B on an α site be "wrong" states. Then for any configuration with given N, N_{AB} and $\mathscr{B}$, the number of "right" states is

$$N_R = N_A{}^\alpha + N_B{}^\beta = \mathscr{B} - N_W \tag{41.67}$$

(where $N_A{}^\alpha$ is the number of A's on α sites, etc.) and the number of RW nearest-neighbor pairs is

$$N_{RW} = \frac{c\mathscr{B}}{2} - N_{AB} \tag{41.68}$$

since every AA or BB pair is an RW pair. Starting with Eq. (41.59) [which is equivalent to Eq. (41.64)], using Eqs. (41.62) and (41.68), we have (with $w > 0$ still)

$$\Delta(\mathscr{H} = 0) = \jmath_m{}^{\mathscr{B}} \exp\left[-\frac{c\mathscr{B}(\varepsilon_{AA} + \varepsilon_{BB})}{4\mathbf{k}T}\right]$$
$$\times \exp\left(-\frac{c\mathscr{B}w}{2\mathbf{k}T}\right) \sum_{N_{RW}=0}^{c\mathscr{B}/2} G(N_{RW},\mathscr{B})e^{wN_{RW}/\mathbf{k}T} \tag{41.69}$$

where $G(N_{RW},\mathscr{B})$ is the number of configurations (irrespective of N) with N_{RW} AB pairs on $\mathscr{B}$ sites. But, since the combinatorial problems are identical, $G(N_{RW},\mathscr{B})$ is also the number of configurations (irrespective of N_W)

with N_{RW} RW pairs on $\mathscr{B}$ sites. Similarly, the number of configurations $g(N_W, N_{RW}, \mathscr{B})$ with N_W, N_{RW}, and $\mathscr{B}$ is the same function of these variables as $g(N, N_{AB}, \mathscr{B})$ is of N, N_{AB}, and $\mathscr{B}$. Hence the sum in Eq. (41.69) can be written as

$$\sum_{N_W=0}^{\mathscr{B}} \sum_{N_{RW}} g(N_W, N_{RW}, \mathscr{B}) e^{wN_{RW}/\mathbf{k}T}$$

and the probability $P(N_W)$ of observing N_W is

$$P(N_W) = \frac{\sum_{N_{RW}} g(N_W, N_{RW}, \mathscr{B}) e^{-(-w)N_{RW}/\mathbf{k}T}}{\sum_{N_W=0}^{\mathscr{B}} \sum_{N_{RW}} g(N_W, N_{RW}, \mathscr{B}) e^{-(-w)N_{RW}/\mathbf{k}T}} \tag{41.70}$$

Thus $P(N)$ for a ferromagnet with w ($w > 0$) is the same function as $P(N_W)$ for an antiferromagnet with $-w$. Hence in the antiferromagnet ($-w$) $P(N_W)$ is symmetrical[1] about $\mathscr{B}/2$, $\bar{N}_W = \mathscr{B}/2$ at all temperatures, $P(N_W)$ has a single extremely sharp peak at $N_W = \mathscr{B}/2$ for $T > T_c$ and $P(N_W)$ has two peaks at $N_W = N_W{}^0 = (\mathscr{B}/2)(1 - M)$ and $N_W = \mathscr{B} - N_W{}^0 = (\mathscr{B}/2) \times (1 + M)$, for $T < T_c$, where M is the spontaneous magnetization in the ferromagnetic case (w) and $N_W{}^0 = N^0$.

Instead of two possible *phases*, we have, in the antiferromagnet with $T < T_c$, two possible "regions," in both of which $I = 0$, but with the two different values of N_W, $N_W{}^0$ and $\mathscr{B} - N_W{}^0$. The two regions differ only with respect to reversal of the labeling of α and β sites (which is arbitrary in the first place). The two regions are physically indistinguishable (but a boundary between them, if they exist together, is in principle detectable) and thus have the same degree of long-range order s, which we now define. However, to be specific, we refer in this definition only to the region in which $N_W = N_W{}^0 \leqslant \mathscr{B}/2$. It is conventional to define s as

$$s = 1 - \frac{2N_W{}^0}{\mathscr{B}} \qquad 0 \leqslant N_W{}^0 \leqslant \frac{\mathscr{B}}{2} \tag{41.71}$$

so that $s = 1$ when all sites are "right" (perfect order) and $s = 0$ when as many sites are "right" as "wrong" (complete disorder). But since $N_W{}^0 = N^0$, as already noted, with this definition of s we have the result that the spontaneous magnetization M in a ferromagnet with w ($w > 0$) is the same as the degree of long-range order s in an antiferromagnet with $-w$. Thus the curve in Fig. 34 refers to M or s, according as we use w or $-w$, respectively. M and s are both zero for $T > T_c$. The critical temperature T_c in the ferromagnet is associated with a first-order phase transition; in the

[1] One special case: $N_W = 0$ (all sites "right") and $N_W = \mathscr{B}$ (all sites "wrong") both correspond to perfect alternate A and B occupation of sites and have the same probability.

antiferromagnet T_c refers to a phase transition of higher order and locates the temperature above which no long-range order exists.

It should perhaps be mentioned at this point that long-range correlation in a ferromagnet ($T < T_c$) has been discussed[1] in terms of the correlation $\overline{s_j s_k}$ between two sites j and k very far apart (in the same phase), where s_j is the spin ($+1$ or -1) at site j and s_k is the spin ($+1$ or -1) at site k. The value of $\overline{s_j s_k}$ is the same in the two phases; therefore we discuss only the phase with $N^0 \leqslant \mathscr{B}/2$ [Eq. (41.66)]. The correlation in a ferromagnet is not a measure of the same type of long-range order defined above [Eq. (41.71)] for an antiferromagnet, for in either one of the two ferromagnetic phases the sites are all equivalent and each one has the same average net spin. Hence, within a phase $N_R = N_W$ and $s = 0$. In a ferromagnet, the two distant sites j and k have independent probabilities for being $+1$ or -1 (which is not the case in the same region of an antiferromagnet), but the two probabilities are the same and lead to a correlation. That is, the probability of a $+1$ spin (A or $\downarrow$) at either site j or k is, from Eq. (41.66), $\rho^0 = (1 - M)/2$. Then we have the four possibilities in Table 12.

TABLE 12. LONG-RANGE CORRELATION

Site j		Site k	
s_j	Probability	s_k	Probability
$+1$	$(1 - M)/2$	$+1$	$(1 - M)/2$
$+1$	$(1 - M)/2$	-1	$(1 + M)/2$
-1	$(1 + M)/2$	$+1$	$(1 - M)/2$
-1	$(1 + M)/2$	-1	$(1 + M)/2$

From this list we find

$$\overline{s_j s_k} = M^2 \tag{41.72}$$

The curve of long-range correlation $\overline{s_j s_k}$ against temperature for a ferromagnet can therefore be deduced from Fig. 34 and will resemble this figure qualitatively (M^2 instead of M).

Thermodynamic Interconnections. Most of our discussion above has concerned the magnetic case. We now wish to use that discussion as a reference point in noting the thermodynamic interconnections between the magnetic problem, the lattice gas, and the binary system. Most of these interconnections will by this time be rather obvious to the reader.

[1] See Kaufman and Onsager, *loc. cit.*, and Newell and Montroll (Gen. Ref.). Short-range correlation is also included in these analyses.

a. Lattice gas. If we substitute Eq. (41.11) in Eq. (41.12) and Eq. (41.30) in Eq. (41.31), we observe that equivalent roles are played by the properties listed in Table 13. The results in Table 13 can also be deduced by a suitable comparison of Eqs. (41.5) and (41.26).

TABLE 13. EQUIVALENT THERMODYNAMIC PROPERTIES

Magnet	*Lattice gas*
No. of spins, $\mathscr{B}$	Volume (or area), $\mathscr{B}$
No. of $\downarrow$ or $+1$ spins, N	No. of molecules, N
$(1 - I)/2$	Density, $\rho = N/\mathscr{B}$
$-2H$	$\mu + \mathbf{k}T \ln j = \mathbf{k}T \ln z$ (z = activity)
$H + \mu_m + \mathbf{k}T \ln j_m$	$-p$

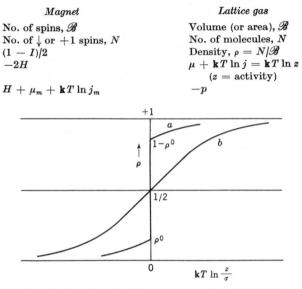

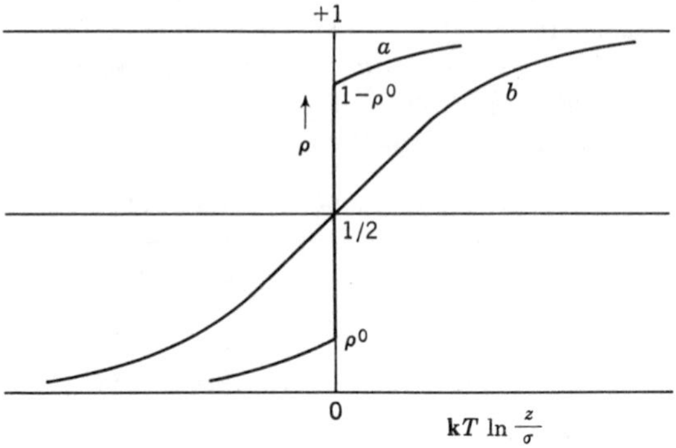

FIG. 36. Density of lattice gas (ρ) versus activity (z).

In Table 13 it is implicit that the same choices are made for ε_{AA}, ε_{BB}, and ε_{AB} in the two cases. As an example of different choices, Lee and Yang[1] used for the magnetic problem $\varepsilon_{AB} = \varepsilon$, $\varepsilon_{AA} = \varepsilon_{BB} = 0$ (and $j_m = 1$) and for the lattice gas, $\varepsilon_{AA} = -2\varepsilon$, $\varepsilon_{AB} = \varepsilon_{BB} = 0$. This gives $w = \varepsilon$ in both problems. Then a comparison of Eqs. (41.12) and (41.21) with Eqs. (41.31) and (41.32) yields Table 14 (we list only the alterations which have to be made in Table 13).

TABLE 14. EQUIVALENT PROPERTIES

(Lee and Yang)

Magnet	*Lattice gas*
$-2H$	$\mu + \mathbf{k}T \ln j + c\varepsilon$
$H + \mu_m$	$-p$

From Table 13 it is apparent that the magnetization curve (Fig. 33) of I versus $\mathscr{H}$ is closely related to a plot of density ρ against $\mathbf{k}T \ln (z/\sigma)$ in a

[1] Lee and Yang, *loc. cit.*

lattice gas as shown in Fig. 36, where $\sigma = \exp[c(\varepsilon_{AA} - \varepsilon_{BB})/2\mathbf{k}T]$. When the lattice gas is an adsorbed phase, Fig. 36 is particularly significant, for in this case the curves represent the directly measurable adsorption isotherms [for example, if a perfect gas is being adsorbed, $\mathbf{k}T \ln z$ is the same as $\mathbf{k}T \ln$ (gas pressure) to within a term which is a function of temperature only]. The "ferromagnetic" curve a in Fig. 36 ($w > 0$, $T < T_c$) corresponds to a condensation of the lattice gas into a condensed phase below the critical temperature. Adsorption isotherms with a phase transition are rather well known experimentally, but the best available illustration is the absorption of hydrogen gas by palladium and certain other metals.[1]

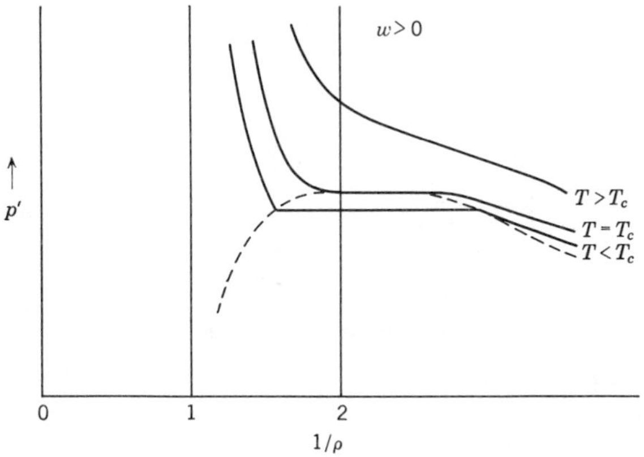

FIG. 37. Equation of state for lattice gas with $w > 0$ showing two-phase region ($T < T_c$).

The equation of state (p versus $1/\rho$) is important in a lattice gas, although the corresponding curve ($H + \mu_m$) versus I is not ordinarily discussed. The equation of state may be deduced from Fig. 36 and Eq. (41.27):

$$\mathscr{B}\, dp = S\, dT + N\, d\mu \tag{41.73}$$

$$p' \equiv p + \frac{c\varepsilon_{BB}}{2} = \mathbf{k}T \int_0^z \rho \frac{dz}{z} \qquad T \text{ const} \tag{41.74}$$

In adsorption work, Eq. (41.74) is the well-known "Gibbs equation" for the calculation (with $\varepsilon_{BB} = \varepsilon_{AB} = 0$) of the surface or spreading pressure p from an adsorption isotherm. The equation of state is shown (schematically) for several temperatures in Fig. 37 when $w > 0$. The dotted curve in Fig. 37 is the two-phase boundary region. It is replotted in the form ρ^0 and $1 - \rho^0$ versus T in Fig. 38. In view of the relation between Fig. 33 and Fig. 36,

[1] See Fowler and Guggenheim (Gen. Ref.).

Fig. 38 is obviously just the spontaneous magnetization curve of Fig. 34 turned on its side.

An important symmetry property follows from Eq. (41.74) and the symmetry of Fig. 36. Let $z < \sigma, \rho(z)$ and $z' > \sigma, \rho(z')$ be symmetrically placed points on the ρ versus $\ln (z/\sigma)$ curve (Fig. 39). That is,

$$\rho(z) + \rho(z') = 1$$

$$-\ln \frac{z}{\sigma} = \ln \frac{z'}{\sigma} \tag{41.75}$$

or

$$zz' = \sigma^2$$

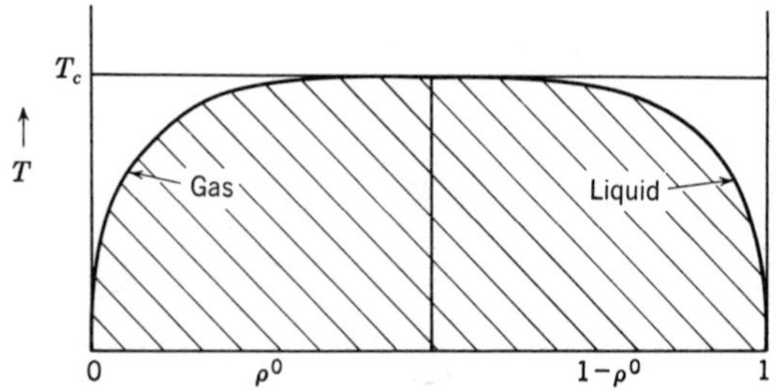

FIG. 38. Two-phase region of lattice gas, density versus temperature.

According to Eq. (41.74),

$$\frac{p'(z)}{\mathbf{k}T} = \int_{-\infty}^{\ln \frac{z}{\sigma}} \rho(z) d \ln \frac{z}{\sigma} \tag{41.76}$$

That is, $p'(z)/\mathbf{k}T$ is equal to the area $ABCD$ in Fig. 39. Also

$$\frac{p'(z')}{\mathbf{k}T} = AJEHID = AJOD + OFGI - BJOC$$

$$= \frac{p'(\sigma)}{\mathbf{k}T} + \ln \frac{z'}{\sigma} - \left[\frac{p'(\sigma)}{\mathbf{k}T} - \frac{p'(z)}{\mathbf{k}T}\right]$$

Hence we have the symmetry property, which will be useful later,

$$\frac{p'(z)}{\mathbf{k}T} = \ln \frac{z}{\sigma} + \frac{p'(\sigma^2/z)}{\mathbf{k}T} \tag{41.77}$$

For a ferro- or antiferromagnet, instead of Eq. (41.73), we have

$$\mathscr{B} d\mu_m = -S_m\, dT - \mathscr{I}\, dH \tag{41.78}$$

and, from Table 13 and Eq. (41.74),

$$H + \mu_m + \mathbf{k}T \ln j_m - \frac{c\varepsilon_{BB}}{2} = \int_H^\infty (I - 1) dH \qquad T \text{ const} \tag{41.79}$$

From Eq. (41.5), we have

$$C_{\mathscr{I}} = \left(\frac{\partial E_m}{\partial T}\right)_{\mathscr{I},\mathscr{B}} = \mathscr{B}\frac{d}{dT}\left(\mathbf{k}T^2\frac{d\ln j_m}{dT}\right) + C_1 \tag{41.80}$$

where C_1 is a configurational heat capacity,

$$C_1 = \left[\frac{\partial}{\partial T}\left(\mathbf{k}T^2\frac{\partial \ln \sum_i e^{-E_i/\mathbf{k}T}}{\partial T}\right)\right]_{N,\mathscr{B}} \tag{41.81}$$

Since E and E_m differ only by obvious terms in j and j_m [see Eqs. (41.9) and

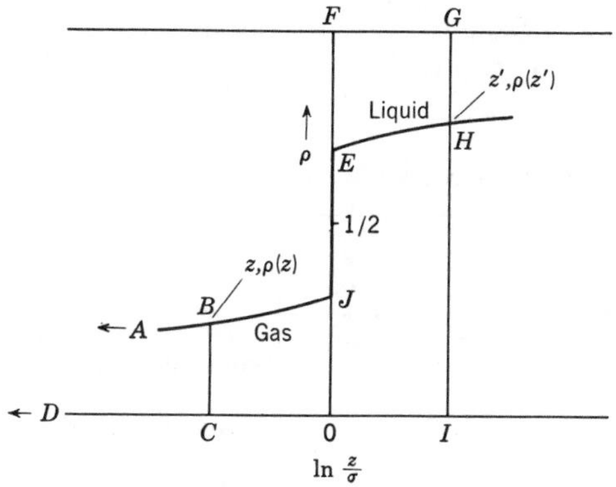

FIG. 39. Density of lattice gas (ρ) versus activity (z), illustrating symmetry property.

(41.28)], we have, from Table 13, for the heat capacity of a lattice gas at constant volume,

$$C_{\mathscr{B}} = \left(\frac{\partial E}{\partial T}\right)_{\mathscr{B},N} = N\frac{d}{dT}\left(\mathbf{k}T^2\frac{d\ln j}{dT}\right) + C_1 \tag{41.82}$$

Similarly, from Eq. (41.15),

$$C_H = \left(\frac{\partial(E_m - H\mathscr{I})}{\partial T}\right)_{H,\mathscr{B}} = \mathscr{B}\frac{d}{dT}\left(\mathbf{k}T^2\frac{d\ln j_m}{dT}\right) + C_2 \tag{41.83}$$

where the configurational heat capacity C_2 is

$$C_2 = \left[\frac{\partial}{\partial T}\left(\mathbf{k}T^2\frac{\partial \ln \sum_N e^{-2HN/\mathbf{k}T}\sum_i e^{-E_i/\mathbf{k}T}}{\partial T}\right)\right]_{H,\mathscr{B}} \tag{41.84}$$

The heat capacity of a lattice gas which corresponds to the heat capacity at constant magnetic field in the magnetic problem is, from Table 13,

$$C_{\mu+\mathbf{k}T\ln j} = \left(\frac{\partial[E - N(\mu + \mathbf{k}T\ln j)]}{\partial T}\right)_{\mu+\mathbf{k}T\ln j,\mathscr{B}} = N\frac{d}{dT}\left(\mathbf{k}T^2\frac{d\ln j}{dT}\right) + C_2 \tag{41.85}$$

where in Eq. (41.84) we replace $-2H$ by $\mu + \mathbf{k}T\ln j$. See also page 106.

As noted in Eq. (41.58), in a lattice gas with $\rho = {}^1/_2$, the two heat capacities $C_{\mathscr{B}}$ and $C_{\mu+\mathbf{k}T\ln j}$ are identical. Also, the configurational heat capacity of a lattice gas with ρ held fixed at ${}^1/_2$ is identical with the configurational heat capacity in the magnetic case with $\mathscr{H} = 0$. In particular, the two heat capacities will behave in the same way at $T = T_c$.

b. Binary system. From Eqs. (41.11), (41.12), (41.45), and (41.47) [or from Eqs. (41.5) and (41.34)] we can construct Table 15, which shows comparable quantities as in Table 13.

TABLE 15. EQUIVALENT THERMODYNAMIC PROPERTIES

Magnet	Lattice gas	Binary system
No. of spins, $\mathscr{B}$	$\mathscr{B}$	Total number of molecules, $N_A + N_B$
No. of $\downarrow$ or $+1$ spins, N	N	Number of A molecules, N_A
$(1 - I)/2$	$\rho = N/\mathscr{B}$	Mole fraction of A, $N_A/(N_A + N_B)$
$-2H$	$\mu + \mathbf{k}T\ln j = \mathbf{k}T\ln z$	$(\mu_A + \mathbf{k}T\ln j_A) - (\mu_B + \mathbf{k}T\ln j_B) = \mathbf{k}T\ln(z_A/z_B)$
$H + \mu_m + \mathbf{k}T\ln j_m$	$-p$	$\mu_B + \mathbf{k}T\ln j_B$

The magnetization curve (Fig. 33) and the "adsorption isotherm" (Fig. 36) become in a binary system a plot of $\mathbf{k}T\ln(z_A/z_B)$ against relative composition, as shown in Fig. 40. To obtain from Fig. 40 the separate curve of μ_B as a function of composition, we make use of Eqs. (41.37) and (41.74), and Table 15:

$$\mathscr{B}d\mu_B = -S_s\,dT - N_A d(\mu_A - \mu_B) \tag{41.86}$$

$$-\mu_B - \mathbf{k}T\ln j_B + \frac{c\varepsilon_{BB}}{2} = \mathbf{k}T\int_{-\infty}^{\ln\frac{(z_A/z_B)}{\sigma}} \frac{N_A}{\mathscr{B}}\,d\ln\frac{(z_A/z_B)}{\sigma} \qquad T\text{ const} \tag{41.87}$$

The result, shown in Fig. 41, corresponds to the equation of state in Fig. 37, plotted, however, upside down and against ρ instead of $1/\rho$.

Experimentally, the vapor pressures p_A and p_B (the vapor mixture is assumed perfect) are frequently the observed quantities. The relation between μ_B and p_B is

$$\mu_B = \mu_B{}^0(T) + \mathbf{k}T \ln p_B \tag{41.88}$$

where $\mu_B{}^0$ is a function of temperature only. From Eq. (41.38), for pure B,

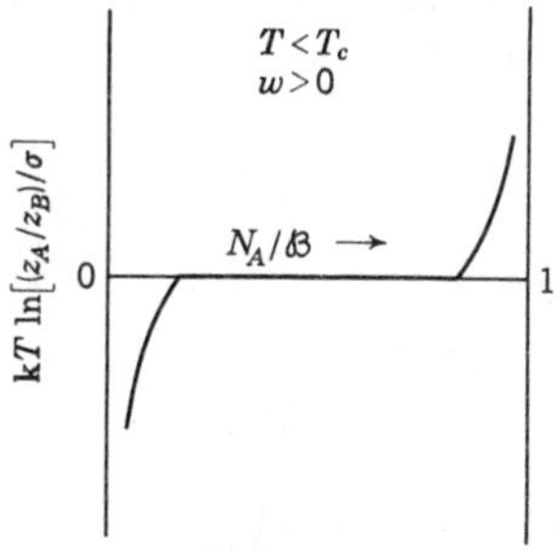

FIG. 40. Analogue of curve a of Figs. 33 and 36 for binary system.

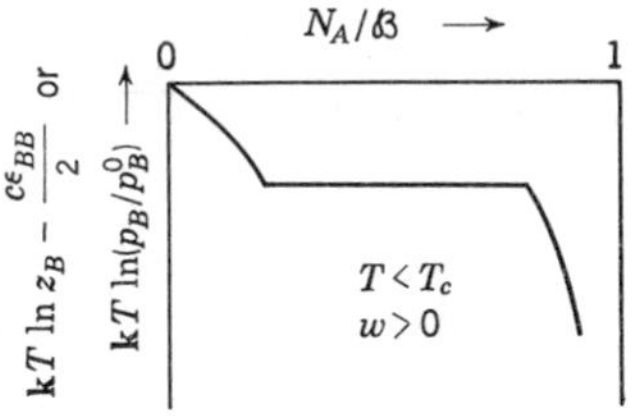

FIG. 41. Chemical potential of B versus mole fraction of A.

$$Q_s(N_B,T) = j_B{}^{N_B} e^{-cN_B\epsilon_{BB}/2\mathbf{k}T} \tag{41.89}$$

and

$$\mu_B = -\mathbf{k}T \ln j_B + \frac{c\epsilon_{BB}}{2} \tag{41.90}$$

$$= \mu_B{}^0(T) + \mathbf{k}T \ln p_B{}^0 \tag{41.91}$$

where $p_B{}^0$ is the vapor pressure of pure B at T. Then

$$\mathbf{k}T \ln z_B - \frac{c\epsilon_{BB}}{2} = \mathbf{k}T \ln \frac{p_B}{p_B{}^0} \tag{41.92}$$

as indicated in Fig. 41. If we use $p_B/p_B{}^0$ as ordinate instead of $\ln (p_B/p_B{}^0)$, Fig. 41 becomes the solid curve in Fig. 42.

The curve of μ_A versus $N_A/\mathcal{B}$ can be deduced from Figs. 40 and 41 or from Fig. 41 together with the Gibbs-Duhem equation:

$$0 = N_A \, d\mu_A + N_B \, d\mu_B \qquad T \text{ const} \tag{41.93}$$

$$-\mu_A - \mathbf{k}T \ln j_A + \frac{c\epsilon_{AA}}{2} = \mathbf{k}T \int_{-\infty}^{\ln z_B} \frac{N_B}{N_A} \, d \ln z_B \qquad T \text{ const} \tag{41.94}$$

The dotted curve in Fig. 42 is a plot of $p_A/p_A{}^0$ against $N_A/\mathscr{B}$, where

$$\mathbf{k}T \ln z_A - \frac{c\varepsilon_{AA}}{2} = \mathbf{k}T \ln \frac{p_A}{p_A{}^0} \tag{41.95}$$

The symmetry in Fig. 42 follows from Eqs. (41.38) and (41.54).

The phase separation which occurs when $T < T_c$ with $w > 0$, as in Figs. 40 to 42, involves of course two condensed phases of different relative compositions. From Figs. 37, 38, and 41 it is clear that Fig. 38 represents here the critical mixing curve of the binary solution. In the shaded area of

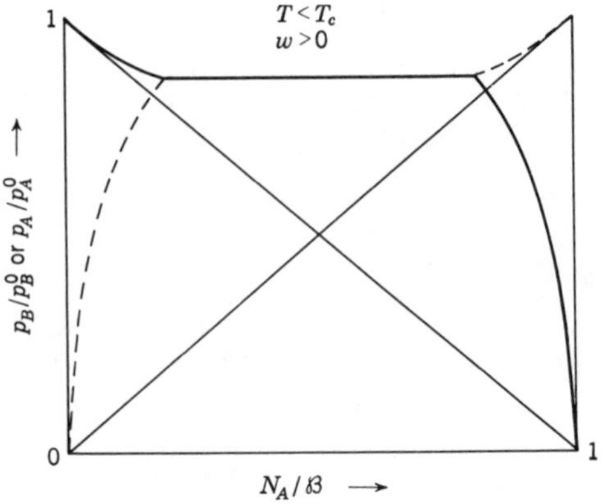

FIG. 42. Vapor pressure of B (solid curve) and A (dashed curve) versus mole fraction of A.

Fig. 38 two phases exist in equilibrium with each other; outside of the shaded area only one phase exists.

The heat capacity at constant N_A and N_B [see Eqs. (41.80) and (41.82)] is

$$C_{N_A,N_B} = \left(\frac{\partial E_s}{\partial T}\right)_{N_A,N_B} = N_A \frac{d}{dT}\left(\mathbf{k}T^2 \frac{d \ln j_A}{dT}\right) + N_B \frac{d}{dT}\left(\mathbf{k}T^2 \frac{d \ln j_B}{dT}\right) + C_1 \tag{41.96}$$

Also, from Eq. (41.83) and Table 15,

$$C_{\mathbf{k}T \ln (z_A/z_B)} = \left(\frac{\partial[E_s - N_A\mathbf{k}T \ln (z_A/z_B)]}{\partial T}\right)_{\mathbf{k}T \ln (z_A/z_B)}$$

$$= N_A \frac{d}{dT}\left(\mathbf{k}T^2 \frac{d \ln j_A}{dT}\right) + N_B \frac{d}{dT}\left(\mathbf{k}T^2 \frac{d \ln j_B}{dT}\right) + C_2 \tag{41.97}$$

where $-2H$ in Eq. (41.84) is replaced by $\mathbf{k}T \ln (z_A/z_B)$. These two heat capacities are the same in a solution with $N_A = N_B$, and the configurational part of the heat capacity in this case ($N_A = N_B$) is the same as the magnetic $\mathscr{H} = 0$ and lattice gas $\rho = 1/2$ configurational heat capacities.

The $w < 0$ case in a binary system is particularly important as this includes the well-known order-disorder transition in alloys (e.g., β brass, an alloy of copper and zinc). Our discussion above of antiferromagnetism obviously applies here. If we consider the case $N_A = N_B$ with the two lattices α and β already defined, the following statements can be made: (1) The heat capacity as a function of temperature is the same as in the case with w reversed in sign (binary solution, $N_A = N_B$, with $w > 0$). (2) Below $T = T_c$ the alloy does not split into two phases but long-range order exists, as defined in Eq. (41.71). The degree of order s in the alloy can be read immediately off the critical mixing curve (see Figs. 34 and 38) for the binary solution with w reversed in sign ($w > 0$).

42. Exact and Formal Methods

The purpose of this section is to introduce, briefly and without carrying through any detailed analysis, the exact and formal methods which have been used in one-, two-, and three-dimensional problems in lattice statistics. The results of the application of these techniques will then be discussed in Secs. 43 to 45. In two and three dimensions, approximate methods are also of considerable interest since the exact solutions are incomplete in these cases. Two of these approximations will be considered in Sec. 46, but not in great detail on account of the availability of excellent summaries by Fowler and Guggenheim (Gen. Ref.) and Guggenheim (Gen. Ref.).

Combinatorial Methods. We use Eq. (41.23) as a starting point and denote the double sum by $\mathscr{S}$:

$$\mathscr{S} = \sum_{N=0}^{\mathscr{B}} \exp\left(-\frac{2\mathscr{H}N}{\mathbf{k}T}\right) \sum_{i=1}^{\Omega} e^{-E_i'/\mathbf{k}T} \tag{42.1}$$

$$E_i' = 2JN_{AB}{}^{i} - \frac{c\mathscr{B}J}{2} \tag{42.2}$$

The first combinatorial formulation of this problem has already been employed in Sec. 41. We write $\mathscr{S}$ as[1]

$$\mathscr{S} = \exp\left(\frac{c\mathscr{B}K}{2}\right) \sum_{N=0}^{\mathscr{B}} \exp\left(-\frac{2\mathscr{H}N}{\mathbf{k}T}\right) \sum_{N_{AB}} g(N,N_{AB},\mathscr{B})e^{-2N_{AB}K} \tag{42.3}$$

where

$$K = \frac{J}{\mathbf{k}T} = \frac{w}{2\mathbf{k}T} \tag{42.4}$$

and $g(N,N_{AB},\mathscr{B})$ is the number of configurations with N_{AB} AB pairs when there are N sites of type A out of a total of $\mathscr{B}$ sites. Thus the problem of

[1] Wakefield, Rushbrooke, and Domb use a function Λ which is related to $\mathscr{S}$ by $\mathscr{S} = \Lambda^{\mathscr{B}} \exp(c\mathscr{B}K/2)$.

finding the thermodynamic properties of the Ising model is reduced to the combinatorial problem of discovering the function $g(N,N_{AB},\mathscr{B})$. This function is known in one dimension (Sec. 43), but has not been found for any two- or three-dimensional problem.

In the special case $\mathscr{H} = 0$,

$$\mathscr{S} = \exp\left(\frac{c\mathscr{B}K}{2}\right) \sum_{N_{AB}=0}^{c\mathscr{B}/2} G(N_{AB},\mathscr{B})e^{-2N_{AB}K} \tag{42.5}$$

where G is defined in Eq. (41.60). The function $G(N_{AB},\mathscr{B})$ has, in effect, been found for a simple square lattice in two dimensions (Sec. 44), as well as in one dimension.

The second combinatorial formulation is due originally to van der Waerden and is restricted to $\mathscr{H} = 0$. We follow the discussion of Newell and Montroll rather closely. We first note that E_i' in Eq. (42.2) can be written as

$$E_i' = \sum_{\text{n.n.}} - s_j s_k J \tag{42.6}$$

where the summation, for any configuration, is over all the nearest-neighbor pairs of sites, jk. It will be recalled that s_j and s_k are the "spins," $+1$ or -1. Equation (42.6) is formally equivalent, as pointed out in connection with Eq. (41.23), to an expression for the configurational energy when each $+1$, $+1$ and -1, -1 nearest-neighbor pair has an energy $-J$ and each $+1$, -1 nearest-neighbor pair has an energy $+J$. Then Eq. (42.1) becomes ($\mathscr{H} = 0$)

$$\begin{aligned}\mathscr{S} &= \sum_{N=0}^{\mathscr{B}} \sum_{i=1}^{\Omega} e^{-E_i'/\mathbf{k}T} \\ &= \sum_{s_1,\ldots,s_{\mathscr{B}}=\pm 1} \exp\left(K \sum_{\text{n.n.}} s_j s_k\right)\end{aligned} \tag{42.7}$$

That is, we sum over the $2^{\mathscr{B}}$ configurations generated by letting each spin be $+1$ or -1, and in the exponent we sum over all $c\mathscr{B}/2$ nearest-neighbor pairs for each of the $2^{\mathscr{B}}$ configurations. Now we observe that the product $s_j s_k$ must be either $+1$ or -1 and in either case

$$e^{Ks_js_k} = (1 + s_j s_k u) \cosh K \tag{42.8}$$

$$u = \tanh K \tag{42.9}$$

Thus Eq. (42.7) can be written as

$$\mathscr{S} = \sum_{s_1,\ldots,s_{\mathscr{B}}} \prod_{\text{n.n.}} e^{Ks_js_k} = (\cosh K)^{c\mathscr{B}/2} \sum_{s_1,\ldots,s_{\mathscr{B}}} \prod_{\text{n.n.}} (1 + s_j s_k u) \tag{42.10}$$

Next, the product is expanded:

$$\mathscr{S} = (\cosh K)^{c\mathscr{B}/2} \sum_{s_1,\ldots,s_{\mathscr{B}}} \left[1 + u \sum_{\text{n.n.}} s_j s_k + u^2 \Sigma (s_j s_k)(s_l s_m) + \cdots\right] \tag{42.11}$$

The coefficient of u^l is a sum of products in which each product corresponds to a different selection of l nearest-neighbor pairs jk out of a total of $c\mathscr{B}/2$. If we represent each of the l nearest-neighbor pairs by a bond, a product such as ($l = 7$)

$$(s_1s_2)(s_2s_6)(s_5s_6)(s_1s_5)(s_3s_4)(s_3s_7)(s_{10}s_{11}) \tag{42.12}$$

can be represented by a graph as in Fig. 43. Each bond (for example, s_1s_2) can occur at most once in a graph (product) such as (42.12) or Fig. 43, and each spin (for example, s_1) can occur at most c times.

When we examine the contribution of a product [such as (42.12)] to the sum over $s_1, \ldots, s_{\mathscr{B}}$ in Eq. (42.11), we observe that only those products in

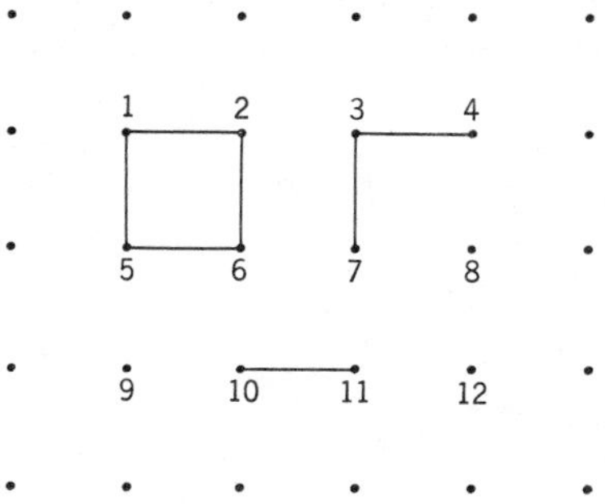

FIG. 43. Graph representing product in Eq. (42.12).

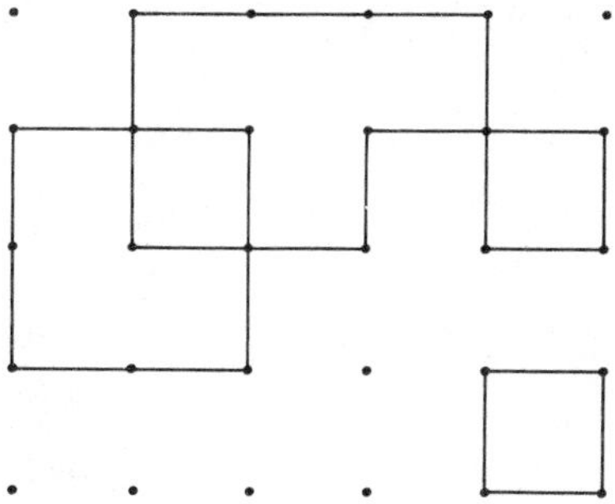

FIG. 44. Closed graph representing a product in Eq. (42.11).

which *every* spin occurs an even number of times will give a nonzero result, for

$$\begin{aligned} \sum_{s_i=\pm 1} s_i^{\,n} &= 0 \qquad n \text{ odd} \\ &= 2 \qquad n \text{ even} \end{aligned} \tag{42.13}$$

For example, the sum over s_4 in (42.12) will give zero. This means that graphs with "loose ends" (e.g., Fig. 43) contribute nothing to $\mathscr{S}$; in fact only "closed graphs" in which each point is connected by an even number of bonds (Fig. 44) are of interest.

In a closed graph (or product) each spin, say s_i, occurs an even number of times and the sum over s_i gives a factor of two [Eq. (42.13)], or a total of $2^{\mathscr{B}}$ for all spins. Hence if we let $\mathscr{G}(l,\mathscr{B})$ be the number of different closed graphs that can be drawn (on $\mathscr{B}$ lattice sites) in each of which there are a total of l bonds, Eq. (42.11) becomes

$$\mathscr{S} = 2^{\mathscr{B}}(\cosh K)^{c\mathscr{B}/2} \sum_{l=0}^{c\mathscr{B}/2} \mathscr{G}(l,\mathscr{B}) \tanh^l K \tag{42.14}$$

where $\mathcal{G}(0,\mathcal{B}) = 1$. The problem has thus been recast in such a way that the combinatorial function $\mathcal{G}(l,\mathcal{B})$ is required in order to compute thermodynamic properties.

Matrix Method.[1] Let us write Eq. (42.1) in the form

$$\mathcal{S} = \sum_{s_1,\ldots,s_{\mathcal{B}}=\pm 1} \exp\left[-\frac{(E_i' + 2\mathcal{H}N)}{\mathbf{k}T}\right] \tag{42.15}$$

Suppose the lattice consists of m layers, each with $\mathcal{B}/m$ sites. Examples are shown in Fig. 45. In one dimension, a "layer" is a single molecule; in two

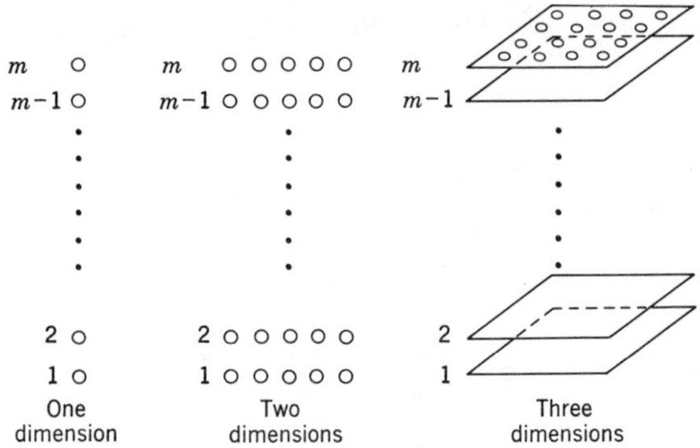

FIG. 45. One-, two-, and three-dimensional lattices consisting of m "layers" each.

dimensions, a "layer" is a row of molecules; and in three dimensions, a "layer" is a conventional layer of molecules. Eventually, of course, we shall be interested in the limit $m,\mathcal{B}/m \to \infty$ (except in one dimension where $m = \mathcal{B}$ and $\mathcal{B} \to \infty$). We now imagine the stack of m layers bent into a ring so that layers 1 and m are next to each other (just as are layers 1 and 2, etc.). This eliminates end effects but does not alter thermodynamic properties (in an infinite crystal). Each of the $\mathcal{B}/m$ spins in a layer can be $+1$ or -1 so that a layer has a total of $2^{\mathcal{B}/m}$ different configurations. Let us number these configurations from 1 to $2^{\mathcal{B}/m}$, using the same order in every layer, and let ν_k be the number of an arbitrary configuration in the kth layer ($1 \leqslant \nu_k \leqslant 2^{\mathcal{B}/m}$). The configuration of the entire lattice is specified then by a set of numbers $\nu_1, \nu_2, \ldots, \nu_m$.

We have already remarked that E_i' in Eq. (42.2) may be regarded formally as the configurational energy when each $+1$, $+1$ and -1, -1 nearest-neighbor pair has an energy $-J$, and each $+1$, -1 pair an energy $+J$.

[1] H. A. Kramers and G. H. Wannier, *Phys. Rev.*, **60**, 252 (1941); E. N. Lassettre and J. P. Howe, *J. Chem. Phys.*, **9**, 747 (1941); E. W. Montroll, *J. Chem. Phys.*, **9**, 706 (1941).

Adopting this convention, in an arbitrary configuration of the lattice with $\nu_1, \nu_2, \ldots, \nu_m$, let $E_{\nu_k\nu_{k+1}}$ be the total interaction energy of all nearest-neighbor pairs between layer k (in state ν_k) and layer $k+1$ (in state ν_{k+1}); and let E_{ν_k} be the sum of (1) the interaction energy of all nearest-neighbor pairs within layer k (in state ν_k) and (2) $2\mathscr{H}N_{\nu_k}$, where N_{ν_k} is the number of $+1$ spins in the kth layer (in the state ν_k). Then, in Eq. (42.15),

$$E_i' + 2\mathscr{H}N = \sum_{k=1}^{m} (E_{\nu_k\nu_{k+1}} + E_{\nu_k})$$

$$= \sum_{k=1}^{m} (E_{\nu_k\nu_{k+1}} + {}^1\!/_2 E_{\nu_k} + {}^1\!/_2 E_{\nu_{k+1}}) \tag{42.16}$$

where $\nu_{m+1} \equiv \nu_1$ (because the layers are in a ring) and Eq. (42.16) is written for reasons of symmetry. Equation (42.15), becomes

$$\mathscr{S} = \sum_{\nu_1, \ldots, \nu_m = 1}^{2^{\mathscr{B}/m}} \exp\left[-\frac{\sum_{k=1}^{m} (E_{\nu_k\nu_{k+1}} + {}^1\!/_2 E_{\nu_k} + {}^1\!/_2 E_{\nu_{k+1}})}{\mathbf{k}T}\right]$$

$$= \sum_{\nu_1, \ldots, \nu_m} P_{\nu_1\nu_2} P_{\nu_2\nu_3} \cdots P_{\nu_{m-1}\nu_m} P_{\nu_m\nu_1} \tag{42.17}$$

where
$$P_{\nu_k\nu_{k+1}} = e^{-(E_{\nu_k\nu_{k+1}} + {}^1/_2 E_{\nu_k} + {}^1/_2 E_{\nu_{k+1}})/\mathbf{k}T} \tag{42.18}$$

The quantities $P_{\nu_k\nu_{k+1}}$ may be considered as the elements of a symmetric matrix with $2^{\mathscr{B}/m}$ rows and columns (see Chap. 2). This point of view is suggested by Eq. (42.17), which resembles a matrix product. That is,

$$\sum_{\nu_2} P_{\nu_1\nu_2} P_{\nu_2\nu_3} = P_{\nu_1\nu_3}{}^2$$

$$\sum_{\nu_2,\nu_3} P_{\nu_1\nu_2} P_{\nu_2\nu_3} P_{\nu_3\nu_4} = \sum_{\nu_3} P_{\nu_1\nu_3}{}^2 P_{\nu_3\nu_4} = P_{\nu_1\nu_4}{}^3$$

$$\cdots\cdots\cdots\cdots$$

$$\sum_{\nu_2, \ldots, \nu_m} P_{\nu_1\nu_2} \cdots P_{\nu_{m-1}\nu_m} P_{\nu_m\nu_1} = P_{\nu_1\nu_1}{}^m$$

so that, in Eq. (42.17),

$$\mathscr{S} = \sum_{\nu_1} P_{\nu_1\nu_1}{}^m = \text{trace}\,\mathscr{P}^m = \lambda_1{}^m + \lambda_2{}^m + \cdots + \lambda_{2^{\mathscr{B}/m}}{}^m \tag{42.19}$$

where $\lambda_1, \lambda_2, \ldots$ are the eigenvalues of the matrix $\mathscr{P} = \|P_{\nu_k\nu_{k+1}}\|$. Since $\mathscr{P}$ is a symmetric matrix with real elements, the eigenvalues are all real [Eq. (8.68)].

Suppose we number the eigenvalues in such a way that $\lambda_1 > \lambda_2 > \lambda_3 > \ldots$.

Then in the thermodynamic limit $m \to \infty$, only the largest eigenvalue contributes appreciably to $\mathscr{S}$:

$$\mathscr{S} = \lambda_1{}^m \left[1 + \left(\frac{\lambda_2}{\lambda_1}\right)^m + \left(\frac{\lambda_3}{\lambda_1}\right)^m + \cdots\right]$$

$$\lim_{m\to\infty} \mathscr{S} = \lambda_1{}^m \tag{42.20}$$

If the largest eigenvalue is, say, α-fold degenerate,

$$\lim_{m\to\infty} \mathscr{S} = \alpha\lambda_1{}^m \tag{42.21}$$

But on taking the logarithm of $\alpha\lambda_1{}^m$, $\ln \alpha$ is negligible (assuming that α is finite when $m \to \infty$) compared to $m \ln \lambda_1$. Incidentally, there is a relation between degeneracy and phase separation, which is discussed by several writers.[1]

We thus see [Eq. (42.20)] that the evaluation of $\mathscr{S}$ can be transformed into the problem of finding the largest eigenvalue of a matrix. This matrix problem is easy to solve in one dimension (Sec. 43); it is very difficult in two dimensions, but a complete solution (with $\mathscr{H} = 0$) was obtained by Onsager (Sec. 44).

Series Expansions. No closed expressions for the thermodynamic properties of a three-dimensional lattice model with any $\mathscr{H}$ or of a two-dimensional model with $\mathscr{H} \neq 0$ have yet been found; therefore series expansions are especially useful in these cases. Details will be given in Secs. 44 and 45. We merely indicate here the general origin of the most important series to be discussed (since the methods are not restricted as to type of lattice or number of dimensions).

Kirkwood[2] pointed out that an exact series expansion for $\ln \mathscr{S}$ at high temperatures can be obtained as follows (when $\mathscr{H} = 0$). If the exponential in

$$\mathscr{S} = \sum_{s_1, \ldots, s_{\mathscr{B}}} e^{-E_i'/\mathbf{k}T}$$

is expanded, one obtains

$$\frac{\mathscr{S}}{2^{\mathscr{B}}} = \frac{1}{2^{\mathscr{B}}} \sum_{s_1, \ldots, s_{\mathscr{B}}} \left[1 - \frac{E_i'}{\mathbf{k}T} + \frac{E_i'^2}{2!(\mathbf{k}T)^2} - \frac{E_i'^3}{3!(\mathbf{k}T)^3} + \cdots\right]$$

$$= 1 - \frac{\overline{E'}}{\mathbf{k}T} + \frac{\overline{E'^2}}{2!(\mathbf{k}T)^2} - \frac{\overline{E'^3}}{3!(\mathbf{k}T)^3} + \cdots \tag{42.22}$$

[1] Lassetre and Howe, *loc. cit.*, J. Ashkin and W. E. Lamb, Jr., *Phys. Rev.*, **64**, 159 (1943); Onsager, *loc. cit.*; Newell and Montroll (Gen. Ref.).

[2] J. G. Kirkwood, *J. Chem. Phys.*, **6**, 70 (1938); W. Opechowski, *Physica*, **4**, 181 (1937).

where
$$\overline{E'^n} = \frac{1}{2^{\mathscr{B}}} \sum_{s_1,\ldots,s_{\mathscr{B}}} E'^n \tag{42.23}$$

This average, it should be noted, is an a priori average (not a Boltzmann average) of $E_i'^n$ over all of the $2^{\mathscr{B}}$ configurations. On taking the logarithm[1] of both sides of Eq. (42.22),

$$\ln \mathscr{S} = \mathscr{B} \ln 2 - \frac{\overline{E'}}{\mathbf{k}T} + \frac{[\overline{E'^2} - (\overline{E'})^2]}{2!(\mathbf{k}T)^2} - \cdots \tag{42.24}$$

These averages can be calculated[2] to give an expansion in powers of $K = J/\mathbf{k}T$, but we omit any details since the same expansion can be found much more easily by the next method.

A second procedure to obtain a high-temperature expansion (with $\mathscr{H} = 0$) is already contained in Eq. (42.14). That is, for small values of l one calculates $\mathscr{G}(l,\mathscr{B})$ explicitly by a consideration of the geometry of the lattice (see Secs. 44 and 45). This gives a series in powers of $u = \tanh K$. An expansion in powers of K is then readily obtained, if desired, by introducing the expansions of $\cosh K$ and $\tanh K$.

In a similar way, Eq. (42.3) can be used to obtain a low-temperature expansion in powers of $e^{-2K}(K > 0)$. We define

$$x = e^{-2K} \qquad y = \exp\left(-\frac{2\mathscr{H}}{\mathbf{k}T}\right) = \frac{e^{-2H/\mathbf{k}T}}{\sigma} \tag{42.25}$$

so that

$$\mathscr{S} \exp\left(-\frac{c\mathscr{B}K}{2}\right) = \sum_{N_{AB}=0}^{c\mathscr{B}/2} [\sum_N g(N,N_{AB},\mathscr{B})y^N]x^{N_{AB}} \tag{42.26}$$

For small values of N_{AB}, we determine the possible values of N and calculate g explicitly from the lattice structure (see Secs. 44 and 45). The result is a series in powers of x with coefficients which are functions of y.

Several variations on these methods will be introduced where needed in Secs. 44 and 45.

Zeros of the Grand Partition Function. In Chap. 5, in connection with the theory of condensation, we discussed the new formal approach of Yang and Lee. This method is applicable here, and in fact the second paper by Lee and Yang[3] is devoted in part to this application.

In the Yang-Lee method, thermodynamic properties are related to the distribution of the zeros of the grand partition function (a polynomial) in the complex plane. Lee and Yang show that in the present problem, the nature

[1] Compare Eqs. (23.41) to (23.43).

[2] Kirkwood, *loc. cit.* See also T. S. Chang, *Proc. Cambridge Phil. Soc.*, **35**, 277 (1939).

[3] Lee and Yang, *loc. cit.*

of this distribution is surprisingly simple (though it is still only known in complete detail in one dimension).

It is convenient to use the lattice gas language here since we have introduced the Yang-Lee procedure in Chap. 5 in terms of the grand partition function. Equation (41.32) can be written as

$$\Xi = \exp\left(-\frac{c\mathscr{B}\varepsilon_{BB}}{2\mathbf{k}T}\right)\sum_{N=0}^{\mathscr{B}}\left(\sum_{N_{AB}} g(N,N_{AB},\mathscr{B})e^{-wN_{AB}/\mathbf{k}T}\right)y^N \tag{42.27}$$

where

$$\sigma y = z \tag{42.28}$$

This definition of y is equivalent[1] (see Table 13) to that in Eq. (42.25). From Eqs. (41.21), (41.23), and (41.32) we note that the relation between Ξ, p, and $\mathscr{S}$ is

$$\Xi = \exp\left(\frac{p\mathscr{B}}{\mathbf{k}T}\right) = \exp\left(-\frac{c\mathscr{B}\varepsilon_{BB}}{2\mathbf{k}T}\right)\exp\left(-\frac{c\mathscr{B}J}{2\mathbf{k}T}\right)\mathscr{S} \tag{42.29}$$

The sum over N in Eq. (42.27) is a polynomial in y of degree $\mathscr{B}$. The coefficients of y^N and $y^{\mathscr{B}-N}$ are equal, in view of Eq. (41.54). Lee and Yang prove that *all the zeros of this polynomial are on the unit circle in the complex y plane if $w \geqslant 0$.* The proof requires a lengthy mathematical detour, and we omit it. One immediate consequence of this theorem of Lee and Yang is that a lattice gas can have only one phase transition (at $y = 1$, or $z = \sigma$).

As a very simple example, consider an equilateral triangle of sites ($\mathscr{B} = 3$). The polynomial in this case is

$$1 + 3x^2y + 3x^2y^2 + y^3 = (1 + y)[1 + (3x^2 - 1)y + y^2]$$
$$x = e^{-w/\mathbf{k}T}$$

If $x = 1$ ($w = 0$), all three zeros of this polynomial are at $y = -1$. If $0 \leqslant x < 1$ ($w > 0$), one zero is at $y = -1$ and the other two are complex but also on the unit circle. If $x > 1$ ($w < 0$), all zeros are on the negative real axis (one at $y = -1$).

The theorem actually proved by Lee and Yang is considerably more general than the statement above: If the interaction energy between any two molecules of a one-component lattice gas is $+\infty$ when the two molecules occupy the same lattice site and $\leqslant 0$ otherwise, then all the zeros of the grand partition function lie on a circle in the complex z plane with center at $z = 0$. In the theorem, it will be noted, no assumption is made about the range of interaction or about the dimensionality, size, structure, or periodicity of the lattice.

Lee and Yang imply that the above theorem applies to a real continuum

[1] Also, it should be observed that we have reversed the meanings of y and z relative to the Lee-Yang convention [because of our use throughout the book of z as the activity; note in Eqs. (44.24), (44.25), and (45.7) that $z \to \rho$ as $\rho \to 0$ when $\varepsilon_{AB} = \varepsilon_{BB} = 0$].

gas, such as that studied in Chap. 5, because a real continuum gas can be considered as the limit of a lattice gas if the distance between nearest neighbor sites is made to approach zero (holding the "size" of the molecules fixed). This conclusion is incorrect, however, for the range of the repulsive part of the "real" intermolecular potential will extend over many sites as the lattice becomes finer grained, and hence the condition in the theorem that the interaction energy be $\leqslant 0$ for a pair of molecules on any two different sites will be violated. Of course, it may turn out that the zeros of the grand partition function still all lie on a circle, but this has yet to be proved.

Returning now to Eq. (42.27), we have, as in Eq. (28.26),

$$\ln \Xi = -\frac{c\mathcal{B}\varepsilon_{BB}}{2\mathbf{k}T} + \sum_{k=1}^{\mathcal{B}} \ln \left(1 - \frac{y}{y_k}\right) \tag{42.30}$$

where the y_k are the zeros of the polynomial in Eq. (42.27). Also, corresponding to Eqs. (28.43) and (28.44),

$$\frac{p'}{\mathbf{k}T} = \int_0^\pi g(\theta) \ln (y^2 - 2y \cos \theta + 1)\, d\theta \tag{42.31}$$

$$\rho = 2y \int_0^\pi \frac{g(\theta)(y - \cos \theta)\, d\theta}{y^2 - 2y \cos \theta + 1} \tag{42.32}$$

where $\mathcal{B}g(\theta)\, d\theta$ is the number of zeros y_k between θ and $\theta + d\theta$ on the unit circle as $\mathcal{B} \to \infty$, and we have used $g(\theta) = g(2\pi - \theta)$.

We have thus converted our problem into one of finding the distribution of zeros of a polynomial on the unit circle.

In conclusion, we might point out the relation here between Eq. (42.31) and a Mayer-type expansion of the pressure in powers of z. Let us write

$$\frac{p'}{\mathbf{k}T} = \sum_{j\geqslant 1} \bar{b}_j z^j = \sum_{j\geqslant 1} \bar{b}_j \sigma^j y^j \tag{42.33}$$

which defines the $\bar{b}_j$. Since this is not a continuum gas, these $\bar{b}_j$ are not the b_j of Chap. 5 but they are analogous quantities.[1] We have also, as in Eq. (28.50),

$$\frac{p'}{\mathbf{k}T} = -\frac{1}{\mathcal{B}} \sum_{k=1}^{\mathcal{B}} \left[\frac{y}{y_k} + \frac{1}{2}\left(\frac{y}{y_k}\right)^2 + \frac{1}{3}\left(\frac{y}{y_k}\right)^3 + \cdots\right] \tag{42.34}$$

This expansion is valid anywhere inside the unit circle. A comparison of Eqs. (42.33) and (42.34) gives

$$\begin{aligned} \bar{b}_j &= -\frac{1}{\mathcal{B}\sigma^j j} \sum_{k=1}^{\mathcal{B}} \left(\frac{1}{y_k}\right)^j \\ &= -\frac{2}{\sigma^j j} \int_0^\pi g(\theta) \cos j\theta\, d\theta \end{aligned} \tag{42.35}$$

since $\qquad y_k^{-j} = \cos j\theta_k - i \sin j\theta_k$

[1] See the next subsection, and Secs. 44 and 45.

Thus the $\bar{b}_j$ are essentially the Fourier coefficients of $g(\theta)$, and the inverse of Eq. (42.35) is

$$g(\theta) = \frac{1}{2\pi} - \frac{1}{\pi} \sum_{j \geqslant 1} j\bar{b}_j \sigma^j \cos j\theta \tag{42.36}$$

Other Methods. Mayer's cluster theory in Chap. 5 can easily be modified to apply to a lattice gas, as has been shown by Fuchs.[1] Cluster sums replace cluster integrals. The analysis is formally very similar to that given in Chap. 5 and will be omitted, except for further brief mention in Secs. 44 and 45.

Similarly, the distribution function and integral equation methods of Chap. 6 are applicable to a lattice gas. This subject has been discussed by Murakami and Ono.[2]

43. One-dimensional Lattice

The one-dimensional problem is worth discussing for several reasons. In the first place, we can illustrate very conveniently the general methods of Sec. 42 since the analysis is relatively simple in one dimension. Second, there are one-dimensional nearest-neighbor physical problems of some interest, for example, adsorption on a linear polymer or protein chain[3] and the elastic properties of fibrous proteins.[4] Finally, a useful approximate method, the Bethe-Guggenheim or quasi-chemical method (Sec. 46), is exact in one dimension so that a one-dimensional discussion is rather illuminating as an introduction to this method.

We now turn to an application of some of the exact methods of Sec. 42 to this problem.

First Combinatorial Method. Ising[5] gave an exact solution to the one-dimensional problem in 1925 using a combinatorial argument equivalent to that to be given here.

The essential problem is to find the function $g(N,N_{AB},\mathscr{B})$ in Eq. (42.3). Consider a row of $\mathscr{B}$ sites, of which N are of type A and $\mathscr{B} - N$ of type B. The question is, how many configurations are possible such that there are N_{AB} nearest-neighbor AB pairs?

We are interested in the case with N, N_{AB}, and $\mathscr{B}$ very large numbers ($\mathscr{B} \to \infty$ holding $N/\mathscr{B}$ and $N_{AB}/\mathscr{B}$ constant). For concreteness, suppose N_{AB} is odd and that the site on the left is of type A:

$$AA|BB|AAA|B|A|BBB|AA|B$$

$$N = 8,\ N_{AB} = 7,\ \mathscr{B} = 15,\ \mathscr{B} - N = 7$$

[1] K. Fuchs, *Proc. Roy. Soc.* (*London*), **A179**, 340 (1942); **A181**, 411 (1943); G. H. Wannier, *Proc. Roy. Soc.* (*London*), **A181**, 409 (1943).

[2] T. Murakami and S. Ono, *Mem. Fac. Eng., Kyushu Univ.*, **12**, 309, 319 (1951).

[3] R. F. Steiner, *J. Chem. Phys.*, **22**, 1458 (1954).

[4] T. L. Hill, *J. Chem. Phys.*, **20**, 1259 (1952).

[5] E. Ising, *Z. Physik*, **31**, 253 (1925).

Then there are $(N_{AB} + 1)/2$ groups of A's, $(N_{AB} + 1)/2$ groups of B's, and the left-hand group is an A group, while the right-hand group is a B group. These remarks follow from the fact that an AB pair occurs at each boundary between an A group and a B group. Now consider the number of ways of arranging NA's in $(N_{AB} + 1)/2$ groups. Each A group must have at least one A site in it; thus the required number of arrangements is the number of ways of assigning the remaining $N - [(N_{AB} + 1)/2]$ A's among the $(N_{AB} + 1)/2$ groups with no restriction on the number of these A's per group. This number is[1]

$$\frac{(N-1)!}{\left(\dfrac{N_{AB}+1}{2}-1\right)!\left(N-\dfrac{N_{AB}+1}{2}\right)!}$$

The corresponding number for the B's is

$$\frac{(\mathscr{B}-N-1)!}{\left(\dfrac{N_{AB}+1}{2}-1\right)!\left(\mathscr{B}-N-\dfrac{N_{AB}+1}{2}\right)!}$$

Then g is twice the product of these two expressions (the factor of two arises because the left group could just as well be B as A). We shall actually use only $\ln g$; thus the factor of 2 is negligible. Also, unity is negligible compared to N, $\mathscr{B} - N$, and N_{AB} above; therefore we have

$$g(N,N_{AB},\mathscr{B}) = \frac{N!(\mathscr{B}-N)!}{(N-{}^1/_2N_{AB})!(\mathscr{B}-N-{}^1/_2N_{AB})![({}^1/_2N_{AB})!]^2} \tag{43.1}$$

The function $G(N_{AB},\mathscr{B})$, defined in Eq. (41.60), can for our purposes be taken as the largest value of g in the sum $\sum\limits_N g$. On setting $\partial \ln g/\partial N = 0$ (using the Stirling approximation), we find that g is a maximum when $N = \mathscr{B}/2$. Substituting $\mathscr{B}/2$ for N in the expression for $\ln g$, we obtain

$$G(N_{AB},\mathscr{B}) = g_{\max} = \frac{\mathscr{B}!}{(\mathscr{B}-N_{AB})!N_{AB}!} \tag{43.2}$$

In the special case $\mathscr{H} = 0$, we have then from Eq. (42.5),

$$\begin{aligned}\mathscr{S} &= \exp(\mathscr{B}K)\sum_{N_{AB}=0}^{\mathscr{B}}\frac{\mathscr{B}!(e^{-2K})^{N_{AB}}}{(\mathscr{B}-N_{AB})!N_{AB}!} \\ &= \exp(\mathscr{B}K)(1+e^{-2K})^{\mathscr{B}} = (2\cosh K)^{\mathscr{B}}\end{aligned} \tag{43.3}$$

[1] See J. E. Mayer and M. G. Mayer, "Statistical Mechanics" (Wiley, New York, 1940), p. 438.

The configurational heat capacity C_2 of Eq. (41.84) is seen from Eqs. (41.11) and (41.23) to be

$$C_2 = \left(\frac{\partial E_{\text{config}}}{\partial T}\right)_{\mathscr{H},\mathscr{B}} \tag{43.4}$$

where

$$E_{\text{config}} = \mathbf{k}T^2 \left(\frac{\partial \ln \mathscr{S}}{\partial T}\right)_{\mathscr{H},\mathscr{B}} \tag{43.5}$$

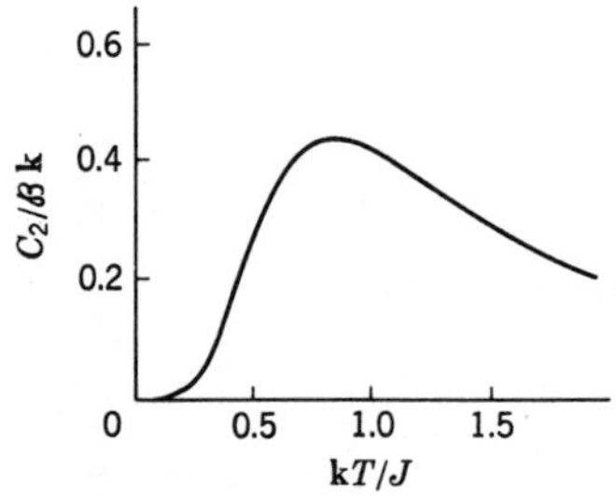

FIG. 46. Configurational heat capacity of one-dimensional Ising model at $\mathscr{H} = 0$.

Thus at $\mathscr{H} = 0$, according to Eq. (43.3),

$$E_{\text{config}} = -\mathscr{B}J \tanh \frac{J}{\mathbf{k}T} \tag{43.6}$$

and

$$\frac{C_2}{\mathscr{B}\mathbf{k}} = \left(\frac{J}{\mathbf{k}T} \operatorname{sech} \frac{J}{\mathbf{k}T}\right)^2 \tag{43.7}$$

C_2 is a smooth function of temperature (Fig. 46) so that there is no indication of a phase transition.

Returning to the more general equation (42.3), we shall use the maximum term in the double sum over N and N_{AB} to determine $\mathscr{S}$. We write

$$\ln \mathscr{S} = \mathscr{B}K + \ln \sum_{N,N_{AB}} t(N,N_{AB})$$

$$= \mathscr{B}K + \ln t_{\max} \tag{43.8}$$

where

$$\ln t = \ln g + N \ln y + N_{AB} \ln x \tag{43.9}$$

To find $\ln t_{\max}$, we set

$$\frac{\partial \ln t}{\partial N} = 0 \qquad \text{and} \qquad \frac{\partial \ln t}{\partial N_{AB}} = 0$$

and obtain

$$\frac{\rho}{1-\rho}\left(\frac{1-\rho-a}{\rho-a}\right)=\frac{1}{y} \tag{43.10}$$

$$\frac{(\rho-a)(1-\rho-a)}{a^2}=\frac{1}{x^2}=e^{2w/\mathbf{k}T} \tag{43.11}$$

where

$$a=\frac{N_{AB}}{2\mathscr{B}} \tag{43.12}$$

Equations (43.10) and (43.11) can be used to determine (see below) N and N_{AB} as functions[1] of x, y, and $\mathscr{B}$.

We note in passing that Eq. (43.11) can be written as [see Eq. (41.17)]

$$\frac{N_{AA}N_{BB}}{N_{AB}^2}=\frac{e^{2w/\mathbf{k}T}}{4} \tag{43.13}$$

This has the form of a chemical equilibrium quotient, a feature which is discussed further in Sec. 46.

Equation (43.11) is a quadratic equation in a. The solution is[2]

$$a=\frac{N_{AB}}{2\mathscr{B}}=\frac{2\rho(1-\rho)}{\beta+1} \tag{43.14}$$

where

$$\beta=[1-4\rho(1-\rho)(1-x^{-2})]^{1/2} \tag{43.15}$$

On using this expression for a, Eq. (43.10) becomes

$$y=\frac{\beta-1+2\rho}{\beta+1-2\rho} \tag{43.16}$$

This is the "adsorption isotherm."

Equations (43.14) and (43.16) can now be employed to eliminate N_{AB} and y from Eq. (43.9) to give, after cancellation of all other terms, the equation of state

$$\frac{p'}{\mathbf{k}T}=\frac{1}{\mathscr{B}}\ln t_{\max}=\ln\frac{\beta+1}{\beta+1-2\rho} \tag{43.17}$$

and

$$\frac{1}{\mathscr{B}}\ln\mathscr{S}=K+\ln\frac{\beta+1}{\beta+1-2\rho} \tag{43.18}$$

This provides us with $\mathscr{S}$ as a function of N, $\mathscr{B}$, and x (instead of the "natural" variables $\mathscr{B}$, x, and y).

[1] These are the most probable (or mean) values of N and N_{AB}, but we do not use a special notation.

[2] In solving this equation, the negative sign on the square root should be taken, as can be seen from the special case $x=1$ (random distribution).

We deduce from Eq. (43.17) that

$$\left(\frac{\partial p'/\mathbf{k}T}{\partial \rho}\right)_{\rho=1/2} = 2x > 0 \qquad \text{for } T > 0 \tag{43.19}$$

and hence that there is no first-order phase change for $T > 0$.

Our next task is to eliminate N from Eq. (43.19). From Eq. (43.10),

$$a = \frac{\rho(1-\rho)(y-1)}{\rho y - (1-\rho)} \tag{43.20}$$

This is substituted for a in Eq. (43.11), with the result[1]

$$\rho = \frac{4y + x^{-2}(y-1)^2 + (y-1)[4yx^{-2} + x^{-4}(y-1)^2]^{1/2}}{2[4y + x^{-2}(y-1)^2]} \tag{43.21}$$

or

$$I = 1 - 2\rho = \frac{e^K \sinh(\mathscr{H}/\mathbf{k}T)}{[e^{-2K} + e^{2K}\sinh^2(\mathscr{H}/\mathbf{k}T)]^{1/2}} \tag{43.22}$$

With this expression for ρ, we find that

$$\beta = \frac{e^K \cosh(\mathscr{H}/\mathbf{k}T)}{[e^{-2K} + e^{2K}\sinh^2(\mathscr{H}/\mathbf{k}T)]^{1/2}} \tag{43.23}$$

and finally, from Eq. (43.18),

$$\frac{1}{\mathscr{B}} \ln \mathscr{S} = \ln \left\{ \exp\left(-\frac{\mathscr{H}}{\mathbf{k}T}\right) \left[e^K \cosh \frac{\mathscr{H}}{\mathbf{k}T} + \left(e^{-2K} + e^{2K} \sinh^2 \frac{\mathscr{H}}{\mathbf{k}T}\right)^{1/2}\right]\right\} \tag{43.24}$$

This reduces, as it should, to Eq. (43.3) when $\mathscr{H} \to 0$.

Incidentally, we can eliminate ρ from Eq. (43.20), with the aid of Eq. (43.22), and find

$$a = \frac{e^{-2K}}{2\left[e^K \cosh \frac{\mathscr{H}}{\mathbf{k}T} + \left(e^{-2K} + e^{2K}\sinh^2 \frac{\mathscr{H}}{\mathbf{k}T}\right)^{1/2}\right]\left(e^{-2K} + e^{2K}\sinh^2\frac{\mathscr{H}}{\mathbf{k}T}\right)^{1/2}} \tag{43.25}$$

As a check, we observe that Eq. (43.22) follows from Eq. (43.24) and the thermodynamic relation

$$I = \frac{\mathbf{k}T}{\mathscr{B}} \frac{\partial \ln \Delta}{\partial \mathscr{H}} = 1 + \frac{\mathbf{k}T}{\mathscr{B}} \frac{\partial \ln \mathscr{S}}{\partial \mathscr{H}} \tag{43.26}$$

From Eq. (43.22) it is apparent that $I \to 0$ as $\mathscr{H} \to 0$ for any finite value of K, which is consistent with the absence of a first-order phase transition for

[1] The positive sign is correct on the square root, as can be seen from the special case $x = 1$.

$T > 0$. Figure 47 presents I versus $\mathscr{H}/\mathbf{k}T$ curves for several values of K. In the lattice-gas language, these are essentially adsorption isotherms. For example, in the case $K = 0$ (or $x = 1$), Eq. (43.21) becomes $\rho = y/(1 + y)$, which is the Langmuir adsorption isotherm.

It may be helpful to the reader if we point out, in conclusion, that Eqs. (43.16) and (43.18) are essentially the same as Eqs. (1012.7) (activity of a lattice gas) and (1012.15) (pressure of a lattice gas) of Fowler and Guggenheim

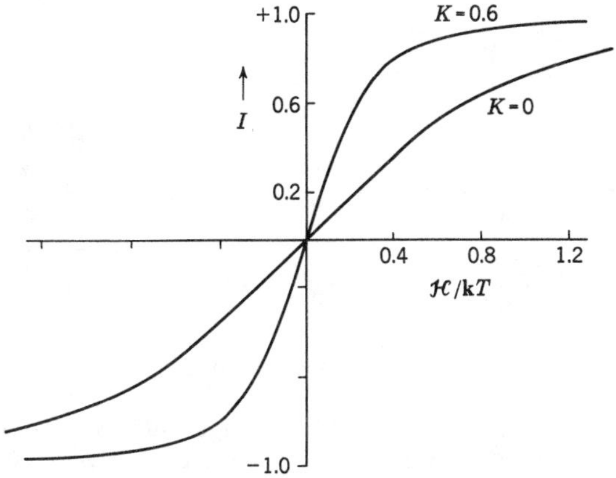

FIG. 47. Magnetization (I) versus magnetic field ($\mathscr{H}/\mathbf{k}T$) for one-dimensional Ising model.

($c = 2$), respectively, since the quasi-chemical assumption (Sec. 46) is exact in one dimension.

Second Combinatorial Method. This applies only when $\mathscr{H} = 0$. Neglecting end effects ($\mathscr{B} \to \infty$), $\mathscr{G}$ in Eq. (42.14) is unity when $l = 0$ or $l = \mathscr{B}$ and zero otherwise. Hence

$$\mathscr{S} = (2 \cosh K)^{\mathscr{B}}(1 + \tanh^{\mathscr{B}} K) \tag{43.27}$$

$$= (2 \cosh K)^{\mathscr{B}} \tag{43.28}$$

since $|\tanh K| < 1$ when $T > 0$, and $\mathscr{B} \to \infty$. This verifies Eq. (43.3).

Matrix Method. In one dimension, ν_k is just s_k, the state of a single spin, and $m = \mathscr{B}$. There are four different P's [Eq. (42.18)]:

$$P_{++} = \exp\left[-\frac{(-J + 2\mathscr{H})}{\mathbf{k}T}\right] \qquad P_{+-} = \exp\left[-\frac{(J + \mathscr{H})}{\mathbf{k}T}\right]$$

$$P_{-+} = \exp\left[-\frac{(J + \mathscr{H})}{\mathbf{k}T}\right] \qquad P_{--} = \exp\left[-\frac{(-J)}{\mathbf{k}T}\right]$$

The matrix

$$\begin{pmatrix} P_{++} & P_{+-} \\ P_{-+} & P_{--} \end{pmatrix}$$

has eigenvalues λ determined by the equation

$$\begin{vmatrix} e^K \exp\left(-\dfrac{2\mathscr{H}}{\mathbf{k}T}\right) - \lambda & e^{-K} \exp\left(-\dfrac{\mathscr{H}}{\mathbf{k}T}\right) \\ e^{-K} \exp\left(-\dfrac{\mathscr{H}}{\mathbf{k}T}\right) & e^K - \lambda \end{vmatrix} = 0 \tag{43.29}$$

We find, on solving for λ,

$$\lambda_1 = \exp\left(-\frac{\mathscr{H}}{\mathbf{k}T}\right)\left[e^K \cosh\frac{\mathscr{H}}{\mathbf{k}T} + \left(e^{-2K} + e^{2K}\sinh^2\frac{\mathscr{H}}{\mathbf{k}T}\right)^{1/2}\right] \tag{43.30}$$

$$\lambda_2 = \exp\left(-\frac{\mathscr{H}}{\mathbf{k}T}\right)\left[e^K \cosh\frac{\mathscr{H}}{\mathbf{k}T} - \left(e^{-2K} + e^{2K}\sinh^2\frac{\mathscr{H}}{\mathbf{k}T}\right)^{1/2}\right] \tag{43.31}$$

When $\mathscr{H} = 0$,

$$\mathscr{S} = \lambda_1^{\mathscr{B}} + \lambda_2^{\mathscr{B}} \tag{43.32}$$

is the same as Eq. (43.27); also, the larger eigenvalue is λ_1 and $\mathscr{S} = \lambda_1^{\mathscr{B}}$ is the same as Eq. (43.28). When $\mathscr{H}$ is not necessarily zero, $\mathscr{S} = \lambda_1^{\mathscr{B}}$ from Eq. (43.30) is identical with Eq. (43.24).

Zeros of the Grand Partition Function. It is of some interest to find the distribution of zeros, $g(\theta)$, of the grand partition function on the unit circle when $K \geqslant 0$. According to Eq. (42.29), $\Xi = 0$ when $\mathscr{S} = 0$; therefore we begin by setting $\mathscr{S} = 0$ in Eq. (43.32), with $\mathscr{B}$ finite for the present. We have then

$$\left\{\cosh\frac{\mathscr{H}}{\mathbf{k}T} + \left(e^{-4K} + \sinh^2\frac{\mathscr{H}}{\mathbf{k}T}\right)^{1/2}\right\}^{\mathscr{B}} + \left\{\cosh\frac{\mathscr{H}}{\mathbf{k}T} - \left(e^{-4K} + \sinh^2\frac{\mathscr{H}}{\mathbf{k}T}\right)^{1/2}\right\}^{\mathscr{B}} = 0 \tag{43.33}$$

We put, as before,

$$\exp\left(-\frac{2\mathscr{H}}{\mathbf{k}T}\right) = y \qquad \text{and} \qquad e^{-2K} = x$$

and then replace y by $e^{i\theta}$, since all zeros of $\mathscr{S}$ occur with y on the unit circle. Equation (43.33) becomes

$$\left\{\left(\cos^2\frac{\theta}{2}\right)^{1/2} + \left(x^2 - \sin^2\frac{\theta}{2}\right)^{1/2}\right\}^{\mathscr{B}} + \left\{\left(\cos^2\frac{\theta}{2}\right)^{1/2} - \left(x^2 - \sin^2\frac{\theta}{2}\right)^{1/2}\right\}^{\mathscr{B}} = 0 \tag{43.34}$$

Our object is to find those values of θ, in terms of x and $\mathscr{B}$, which satisfy Eq. (43.34). Because of the way $\mathscr{B}$ occurs in Eq. (43.34), it would be very convenient if the two $(\quad)^{1/2}$ terms in $\{\quad\}$ could be combined. This would be possible, for example, if we could find a function $f(x)$ and a quantity γ such that

$$\cos^2 \frac{\theta}{2} = f(x) \cos^2 \gamma \tag{43.35}$$

and

$$x^2 - \sin^2 \frac{\theta}{2} = -f(x) \sin^2 \gamma \tag{43.36}$$

for in this case the first $\{\quad\} = f^{1/2}e^{i\gamma}$ and the second $\{\quad\} = f^{1/2}e^{-i\gamma}$. This suggestion is in fact realizable, for on subtracting Eq. (43.36) from Eq. (43.35),

$$f(x) = 1 - x^2$$

Then Eq. (43.34) simplifies to

$$\{(1 - x^2)^{1/2}e^{i\gamma}\}^{\mathscr{B}} + \{(1 - x^2)^{1/2}e^{-i\gamma}\}^{\mathscr{B}} = 0$$

or

$$\cos \gamma\mathscr{B} = 0$$

The zeros of $\cos \gamma\mathscr{B}$ are

$$\gamma = \pm \frac{(2j - 1)\pi}{2\mathscr{B}} \qquad j = 1, 2, \ldots \tag{43.37}$$

Each of these values of γ, when substituted in Eq. (43.35) or in the equivalent equation

$$\cos \theta = -x^2 + (1 - x^2) \cos 2\gamma \tag{43.38}$$

gives a value of θ which satisfies Eq. (43.34). However, no new values of θ arise from the negative sign in Eq. (43.37) or from $j > (\mathscr{B} + 1)/2$; therefore we can write

$$\cos \theta = -x^2 + (1 - x^2) \cos \frac{(2j - 1)\pi}{\mathscr{B}} \qquad 1 \leqslant j \leqslant \frac{1}{2}(\mathscr{B} + 1) \tag{43.39}$$

as the equation locating the $\mathscr{B}$ zeros of the grand partition function on the unit circle.

As $\mathscr{B} \to \infty$, the distribution of zeros becomes continuous. As j runs from 1 to $\mathscr{B}/2$, $\cos \theta$ runs from $1 - 2x^2$ $(0 \leqslant x \leqslant 1)$ to -1 and θ from $\pm \cos^{-1} (1 - 2x^2)$ to $\pm \pi$. Thus there is a "horseshoe" of zeros on the unit circle with the opening along the real axis (Fig. 48). The zeros close in on the real axis only when $x = 0$ (that is, at $T = 0$). This confirms the fact that there is no phase transition in a one-dimensional ferromagnet.

The number of zeros between j and $j + dj$ [treating j as continuous in Eq. (43.39) when $\mathscr{B} \to \infty$] is $2dj$ and the number between θ and $\theta + d\theta$ is

$$\mathscr{B}g(\theta)\,d\theta = \frac{dj}{d\theta}\,d\theta \tag{43.40}$$

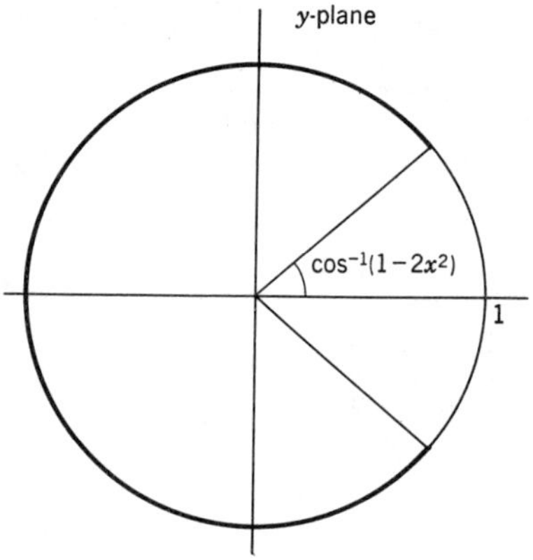

FIG. 48. Zeros of the grand partition function on the unit circle in the y plane for a one-dimensional Ising model.

We find $dj/d\theta$ from Eq. (43.39), and hence

$$\begin{aligned} g(\theta) &= \frac{1}{2\pi}\frac{\sin(\theta/2)}{[\sin^2(\theta/2) - x^2]^{1/2}} \qquad -1 \leqslant \cos\theta \leqslant 1 - 2x^2 \\ &= 0 \quad 1 \geqslant \cos\theta \geqslant 1 - 2x^2 \end{aligned} \tag{43.41}$$

When $x = 1$ ($T = \infty$, or no interaction), $g(\theta)$ is a Dirac δ function at $\theta = \pi$ ($y = -1$). As T decreases, the distribution of zeros broadens and becomes completely uniform ($g = 1/2\pi$) at $x = 0$ ($T = 0$). Figure 49 shows the form of $g(\theta)$ for an intermediate value of x, $x = 1/\sqrt{2}$.

It is easy to verify that $g(\theta)$ in Eq. (43.41) satisfies

$$\int_0^\pi g(\theta)\,d\theta = 1/2 \tag{43.42}$$

as it should. Also, let us check Eq. (42.31). We have

$$\frac{p'}{\mathbf{k}T} = \frac{1}{\mathscr{B}} \ln \Xi + \frac{\varepsilon_{BB}}{\mathbf{k}T} = -K + \frac{1}{\mathscr{B}} \ln \mathscr{S}$$

$$= \int_0^\pi g(\theta) \ln (y^2 - 2y \cos \theta + 1)\, d\theta \tag{43.43}$$

If we substitute Eq. (43.41) into Eq. (43.43) and change variables from θ to r, where

$$\cos \theta = -1 + 2(1 - x^2)r^2$$

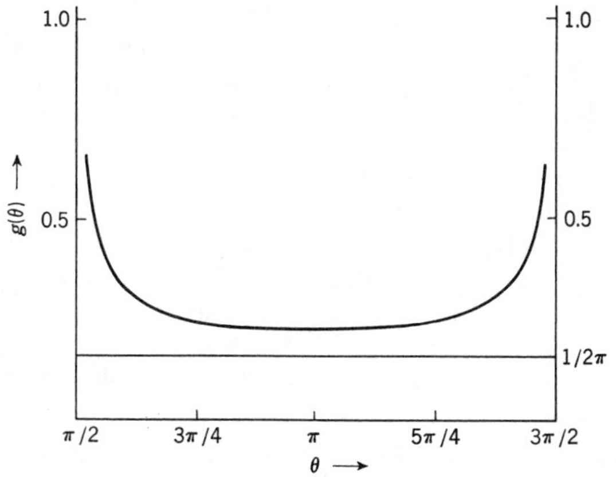

FIG. 49. Distribution of zeros of grand partition function on unit circle for $x = 1/\sqrt{2}$ (one-dimensional Ising model).

we obtain

$$-K + \frac{1}{\mathscr{B}} \ln \mathscr{S} = \frac{2}{\pi} \ln (1 + y) \int_0^1 \frac{dr}{(1 - r^2)^{1/2}} + \frac{1}{\pi} \int_0^1 \frac{\ln (1 - \tau r^2) dr}{(1 - r^2)^{1/2}} \tag{43.44}$$

with

$$\tau = \frac{4y(1 - x^2)}{(1 + y)^2} < 1$$

The second integral is[1]

$$\pi \ln \left[\frac{1 + (1 - \tau)^{1/2}}{2} \right]$$

so that

$$-K + \frac{1}{\mathscr{B}} \ln \mathscr{S} = \ln \left\{ \frac{1 + y + [(1 - y)^2 + 4yx^2]^{1/2}}{2} \right\} \tag{43.45}$$

This is easily seen to be identical with Eq. (43.24).

[1] D. Bierens de Haan, "Nouvelles tables d'intégrales définies" (P. Engels, Leyden, 1867).

44. Two-dimensional Lattice

We restrict the discussion here to a summary of the results found by Kramers and Wannier and Onsager[1] for a two-dimensional square[2] lattice, supplemented somewhat by series expansions, etc. Other two-dimensional lattices are considered by Newell and Montroll. Onsager used the matrix method and was able, after a very long and sophisticated argument, to obtain the solution to this problem in closed form, when $\mathscr{H} = 0$. Kaufman[3] succeeded later in shortening the algebra considerably and Kac and Ward[4] have recently shown that the second combinatorial method [Eq. (42.14)] can be pushed through to give the same solution.

Newell and Montroll (Gen. Ref.) give an excellent survey of the more mathematical aspects of the present problem, to which we refer the reader. We avoid complicated mathematics here by simply quoting results where necessary.

Duality Theorem and Critical Temperature. Kramers and Wannier[5] were the first to locate exactly the critical temperature in the square-lattice problem using essentially the following argument.[6] We observe in Fig. 50 that if we bisect every AB bond by a dotted line, we obtain a closed graph (of the type shown in Fig. 44) on the lattice of X's, which is called the dual lattice of the original AB lattice. The dual lattice is of the same type (square) as the original lattice. We observe further that (1) the number of bonds in the closed graph of the dual lattice is equal to the number of AB nearest-neighbor pairs in the original lattice and that (2) there is a one-to-one correspondence between closed graphs in the dual lattice and possible arrangements of AB pairs in the original lattice. In other words, the two functions $\mathscr{G}(l,\mathscr{B})$ and $G(l,\mathscr{B})$ in Eqs. (42.5) and (42.14) are identical for a square lattice; thus we have from these equations ($c = 4$)

$$\mathscr{S}(K) = 2^{\mathscr{B}}(\cosh K)^{2\mathscr{B}} \sum_{l=0}^{2\mathscr{B}} \mathscr{G}(l,\mathscr{B}) \tanh^l K \tag{44.1}$$

$$= \exp(2\mathscr{B}K) \sum_{l=0}^{2\mathscr{B}} \mathscr{G}(l,\mathscr{B}) e^{-2lK} \tag{44.2}$$

Now define $K^*(K)$ by

$$\tanh K^* = e^{-2K} \tag{44.3}$$

[1] Kramers and Wannier, *loc. cit.*; Onsager, *loc. cit.*

[2] Actually Onsager studied a "rectangular" lattice with different interactions in different directions, but we omit this generalization except in Fig. 51.

[3] B. Kaufman, *Phys. Rev.*, **76**, 1232 (1949).

[4] M. Kac and J. C. Ward, *Phys. Rev.*, **88**, 1332 (1952).

[5] Kramers and Wannier, *loc. cit.*

[6] Newell and Montroll (Gen. Ref.).

and Eq. (44.2) becomes

$$\mathscr{S}(K)\exp(-2\mathscr{B}K) = \sum_{l=0}^{2\mathscr{B}} \mathscr{G}(l,\mathscr{B}) \tanh^l K^* = \frac{\mathscr{S}(K^*)}{2^{\mathscr{B}}(\cosh K^*)^{2\mathscr{B}}} \tag{44.4}$$

Equation (44.4) relates $\mathscr{S}$ at any temperature $T_1 = J/\mathbf{k}K$ to $\mathscr{S}$ at a second temperature $T_2 = J/\mathbf{k}K^*$, where $K^*(K)$ is given by Eq. (44.3). Thus if there is a singularity in $\mathscr{S}$ at T_1, there must also be a singularity in $\mathscr{S}$ at T_2. If we assume that a singularity in $\mathscr{S}$ occurs at only one temperature (as

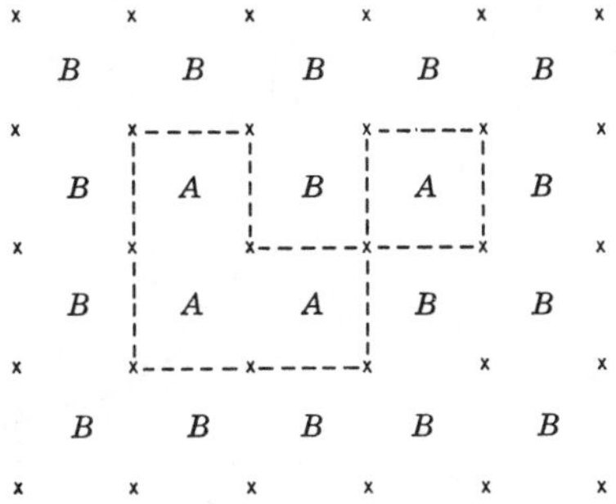

FIG. 50. Closed graph in the dual lattice (X) of the AB lattice.

shown by Lee and Yang when $J > 0$; see Sec. 42), then $T_1 = T_2$, $K = K^*$, and

$$\tanh K = e^{-2K} \tag{44.5}$$

This equation determines the critical temperature as

$$K_c = \frac{J}{\mathbf{k}T_c} = 0.4407 \cdots = {}^1/_2 \sinh^{-1} 1$$
$$x_c = e^{-2K_c} = \sqrt{2} - 1 = 0.4142 \cdots \tag{44.6}$$

This result was later confirmed by Onsager in his more complete investigation.

Partition Function and Heat Capacity. For $\mathscr{S}$ in an infinite crystal with $\mathscr{H} = 0$, Onsager obtained from Eq. (42.20)

$$\frac{1}{\mathscr{B}} \ln \mathscr{S} = \ln(2\cosh 2K) + \frac{1}{2\pi}\int_0^{\pi} \ln\left\{\frac{1}{2}[1 + (1 - k_1^2 \sin^2\varphi)^{1/2}]\right\} d\varphi \tag{44.7}$$

where
$$k_1 = \frac{2\sinh 2K}{\cosh^2 2K} = \frac{4x(1-x^2)}{(1+x^2)^2}$$

For the pressure of the lattice gas, we have then, from Eq. (42.29),

$$\frac{p'}{\mathbf{k}T} = \ln(1 + x^2) + \frac{1}{2\pi}\int_0^\pi \ln\left\{\frac{1}{2}[1 + (1 - k_1^2 \sin^2\varphi)^{1/2}]\right\} d\varphi \tag{44.8}$$

This equation provides the pressure as a function of temperature (x) at constant density ($\rho = 1/2$). With $J > 0$, the system splits into two phases below $T = T_c$ (see below), and $p'(x)$ is then the vapor pressure of the condensed phase as a function of temperature.

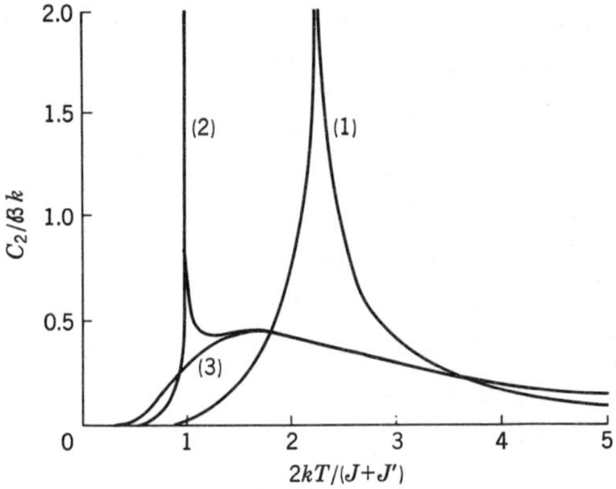

FIG. 51. The heat capacity of two-dimensional Ising model at $\mathcal{H} = 0$. (1) $J'/J = 1$; (2) $J'/J = 1/100$; (3) $J'/J = 0$.

Differentiation of Eq. (44.7) with respect to temperature gives the configurational energy E_{config} of Eq. (43.5). One finds after some cancellation

$$E_{\text{config}} = -\mathcal{B}J \coth 2K \left[1 + \frac{2}{\pi}(2\tanh^2 2K - 1)K_1(k_1)\right] \tag{44.9}$$

where $K_1(k_1)$ is the complete elliptic integral of the first kind

$$K_1(k_1) = \int_0^{\pi/2} (1 - k_1^2 \sin^2\varphi)^{-1/2}\, d\varphi \tag{44.10}$$

At $T = T_c$, $k_1 = 1$. On expanding the integrand in Eq. (44.10) about $T = T_c$ and $\varphi = \pi/2$, we find that $K_1(k_1)$ behaves as $\ln|T - T_c|$ near $T = T_c$. But the coefficient of $K_1(k_1)$ in Eq. (44.9) is easily seen to be linear in $T - T_c$ near the critical point so that E_{config} is continuous and equal to $-\mathcal{B}J \coth 2K_c$ at $T = T_c$.

On further differentiation of Eq. (44.9) to obtain the configurational heat capacity C_2 [see Eq. (43.4)], we find from the term $(T - T_c)\ln|T - T_c|$ just mentioned that C_2 is proportional to $\ln|T - T_c|$ near $T = T_c$. Thus the heat capacity has a logarithmic singularity at the critical point. The heat capacity is shown in Fig. 51. Included in the figure is the heat-capacity curve obtained by Onsager in the more complicated case: (horizontal interaction, J')/(vertical interaction, J) = 1/100. The symmetrical square lattice has $J'/J = 1$ and the one-dimensional lattice [Eq. (43.7)] has $J'/J = 0$.

It should be recalled from Sec. 41 that the heat-capacity curve of Fig. 51 ($J'/J = 1$) refers to a ferromagnet with $\mathscr{H} = 0$ or $I = 0$, to a lattice gas or adsorbed phase at fixed density $\rho = {}^1/_2$, to a binary solution with $N_A = N_B$,

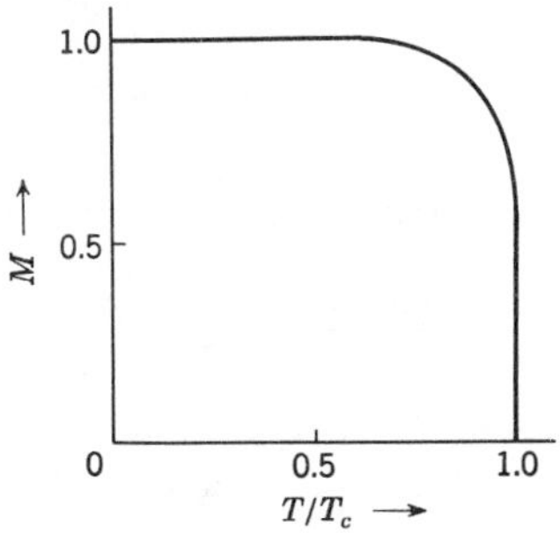

FIG. 52. Spontaneous magnetization (M) of two-dimensional (square) Ising model versus temperature.

or to an alloy (with an order-disorder transition) of composition $N_A = N_B$—all on a square lattice in two dimensions.

Spontaneous Magnetization. The spontaneous magnetization M [Eq. (41.57)] has been found by Onsager and Kaufman[1] and by Yang[2] to be

$$M = (1 - \sinh^{-4} 2K)^{1/8} \tag{44.11}$$

This can also be written as

$$M = \left[\frac{(1 + x^2)(1 - 6x^2 + x^4)^{1/2}}{(1 - x^2)^2}\right]^{1/4} \tag{44.12}$$

Figure 52 shows M versus T/T_c.

As pointed out in Sec. 41, the spontaneous magnetization curve is also essentially the phase boundary curve in a lattice gas or in a binary solution. In the lattice gas, $\rho^0 = (1 - M)/2$ is the density of the gas phase and $1 - \rho^0$

[1] Reported by Onsager in *Nuovo cimento*, **6** (suppl. 2), 261 (1949), without proof.

[2] C. N. Yang, *Phys. Rev.*, **85**, 808 (1952).

is the density of the liquid phase in the two-phase equilibrium (see Fig. 38), where

$$\rho^0(x) = \frac{1}{2} - \frac{1}{2}\left[\frac{(1+x^2)(1-6x^2+x^4)^{1/2}}{(1-x^2)^2}\right]^{1/4} \tag{44.13}$$

The two-phase boundary is shown in Fig. 53.

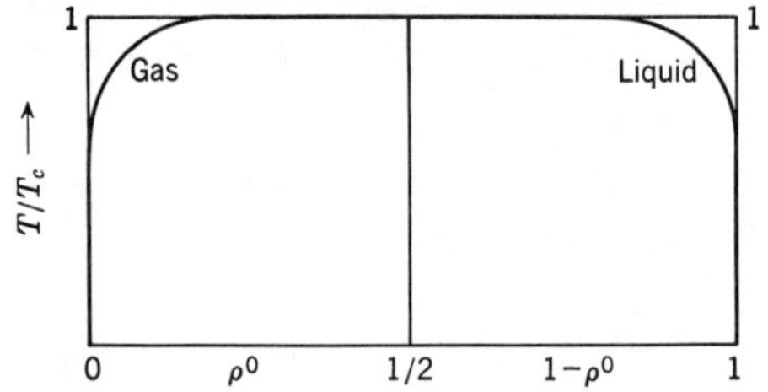

FIG. 53. Two-phase boundary as a function of temperature for two-dimensional (square) lattice gas.

The two-phase boundary can also be presented as a function of p' by using Eq. (44.8) to eliminate x. The result is shown in Fig. 54.

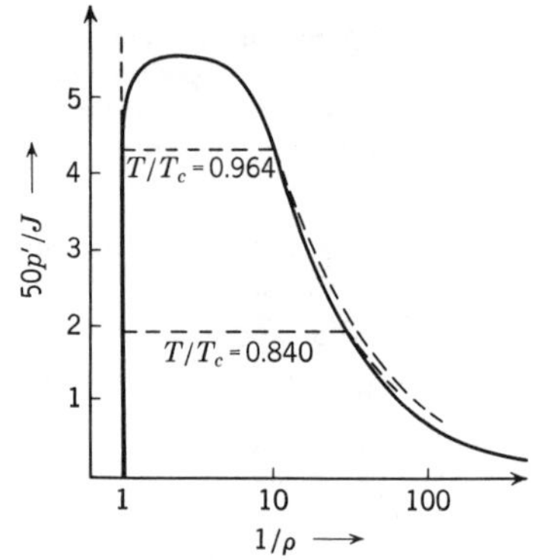

FIG. 54. Two-phase boundary as a function of pressure for two-dimensional (square) lattice gas. Two liquid-gas pressure-volume isotherms are included.

By expanding [ ] in Eq. (44.12) about $x_c = \sqrt{2} - 1$, we find near $T = T_c$

$$M = \rho_L - \rho_G = [4(\sqrt{2}+2)(x_c - x)]^{1/8} = \left[8\sqrt{2}\,\frac{J}{\mathbf{k}T_c^2}(T_c - T)\right]^{1/8} \tag{44.14}$$

$$\rho_G \equiv \rho^0 \qquad \rho_L \equiv 1 - \rho^0$$

Hence in Fig. 53, $T_c - T$ is proportional to $(\rho_L - \rho_G)^8$ near $T = T_c$. Real three-dimensional gas-liquid systems, obeying the law of corresponding states, are found empirically[1] to obey $T_c - T = \text{constant} \times (\rho_L - \rho_G)^3$. In Fig. 54, $p - p_c$ is also proportional to $(\rho_L - \rho_G)^8$ near $p = p_c$, for

$$p - p_c = \left(\frac{\partial p}{\partial T}\right)_c (T - T_c) \tag{44.15}$$

where, from Eqs. (42.29) and (43.5),

$$\left(\frac{\partial p}{\partial T}\right)_c = \frac{1}{\mathscr{B}}\left(\frac{E_{\text{config}}}{T.}\right)_c + \mathbf{k}\left(\frac{1}{\mathscr{B}}\ln \mathscr{S}\right)_c \tag{44.16}$$

This quantity is continuous at $T = T_c$.

Figure 52 for M is also a plot of the degree of order s in an antiferromagnet or in an alloy with an order-disorder transition. Further, M^2 is the long-range correlation in the ferromagnet.

Cluster Expansion of Pressure. It was mentioned at the end of Sec. 42 that Fuchs[2] has extended Mayer's cluster method to the lattice problem. We shall use the alternative but equivalent method of Eqs. (23.41) to (23.43), to avoid any new analysis, to obtain the first few terms in the expansion of the pressure in powers of z. In this discussion, we are of course no longer restricted to $\mathscr{H} = 0$ and $\rho = {}^1/_2$.

We have, from Eq. (42.27),

$$\begin{aligned}\frac{p'}{\mathbf{k}T} &= \frac{1}{\mathscr{B}} \ln\left[\Xi \exp\frac{c\mathscr{B}\varepsilon_{BB}}{2\mathbf{k}T}\right] \\ &= \frac{1}{\mathscr{B}} \ln \sum_{N=0}^{\mathscr{B}} \frac{z^N \bar{Z}_N}{N!}\end{aligned} \tag{44.17}$$

where

$$\bar{Z}_N = N!\sigma^{-N} \sum_{N_{AB}} g(N, N_{AB}, \mathscr{B}) x^{N_{AB}} \tag{44.18}$$

Also, as in Eq. (42.33), we write

$$\frac{p'}{\mathbf{k}T} = \sum_{j \geqslant 1} \bar{b}_j z^j \tag{44.19}$$

Then, just as in Eq. (23.43), we obtain

$$\begin{aligned}1!\mathscr{B}\bar{b}_1 &= \bar{Z}_1 \\ 2!\mathscr{B}\bar{b}_2 &= \bar{Z}_2 - \bar{Z}_1^2 \\ 3!\mathscr{B}\bar{b}_3 &= \bar{Z}_3 - 3\bar{Z}_1\bar{Z}_2 + 2\bar{Z}_1^3 \\ 4!\mathscr{B}\bar{b}_4 &= \bar{Z}_4 - 4\bar{Z}_1\bar{Z}_3 - 3\bar{Z}_2^2 + 12\bar{Z}_1^2\bar{Z}_2 - 6\bar{Z}_1^4 \\ &\text{etc.}\end{aligned} \tag{44.20}$$

[1] E. A. Guggenheim, *J. Chem. Phys.*, **13**, 253 (1945).

[2] Fuchs, *loc. cit.* See also Lee and Yang, *loc. cit.*, and Domb, *loc. cit.*

To find $\bar{Z}_1$ we put $N = 1$ in Eq. (44.18). The only possible value of N_{AB} is c, and $g(1,c,\mathscr{B}) = \mathscr{B}$, since the single molecule can be on any one of the $\mathscr{B}$ sites. Then

$$\bar{Z}_1 = \mathscr{B}\sigma^{-1}x^c \qquad \bar{b}_1 = \sigma^{-1}x^c \tag{44.21}$$

When we put $N = 2$ in Eq. (44.18), we can have $N_{AB} = 2c$ (two molecules not on nearest-neighbor sites) or $N_{AB} = 2c - 2$ (two molecules on nearest-neighbor sites). In the former case, $g(2,2c,\mathscr{B}) = \mathscr{B}(\mathscr{B} - c - 1)/2$ since the first molecule can be on any of $\mathscr{B}$ sites, but $c + 1$ sites about the first molecule are always excluded to the second. In the latter case, $g(2,2c - 2,\mathscr{B}) = \mathscr{B}c/2$ since the second molecule must be on a site nearest neighbor to the first molecule. Thus

$$\bar{Z}_2 = \mathscr{B}\sigma^{-2}[(\mathscr{B} - c - 1)x^{2c} + cx^{2c-2}]$$

$$= \mathscr{B}(\sigma^{-1}x^c)^2(\mathscr{B} - c - 1 + cx^{-2})$$

$$\bar{b}_2 = (\sigma^{-1}x^c)^2 \frac{(cx^{-2} - c - 1)}{2} \tag{44.22}$$

so that

$$\frac{p'}{\mathbf{k}T} = \frac{x^c z}{\sigma} + \left(\frac{cx^{-2} - c - 1}{2}\right)\left(\frac{x^c z}{\sigma}\right)^2 + \cdots \tag{44.23}$$

Each lattice type must be considered separately to go beyond $\bar{b}_2$ and the counting of course gets more complicated. In the special case[1] of a square lattice

$$\frac{p'(z)}{\mathbf{k}T} = \left(\frac{x^4 z}{\sigma}\right) + \left(2x^{-2} - \frac{5}{2}\right)\left(\frac{x^4 z}{\sigma}\right)^2 + \left(6x^{-4} - 16x^{-2} + \frac{31}{3}\right)\left(\frac{x^4 z}{\sigma}\right)^3$$

$$+ \left(x^{-8} + 18x^{-6} - 85x^{-4} + 118x^{-2} - \frac{209}{4}\right)\left(\frac{x^4 z}{\sigma}\right)^4 + \cdots \qquad z < \sigma \text{ if } x < 1 \tag{44.24}$$

When $x < 1$, all the zeros of the grand partition function are on the circle $|y| = 1$ in the complex y plane or on $|z| = \sigma$ in the complex z plane. Hence Eq. (44.24) converges for all $z < \sigma$; a first-order phase transition occurs at $z = \sigma$. Equation (44.24) has therefore been used to calculate the dotted p-v curves in Fig. 54 in the gas region, by combining it with

$$\rho = z\frac{\partial}{\partial z}\left(\frac{p}{\mathbf{k}T}\right) \tag{44.25}$$

[1] Lee and Yang, *loc. cit.* See also the next subsection on series.

Lee and Yang give equations for the special case $\varepsilon_{AB} = \varepsilon_{BB} = 0$ so that $p' = p$, $x = \exp(\varepsilon_{AA}/2\mathbf{k}T)$, and $\sigma = x^4$.

Equation (44.24) can also be used, after suitable modification, in the liquid region, because of the symmetry property in Eq. (41.77). Equations (41.77) and (44.24) give

$$\frac{p'(z)}{\mathbf{k}T} = \ln\frac{z}{\sigma} + x^8\left(\frac{x^4z}{\sigma}\right)^{-1} + \left(2x^{14} - \frac{5}{2}x^{16}\right)\left(\frac{x^4z}{\sigma}\right)^{-2}$$
$$+ \left(6x^{20} - 16x^{22} + \frac{31}{3}x^{24}\right)\left(\frac{x^4z}{\sigma}\right)^{-3}$$
$$+ \left(x^{24} + 18x^{26} - 85x^{28} + 118x^{30} - \frac{209}{4}x^{32}\right)\left(\frac{x^4z}{\sigma}\right)^{-4} + \cdots$$
$$z > \sigma \text{ if } x < 1 \qquad (44.26)$$

This series obviously converges for $z > \sigma$ if $x < 1$. The liquid isotherms in Fig. 54 follow from Eqs. (44.25) and (44.26).

Low-temperature Series Expansions. We consider first the ferromagnetic case $x < 1$, using Eq. (42.26) as the starting point. When $N_{AB} = 0$, $N = 0$ or $N = \mathscr{B}$, and in either case $g = 1$. When $N_{AB} = 4$, $N = 1$ or $N = \mathscr{B} - 1$, and in either case $g = \mathscr{B}$. When $N_{AB} = 6$, $N = 2$ (two nearest-neighbor sites occupied) or $N = \mathscr{B} - 2$, and in either case $g = 2\mathscr{B}$ [see the discussion above Eq. (44.22)]. For $N_{AB} = 8$, there are four possibilities:

○ ↔ ○ (separate)	○○○	○○ ○	○○ ○○
$N = 2, \mathscr{B} - 2$	$N = 3, \mathscr{B} - 3$	$N = 3, \mathscr{B} - 3$	$N = 4, \mathscr{B} - 4$
$g = {}^1/_2\mathscr{B}(\mathscr{B} - 5)$	$g = 2\mathscr{B}$	$g = 4\mathscr{B}$	$g = \mathscr{B}$

Thus, we have so far

$$\mathscr{S} \exp(-2\mathscr{B}K) = (1 + y^{\mathscr{B}}) + \mathscr{B}(y + y^{\mathscr{B}-1})x^4 + 2\mathscr{B}(y^2 + y^{\mathscr{B}-2})x^6$$
$$+ [{}^1/_2\mathscr{B}(\mathscr{B} - 5)(y^2 + y^{\mathscr{B}-2}) + 6\mathscr{B}(y^3 + y^{\mathscr{B}-3})$$
$$+ \mathscr{B}(y^4 + y^{\mathscr{B}-4})]x^8 + \cdots \qquad (44.27)$$

Now suppose $y < 1$. Then $y^{\mathscr{B}}$, $y^{\mathscr{B}-1}$, etc., are negligible and

$$\mathscr{S} \exp(-2\mathscr{B}K) = 1 + \mathscr{B}yx^4 + 2\mathscr{B}y^2x^6 + [{}^1/_2\mathscr{B}(\mathscr{B} - 5)y^2$$
$$+ 6\mathscr{B}y^3 + By^4]x^8 + \cdots \qquad y < 1 \qquad (44.28)$$

On extracting the $\mathscr{B}$th root using the binomial theorem, we obtain easily the terms through x^8 in the following series (the higher terms having been found by Domb[1]):

$$\mathscr{S}^{1/\mathscr{B}}e^{-2K} = \Lambda(y,x) = 1 + yx^4 + 2y^2x^6 + (-2y^2 + 6y^3 + y^4)x^8$$
$$+ (-14y^3 + 18y^4 + 8y^5 + 2y^6)x^{10}$$
$$+ (8y^3 - 77y^4 + 44y^5 + 40y^6 + 22y^7 + 6y^8 + y^9)x^{12}$$
$$+ (98y^4 - 370y^5 + 40y^6 + 138y^7 + 134y^8 + 72y^9 + 30y^{10}$$
$$+ 8y^{11} + 2y^{12})x^{14} + (-40y^4 + \cdots)x^{16} + \cdots \qquad y < 1 \quad (44.29)$$

where in general $\Lambda(y,x)$ is defined by[2]

$$\Lambda(y,x) = \mathscr{S}^{1/\mathscr{B}}e^{-cK/2} = e^{p'/\mathbf{k}T} \tag{44.30}$$

To obtain $\Lambda(1,x)$ it is legitimate to set $y = 1$ in Eq. (44.29), for according to Eq. (44.27) this errs only by omitting the inconsequential factor $2^{1/\mathscr{B}}$:

$$\Lambda(1,x) = 1 + x^4 + 2x^6 + 5x^8 + 14x^{10} + 44x^{12} + 152x^{14} + \cdots \tag{44.31}$$

or

$$\ln \Lambda(1,x) = x^4 + 2x^6 + {}^9\!/_2x^8 + 12x^{10} + {}^{112}\!/_3x^{12} + 130x^{14} + \cdots \tag{44.32}$$

which is the expansion of Eq. (44.8).

When $y > 1$, $\Lambda(y,x)$ may be found directly from Eq. (44.27) by neglecting 1, y, y^2, etc., compared to $y^{\mathscr{B}}$, $y^{\mathscr{B}-1}$, etc., or from the symmetry property, Eq. (41.77), which may be written, in view of Eq. (44.30), as

$$\Lambda(y,x) = y\Lambda\left(\frac{1}{y}, x\right) \tag{44.33}$$

From Eqs. (44.29) and (44.33),

$$\Lambda(y,x) = y[1 + y^{-1}x^4 + 2y^{-2}x^6 + (-2y^{-2} + 6y^{-3} + y^{-4})x^8 + \cdots]$$
$$y > 1 \qquad (44.34)$$

From Eqs. (41.57) and (43.26), the spontaneous magnetization M is

$$M = 1 - 2\left[\frac{y}{\Lambda}\frac{\partial \Lambda}{\partial y}\right]_{y\to 1-} \tag{44.35}$$

[1] Domb, *loc. cit.* Domb used a matrix method to obtain his series.

[2] See footnote to Eq. (42.3), and Eq. (42.29). The function $\Lambda(y,x)$ is of course not to be confused with the constant Λ of Chaps. 5 and 6.

Using Eqs. (44.29) and (44.31), we find

$$M = 1 - 2x^4 - 8x^6 - 34x^8 - 152x^{10} - 714x^{12} - 3472x^{14} - \cdots \quad (44.36)$$

This is the expansion of Eq. (44.12).

Let us rearrange the terms in Eq. (44.29) to give an expansion[1] in powers of y:

$$\Lambda(y,x) = e^{p'/kT} = 1 + x^4y + (2x^6 - 2x^8)y^2 + (6x^8 - 14x^{10} + 8x^{12})y^3 + (x^8 + 18x^{10} - 77x^{12} + 98x^{14} - 40x^{16})y^4 + \cdots \qquad y < 1 \quad (44.37)$$

Incidentally, the coefficient of each power of y above the first is zero when $x = 1$ because $\Lambda(y,1) = 1 + y$. This follows from Eqs. (41.52) and (42.26):

$$\Lambda(y,1)^{\mathscr{B}} = \sum_N [\sum_{N_{AB}} g(N,N_{AB},\mathscr{B})]y^N = \sum_{N=0}^{\mathscr{B}} \frac{\mathscr{B}!y^N}{N!(\mathscr{B}-N)!} = (1+y)^{\mathscr{B}} \quad (44.38)$$

Now if we take the logarithm of both sides of Eq. (44.37), we again find Eq. (44.24). Similarly, if we rearrange the terms in Eq. (44.34) to give an expansion in powers of y^{-1}, and then take the logarithm, the result is Eq. (44.26).

We turn now to the antiferromagnetic case ($x > 1$) at low temperatures. Since the stable state[2] at $T = 0$ is the one with A and B sites alternating, we count configurations for values of N_{AB} near $N_{AB} = 2\mathscr{B}$ instead of near $N_{AB} = 0$, as in Eq. (44.27). Otherwise the procedure is the same. For example, for $N_{AB} = 2\mathscr{B}$, the only possible value of N is $\mathscr{B}/2$ and $g = 2$ (there are two ways of filling the $\mathscr{B}$ sites with alternate A's and B's). Next, we can obtain $N_{AB} = 2\mathscr{B} - 4$ with either $N = (\mathscr{B}/2) - 1$ or $N = (\mathscr{B}/2) + 1$ (by removing an A or a B from a "perfect" configuration and replacing it by a B or an A, respectively), and in both cases $g = 2(\mathscr{B}/2)$. The next smaller value of N_{AB} is $2\mathscr{B} - 6$, which arises with $N = \mathscr{B}/2$ by exchanging an A and a B which are nearest neighbors in a "perfect" configuration. The value of g is $2(2\mathscr{B})$. Thus we have

$$\mathscr{S} \exp(-2\mathscr{B}K) = \Lambda(y,x)^{\mathscr{B}} = 2y^{\mathscr{B}/2}x^{2\mathscr{B}} + \mathscr{B}(y^{\frac{\mathscr{B}}{2}-1} + y^{\frac{\mathscr{B}}{2}+1})x^{2\mathscr{B}-4} + 4\mathscr{B}y^{\mathscr{B}/2}x^{2\mathscr{B}-6} + \cdots$$

$$= 2y^{\mathscr{B}/2}x^{2\mathscr{B}}\left[1 + \frac{\mathscr{B}}{2}(y^{-1} + y)x^{-4} + 2\mathscr{B}x^{-6} + \cdots\right] \quad (44.39)$$

[1] Domb gives this series through the term in y^8.

[2] Provided $|\mathscr{H}|$ is not too large: $-cJ > |\mathscr{H}|$. For large $|\mathscr{H}|$ the ferromagnetic expansions are still useful; see Wakefield, *loc. cit.*

This process can be continued. On extracting the $\mathscr{B}$th root (we omit the factor $2^{1/\mathscr{B}}$),

$$\begin{aligned}\Lambda(y,x) = y^{1/2}x^2\{1 &+ \tfrac{1}{2}(y^{-1} + y)x^{-4} + 2x^{-6} \\ &+ [-\tfrac{1}{8}(y^{-2} + y^2) + 3(y^{-1} + y) - \tfrac{3}{4}]x^{-8} \\ &+ [2(y^{-2} + y^2) - 3(y^{-1} + y) + 16]x^{-10} \\ &+ [\tfrac{9}{16}(y^{-3} + y^3) - \tfrac{9}{2}(y^{-2} + y^2) + \tfrac{591}{16}(y^{-1} + y) - 22]x^{-12} \\ &+ [-3(y^{-3} + y^3) + \tfrac{191}{4}(y^{-2} + y^2) - 73(y^{-1} + y) + \tfrac{417}{2}]x^{-14} \\ &+ \cdots\} \end{aligned} \tag{44.40}$$

This expansion is obviously equally valid for $y > 1$ and $y < 1$; Eq. (44.33) does not lead to anything new. When $y = 1$,

$$\Lambda(1,x) = x^2(1 + x^{-4} + 2x^{-6} + 5x^{-8} + 14x^{-10} + 44x^{-12} + 152x^{-14} + \cdots) \tag{44.41}$$

These coefficients are the same as in Eq. (44.31). This is a consequence of the symmetry property, Eq. (41.63). In present notation, using Eq. (41.62) and writing $N_{AB}^* = \dfrac{c\mathscr{B}}{2} - N_{AB}$,

$$\begin{aligned}\Lambda(1,x)^{\mathscr{B}} &= \sum_{N_{AB}} G(N_{AB},\mathscr{B})e^{-2N_{AB}K} \\ &= \sum_{N_{AB}^*} G(N_{AB}^*,\mathscr{B}) \exp\left[2K\left(N_{AB}^* - \frac{c\mathscr{B}}{2}\right)\right]\end{aligned}$$

or

$$\Lambda(1,x) = x^{c/2}\Lambda\left(1,\frac{1}{x}\right) \tag{44.42}$$

We find easily from Eq. (44.40) that

$$\left(\frac{y}{\Lambda}\frac{\partial\Lambda}{\partial y}\right)_{y\to 1-} = \frac{1}{2}$$

Therefore $M = 0$ as expected in an antiferromagnet.

High-temperature Series Expansions. As $T \to \infty$, $x \to 1$ for both the ferromagnet ($x < 1$) and the antiferromagnet ($x > 1$), so it is possible to give a single discussion valid in both instances.

Consider first the special case $y = 1$ ($\mathscr{H} = 0$). The straightforward procedure here is to count closed graphs and use Eq. (42.14). However, with a square lattice, this high-temperature expansion follows immediately from the low-temperature expansion in Eq. (44.31), because $\mathscr{G}(l,\mathscr{B}) = G(l,\mathscr{B})$, as in Eqs. (44.1) and (44.2). Explicitly, since

$$x = e^{-2K} \qquad \text{and} \qquad \tanh K = u = \frac{1-x}{1+x}$$

we have

$$\Lambda(1,x) = \frac{(1+x)^2}{2} \Lambda(1,u) \tag{44.43}$$

Hence, from Eq. (44.31), the high-temperature expansion is

$$\Lambda(1,x) = \frac{(1+x)^2}{2} (1 + u^4 + 2u^6 + 5u^8 + 14u^{10} + 44u^{12} + 152u^{14} + \cdots) \tag{44.44}$$

or

$$\mathscr{S}^{1/\mathscr{B}} = 2 (\cosh^2 K)(1 + u^4 + 2u^6 + 5u^8 + \cdots) \tag{44.45}$$

If we expand $\cosh K$ and u in powers of K, Eq. (44.45) becomes

$$\mathscr{S}^{1/\mathscr{B}} = 2 \left(1 + K^2 + \frac{4}{3} K^4 + \frac{77}{45} K^6 + \frac{1009}{315} K^8 + \cdots\right) \tag{44.46}$$

or

$$\frac{1}{\mathscr{B}} \ln \mathscr{S} = \ln 2 + K^2 + \frac{5}{6} K^4 + \frac{32}{45} K^6 + \frac{425}{252} K^8 + \cdots \tag{44.47}$$

Equation (44.47) is in the form of the Kirkwood-Opechowski expansion, Eq. (42.24).

We turn now to a method due to Domb[1] which leads to a high-temperature expansion when $y \neq 1$. We introduce the variable $t = 1 - x^2$, and seek an expansion in powers of t. If we set $x^2 = 1 - t$ in Eq. (44.29), the coefficient of t is seen to be

$$-2y + 2y^2 - 2y^3 + 2y^4 - \cdots$$

which is apparently $-2y/(1 + y)$. Similarly, the coefficient of t^2 is

$$y - 6y^2 + 16y^3 - 31y^4 + \cdots$$

which appears to be $(y - 3y^2 + y^3)/(1 + y)^3$. Domb therefore surmised that $\Lambda(y,x)$ could be expressed in the form

$$\Lambda(y,x) = 1 + y + \sum_{r \geqslant 1} \frac{\varphi_r(y) t^r}{(1+y)^{2r-1}} \tag{44.48}$$

where φ_r is a polynomial of degree not exceeding $2r$. With this assumption and noting a symmetry property in the φ_r's [see Eq. (44.49)], Domb was able to find the φ_r's through $r = 9$. The first few of these are:

$$\begin{aligned} \varphi_1 &= -2y \\ \varphi_2 &= y - 3y^2 + y^3 \\ \varphi_3 &= 6y^2 - 14y^3 + 6y^4 \\ \varphi_4 &= -2y^2 + 42y^3 - 84y^4 + 42y^5 - 2y^6 \end{aligned} \tag{44.49}$$

[1] Domb, *loc. cit.*

Equations (44.48) and (44.49) can be checked by putting $y = 1$, replacing t by $1 - \left(\frac{1-x}{1+x}\right)^2$, multiplying the resulting series by $(1 + x)^2/2$, and noting that Eq. (44.31) is then recovered, as required by Eq. (44.43). Equations (44.48) and (44.49) are valid for both $y \leqslant 1$ and $y \geqslant 1$; the symmetry property, Eq. (44.33), gives nothing new.

Other Methods. The distribution, $g(\theta)$, of zeros of the grand partition function is unknown here, although some of the properties of this distribution are available[1] from Onsager's results.

The combinatorial function $G(N_{AB},\mathscr{B})^{1/\mathscr{B}}$ can, on the other hand, be computed numerically from Onsager's equation for $\mathscr{S}$, Eq. (44.7). This has been

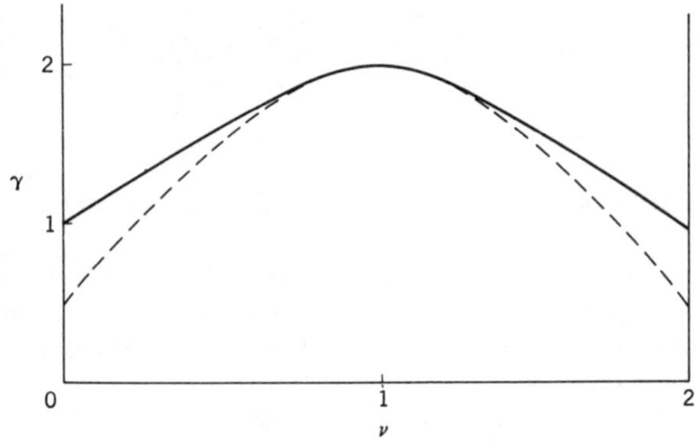

FIG. 55. Combinatorial factor $\gamma(\nu)$ for two-dimensional (square) Ising model. Solid curve, exact; dashed curve, quasi-chemical approximation (Sec. 46).

done by Prigogine, Mathot-Sarolea, and van Hove.[2] In the notation of these authors, let

$$\nu = \frac{N_{AB}}{\mathscr{B}} \qquad \gamma(\nu) = G(N_{AB},\mathscr{B})^{1/\mathscr{B}}$$

$$\lambda = \mathscr{S}^{1/\mathscr{B}}$$

We replace the sum in Eq. (42.5) by its maximum term, and this equation becomes

$$\lambda(K) = e^{2K}\gamma(\nu(K))e^{-2\nu(K)K}, \tag{44.50}$$

where $\nu(K)$ is determined by the relation

$$\frac{d \ln \gamma(\nu)}{d\nu} = 2K \tag{44.51}$$

[1] Lee and Yang, *loc. cit.*

[2] I. Prigogine, L. Mathot-Sarolea, and L. van Hove, *Trans. Faraday Soc.*, **48**, 485 (1952).

Differentiating Eq. (44.50),

$$\frac{d \ln \lambda}{dK} = 2 + \frac{d \ln \gamma}{d\nu}\frac{d\nu}{dK} - 2\nu - 2K\frac{d\nu}{dK} = 2(1-\nu) \qquad (44.52)$$

Equation (44.52) is then substituted in Eq. (44.50) with the result

$$\gamma(K) = \lambda(K) \exp\left(-K\frac{d \ln \lambda}{dK}\right) \qquad (44.53)$$

Equation (44.7) gives expressions for $\lambda(K)$ and $d \ln \lambda/dK$. Hence from Eqs. (44.52) and (44.53), we can compute $\nu(K)$ and $\gamma(K)$, and finally $\gamma(\nu)$. The function $\gamma(\nu)$ is plotted in Fig. 55. The minimum values are

$$G(0,\mathscr{B})^{1/\mathscr{B}} = G(2\mathscr{B},\mathscr{B})^{1/\mathscr{B}} = 1^{1/\mathscr{B}} = 1$$

and the maximum value is

$$G(\mathscr{B},\mathscr{B})^{1/\mathscr{B}} = [\sum_{N_{AB}} G(N_{AB},\mathscr{B})]^{1/\mathscr{B}} = (2^{\mathscr{B}})^{1/\mathscr{B}} = 2$$

since the sum can be replaced by its maximum term, $G(\mathscr{B},\mathscr{B})$.

45. Three-dimensional Lattice

We limit ourselves here to the simple cubic lattice as an illustration.[1] The only exact results available are in the form of series expansions.[2] We follow closely the discussion above of series for the square lattice, and hence omit most details.

Low-temperature Series Expansions. In place of Eq. (44.29), we have[3]

$$\begin{aligned}\mathscr{S}^{1/\mathscr{B}}e^{-3K} = \Lambda(y,x) &= 1 + yx^6 + 3y^2x^{10} - 3y^2x^{12} + 15y^3x^{14} \\ &+ (-33y^3 + 3y^4)x^{16} + (18y^3 + 83y^4)x^{18} \\ &+ (-309y^4 + 48y^5)x^{20} + (360y^4 + 429y^5 + 18y^6)x^{22} \\ &+ (-137y^4 + \cdots)x^{24} + \cdots \qquad y < 1 \qquad (45.1)\end{aligned}$$

On putting $y = 1$ in Eq. (45.1),

$$\begin{aligned}\Lambda(1,x) = 1 + x^6 + 3x^{10} - 3x^{12} + 15x^{14} - 30x^{16} + 101x^{18} \\ - 261x^{20} + 807x^{22} + \cdots \qquad (45.2)\end{aligned}$$

and $$\begin{aligned}\ln \Lambda(1,x) = x^6 + 3x^{10} - {}^7/_2\, x^{12} + 15x^{14} - 33x^{16} + {}^{313}/_3\, x^{18} \\ - {}^{561}/_2\, x^{20} + 849x^{22} + \cdots \qquad (45.3)\end{aligned}$$

From Eqs. (44.33) and (45.1), we obtain, when $y > 1$,

$$\Lambda(y,x) = y[1 + y^{-1}x^6 + 3y^{-2}x^{10} - 3y^{-2}x^{12} + \cdots] \qquad y > 1 \qquad (45.4)$$

[1] For other lattices, see Newell and Montroll (Gen. Ref.) and Rushbrooke (Gen. Ref., 1952).

[2] See, for example, Wakefield, *loc. cit.*, and Rushbrooke (Gen. Ref., 1952).

[3] Wakefield gives terms through x^{28}.

Equations (44.35) and (45.1) give for the spontaneous magnetization[1] M,

$$M = 1 - 2x^6 - 12x^{10} + 14x^{12} - 90x^{14} + 192x^{16} - 792x^{18} + 2148x^{20} - 7716x^{22} + \cdots \quad (45.5)$$

The terms in Eq. (45.1) can be rearranged to give an expansion[2] in powers of y:

$$\Lambda(y,x) = e^{p'/\mathbf{k}T} = 1 + x^6 y + (3x^{10} - 3x^{12})y^2 + (15x^{14} - 33x^{16} + 18x^{18})y^3 + (3x^{16} + 83x^{18} - 309x^{20} + 360x^{22} - 137x^{24})y^4 + \cdots \qquad y < 1 \quad (45.6)$$

or

$$\ln \Lambda = \frac{p'}{\mathbf{k}T} = \frac{x^6 z}{\sigma} + \left(3x^{-2} - \frac{7}{2}\right)\left(\frac{x^6 z}{\sigma}\right)^2 + \left(15x^{-4} - \frac{69}{2}x^{-2} + \frac{119}{6}\right)\left(\frac{x^6 z}{\sigma}\right)^3 + \left(3x^{-8} + 83x^{-6} - 321x^{-4} + \frac{775}{2}x^{-2} - \frac{611}{4}\right)\left(\frac{x^6 z}{\sigma}\right)^4 + \cdots \qquad z < \sigma \text{ if } x < 1 \quad (45.7)$$

This is the "cluster" expansion of the equation of state of the lattice gas, valid to the condensation point $z = \sigma$. In the liquid region, from Eqs. (41.77) and (45.7), we have

$$\frac{p'(z)}{\mathbf{k}T} = \ln\frac{z}{\sigma} + x^{12}\left(\frac{x^6 z}{\sigma}\right)^{-1} + \left(3x^{22} - \frac{7}{2}x^{24}\right)\left(\frac{x^6 z}{\sigma}\right)^{-2} + \left(15x^{32} - \frac{69}{2}x^{34} + \frac{119}{6}x^{36}\right)\left(\frac{x^6 z}{\sigma}\right)^{-3} + \left(3x^{40} + 83x^{42} - 321x^{44} + \frac{775}{2}x^{46} - \frac{611}{4}x^{48}\right)\left(\frac{x^6 z}{\sigma}\right)^{-4} + \cdots \qquad z > \sigma \text{ if } x < 1 \quad (45.8)$$

In the special case $\varepsilon_{AB} = \varepsilon_{BB} = 0$: $p' = p$, $x = \exp(\varepsilon_{AA}/2\mathbf{k}T)$, $\sigma = x^6$.

For the antiferromagnet ($x > 1$) at low temperatures, corresponding to Eq. (44.40), we obtain[3]

$$\Lambda(y,x) = y^{1/2}x^3\{1 + {}^1/_2(y^{-1} + y)x^{-6} + 3x^{-10} - {}^1/_8(y^{-2} + 22 + y^2)x^{-12} + {}^{15}/_2(y^{-1} + y)x^{-14} - {}^1/_2(33y^{-1} - 6 + 33y)x^{-16} + \cdots\} \quad (45.9)$$

and, when $y = 1$,

$$\Lambda(1,x) = x^3(1 + x^{-6} + 3x^{-10} - 3x^{-12} + 15x^{-14} - 30x^{-16} + \cdots) \quad (45.10)$$

in agreement with Eqs. (44.42) and (45.2).

[1] See Rushbrooke (Gen. Ref., 1952) for terms to x^{28}.
[2] Wakefield gives terms through y^7.
[3] Wakefield gives terms through x^{-28}.

High-temperature Series Expansions. In the special case $y = 1$ ($\mathscr{H} = 0$), we must count closed graphs and use Eq. (42.14), for there is no duality theorem or symmetry property available such as Eq. (44.43). When $l = 0$ in Eq. (42.14), $\mathscr{G} = 1$. When $l = 4$, $\mathscr{G} = 12\mathscr{B}/4 = 3\mathscr{B}$, since each lattice site belongs to 12 squares of 4 bonds, but each square is counted four times if we include every lattice site. When $l = 6$, there are three kinds of closed graph, as shown in Fig. 56. Thus

$$\mathscr{S} = 2^{\mathscr{B}}(\cosh K)^{3\mathscr{B}}(1 + 3\mathscr{B}u^4 + 22\mathscr{B}u^6 + \cdots)$$

On taking the $\mathscr{B}$th root and including further terms,[1]

$$\mathscr{S}^{1/\mathscr{B}} = 2\,(\cosh^3 K)(1 + 3u^4 + 22u^6 + 192u^8 + 2070u^{10} + 24{,}943u^{12} + \cdots) \qquad (45.11)$$

or

$$\Lambda(1,x) = \frac{(1+x)^3}{4}(1 + 3u^4 + 22u^6 + \cdots) \qquad (45.12)$$

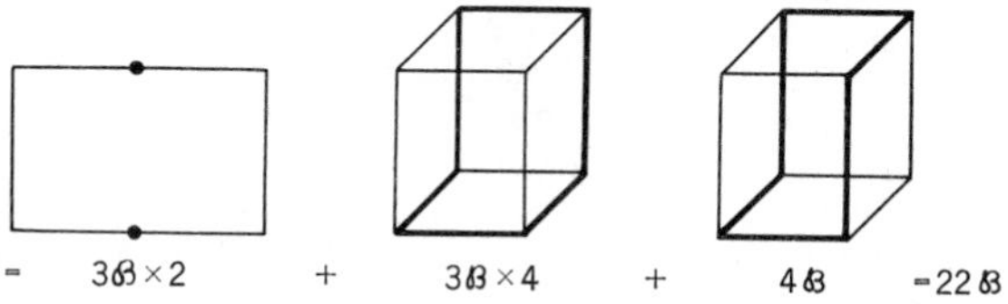

FIG. 56. Closed graphs in the simple cubic lattice.

If we expand cosh K and u in powers of K, and take logarithms, Eq. (45.11) becomes[2]

$$\frac{1}{\mathscr{B}}\ln \mathscr{S} = \ln 2 + \frac{3}{2}K^2 + \frac{11}{4}K^4 + \frac{271}{15}K^6 + \cdots \qquad (45.13)$$

This is in the Kirkwood-Opechowski form, Eq. (42.24).

Domb's equation, Eq. (44.48), for a high-temperature expansion when $y \neq 1$ was also employed by Wakefield for the simple cubic lattice. Wakefield gives the φ_r's through $r = 7$; the first few are

$$\begin{aligned}
\varphi_1 &= -3y \\
\varphi_2 &= 3y - 6y^2 + 3y^3 \\
\varphi_3 &= -y + 25y^2 - 49y^3 + 25y^4 - y^5 \\
\varphi_4 &= -30y^2 + 273y^3 - 486y^4 + 273y^5 - 30y^6
\end{aligned} \qquad (45.14)$$

[1] See Wakefield, *loc. cit.*, and M. Kurata, R. Kikuchi, and T. Watari, *J. Chem. Phys.*, **21**, 434 (1953).

[2] Wakefield gives terms through K^{12}.

Wakefield derived, in addition, an alternative expansion, for $y \neq 1$, in powers of u rather than t. We write Eq. (44.48) in the form

$$\Lambda(y,x) = \frac{(1+x)^3}{4} \cdot \frac{(1+u)^3}{2} \left[1 + y + \sum_{r \geqslant 1} \frac{\varphi_r(y) t^r}{(1+y)^{2r-1}} \right]$$

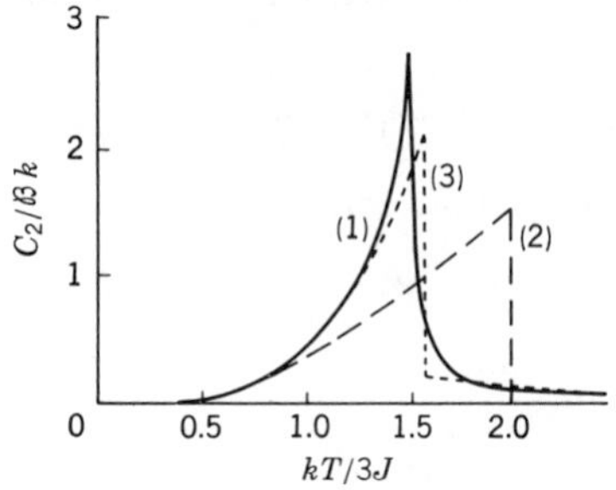

FIG. 57. Heat capacity of simple cubic Ising model at $\mathscr{H} = 0$ according to (1) Wakefield, (2) the Bragg-Williams approximation, and (3) the quasi-chemical approximation.

and then replace t by $4u/(1+u)^2$. Collecting coefficients of the various powers of u, we find[1]

$$\Lambda(y,x) = \frac{(1+x)^3}{4} \cdot \frac{(1+y)}{2} \sum_{r \geqslant 0} \frac{\psi_r(y) u^r}{(1+y)^{2r}} \tag{45.15}$$

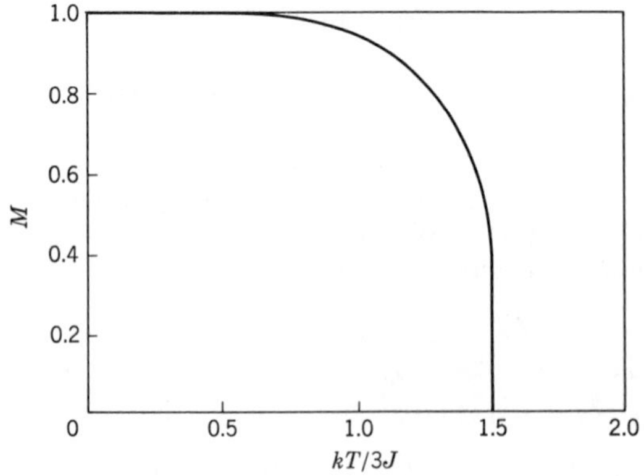

FIG. 58. Spontaneous magnetization of simple cubic Ising model according to Wakefield.

[1] Wakefield gives ψ_r through $r = 7$.

$$
\begin{aligned}
\psi_0 &= 1 \\
\psi_1 &= 3 - 6y + 3y^2 \\
\psi_2 &= 3 + 48y - 102y^2 + 48y^3 + 3y^4 \\
\psi_3 &= 1 - 106y + 1615y^2 - 3020y^3 + 1615y^4 - 106y^5 + y^6
\end{aligned}
\tag{45.16}
$$

Thermodynamic Properties. Wakefield[1] has estimated the remainders in the series necessary to compute thermodynamic functions for the simple cubic lattice. The thermodynamic properties that he deduced using these remainders are not exact, but they are certainly the best available for any three-dimensional lattice. He finds a critical point at $x_c = 0.641$; a configurational energy continuous at x_c; and a configurational heat capacity with a logarithmic singularity at x_c. In Fig. 57 the heat capacity is plotted, and in Fig. 58 the spontaneous magnetization. The break in the magnetization curve at $M = 0.41$ is presumably due to the numerical approximations introduced in the calculations.

46. Approximate Methods

Our aim in this chapter has been to outline the present status of exact treatments of nearest-neighbor lattice problems. Other books, particularly those of Fowler and Guggenheim and of Guggenheim, discuss approximate methods in great detail. It would seem appropriate for purposes of orientation, however, to conclude this chapter with a very brief outline of the two most useful approximate methods, the Bragg-Williams approximation and the Bethe-Guggenheim (or quasi-chemical) approximation. The more refined methods[2] of Kramers and Wannier and of Kikuchi have been discussed recently by Kurata, Kikuchi, and Watari.[3]

Bragg-Williams Approximation. In Eq. (42.26),

$$
\mathcal{S} \exp\left(-\frac{c\mathcal{B}K}{2}\right) = \sum_N [\sum_{N_{AB}} g(N,N_{AB},\mathcal{B})x^{N_{AB}}]y^N \tag{46.1}
$$

we make the approximation of collecting together all configurations with the same N, and use a single average N_{AB} for this set of configurations:

$$
\sum_{N_{AB}} g(N,N_{AB},\mathcal{B})x^{N_{AB}} \cong \frac{\mathcal{B}!x^{\bar{N}_{AB}}}{N!(\mathcal{B}-N)!} \tag{46.2}
$$

Furthermore, we choose for the value of $\bar{N}_{AB}$ the average number of AB pairs in a completely *random* distribution of N sites of type A on the total of

[1] Wakefield, *loc. cit.*

[2] H. A. Kramers and G. H. Wannier, *Phys. Rev.*, **60**, 263 (1941); R. Kikuchi, *Phys. Rev.*, **81**, 988 (1951); *J. Chem. Phys.*, **19**, 1230 (1951).

[3] Kurata, Kikuchi, and Watari, *loc. cit.* See also J. Hijmans and J. DeBoer, *Physica*, **21**, 471, 485, 499 (1955).

$\mathscr{B}$ sites.[1] The average number of B sites next to an A site is, on this basis, $c(\mathscr{B} - N)/\mathscr{B}$, so that $\bar{N}_{AB} = cN(\mathscr{B} - \hat{N})/\mathscr{B}$. Hence, in the Bragg-Williams approximation,

$$\mathscr{S} \exp\left(-\frac{c\mathscr{B}K}{2}\right) = \sum_{N=0}^{\mathscr{B}} \frac{\mathscr{B}!}{N!(\mathscr{B} - N)!} y^N x^{cN(\mathscr{B}-N)/\mathscr{B}} \tag{46.3}$$

$$= \sum_{N=0}^{\mathscr{B}} \frac{\mathscr{B}!}{N!(\mathscr{B} - N)!} \left(\frac{zx^c}{\sigma}\right)^N x^{-cN^2/\mathscr{B}} \tag{46.4}$$

In the special case $\varepsilon_{AB} = \varepsilon_{BB} = 0$, $\sigma = x^c$.

The summation in Eq. (46.4) is difficult; therefore we replace the sum by its maximum term.[2] Let $t(N)$ be the summand in Eq. (46.4). We find easily from $\partial \ln t/\partial N = 0$ that N_{max} or $\bar{N}$ is given by

$$\frac{zx^c}{\sigma} = \frac{\rho}{1-\rho} x^{2c\rho} \qquad \rho = \frac{\bar{N}}{\mathscr{B}} \tag{46.5}$$

and hence that

$$\frac{1}{\mathscr{B}} \ln \mathscr{S} - \frac{cK}{2} = \frac{1}{\mathscr{B}} \ln t_{\text{max}} = \frac{p'}{\mathbf{k}T} = \ln \frac{x^{c\rho^2}}{1-\rho} \tag{46.6}$$

Equations (46.5) and (46.6) give essentially the chemical potential (or adsorption isotherm) and pressure of the lattice gas as functions of ρ and T. It is easy to see that Eq. (46.5) has the symmetry of Fig. 36; thus if a critical point exists, it is at $\rho = 1/2$ and $z = \sigma$.

From Eq. (46.6), we obtain

$$\frac{\partial(p'/\mathbf{k}T)}{\partial \rho} = 2c\rho \ln x + \frac{1}{1-\rho} \tag{46.7}$$

When $w/\mathbf{k}T > 2/c$, $\partial(p'/\mathbf{k}T)/\partial\rho$ is negative at $\rho = 1/2$; hence there is a loop in the equation of state and the lattice gas has a first-order phase transition. On setting the derivative in Eq. (46.7) equal to zero, we find that the critical point is located at

$$\rho_c = {}^1/_2$$

$$x_c = e^{-2/c} \qquad \frac{w}{\mathbf{k}T_c} = \frac{2}{c} \tag{46.8}$$

$$z_c = \sigma \qquad y_c = 1$$

We obtain a loop instead of a flat region in the equation of state because[3] of our use of the "maximum" term only in Eq. (46.4). The complete sum

[1] This will make the present approximation rigorous in the limit $T \to \infty$. See below.

[2] This is equivalent to using the canonical ensemble in the first place. See Sec. 20.

[3] See Sec. 28, T. L. Hill, *J. Phys. Chem.*, **57**, 324 (1953), and Appendix 9.

would not give a loop. Actually, the "maximum" consists of two maxima and a minimum in the region of the loop.

If we put $\rho = {}^1/_2$ in Eq. (46.6), we find

$$\frac{1}{\mathscr{B}} \ln \mathscr{S} = \ln 2$$

This result is the same as that (for $\mathscr{H} = 0$) in Eqs. (44.47) and (45.13) in the limit $K \to 0$ ($T \to \infty$).

The polynomial in y in Eq. (46.3) is included in the Lee-Yang theorem on zeros of the grand partition function. When $x < 1$, all the zeros of this polynomial are on the unit circle in the complex y plane. The distribution of zeros $g(\theta)$ on the unit circle has been found recently by Katsura.[1]

In conclusion, we consider the order-disorder problem, $x > 1$ and $\rho = {}^1/_2$. According to Eq. (46.5), $y = z/\sigma = 1$ at $\rho = {}^1/_2$. When we put $y = 1$ and $x > 1$ in Eq. (46.3), the summand has a single maximum at $N = \mathscr{B}/2$. Hence we can use Eq. (46.6) without concern about phase separation. Equations (43.5) and (46.6) then give E_{config} and C_2 identically zero (equal numbers of like and unlike nearest-neighbor pairs at all temperatures).

On the other hand, we have included the well-known Bragg-Williams heat-capacity curve[2] for the order-disorder problem in Fig. 57, and this heat capacity is not zero everywhere. The reason for the apparent discrepancy is that there are two different "Bragg-Williams" (and quasi-chemical) treatments of this problem, which have to be distinguished. One is the theory we have outlined above. The second method is slightly more refined. Instead of collecting together *all* configurations with the same N ($N = \mathscr{B}/2$ here), as in Eq. (46.2), we introduce a long-range order parameter[3] s and collect the $h(s)$ configurations with the same N and s:

$$\sum_{N_{AB}} g\left(\frac{\mathscr{B}}{2}, N_{AB}, \mathscr{B}\right) x^{N_{AB}} \cong \sum_s h(s) x^{\overline{N_{AB}(s)}} \tag{46.9}$$

$$\sum_s h(s) = \frac{\mathscr{B}!}{[(\mathscr{B}/2)!]^2} \tag{46.10}$$

A separate random distribution (Bragg-Williams approximation) of A and B sites is then assumed for configurations with each value of s, in computing $\overline{N_{AB}(s)}$. The sum over s in Eq. (46.9) is replaced by its maximum term, for computational purposes. The value of s giving the maximum term is called the "equilibrium" value of s. At high temperatures, the equilibrium value

[1] S. Katsura, *J. Chem. Phys.*, **22**, 1277 (1954).

[2] Fowler and Guggenheim (Gen. Ref.).

[3] For any configuration, let N_W be the number of "wrong" states, as in Eq. (41.67); this limits the discussion to lattices divisible into α and β sites. Then for this configuration, s is defined as $s = 1 - (2N_W/\mathscr{B})$; see Eq. (41.71).

of s turns out to be $s = 0$, and Eq. (46.9) reduces to the simpler treatment, Eq. (46.2); hence the heat capacity is zero at high temperatures. But the inclusion of the parameter s improves the theory at low temperatures, and leads to the unsymmetrical heat-capacity curve of Fig. 57. For further details, the reader is referred to Fowler and Guggenheim, and Guggenheim.

It is to be emphasized that the rigorous theory of Secs. 42 to 45 puts high and low temperatures on an equal footing, and leads to relatively symmetrical heat-capacity curves in the neighborhood of T_c (Figs. 51 and 57).

Quasi-chemical (Bethe-Guggenheim) Approximation. We use the following argument to make a reasonable guess at $g(N,N_{AB},\mathscr{B})$. For given values of N, N_{AB}, and $\mathscr{B}$, we have [Eq. (41.17)]:

$$\text{Number of } AA \text{ pairs} = N_{AA} = \frac{cN}{2} - \frac{N_{AB}}{2}$$

$$\text{Number of } BB \text{ pairs} = N_{BB} = \frac{c(\mathscr{B} - N)}{2} - \frac{N_{AB}}{2}$$

$$\text{Number of } AB \text{ pairs} = \frac{N_{AB}}{2}$$

$$\text{Number of } BA \text{ pairs} = \frac{N_{AB}}{2}$$

$$\text{Total number of pairs} = \frac{c\mathscr{B}}{2}$$

Now we assume that these pairs are independent of each other and that the desired number of configurations $g(N,N_{AB},\mathscr{B})$ is equal, apart from a normalization constant, to the number of ways, ω, of dividing up the $c\mathscr{B}/2$ pairs. That is,

$$\omega = \frac{\left(\frac{c\mathscr{B}}{2}\right)!}{\left(\frac{cN}{2} - \frac{N_{AB}}{2}\right)! \left[\frac{c(\mathscr{B} - N)}{2} - \frac{N_{AB}}{2}\right]! \left(\frac{N_{AB}}{2}!\right)^2} \tag{46.11}$$

$$g = \text{const} \times \omega \tag{46.12}$$

To find the constant in Eq. (46.12), we use

$$\sum_{N_{AB}} g(N,N_{AB},\mathscr{B}) = \frac{\mathscr{B}!}{N!(\mathscr{B} - N)!} = \text{const} \times \sum_{N_{AB}} \omega(N,N_{AB},\mathscr{B}) \tag{46.13}$$

The ω sum is replaced by the largest term in the sum. From $\partial \ln \omega/\partial N_{AB} = 0$ we find

$$\frac{N_{AB}(\max)}{2} = \frac{cN(\mathscr{B} - N)}{2\mathscr{B}} \tag{46.14}$$

and
$$\sum_{N_{AB}} \omega = \omega_{\max} = \left[\frac{\mathscr{B}!}{N!(\mathscr{B}-N)!}\right]^{c} \tag{46.15}$$

Hence
$$\text{const} = \left[\frac{\mathscr{B}!}{N!(\mathscr{B}-N)!}\right]^{1-c} \tag{46.16}$$

We note, incidentally, that ω can also be written in the form

$$\omega = \left[\frac{\mathscr{B}!}{\left(N - \dfrac{N_{AB}}{c}\right)!\left(\mathscr{B} - N - \dfrac{N_{AB}}{c}\right)!\left(\dfrac{N_{AB}}{c}!\right)^{2}}\right]^{c/2} \tag{46.17}$$

From Eqs. (46.12) and (46.16), we observe that ω greatly overcounts the actual number of configurations g. There are very many divisions of the total number $c\mathscr{B}/2$ of pairs into AA, BB, AB, and BA pairs which are actually impossible configurations because the pairs are in fact *not* independent of each other but overlap:

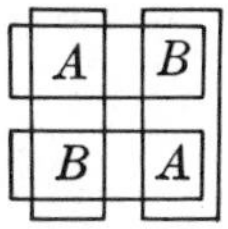

As can be seen by comparing Eqs. (43.1) and (46.12), the present approximation does, however, turn out to be exact for a one-dimensional lattice: the ω counting is overdone in this case by the same factor for each N_{AB}, but is otherwise correct.

We now substitute Eq. (46.12) for g in Eq. (46.1). The summation is difficult; therefore we let $t(N,N_{AB})$ be the general term and proceed to find the maximum term, just as in Eqs. (43.8) to (43.12). We obtain

$$\left(\frac{\rho}{1-\rho}\right)^{c-1}\left(\frac{1-\rho-a}{\rho-a}\right)^{c/2} = \frac{1}{y} \tag{46.18}$$

and
$$\frac{(\rho-a)(1-\rho-a)}{a^2} = \frac{1}{x^2} = e^{2w/\mathbf{k}T} \tag{46.19}$$

where
$$a = \frac{N_{AB}}{c\mathscr{B}} \tag{46.20}$$

Equations (46.18) and (46.19) determine ρ and a as functions of x, y, and $\mathscr{B}$.

Equation (46.19) can be rewritten as

$$\frac{N_{AA}N_{BB}}{N_{AB}^{2}} = \frac{e^{2w/\mathbf{k}T}}{4} \tag{46.21}$$

This resembles a chemical equilibrium quotient for the process

$$2AB \rightarrow AA + BB$$

$$\Delta\varepsilon = \varepsilon_{AA} + \varepsilon_{BB} - 2\varepsilon_{AB} = -2w$$

The "equilibrium constant" is $[\exp(-\Delta\varepsilon/\mathbf{k}T)]/4$, the factor of 4 arising because AA and BB have symmetry numbers of 2. In the Bragg-Williams approximation, it is easy to see from $N_{AB} = cN(\mathscr{B} - N)/\mathscr{B}$ and Eq. (41.17) that

$$\frac{N_{AA}N_{BB}}{N_{AB}^2} = \frac{1}{4} \tag{46.22}$$

which corresponds to a completely random distribution. Equation (46.21) is obviously an improvement over Eq. (46.22) and reduces to Eq. (46.22) as $T \rightarrow \infty$. Historically, Guggenheim adopted Eq. (46.21) rather than Eq. (46.12) as the starting point in the quasi-chemical treatment. He later showed that the quasi-chemical approximation and the so-called Bethe approximation (which we shall not discuss explicitly) are identical.

Equation (46.19) can be solved for a:

$$a = \frac{N_{AB}}{c\mathscr{B}} = \frac{2\rho(1-\rho)}{\beta + 1} \tag{46.23}$$

where
$$\beta = [1 - 4\rho(1-\rho)(1 - x^{-2})]^{1/2} \tag{46.24}$$

Using this in Eq. (46.18), we find for the chemical potential or adsorption isotherm

$$y = \left[\frac{(\beta - 1 + 2\rho)(1-\rho)}{(\beta + 1 - 2\rho)\rho}\right]^{c/2} \frac{\rho}{1-\rho} \tag{46.25}$$

Finally, with the aid of Eqs. (46.23) and (46.25), the equation of state turns out to be

$$\frac{1}{\mathscr{B}} \ln \mathscr{S} - \frac{cK}{2} = \frac{1}{\mathscr{B}} \ln t_{\max} = \frac{p'}{\mathbf{k}T} = \ln\left\{\left[\frac{(\beta+1)(1-\rho)}{\beta + 1 - 2\rho}\right]^{c/2} \frac{1}{1-\rho}\right\} \tag{46.26}$$

Equations (46.25) and (46.26) reduce to Eqs. (46.5) and (46.6) when $T \rightarrow \infty$.

Equation (46.25) has the symmetry property (Fig. 36)

$$y(\rho)y(1-\rho) = y^2(1/2) = 1$$

and hence $\rho_c = 1/2$ and $y_c = 1$ if a critical point exists.

From Eq. (46.26), we have

$$\left(\frac{\partial p'/\mathbf{k}T}{\partial \rho}\right)_{\rho=1/2} = c(x - 1) + 2 \tag{46.27}$$

This slope is negative if $x < (c - 2)/c$. Hence a loop and first-order phase change are predicted. The critical point is at

$$\rho_c = 1/2$$

$$x_c = \frac{(c-2)}{c} \qquad \frac{w}{\mathbf{k}T_c} = \ln \frac{c}{(c-2)} \tag{46.28}$$

$$z_c = \sigma \qquad y_c = 1$$

The remarks, following Eq. (46.8), about the loop in the equation of state and the zeros of the grand partition function apply here as well, of course.

The critical temperatures, discussed above, for square and cubic lattices are collected in Table 16.

TABLE 16. CRITICAL TEMPERATURES

	Square lattice $c = 4$	Cubic lattice $c = 6$
Exact.............	$x_c = \exp(-w/\mathbf{k}T_c)$ $= .4142$	$\sim .641$
Quasi-chemical......	.5000	.6667
Bragg-Williams	.6065	.7165

The "exact" value for the cubic lattice is Wakefield's, mentioned at the end of Sec. 45.

We turn now to the order-disorder problem, $x > 1$ at $\rho = {}^1/_2$ (or $y = 1$). There is a single maximum term in the sum in Eq. (46.1) when $y = 1$ and $x > x_c$. Therefore we put $\rho = {}^1/_2$ in Eq. (46.26) and find

$$\frac{1}{\mathscr{B}} \ln \mathscr{S} = \ln [2(\cosh K)^{c/2}] \tag{46.29}$$

$$E_{\text{config}} = -\frac{\mathscr{B}Jc}{2} \tanh K \tag{46.30}$$

$$\frac{C_2}{(c\mathscr{B}\mathbf{k}/2)} = K^2 \operatorname{sech}^2 K \tag{46.31}$$

Equation (46.31) gives a smooth curve (Fig. 46) for the configurational heat capacity against temperature, with no hint of a singularity. But, again, there is in addition to the above a *second* quasi-chemical treatment of the order-disorder problem: An ordering parameter s is introduced as in Eqs. (46.9) and (46.10), and a separate quasi-chemical pair equilibrium is assumed within each group of configurations [$h(s)$ in number] with the same value of

s. This leads to an improved result at low temperatures but not at high temperatures ($s = 0$), just as in the Bragg-Williams theory. This more refined heat-capacity curve is included in Fig. 57.

In Fig. 55, we have plotted the rigorous $G(N_{AB},\mathscr{B})^{1/\mathscr{B}}$ against $N_{AB}/\mathscr{B}$ for a square lattice ($c = 4$). It is of interest to compare this curve with the corresponding quasi-chemical result. We have

$$G(N_{AB},\mathscr{B}) = \sum_{N} g(N,N_{AB},\mathscr{B})$$
$$= g\left(\frac{\mathscr{B}}{2},N_{AB},\mathscr{B}\right)$$

since $N = \mathscr{B}/2$ gives the maximum term in the sum. Thus from Eq. (46.12) we get

$$G(N_{AB},\mathscr{B}) = \frac{(c\mathscr{B}/2)!\,2^{\mathscr{B}\left(1-\frac{c}{2}\right)}}{[(c\mathscr{B}/2) - N_{AB}]!\,N_{AB}!} \tag{46.32}$$

This function behaves properly[1] (Fig. 55, $c = 4$) near $N_{AB} = \mathscr{B}$ but not otherwise.

A useful feature of the quasi-chemical method is that it can be generalized[2] to larger basic groups of sites than pairs. For example, in a square lattice, squares of four sites, in their different manners of occupation, can be assumed independent of each other in constructing an approximate $g(N,N_{AB},\mathscr{B})$. In principle, by choosing larger and larger basic units, one must converge on the correct result; but unfortunately this is not a practical procedure, for the convergence is very slow. The generalization to larger groups of sites than pairs obviously makes possible[3] the inclusion of second and higher neighbor interactions. For example, with a square of four sites:

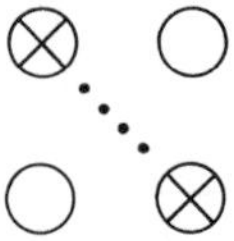

This refinement does not seem to be important in the theory of regular solutions,[4] but it may be important in adsorption, since intersite distances

[1] Prigogine et al., *loc. cit.*

[2] E. A. Guggenheim, *Proc. Roy. Soc.* (*London*), **A183**, 213 (1944); E. A. Guggenheim (Gen. Ref.); E. A. Guggenheim and M. L. McGlashan, *Proc. Roy. Soc.* (*London*), **A206**, 335 (1951); *Trans. Faraday Soc.*, **47**, 932 (1951); C. N. Yang, *J. Chem. Phys.*, **13**, 66 (1945); Y. Y. Li, *J. Chem. Phys.*, **17**, 447 (1949); T. L. Hill, *J. Chem. Phys.*, **18**, 988 (1950); R. Kikuchi, *Phys. Rev.*, **81**, 988 (1951).

[3] Hill, *loc. cit.*; Guggenheim and McGlashan, *loc. cit.* The discrepancies between these two papers are due to an error in the latter. See the correction by Guggenheim and McGlashan, *Trans. Faraday Soc.*, **48**, 773 (1952).

[4] Guggenheim and McGlashan, *loc. cit.*

are predetermined by the structure of the solid surface in "localized" adsorption.

Incidentally, the quasi-chemical "approximation" is exact in one dimension not only for nearest-neighbor systems but also for arbitrary higher neighbor systems.[1]

GENERAL REFERENCES

Fowler, R. H., and E. A. Guggenheim, "Statistical Thermodynamics" (Cambridge, London, 1939).

Guggenheim, E. A., "Mixtures" (Oxford, London, 1952).

ter Haar, D., "Elements of Statistical Mechanics" (Rinehart, New York, 1954).

Newell, G. F., and E. W. Montroll, *Revs. Mod. Phys.*, **25**, 353 (1953).

Nix, F. C., and W. Shockley, *Revs. Mod. Phys.*, **10**, 1 (1938).

Rushbrooke, G. S., "Statistical Mechanics" (Oxford, London, 1949).

Rushbrooke, G. S., *Compt. rend. réunion ann. union intern. phys.* (Paris, 1952), p. 177.

Wannier, G. H., *Revs. Mod. Phys.*, **17**, 50 (1945).

Zimm, B. H., R. A. Oriani, and J. D. Hoffman, *Ann. Rev. Phys. Chem.*, **4**, 207 (1953).

---

[1] H. N. V. Temperley, *Proc. Cambridge Phil. Soc.*, **40**, 239 (1944); G. S. Rushbrooke and H. D. Ursell, *Proc. Cambridge Phil. Soc.*, **44**, 263 (1948).

CHAPTER 8

LATTICE THEORIES OF THE LIQUID AND SOLID STATES

In view of the mathematical difficulties encountered in attempting to treat the liquid state by the methods described in Chaps. 5 and 6, it is natural that more or less intuitive approaches to this problem should also have been tried. So-called free-volume and hole theories[1] have been outstanding in this connection; qualitative or semiquantitative agreement with experimental thermodynamic properties has been achieved using equations presenting only relatively minor mathematical difficulties.

A former serious objection to lattice theories was the feature that, although a physically plausible model might be adopted as a starting point, the connection between the model and rigorous statistical mechanics was rather obscure. An interesting recent development in this field has been the deduction of the important lattice theories from general principles of statistical mechanics using well-defined approximations (Secs. 48 to 50).

The general configuration integral Z [Eq. (48.2)] with which we shall start the rigorous discussion below is of course valid for gas, liquid, and solid states. However, an important point (amplified in Secs. 48 to 50), which should be kept in mind throughout this chapter, is that the numerous approximations introduced in the analysis of Z before numerical results are finally reached are of such a nature that the liquid state is, in fact, removed from the discussion. Although it is conventional to call the approximate theories discussed below free-volume or hole theories of the liquid state, they are actually theories of the solid state. Their application to liquids involves the additional approximation that the liquid state can be represented by a model of the solid state. The first-order phase transition and critical properties conventionally attributed to a liquid-gas system in these approximations really refer to a phase transition (sublimation) in a system constrained

[1] We arbitrarily adopt the following nomenclature: a "lattice theory" is (1) a "free-volume theory" if the number of cells of the lattice $\mathscr{B}$ is chosen equal to the number of molecules N, and is (2) a "hole theory" if $\mathscr{B} \geqslant N$ and no cell is ever occupied by more than one molecule.

to the solid and gaseous states.[1] It is, therefore, beyond the scope of these theories to discuss fusion, for example.

Fowler and Guggenheim[2] discuss in detail the earlier work on free-volume theories. For this reason we shall emphasize developments of more recent origin. However, in Sec. 47 we give a brief preliminary discussion which will serve to acquaint the reader with the qualitative significance of the terms "free volume" and "communal entropy." A deductive discussion of the subject is then given in Secs. 48 and 49. Hole theories are considered in Sec. 50.

Lattice theories have been extended,[3] principally by Prigogine and Guggenheim, to mixtures and to surfaces, but we shall not consider these topics here.

47. Communal Entropy and Free Volume

Communal Entropy. Free-volume models use the crystalline state, rather than the gaseous state, as the point of departure. In a crystal, each molecule is confined in a "cell" or "cage" formed by its nearest neighbors, and is only very rarely involved in excursions outside of the cell. At the other extreme, in a dilute gas, each molecule is quite free to wander over the *entire* volume V of the container. Because of this additional freedom, gas molecules are said to have "communal entropy" not possessed by molecules in a crystal. The liquid state is intermediate in nature, and it is not at all obvious to what extent the liquid state possesses communal entropy (see Sec. 48). Originally, Hirschfelder, Stevenson, and Eyring[4] assumed that the liquid state had essentially the complete communal entropy and that the communal entropy therefore appeared on melting as a large part of the entropy of fusion. This view was later criticized by O. K. Rice.[5] It can safely be said that the situation with regard to the communal entropy in the liquid state remains obscure.

Communal entropy will be defined rigorously in Sec. 48, but it will no doubt be helpful to the reader if we give at this point the simplest possible explicit illustration of the concept. Suppose that we have a system of N monatomic molecules, without intermolecular forces, in a volume V. Then (Appendix 2, perfect gas)

$$A = -\mathbf{k}T \ln Q = -\mathbf{k}T \ln \frac{V^N}{N!\Lambda^{3N}} = -N\mathbf{k}T \ln \frac{ve}{\Lambda^3} \qquad (47.1)$$

[1] Further, the gaseous state here is not "normal" in that it also has lattice characteristics.

[2] Fowler and Guggenheim (Gen. Ref.).

[3] See, for example, E. A. Guggenheim, "Mixtures" (Oxford, London, 1952); I. Prigogine and V. Mathot, *J. Chem. Phys.*, **20**, 49 (1950); I. Prigogine and G. Garikian, *Physica*, **16**, 236 (1950); I. Sarolea, *J. Chem. Phys.*, **21**, 182 (1953); J. Nasielski, *J. Chem. Phys.*, **21**, 184 (1953); J. A. Pople, *Trans. Faraday Soc.*, **49**, 591 (1953); Z. Salsburg and J. G. Kirkwood, *J. Chem. Phys.*, **21**, 2169 (1953).

[4] J. Hirschfelder, D. Stevenson, and H. Eyring, *J. Chem. Phys.*, **5**, 896 (1937).

[5] O. K. Rice, *J. Chem. Phys.*, **6**, 476 (1938).

and
$$S = N\mathbf{k} \ln \frac{ve^{3/2}}{\Lambda^3} + N\mathbf{k} \tag{47.2}$$

On the other hand, suppose, by the use of hypothetical partitions, that the volume V is divided up into N cells, $\Delta_1, \Delta_2, \ldots, \Delta_N$, each of volume $v = V/N$, and that each cell is occupied by a molecule which is restricted to move inside the cell. Again we assume that there are no intermolecular forces. From the communal-entropy point of view this situation resembles that in a crystal. The configuration integral is now v^N instead of V^N and also the factor $1/N!$ in Eq. (47.1) is omitted since the molecules are now distinguishable (the cells can be labeled). Hence

$$A = -\mathbf{k}T \ln Q = -\mathbf{k}T \ln \frac{v^N}{\Lambda^{3N}} = -N\mathbf{k}T \ln \frac{v}{\Lambda^3} \tag{47.3}$$

and
$$S = N\mathbf{k} \ln \frac{ve^{3/2}}{\Lambda^3} \tag{47.4}$$

Equations (47.1) and (47.2) may be compared with Eqs. (47.3) and (47.4). It is clear that if we start with the system divided up into cells and then remove the partitions, the system acquires in this process the additional "communal" entropy $\Delta S = N\mathbf{k}$.

Free Volume. Now suppose that we take into account, very roughly, the intermolecular forces in the cell model above by assuming that each molecule moves in a field of constant potential χ within its cell, where χ arises from the (averaged) interaction of a given molecule with all the other molecules of the system. The only effect this will have on thermodynamic properties is to raise the energy of the system by $N\chi/2$. Equation (47.3) becomes

$$A = -N\mathbf{k}T \ln \frac{v}{\Lambda^3} + \frac{N\chi}{2} \tag{47.5}$$

As a next step, suppose the interaction potential between a given molecule and all other molecules is not assumed constant but rather is a function of position in the cell, $\chi(\mathbf{r})$, where the origin $\mathbf{r} = 0$ is located at the minimum in χ. For example, owing to intermolecular repulsions, we might expect χ to become very large in a condensed phase when the confined molecule approaches its neighbors at the edge of its cell. The probability of observing the central molecule in a given element of volume is no longer uniform throughout the cell but must be obtained from a Boltzmann factor. As a result of this, the "effective" or "free" volume through which the central molecule can move is reduced to a value less than v. In fact, the cell configuration integral leading to Eq. (47.3),

$$v = \int_\Delta d\mathbf{r}$$

must now be replaced by

$$v_f = \int_\Delta e^{-[\chi(\mathbf{r}) - \chi(0)]/\mathbf{k}T} d\mathbf{r} \tag{47.6}$$

where v_f is the "free" or "effective" volume. Equation (47.5) is then modified to read

$$A = -N\mathbf{k}T \ln \frac{v_f}{\Lambda^3} + \frac{N\chi(0)}{2} \tag{47.7}$$

In this picture we would expect $\chi(0)$ to be a function of ρ or v, and v_f to be a function of v and T.

Equation (47.7) is appropriate for a crystal as a crude approximation; for a liquid, in view of Eq. (47.1), there is a question as to whether v_f should not be replaced by $v_f e$ (as was done in the early theories[1] of Eyring and coworkers and of Lennard-Jones and Devonshire). However, to be more general, let us write

$$A = -N\mathbf{k}T \ln \frac{v_f \sigma}{\Lambda^3} + \frac{N\chi(0)}{2} \tag{47.8}$$

where $\sigma = 1$ for a crystal, $\sigma = e$ for a dilute gas, and σ has some intermediate, unspecified (as yet) value in the liquid state.

The purpose of the above discussion has been to introduce the term "free volume" in its historical and intuitive sense. As in the case of communal entropy, the free volume can be given a precise definition, and this is done in Sec. 48.

48. General Free-volume Theory

In this section we start with a rigorous formulation of free-volume theory, applicable to liquid, solid, and gaseous states, which separates the problem formally into a "free-volume" part and a "communal free-energy" part. We then continue the discussion by introducing certain approximations into the treatment of the free volume and communal free energy, respectively. The Lennard-Jones and Devonshire theory arises from this argument in a clear-cut way, and this theory is discussed in more detail in Sec. 49.

Rigorous Treatment.[2] We consider a one-component, monatomic, classical system of N molecules in a volume V with potential energy a sum of pair potentials. Using the canonical ensemble, we have

$$A = -\mathbf{k}T \ln Q \tag{48.1}$$

$$Q = \frac{Z}{N!\Lambda^{3N}} \tag{48.2}$$

[1] See Fowler and Guggenheim (Gen. Ref.).

[2] J. G. Kirkwood, *J. Chem. Phys.*, **18**, 380 (1950).

as in Eq. (22.4), for example. We now divide the volume V into an imaginary lattice of N cells $\Delta_1, \Delta_2, \ldots, \Delta_N$, each of volume $v = V/N$. The configuration integral when molecule 1 is assigned to cell Δ_{l_1}, molecule 2 to cell Δ_{l_2}, etc., is

$$\int_{\Delta_{l_1}} \cdots \int_{\Delta_{l_N}} e^{-U(\mathbf{r}_1, \ldots, \mathbf{r}_N)/\mathbf{k}T} \, d\mathbf{r}_1 \cdots d\mathbf{r}_N$$

where the integration over $\mathbf{r}_i$ is extended through the cell Δ_{l_i}, etc. In the complete configuration integral Z the integration over each $\mathbf{r}_i$ must be carried out through the entire volume V and therefore over all cells $\Delta_1, \Delta_2, \ldots, \Delta_N$; thus we have

$$Z = \sum_{l_1=1}^{N} \cdots \sum_{l_N=1}^{N} \int_{\Delta_{l_1}} \cdots \int_{\Delta_{l_N}} e^{-U/\mathbf{k}T} \, d\mathbf{r}_1 \cdots d\mathbf{r}_N \tag{48.3}$$

Z is thus a sum of N^N integrals.

Obviously, the number of cells need not be chosen equal to the number of molecules (see Sec. 50, for example), but it is convenient to do so here. Also, the geometry of the lattice is quite arbitrary in Eq. (48.3), but, again, for present purposes, it is advantageous to choose the lattice geometry to conform with the stable lattice of the system in the crystalline state. This ensures that at high enough densities there will be one molecule in each cell. For molecules interacting through central forces, the face-centered cubic structure is presumably most stable; the cells in this case are dodecahedra.

Now let $Z^{(m_1 \cdots m_N)}$ be the configuration integral when m_1 molecules are in cell 1, . . . , and m_N molecules are in cell N. That is,

$$Z^{(m_1 \cdots m_N)} = \underbrace{\int_{\Delta_1} \cdots \int_{\Delta_1}}_{m_1} \underbrace{\int_{\Delta_2} \cdots \int_{\Delta_2}}_{m_2} \cdots \underbrace{\int_{\Delta_N} \cdots \int_{\Delta_N}}_{m_N} e^{-U/\mathbf{k}T} \, d\mathbf{r}_1 \cdots d\mathbf{r}_N \tag{48.4}$$

where, say, $\mathbf{r}_1, \ldots, \mathbf{r}_{m_1}$ are to be integrated over Δ_1, $\mathbf{r}_{m_1+1}, \ldots, \mathbf{r}_{m_1+m_2}$ over Δ_2, etc. Then, since the integrals in Eq. (48.3) are all of the type in Eq. (48.4), we have

$$Z = \sum_{\substack{m_1, \ldots, m_N = 0 \\ (\sum_{s=1}^{N} m_s = N)}}^{N} \frac{N!}{m_1! m_2! \cdots m_N!} Z^{(m_1 \cdots m_N)} \tag{48.5}$$

for there are $N!/m_1! \cdots m_N!$ ways of assigning m_1 molecules to cell 1, m_2 to cell 2, etc.

Of special interest is the integral corresponding to single occupancy of each cell. We denote this by

$$Z^{(1)} = Z^{(1 \cdots 1)} = \int_{\Delta_1} \int_{\Delta_2} \cdots \int_{\Delta_N} e^{-U/\mathbf{k}T} \, d\mathbf{r}_1 \cdots d\mathbf{r}_N \tag{48.6}$$

where, say, $\mathbf{r}_1$ is integrated over Δ_1, $\mathbf{r}_2$ over Δ_2, etc.

Equation (48.5) can now be rewritten as

$$Z = N!Z^{(1)}\sigma^N \tag{48.7}$$

where σ^N is defined by the relation

$$\sigma^N = \sum_{\substack{m_1,\ldots,m_N=0 \\ (\sum_{s=1}^{N} m_s = N)}}^{N} \frac{1}{\prod_{s=1}^{N} m_s!} \frac{Z^{(m_1 \cdots m_N)}}{Z^{(1)}} \tag{48.8}$$

If, as in the crystalline state at high densities, intermolecular repulsive forces exclude multiple occupancy of any of the cells, all of the $Z^{(m_1 \cdots m_N)}$ are zero except $Z^{(1)}$ and therefore $\sigma = 1$. Then, from either Eq. (48.5) or (48.7),

$$Z = N!Z^{(1)} \tag{48.9}$$

and

$$Q = \frac{Z^{(1)}}{\Lambda^{3N}} \tag{48.10}$$

This is the same partition function we would have if we restrained the molecules to single occupancy of cells and, *in addition*, never permitted a molecule to escape from its cell [so that the molecules are, in principle, distinguishable, as in Eq. (47.3)]. At the other extreme, as the density tends to zero, all $Z^{(m_1 \cdots m_N)}$, including $Z^{(1)}$, approach v^N. Then, from Eq. (48.8),

$$\sigma^N \to \frac{1}{N!} N^N = e^N$$

or

$$\sigma \to e$$

Thus σ, as defined in Eq. (48.8), has the same limiting values as, and may be considered the rigorous analogue of, the empirical σ introduced in Eq. (47.8).

With the restraint of single occupancy (taking molecule 1 in cell 1, 2 in cell 2, etc., for concreteness), the probability that particle 1 is in $d\mathbf{r}_1$ at $\mathbf{r}_1$, 2 in $d\mathbf{r}_2$ at $\mathbf{r}_2$, etc., is

$$P(\mathbf{r}_1, \ldots, \mathbf{r}_N)\, d\mathbf{r}_1 \cdots d\mathbf{r}_N = \frac{e^{-U/\mathbf{k}T}\, d\mathbf{r}_1 \cdots d\mathbf{r}_N}{Z^{(1)}} \tag{48.11}$$

which defines P as a probability density. Now let us take the logarithm of both sides of $P = [\exp(-U/\mathbf{k}T)]/Z^{(1)}$, multiply by $P\, d\mathbf{r}_1 \cdots d\mathbf{r}_N$ and integrate. The result is

$$-\mathbf{k}T \ln Z^{(1)} = \int_{\Delta_1} \cdots \int_{\Delta_N} PU\, d\mathbf{r}_1 \cdots d\mathbf{r}_N + \mathbf{k}T \int_{\Delta_1} \cdots \int_{\Delta_N} P \ln P\, d\mathbf{r}_1 \cdots d\mathbf{r}_N \tag{48.12}$$

Equation (48.12) becomes

$$A^{(1)} = E^{(1)} - TS^{(1)} \tag{48.13}$$

if we define

$$A^{(1)} = -\mathbf{k}T \ln Z^{(1)} \tag{48.14}$$

$$E^{(1)} = \int_{\Delta_1} \cdots \int_{\Delta_N} PU \, d\mathbf{r}_1 \cdots d\mathbf{r}_N \tag{48.15}$$

$$S^{(1)} = -\mathbf{k} \int_{\Delta_1} \cdots \int_{\Delta_N} P \ln P \, d\mathbf{r}_1 \cdots d\mathbf{r}_N \tag{48.16}$$

$A^{(1)}$, $E^{(1)}$ and $S^{(1)}$ are the configurational (i.e., the kinetic energy contributions are omitted) Helmholtz free energy, energy, and entropy, respectively, of a system restrained to single occupancy of cells.

We define further an energy $\bar{\mathrm{E}}$ by the equation

$$E^{(1)} = \frac{N\bar{\mathrm{E}}}{2} \tag{48.17}$$

so that $\bar{\mathrm{E}}$ is the average potential energy of interaction of a specified molecule with all other molecules in the system, again subject to the restraint of single occupancy of all cells. Finally, v_f is defined by the equation

$$S^{(1)} = N\mathbf{k} \ln v_f \tag{48.18}$$

That is, the free volume v_f is defined formally in such a way that $\mathbf{k} \ln v_f$ is the configurational entropy per molecule in a system restrained to single occupancy of cells.

If we use

$$-\mathbf{k}T \ln Z^{(1)} = \frac{N\bar{\mathrm{E}}}{2} - N\mathbf{k}T \ln v_f \tag{48.19}$$

to eliminate $Z^{(1)}$ from the expression for Z, Eq. (48.7), we find that Eq. (48.1) becomes

$$A = -N\mathbf{k}T \ln \frac{v_f \sigma}{\Lambda^3} + \frac{N\bar{\mathrm{E}}}{2} \tag{48.20}$$

Equation (48.20) is the rigorous analogue of Eq. (47.8). Aside from the formal similarity, the resemblance between the two equations will become more obvious in certain approximations to be considered below.

All the thermodynamic properties of the system follow from Eq. (48.20), in which, it should be noted, v_f, σ, and $\bar{\mathrm{E}}$ are all, in the absence of approximations, functions of ρ and T or v and T. The contribution $-N\mathbf{k}T \ln \sigma$ to A in Eq. (48.20) may be called, appropriately, the communal free energy, so that

$$N\mathbf{k} \ln \sigma + N\mathbf{k}T \left(\frac{\partial \ln \sigma}{\partial T} \right)_v \tag{48.21}$$

is the communal entropy.

To make further progress, we consider the free volume and communal free energy separately, and introduce approximations.

Free Volume.[1] As an approximation, let us assume that the probability density $P(\mathbf{r}_1, \ldots, \mathbf{r}_N)$ may be written as a product of independent single particle ("singlet") probability densities. That is,

$$P = \prod_{s=1}^{N} \varphi(\mathbf{r}_s) \tag{48.22}$$

where the center of the cell Δ_s is chosen as the origin for $\mathbf{r}_s$. Clearly, $\varphi(\mathbf{r}_s)\, d\mathbf{r}_s$ is the probability of observing molecule s in $d\mathbf{r}_s$ at $\mathbf{r}_s$ (in the cell Δ_s) and, since

$$\int_{\Delta_1} \cdots \int_{\Delta_N} P\, d\mathbf{r}_1 \cdots d\mathbf{r}_N = 1 \tag{48.23}$$

we have

$$\int_{\Delta} \varphi(\mathbf{r})\, d\mathbf{r} = 1 \tag{48.24}$$

Equation (48.22) is analogous to the Hartree approximation in the solution of the Schrödinger equation for several particles.

We now find the best possible P, of the form in Eq. (48.22), by minimizing the configurational free energy $A^{(1)}$. As part of the approximation here, it should be observed that the communal free energy is being excluded from the minimization procedure.[2] We substitute Eq. (48.22) in Eqs. (48.15) and (48.16):

$$\begin{aligned} S^{(1)} &= -\mathbf{k} \int_{\Delta_1} \cdots \int_{\Delta_N} [\prod_{s=1}^{N} \varphi(\mathbf{r}_s)][\sum_{s=1}^{N} \ln \varphi(\mathbf{r}_s)]\, d\mathbf{r}_1 \cdots d\mathbf{r}_N \\ &= -N\mathbf{k} \int_{\Delta} \varphi(\mathbf{r}) \ln \varphi(\mathbf{r})\, d\mathbf{r} \end{aligned} \tag{48.25}$$

$$\begin{aligned} E^{(1)} &= \int_{\Delta_1} \cdots \int_{\Delta_N} [\prod_{s=1}^{N} \varphi(\mathbf{r}_s)][\sum_{1 \leqslant i < j \leqslant N} u(r_{ij})]\, d\mathbf{r}_1 \cdots d\mathbf{r}_N \\ &= \sum_{1 \leqslant i < j \leqslant N} \int_{\Delta_i} \int_{\Delta_j} \varphi(\mathbf{r}_i)\varphi(\mathbf{r}_j) u(r_{ij})\, d\mathbf{r}_i\, d\mathbf{r}_j \end{aligned} \tag{48.26}$$

All molecules are equivalent in Eq. (48.26); thus of the $N(N-1)/2$ terms we retain explicitly those $N-1$ terms involving the interaction of, say, molecule 1 with other molecules, and then multiply this sum by $N/2$:

$$E^{(1)} = \frac{N}{2} \sum_{j=2}^{N} \int_{\Delta_1} \int_{\Delta_j} \varphi(\mathbf{r}_1)\varphi(\mathbf{r}_j) u(\mathbf{r}_{1j})\, d\mathbf{r}_1\, d\mathbf{r}_j \tag{48.27}$$

[1] J. G. Kirkwood, *loc. cit.*

[2] That is, a better approximation would be to minimize A rather than $A^{(1)}$, subject to Eq. (48.22). See the subsection below on the communal free energy.

Let $\mathbf{R}_{1j}$ be the vector joining the origins of cells j and 1 (Fig. 59) so that $\mathbf{r}_{1j} = |\mathbf{R}_{1j} + \mathbf{r}_1 - \mathbf{r}_j|$. If we replace $\mathbf{r}_1$ by $\mathbf{r}$ and $\mathbf{r}_j$ by $\mathbf{r}'$ in Eq. (48.27), $E^{(1)}$ can be expressed as

$$E^{(1)} = \frac{N\bar{\text{E}}}{2} = \frac{N}{2}\int_\Delta \int_{\Delta'} \varphi(\mathbf{r})\varphi(\mathbf{r}')E(\mathbf{r}-\mathbf{r}')\,d\mathbf{r}\,d\mathbf{r}' \tag{48.28}$$

where

$$E(\mathbf{r}) = \sum_{j=2}^{N} u(\mathbf{R}_{1j} + \mathbf{r}) \tag{48.29}$$

Equations (48.25) and (48.28) are substituted in Eq. (48.13), giving

$$\frac{A^{(1)}}{N\mathbf{k}T} = \int_\Delta \varphi(\mathbf{r}) \ln \varphi(\mathbf{r})\,d\mathbf{r} + \frac{1}{2\mathbf{k}T}\int_\Delta \int_{\Delta'} \varphi(\mathbf{r})\varphi(\mathbf{r}')E(\mathbf{r}-\mathbf{r}')\,d\mathbf{r}\,d\mathbf{r}' \tag{48.30}$$

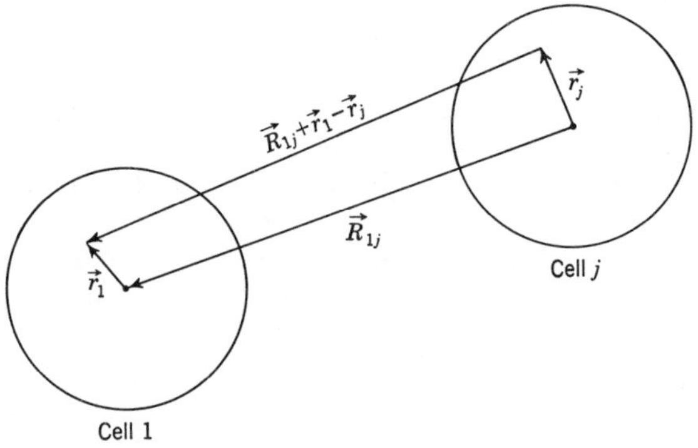

FIG. 59. Vectors locating molecules in cells.

Regarding $A^{(1)}/N\mathbf{k}T$ as a functional of φ, we obtain for the extremalization condition (N,V,T constant)

$$\frac{\delta A^{(1)}}{N\mathbf{k}T} = \int_\Delta \delta\varphi\,d\mathbf{r} + \int_\Delta \delta\varphi \ln\varphi\,d\mathbf{r} + \frac{1}{2\mathbf{k}T}\int_\Delta\int_{\Delta'}[\varphi(\mathbf{r})\,\delta\varphi(\mathbf{r}') + \varphi(\mathbf{r}')\,\delta\varphi(\mathbf{r})]E(\mathbf{r}-\mathbf{r}')\,d\mathbf{r}\,d\mathbf{r}'$$

$$= \int_\Delta \left[\ln\varphi(\mathbf{r}) + \frac{1}{\mathbf{k}T}\int_\Delta \varphi(\mathbf{r}')E(\mathbf{r}-\mathbf{r}')\,d\mathbf{r}'\right]\delta\varphi(\mathbf{r})\,d\mathbf{r} \tag{48.31}$$

$$\int_\Delta \delta\varphi\,d\mathbf{r} = 0 \tag{48.32}$$

The restraint, Eq. (48.32), follows from Eq. (48.24). For the Lagrange multiplier in combining Eqs. (48.31) and (48.32) it proves convenient to use $(\alpha + \bar{E})/\mathbf{k}T$. Then we find

$$\ln \varphi(\mathbf{r}) + \frac{1}{\mathbf{k}T}\int_{\Delta'} \varphi(\mathbf{r}')E(\mathbf{r} - \mathbf{r}')\, d\mathbf{r}' = \frac{\alpha + \bar{E}}{\mathbf{k}T} \tag{48.33}$$

Before discussing this equation, let us replace $\varphi(\mathbf{r})$ by a new function $\psi(\mathbf{r})$ defined by

$$\begin{aligned} \psi(\mathbf{r}) &= -\,\mathbf{k}T \ln \varphi(\mathbf{r}) + \alpha \\ \varphi(\mathbf{r}) &= e^{[\alpha - \psi(\mathbf{r})]/\mathbf{k}T} \end{aligned} \tag{48.34}$$

From Eq. (48.24),

$$e^{-\alpha/\mathbf{k}T} = \int_{\Delta} e^{-\psi(\mathbf{r})/\mathbf{k}T}\, d\mathbf{r} \tag{48.35}$$

$$\varphi(\mathbf{r}) = \frac{e^{-\psi(\mathbf{r})/\mathbf{k}T}}{\displaystyle\int_{\Delta} e^{-\psi(\mathbf{r})/\mathbf{k}T}\, d\mathbf{r}} \tag{48.36}$$

Putting Eq. (48.34) in Eq. (48.33),

$$\psi(\mathbf{r}) = e^{\alpha/\mathbf{k}T}\int_{\Delta'} e^{-\psi(\mathbf{r}')/\mathbf{k}T}\, w(\mathbf{r} - \mathbf{r}')\, d\mathbf{r}' \tag{48.37}$$

where

$$w(\mathbf{r}) = E(\mathbf{r}) - \bar{E} \tag{48.38}$$

With the understanding that $\exp(-\alpha/\mathbf{k}T)$ in Eq. (48.37)[1] is to be replaced by the integral in Eq. (48.35), Eq. (48.37) is an integral equation in $\psi(\mathbf{r})$, the solution of which[2] corresponds to the lowest possible $A^{(1)}$ for a P of the form in Eq. (48.22).

Let us take the gradient with respect to $\mathbf{r}$ in Eq. (48.33) [which is equivalent to Eq. (48.37)]:

$$-\,\mathbf{k}T\nabla_{\mathbf{r}} \ln \varphi(\mathbf{r}) = \int_{\Delta'} \varphi(\mathbf{r}')\Big[\sum_{j=2}^{N} \nabla_{\mathbf{r}} u(\mathbf{R}_{1j} + \mathbf{r} - \mathbf{r}')\Big]\, d\mathbf{r}' \tag{48.39}$$

It is interesting to verify that Eq. (48.39) also follows from the Born-Green-Yvon integrodifferential equation appropriate to the present problem. To see this, we consider the distribution function [see Eq. (29.5)]

$$P^{(n)}(\mathbf{r}_1, \ldots, \mathbf{r}_n) = \frac{\displaystyle\int_{\Delta_{n+1}} \cdots \int_{\Delta_N} e^{-U/\mathbf{k}T}\, d\mathbf{r}_{n+1} \cdots d\mathbf{r}_N}{Z^{(1)}} \tag{48.40}$$

[1] Note that $\exp(\alpha/\mathbf{k}T)$ occurs also in $w(\mathbf{r} - \mathbf{r}')$; see Eqs. (48.38), (48.28), and (48.34).

[2] The exact solution for a system of hard spheres has been found by W. W. Wood, *J. Chem. Phys.*, **20**, 1334 (1952).

for a system restrained to single occupancy of cells. Then

$$-\mathbf{k}T\nabla_1 \ln P^{(n)}(\mathbf{r}_1, \ldots, \mathbf{r}_n) = \frac{\int_{\Delta_{n+1}} \cdots \int_{\Delta_N} e^{-U/\mathbf{k}T} \nabla_1 U \, d\mathbf{r}_{n+1} \cdots d\mathbf{r}_N}{\int_{\Delta_{n+1}} \cdots \int_{\Delta_N} e^{-U/\mathbf{k}T} \, d\mathbf{r}_{n+1} \cdots d\mathbf{r}_N} \tag{48.41}$$

Just as in Eq. (33.3), this reduces to

$$-\mathbf{k}T\nabla_1 \ln P^{(n)}(\mathbf{r}_1, \ldots, \mathbf{r}_n) = \sum_{i=2}^{n} \nabla_1 u(r_{1i}) + \frac{1}{P^{(n)}} \sum_{i=n+1}^{N} \int_{\Delta_i} \nabla_1 u(r_{1i}) P^{(n+1)}(\mathbf{r}_1, \ldots, \mathbf{r}_n, \mathbf{r}_i) \, d\mathbf{r}_i \tag{48.42}$$

In the case $n = 1$,

$$-\mathbf{k}T\nabla_1 \ln P^{(1)}(\mathbf{r}_1) = \frac{1}{P^{(1)}} \sum_{i=2}^{N} \int_{\Delta_i} \nabla_1 u(r_{1i}) P^{(2)}(\mathbf{r}_1,\mathbf{r}_i) \, d\mathbf{r}_i \tag{48.43}$$

This equation is exact for a system of singly occupied cells. Now let us introduce the approximation

$$P^{(2)}(\mathbf{r}_1,\mathbf{r}_i) = P^{(1)}(\mathbf{r}_1)P^{(1)}(\mathbf{r}_i) \tag{48.44}$$

This is equivalent to Eq. (48.22), for

P [Eq. (48.11)] $= P^{(N)}$ [Eq. (48.40)] $\qquad \varphi$ [Eq. (48.22)] $= P^{(1)}$ [Eq. (48.44)]

Equations (48.43) and (48.44) give

$$-\mathbf{k}T\nabla_{\mathbf{r}} \ln P^{(1)}(\mathbf{r}) = \int_{\Delta'} P^{(1)}(\mathbf{r}')\Big[\sum_{i=2}^{N} \nabla_{\mathbf{r}} u(\mathbf{R}_{1i} + \mathbf{r} - \mathbf{r}')\Big] \, d\mathbf{r}' \tag{48.45}$$

which is the same as Eq. (48.39).

The solution, $\psi(\mathbf{r})$, of Eq. (48.37) can now be used to obtain the Helmholtz free energy A in Eq. (48.20), to the present degree of approximation. In Eq. (48.20), $\bar{\mathrm{E}}$ is given by Eqs. (48.28) and (48.36), while

$$\ln v_f = \frac{S^{(1)}}{N\mathbf{k}} = \frac{\int_{\Delta} e^{-\psi(\mathbf{r})/\mathbf{k}T} \left[\dfrac{\psi(\mathbf{r})}{\mathbf{k}T} + \ln \int_{\Delta} e^{-\psi(\mathbf{r})/\mathbf{k}T} \, d\mathbf{r}\right] d\mathbf{r}}{\int_{\Delta} e^{-\psi(\mathbf{r})/\mathbf{k}T} \, d\mathbf{r}}$$

With the aid of Eqs. (48.37) and (48.28), this reduces to

$$v_f = \int_{\Delta} e^{-\psi(\mathbf{r})/\mathbf{k}T} \, d\mathbf{r} \tag{48.46}$$

Thus, v_f, in this approximation, has the formal appearance of a "cell configuration integral," with $\psi(\mathbf{r})$ playing the role of the potential (of mean force) in a Boltzmann factor. The quantity σ in Eq. (48.20) should be considered a constant here if we are to be self-consistent, for it was in effect so considered in omitting it from the minimization procedure.

It is natural to try an iteration method for the solution of Eq. (48.37). As a zeroth approximation, let us take $\varphi(\mathbf{r})$ sharply peaked at the origin, that is, $\varphi_0(\mathbf{r}) = \delta(\mathbf{r})$. If we now substitute $\delta(\mathbf{r})$ for $\varphi(\mathbf{r})$ in Eq. (48.37), we obtain as a first approximation

$$\psi_1(\mathbf{r}) = \int_{\Delta'} \delta(\mathbf{r}')w_0(\mathbf{r} - \mathbf{r}')\,d\mathbf{r}' = w_0(\mathbf{r}) \tag{48.47}$$

$$v_{f(1)} = \int_{\Delta} e^{-\psi_1(\mathbf{r})/\mathbf{k}T}\,d\mathbf{r} \tag{48.48}$$

where

$$w_0(\mathbf{r}) = E(\mathbf{r}) - \bar{\mathrm{E}}_0 \tag{48.49}$$

From Eq. (48.28),

$$\bar{\mathrm{E}}_0 = \int_{\Delta}\int_{\Delta'} \delta(\mathbf{r})\delta(\mathbf{r}')E(\mathbf{r} - \mathbf{r}')\,d\mathbf{r}\,d\mathbf{r}' = E(0) \tag{48.50}$$

therefore

$$\psi_1(\mathbf{r}) = w_0(\mathbf{r}) = E(\mathbf{r}) - E(0) = \sum_{j=2}^{N} [u(\mathbf{R}_{1j} + \mathbf{r}) - u(\mathbf{R}_{1j})] \tag{48.51}$$

That is, $\psi_1 = w_0(\mathbf{r})$ is the total potential energy of interaction between molecule 1 at $\mathbf{r}$ and all other molecules at the centers of their cells *relative to* the same potential energy with molecule 1 also at the center of its cell. If the sum in Eq. (48.51) is extended only over nearest-neighbor cells, and if the nearest-neighbor sum is replaced by an integral over a sphere of radius equal to the distance between the centers of nearest-neighbor cells, $\psi_1 = w_0(\mathbf{r})$ becomes identical with the potential of the Lennard-Jones and Devonshire theory (see Sec. 49).

To achieve an improvement over Eqs. (48.47) to (48.51), a second approximation to $\psi(\mathbf{r})$ might be obtained from

$$\psi_2(\mathbf{r}) = \frac{\int_{\Delta'} e^{-\psi_1(\mathbf{r}')/\mathbf{k}T} w_0(\mathbf{r} - \mathbf{r}')\,d\mathbf{r}'}{\int_{\Delta} e^{-\psi_1(\mathbf{r})/\mathbf{k}T}\,d\mathbf{r}} \tag{48.52}$$

If convergent, a continuation of this iteration procedure could be employed to obtain a numerical solution to the integral equation, Eq. (48.37), leading to a refinement of the Lennard-Jones and Devonshire theory. Whether this would be worthwhile without including the communal-entropy problem is open to question.

In this subsection we have introduced two fundamental approximations [aside from the purely mathematical approximations of Eqs. (48.47) to (48.52)]: (1) the assumption that the single-occupation probability density $P(\mathbf{r}_1, \ldots, \mathbf{r}_N)$ can be written as a product of singlet probability densities $\varphi(\mathbf{r}_s)$ [Eq. (48.22)]; and (2) the omission of the communal free energy on minimizing the free energy of the system. The second assumption implies that σ is a constant, as has been pointed out. As a consequence of these approximations, the singlet probability density [see Eq. (29.8)] of a particle (properly normalized) must satisfy Eq. (48.33) within each cell, for the single-occupation singlet density $\varphi(\mathbf{r}_s)$ is just the singlet density of the system in this case.[1] Now in the liquid state the singlet density of the system is a constant, but φ = constant is not a solution of Eq. (48.33) except in the limit of zero density. Hence, we must conclude that the two approximations mentioned above have the effect of excluding the liquid state from consideration. The singlet probability density of the system (φ) has, in fact, the periodicity of the lattice and hence has long-range order and is appropriate to a crystal.[2] It should also be remarked that the "gaseous" state obtained at low enough densities is not "normal"; it is a quasi-lattice gas, with no second virial coefficient, for example (see Sec. 49).

Communal Free Energy.[3] Equation (48.8) defines σ, a quantity which makes the contribution $-N\mathbf{k}T \ln \sigma$ (the communal free energy) to the Helmholtz free energy of the system [Eq. (48.20)]. We present here an approximate theory for the calculation of σ.

We first make the approximation of writing the quotient $Z^{(m_1 \cdots m_N)}/Z^{(1)}$ in Eq. (48.8) as a product of factors ω'_{m_s}:

$$\frac{Z^{(m_1 \cdots m_N)}}{Z^{(1)}} = \prod_{s=1}^{N} \omega'_{m_s} \tag{48.53}$$

That is, ω'_1 refers to a singly occupied cell, ω'_2 to a doubly occupied cell, etc. Equation (48.53) assumes that $Z^{(m_1 \cdots m_N)}/Z^{(1)}$ depends only on the number (n_0) of vacant cells, the number (n_1) with single occupancy, the number (n_2) with double occupancy, etc., whereas this quotient actually depends in general not only on $n_0, n_1, \ldots, n_N$ but also on the detailed configuration of

[1] That is, in general, the singlet density [derived from Z; Eq. (29.8)] of the actual system and the singlet density [derived from $Z^{(1)}$; Eq. (48.40)] in the single-occupation system, both normalized in the same way, are different and the difference is contained implicitly in σ [see Eqs. (48.5), (48.7), and (48.8)]. But when σ is constant (as in the present approximation), the two singlet densities must be the same.

[2] Although there are short-range remnants of crystalline order remaining in the liquid state, the over-all fixed lattice of the crystal no longer has any special significance, nor has the origin $\mathbf{r}_s = 0$ of each cell: there is no long-range order.

[3] J. A. Pople, *Phil. Mag.*, **42**, 459 (1951); P. Janssens and I. Prigogine, *Physica*, **16**, 895 (1950).

cells of different kinds of occupancy. In a crystal, only a few cells will have nonsingle occupancy and each of these cells will almost always be surrounded entirely by singly occupied cells. Thus the nonsingle cells will not "interact" and $Z^{(m_1 \cdots m_N)}$ will have the same value for almost all configurations consistent with the set $n_0, n_1, \ldots, n_N$. In other words, the approximation in Eq. (48.53) should be accurate in a crystal. It is also accurate in a gas at low densities when the cells are large and effectively independent. Pople therefore suggests that it seems reasonable to use Eq. (48.53) as an approximation at any density. However, this would appear to exclude the liquid state from Pople's theory, for the liquid state is a high-density state in which

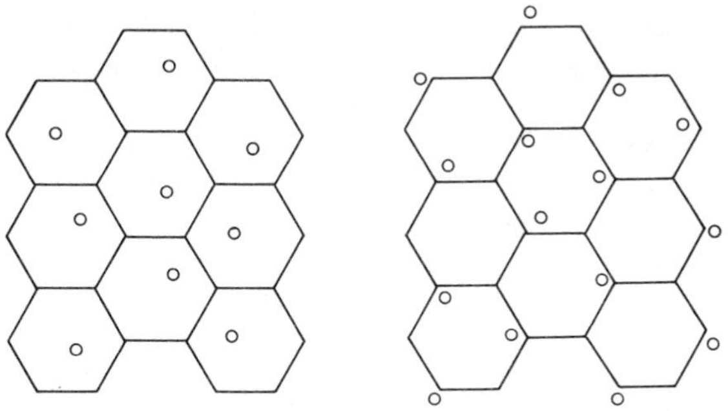

FIG. 60. Two configurations with same degree of local order but with occasional multiple occupancy in one case (*b*).

many cells can have nonsingle occupancy [and therefore Eq. (48.53) would be a poor approximation]. Thus, in Fig. 60*a*, we have (in two dimensions) a crystalline configuration with $\mathbf{r}_s = 0$ as the equilibrium point in each cell. Figure 60*a* could also represent a region of local order in the liquid state; but equally likely, in the liquid state, is, say, the region of Fig. 60*b*, which is just as ordered (locally) as Fig. 60*a* but has several cells with nonsingle occupation. From the special case $m_1 = m_2 = \cdots = m_N = 1$ in Eq. (48.53), we conclude that $\omega'_1 = 1$. Now define $\omega_s (s \neq 0)$ as $\omega'_s \omega_0'^{s-1}$. Then

$$\frac{Z^{(m_1 \cdots m_N)}}{Z^{(1)}} = \prod_{s=1}^{N} \omega'_{m_s} = \prod_{s=0}^{N} \omega_s'^{n_s} = \omega_0'^{n_0} \prod_{s=2}^{N} \omega_s^{n_s} \omega_0'^{n_s(1-s)} \tag{48.54}$$

On using the relations for the number of cells and molecules,

$$\sum_{s=0}^{N} n_s = N \tag{48.55}$$

$$\sum_{s=1}^{N} s n_s = N \tag{48.56}$$

Equation (48.54) becomes

$$\frac{Z^{(m_1 \cdots m_N)}}{Z^{(1)}} = \prod_{s=2}^{N} \omega_s^{n_s}$$

$$= \prod_{s=0}^{N} \omega_s^{n_s} \qquad \omega_0 = \omega_1 = 1 \tag{48.57}$$

If we take $n_0 = n_2 = 1$ and $n_1 = N - 2$,

$$\frac{Z^{(0211 \cdots)}}{Z^{(1)}} = \omega_2 \tag{48.58}$$

Thus ω_2 may be described as the factor by which $Z^{(1)}$ is altered if, starting with every cell singly occupied, we remove a particle from its cell and place it in another cell, leaving a doubly occupied cell and a hole.

Since we are using Eq. (48.57) as an approximation to $Z^{(m_1 \cdots m_N)}/Z^{(1)}$, it is desirable to collect terms in Eq. (48.8) corresponding to the set $n_0, n_1, \ldots n_N$ rather than $m_1, m_2, \ldots, m_N$. There are

$$\frac{N!}{\prod_{i=1}^{N} m_i!} = \frac{N!}{\prod_{s=0}^{N} (s!)^{n_s}}$$

ways of distributing N molecules among N cells such that m_1 are in cell 1, m_2 in cell 2, etc. (and altogether n_s cells contain s molecules). There are also

$$\frac{N!}{\prod_{s=0}^{N} n_s!}$$

different ways of assigning cells into groups of $n_0, n_1, \ldots, n_N$. Hence the total number of ways of distributing N molecules among N cells in such a way that n_s cells contain s molecules is

$$\frac{(N!)^2}{\prod_{s=0}^{N} n_s!(s!)^{n_s}}$$

Thus, Eq. (48.8) becomes, using Eq. (48.57),

$$N!\sigma^N = \sum_{n_0, \ldots, n_N}^{*} \frac{(N!)^2}{\prod_{s=0}^{N} n_s!(s!)^{n_s}} \prod_{s=0}^{N} \omega_s^{n_s}$$

or

$$\sigma^N = \sum_{n_0, \ldots, n_N}^{*} \frac{N!}{\prod_{s=0}^{N} n_s!} \prod_{s=0}^{N} \left(\frac{\omega_s}{s!}\right)^{n_s} = \sum_{n_0, \ldots, n_N}^{*} t(n_0, \ldots, n_N) \tag{48.59}$$

where Σ^* means that the sets $n_0, \ldots, n_N$ are to be restricted by Eqs. (48.55) and (48.56).

The next step is to replace the sum $\Sigma^* t(n_0, \ldots, n_N)$ in Eq. (48.59) by its maximum term. This requires maximizing $\ln t(n_0, \ldots, n_N)$ with the aid of Lagrange multipliers, because of the restrictions on the sum. We find that the maximum term occurs when

$$n_s = \frac{N\lambda\mu^s\omega_s}{s!} \qquad s = 0, 1, \ldots, N \tag{48.60}$$

where λ and μ are determined by

$$\sum_{s=0}^{N} n_s = \sum_{s=0}^{N} \frac{N\lambda\mu^s\omega_s}{s!} = N \tag{48.61}$$

$$\sum_{s=1}^{N} sn_s = \sum_{s=1}^{N} \frac{N\lambda\mu^s s\omega_s}{s!} = N \tag{48.62}$$

If we define

$$f(x) = \sum_{s=0}^{N} \frac{\omega_s}{s!} x^s \tag{48.63}$$

Equations (48.61) and (48.62) may be written

$$\frac{1}{\lambda} = f(\mu) = \mu f'(\mu) \tag{48.64}$$

Thus if the ω_s are given, we are to find that μ which satisfies $f(\mu) = \mu f'(\mu)$, and then obtain λ from $\lambda = 1/f(\mu)$. Finally, using Eq. (48.60),

$$\sigma = t_{\max}{}^{1/N} = \frac{1}{\lambda\mu} = \frac{f(\mu)}{\mu} = f'(\mu) \tag{48.65}$$

If we assume that $\omega_3, \omega_4, \ldots$ vanish, Eq. (48.65) reduces to

$$\sigma = \frac{1 + \mu + (\omega_2/2)\mu^2}{\mu} = 1 + \omega_2\mu$$

or

$$\mu = \sqrt{2/\omega_2}$$

$$\sigma = 1 + \sqrt{2\omega_2} \tag{48.66}$$

This is the solution we get if we impose, in addition to Eq. (48.53), the limitation that no cell can be occupied by more than two molecules.

From Eq. (48.58) we see that ω_2 ranges from 1 at low densities to zero at high densities. Now at low densities $\ln \sigma = \ln e = 1$, while Eq. (48.66) gives $\ln \sigma = \ln (1 + \sqrt{2}) = 0.88$. Hence, at low densities, 88 per cent of the communal entropy is accounted for[1] by double occupation of cells. It is also possible to deduce the low-density equation of state from Eqs. (48.1), (48.2), and (48.7):

$$A = -\mathbf{k}T \ln \frac{Z^{(1)}\sigma^N}{\Lambda^{3N}} \tag{48.67}$$

$$-p = \left(\frac{\partial A}{\partial V}\right)_{N,T} \tag{48.68}$$

Since we can neglect interactions between particles in different cells in the low-density limit, $Z^{(1)} = v^N$. Also, from Eq. (48.58),

$$\begin{aligned} \omega_2 &= \frac{v^{N-2}\displaystyle\iint_\Delta e^{-u(r_{12})/\mathbf{k}T}\, d\mathbf{r}_1\, d\mathbf{r}_2}{v^N} \\ &= \frac{1}{v^2}\iint_\Delta [1 + e^{-u(r_{12})/\mathbf{k}T} - 1]\, d\mathbf{r}_1\, d\mathbf{r}_2 \\ &= 1 + \rho\int_0^\infty [e^{-u(r)/\mathbf{k}T} - 1]4\pi r^2\, dr \end{aligned} \tag{48.69}$$

Then we find

$$\frac{pv}{\mathbf{k}T} = 1 - \left(\frac{\sqrt{2}}{1+\sqrt{2}}\right)\frac{\rho}{2}\int_0^\infty [e^{-u(r)/\mathbf{k}T} - 1]4\pi r^2\, dr \tag{48.70}$$

From this equation we see that the theory allowing for double occupation of cells gives a second virial coefficient equal to $\sqrt{2}/(1+\sqrt{2})$ or 58.6 per cent of the correct value. We shall find in Sec. 49 that the second virial coefficient is zero if every cell is forced to be singly occupied. Thus, double occupation corrects the major part of this second virial coefficient deficiency of "single-occupation" cell theories. In the crystalline state, the values of ω_3, ω_4, etc., will be much smaller relative to ω_2 and Eq. (48.66) should become very accurate.

We now derive an approximate expression for ω_2 not restricted to low densities. We do this by treating $Z^{(0211\cdots)}$ in essentially the same way as $Z^{(1)}$. Suppose cell 1 is empty and molecules 1 and 2 are both in cell 2, which is far from cell 1. Then

$$Z^{(2)} \equiv Z^{(0211\cdots)} = \int_{\Delta_2}\int_{\Delta_2}\int_{\Delta_3}\cdots\int_{\Delta_N} e^{-U/\mathbf{k}T}\, d\mathbf{r}_1 \cdots d\mathbf{r}_N \tag{48.71}$$

and

$$P'(\mathbf{r}_1, \ldots, \mathbf{r}_N) = \frac{e^{-U/\mathbf{k}T}}{Z^{(2)}} \tag{48.72}$$

[1] Recall that Eq. (48.53) is accurate at low densities.

where P' is a probability density. Now as an approximation we write P' in the form

$$P' = \Phi(\mathbf{r}_1,\mathbf{r}_2) \prod_{s=3}^{N} \varphi(\mathbf{r}_s) \tag{48.73}$$

$$\int_\Delta \varphi(\mathbf{r})\, d\mathbf{r} = 1 \tag{48.74}$$

$$\iint_\Delta \Phi(\mathbf{r}_1,\mathbf{r}_2)\, d\mathbf{r}_1\, d\mathbf{r}_2 = 1 \tag{48.75}$$

and find those functions Φ and φ which minimize

$$A^{(2)} \equiv -\mathbf{k}T \ln Z^{(2)} \tag{48.76}$$

$$= E^{(2)} - TS^{(2)} \tag{48.77}$$

where
$$E^{(2)} = \int_{\Delta_2}\int_{\Delta_2}\int_{\Delta_3} \cdots \int_{\Delta_N} P'U\, d\mathbf{r}_1 \cdots d\mathbf{r}_N \tag{48.78}$$

and

$$S^{(2)} = -\mathbf{k}\int_{\Delta_2}\int_{\Delta_2}\int_{\Delta_3} \cdots \int_{\Delta_N} P' \ln P'\, d\mathbf{r}_1 \cdots d\mathbf{r}_N$$
$$= -(N-2)\mathbf{k}\int_\Delta \varphi(\mathbf{r}) \ln \varphi(\mathbf{r})\, d\mathbf{r} - \mathbf{k}\iint_\Delta \Phi(\mathbf{r}_1,\mathbf{r}_2) \ln \Phi(\mathbf{r}_1,\mathbf{r}_2)\, d\mathbf{r}_1\, d\mathbf{r}_2 \tag{48.79}$$

Equations (48.77) to (48.79) follow from Eqs. (48.72) and (48.76). Equation (48.78) can be written more explicitly as follows:

$$E^{(2)} = \int_{\Delta_2}\int_{\Delta_2}\int_{\Delta_3} \cdots \int_{\Delta_N} \Phi(\mathbf{r}_1,\mathbf{r}_2)\varphi(\mathbf{r}_3) \cdots \varphi(\mathbf{r}_N)[\sum_{1 \leqslant i<j \leqslant N} u(r_{ij})]d\mathbf{r}_1 \cdots d\mathbf{r}_N$$
$$= \iint_\Delta \Phi(\mathbf{r}_1,\mathbf{r}_2)u(r_{12})\, d\mathbf{r}_1\, d\mathbf{r}_2 + \sum_{i=1}^{2}\sum_{j=3}^{N} \int_{\Delta_2}\int_{\Delta_2}\int_{\Delta_j} \Phi(\mathbf{r}_1,\mathbf{r}_2)\varphi(\mathbf{r}_j)u(r_{ij})d\mathbf{r}_1\, d\mathbf{r}_2\, d\mathbf{r}_j$$
$$+ \sum_{3 \leqslant i<j \leqslant N} \int_{\Delta_i}\int_{\Delta_j} \varphi(\mathbf{r}_i)\varphi(\mathbf{r}_j)u(r_{ij})\, d\mathbf{r}_i\, d\mathbf{r}_j \tag{48.80}$$

All the $N - 2$ molecules in the last sum in Eq. (48.80) are not equivalent (some are near cells 1 and 2); thus it is more convenient to replace this sum by

$$\sum_{1 \leqslant i<j \leqslant N} \int_{\Delta_i}\int_{\Delta_j} \varphi(\mathbf{r}_i)\varphi(\mathbf{r}_j)u(r_{ij})\, d\mathbf{r}_i\, d\mathbf{r}_j - 2\sum_{j=2}^{N} \int_{\Delta_1}\int_{\Delta_j} \varphi(\mathbf{r}_1)\varphi(\mathbf{r}_j)u(r_{1j})\, d\mathbf{r}_1\, d\mathbf{r}_j \tag{48.81}$$

In (48.81) we have written the appropriate expression for every cell singly occupied and then subtracted the (equal) interactions of molecules 1 (in cell 1) and 2 (in cell 2) with their neighbors [the $u(r_{12})$ interaction is subtracted twice but this does not matter since cells 1 and 2 are far apart]. Equation (48.80) becomes

$$E^{(2)} = \iint\limits_{\Delta} \Phi(\mathbf{r}_1,\mathbf{r}_2)u(r_{12})\, d\mathbf{r}_1\, d\mathbf{r}_2 + \sum_{i=1}^{2} \int_{\Delta_2}\int_{\Delta_2}\int_{\Delta} \Phi(\mathbf{r}_1,\mathbf{r}_2)\varphi(\mathbf{r})E'(\mathbf{r}_i - \mathbf{r})\, d\mathbf{r}_1\, d\mathbf{r}_2\, d\mathbf{r}$$

$$+ \left(\frac{N}{2} - 2\right)\int_{\Delta}\int_{\Delta'} \varphi(\mathbf{r})\varphi(\mathbf{r}')E(\mathbf{r} - \mathbf{r}')\, d\mathbf{r}\, d\mathbf{r}' \quad (48.82)$$

where

$$E'(\mathbf{r}_i) = \sum_{j=3}^{N} u(\mathbf{R}_{2j} + \mathbf{r}_i) \qquad i = 1, 2 \quad (48.83)$$

$$= E(\mathbf{r}_i) \qquad i = 1, 2 \quad (48.84)$$

since the empty cell 1 is far from cell 2.

Equations (48.79) and (48.82) are substituted in Eq. (48.77), and the variation $\delta A^{(2)}/\mathbf{k}T$ with respect to both φ and Φ is taken, with Eqs. (48.74) and (48.75) as restraints. Except for negligible terms of order $1/N$, Eq. (48.33) is again obtained as the integral equation determining $\varphi(\mathbf{r})$. The Φ equation is

$$\mathbf{k}T \ln \Phi(\mathbf{r}_1,\mathbf{r}_2) + u(r_{12}) + \sum_{i=1}^{2} \int_{\Delta} \varphi(\mathbf{r})E(\mathbf{r}_i - \mathbf{r})\, d\mathbf{r} = \varepsilon + E^0 \quad (48.85)$$

where $\varepsilon + E^0$ is used as the undetermined multiplier and

$$E^0 \equiv \iint\limits_{\Delta} \Phi(\mathbf{r}_1,\mathbf{r}_2)u(r_{12})\, d\mathbf{r}_1\, d\mathbf{r}_2 + \sum_{i=1}^{2} \int_{\Delta_2}\int_{\Delta_2}\int_{\Delta} \Phi(\mathbf{r}_1,\mathbf{r}_2)\varphi(\mathbf{r})E(\mathbf{r}_i - \mathbf{r})\, d\mathbf{r}_1\, d\mathbf{r}_2\, d\mathbf{r} \quad (48.86)$$

E^0 is the interaction energy of molecules 1 and 2 with other molecules and with each other.

We define $\Psi(\mathbf{r}_1,\mathbf{r}_2)$ by

$$\Psi = -\mathbf{k}T \ln \Phi + \varepsilon \quad (48.87)$$

Then

$$\Phi = e^{(\varepsilon - \Psi)/\mathbf{k}T} \quad (48.88)$$

and

$$e^{-\varepsilon/\mathbf{k}T} = \iint\limits_{\Delta} e^{-\Psi/\mathbf{k}T}\, d\mathbf{r}_1\, d\mathbf{r}_2 \quad (48.89)$$

Equation (48.85) becomes

$$\Psi(\mathbf{r}_1,\mathbf{r}_2) = u(r_{12}) + \sum_{i=1}^{2} \int_{\Delta} \varphi(\mathbf{r}) \left[E(\mathbf{r}_i - \mathbf{r}) - \frac{E^0}{2}\right] d\mathbf{r} \quad (48.90)$$

In Eq. (48.77) we now have

$$E^{(2)} = \left(\frac{N}{2} - 2\right)\bar{\mathrm{E}} + E^0 \quad (48.91)$$

and, from Eqs. (48.79) and (48.86) to (48.90), we find

$$S^{(2)} = (N - 2)\mathbf{k} \ln v_f + \mathbf{k} \ln v_f^{(2)} \tag{48.92}$$

where

$$v_f^{(2)} = \iint_\Delta e^{-\Psi/\mathbf{k}T}\, d\mathbf{r}_1\, d\mathbf{r}_2 \tag{48.93}$$

As a zeroth approximation, let us take

$$E^0 = 2E(0)$$

and

$$\varphi(\mathbf{r}) = \delta(\mathbf{r})$$

Then

$$E_0^{(2)} = \left(\frac{N}{2} - 2\right) E(0) + 2E(0) = \frac{N}{2} E(0) \tag{48.94}$$

As a first approximation, from Eq. (48.90),

$$\begin{aligned} \Psi_1(\mathbf{r}_1,\mathbf{r}_2) &= u(r_{12}) + [E(\mathbf{r}_1) - E(0)] + [E(\mathbf{r}_2) - E(0)] \\ &= u(r_{12}) + \psi_1(\mathbf{r}_1) + \psi_1(\mathbf{r}_2) \end{aligned} \tag{48.95}$$

and

$$v_{f(1)}^{(2)} = \iint_\Delta e^{-\Psi_1/\mathbf{k}T}\, d\mathbf{r}_1\, d\mathbf{r}_2 \tag{48.96}$$

Finally, to this approximation, we have

$$\omega_2 = \frac{Z^{(2)}}{Z^{(1)}} = \frac{e^{-A^{(2)}/\mathbf{k}T}}{e^{-A^{(1)}/\mathbf{k}T}} = \frac{e^{-NE(0)/2\mathbf{k}T} v_{f(1)}^{N-2} v_{f(1)}^{(2)}}{e^{-NE(0)/2\mathbf{k}T} v_{f(1)}^{N}} = \frac{v_{f(1)}^{(2)}}{v_{f(1)}^{2}} \tag{48.97}$$

For molecule 2 in a fixed position $\mathbf{r}_2$ in the cell, the integration over $\mathbf{r}_1$ in the numerator of Eq. (48.97) moves molecule 1 throughout the cell. As far as $u(r_{12})$ is concerned [see Eqs. (48.95) and (48.96)], let us replace, as a further approximation, this wandering of molecule 1 over the entire cell by a fixed "average" position for molecule 1, namely, the cell center. That is, we replace $u(r_{12})$ by $u(r_2)$ in Eq. (48.96). Then Eq. (48.97) becomes

$$\omega_2 = \frac{v_{f(1)}^*}{v_{f(1)}} \tag{48.98}$$

where

$$v_{f(1)}^* = \int_\Delta e^{-[\psi_1(\mathbf{r}) + u(\mathbf{r})]/\mathbf{k}T}\, d\mathbf{r} \tag{48.99}$$

The quantity $v_{f(1)}^*$ is the free volume of a molecule in a cell with its center already occupied by another molecule, whereas in $v_{f(1)}$ the center is not so occupied.

Buehler et al.[1] have calculated $v_{f(1)}$ and $v^*_{f(1)}$ for a system of hard spheres, and we may use their results to compute

$$\ln \sigma = -\left(\frac{A}{N\mathbf{k}T}\right)_{\text{communal}} = \ln\left[1 + \sqrt{\frac{2v^*_{f(1)}}{v_{f(1)}}}\right] \tag{48.100}$$

as presented in[2] Table 17 (in which v_0 is the least volume per molecule into which the spheres of diameter r_0 can be packed). In the case of hard spheres, $\ln \sigma$ is independent of temperature; thus $N\mathbf{k} \ln \sigma$ is the communal entropy [see Eq. (48.21)].

TABLE 17. THE APPROXIMATE COMMUNAL FREE ENERGY OF A SYSTEM OF HARD SPHERES

$\frac{v}{v_0}$	$\frac{v_{f(1)}}{48r_0^3}$	$\frac{v^*_{f(1)}}{48r_0^3}$	ω_2	$\ln \sigma$
1.000				
2.828	.0102			
4.887	.0550	.0018	.032	.044
7.761	.1143	.0272	.238	.290
10.0	.1473	.0600	.407	.454
15.0	.2210	.1337	.605	.618
20.0	.2946	.2073	.704	.691

Examination of Table 17 reveals that the communal entropy does not become appreciable until the total volume is about five times the volume of the close-packed system of spheres. It seems probable that the results would be similar if other, more accurate, intermolecular potentials were used. Even taking into account the various approximations that have been introduced, the general conclusion would appear to be that the communal entropy is practically zero in the crystalline state and very small in a considerably expanded system constrained, however, to retain the lattice geometry and long-range order of the crystalline state. But no conclusion can be drawn about the communal entropy of the liquid state from the argument of this subsection.

It should be emphasized, finally, that *separate* minimization procedures have been introduced in connection with the free volume and communal free energy (when no more than double occupation of a cell is allowed). It would clearly be an improvement to minimize, instead, the combined free energy A of the system, using Eqs. (48.20), (48.30), (48.58), (48.66), (48.79), and (48.82). This can be done, but the resulting equations are rather complicated.

[1] R. J. Buehler, R. H. Wentorf, J. O. Hirschfelder, and C. F. Curtiss, *J. Chem. Phys.*, **19**, 61 (1951).

[2] From Pople, *loc. cit.*

Modification for Fluids. Without any approximations, Eq. (48.20) of free-volume theory is equally valid for gas, liquid, and solid states. In a fluid (gas or liquid), the singlet density $\rho^{(1)}$ must be constant, and it is easy to see qualitatively how this is possible. Consider any cell of the lattice. There is a certain probability p_0 that this cell will be empty, a probability p_1 that it will be singly occupied, etc. The singlet density inside this cell is of course zero when the cell is empty. Let it be $\rho_1^{(1)}, \rho_2^{(1)}, \ldots$ when the cell is singly, doubly, etc., occupied. At liquid densities (as an example), $\rho_1^{(1)}$ will presumably have a rather high maximum at the *center* of the cell $\mathbf{r} = 0$ and fall off very rapidly near the edges of the cell. On the other hand, $\rho_2^{(1)}, \rho_3^{(1)}$, etc., will tend to be largest near the *periphery* of the cell because of the intermolecular repulsion between the molecules in the cell at small distances apart. In any case, when these different singlet densities are combined,

$$\rho^{(1)} = p_1\rho_1^{(1)} + p_2\rho_2^{(1)} + p_3\rho_3^{(1)} + \cdots \tag{48.101}$$

and the net or total $\rho^{(1)}$ will be independent of $\mathbf{r}$ for values of N, V, and T appropriate to the gas or liquid states (in fact, $\rho^{(1)} = N/V$). In a crystal, $\rho_1^{(1)}$ predominates and $\rho^{(1)}$ will have a maximum at the center of each cell.

Thus the constancy of $\rho^{(1)}$ will automatically take care of itself in a rigorous free-volume theory of a fluid state by a rather delicate balancing of $\rho_1^{(1)}$, $\rho_2^{(1)}, \ldots$ and $p_1, p_2, \ldots$ in Eq. (48.101). However, as soon as approximations are introduced [e.g., in Eqs. (48.22) and (48.73)], we may safely assume that the balance in Eq. (48.101) will be destroyed and that $\rho^{(1)}$ will have the periodicity of the lattice at all densities (except zero). To formulate an approximate free-volume theory strictly applicable to fluid states, it is therefore necessary to introduce a new restraint *forcing* the net $\rho^{(1)}$ to be constant.[1]

As an example, we may refer to the minimization of A discussed in the last paragraph of the preceding subsection. Only double occupation of a cell is allowed; therefore in this case the full burden of "uniformizing" $\rho_1^{(1)}$ falls on $\rho_2^{(1)}$. Equation (48.60) gives the most probable number of cells containing s molecules. From Eqs. (48.60), (48.64), and (48.66), we find that

$$p_0 = p_2 = \frac{\sqrt{\omega_2}}{\sqrt{2} + 2\sqrt{\omega_2}} \tag{48.102}$$

$$p_1 = \frac{\sqrt{2}}{\sqrt{2} + 2\sqrt{\omega_2}} \tag{48.103}$$

[1] In approximate theories of fluids using the method of distribution functions (Chap. 6) essentially the same step is taken, but since $\rho^{(1)}$ occurs explicitly in the equations it is simply necessary to replace $\rho^{(1)}$ by N/V [see, for example, Eqs. (33.6) and (33.8)]. As far as fluids are concerned, this is, in fact, a rather important advantage possessed by the distribution function method relative to lattice theories.

Also
$$\rho_1^{(1)}(\mathbf{r}) = \varphi(\mathbf{r}) \tag{48.104}$$

$$\rho_2^{(1)}(\mathbf{r}) = 2\int_\Delta \Phi(\mathbf{r},\mathbf{r}')\, d\mathbf{r}' \tag{48.105}$$

Thus, if we wish to consider fluid states, A should be minimized subject to the *additional* [see Eqs. (48.74) and (48.75)] restraint

$$\frac{N}{V} = \frac{\sqrt{2}\varphi(\mathbf{r})}{\sqrt{2} + 2\sqrt{\omega_2}} + \frac{2\sqrt{\omega_2}\int_\Delta \Phi(\mathbf{r},\mathbf{r}')\, d\mathbf{r}'}{\sqrt{2} + 2\sqrt{\omega_2}} \tag{48.106}$$

We might recall, however, that Eq. (48.53) is not a good approximation for liquids. Hole theories seem more promising in this respect (Sec. 50).

With the modification introduced here, a theory of the gas-liquid transition can be developed. It might appear offhand that a theory of the liquid-solid transition could also be obtained by omitting the restraint $\rho^{(1)} = N/V$ in Eq. (48.101) for the solid state but including it for the liquid state. However, this will lead at every density to a solid state with a lower Helmholtz free energy than the liquid state if a method of free energy minimization is used.

49. The Lennard-Jones and Devonshire Theory

The model on which the Lennard-Jones and Devonshire (LJD) theory is based has been discussed briefly above in connection with Eqs. (48.47) to (48.51). These equations were obtained by making certain approximations in the general free-volume theory of Sec. 48. Historically, of course, the LJD theory was developed[1] a number of years before the general theory.

In the LJD model, the cells are singly occupied. Each molecule moves within its cell in the potential field [see Eq. (48.51)] of its neighbors fixed at the centers of their cells. To simplify the problem, the c nearest neighbors are treated as uniformly "smeared" over a spherical surface of radius a, where a is the distance between the centers of nearest-neighbor cells. Then

$$a^3 = \gamma v \tag{49.1}$$

where γ is a numerical constant depending on the geometry of the lattice. For a face-centered cubic lattice, $\gamma = \sqrt{2}$ and $c = 12$. Our task is to calculate

$$v_{f(1)} = \int_\Delta e^{-\psi_1(\mathbf{r})/\mathbf{k}T} 4\pi r^2\, dr \tag{49.2}$$

where
$$\psi_1(\mathbf{r}) = E(\mathbf{r}) - E(0) \tag{49.3}$$

[1] J. E. Lennard-Jones and A. F. Devonshire, *Proc. Roy. Soc.* (*London*), **A163**, 53 (1937); **A165**, 1 (1938).

In Fig. 61, the central molecule is at P, a distance r from the center of the cell. The area of the ring shown on the surface of the sphere is

$$2\pi a^2 \sin\theta \, d\theta$$

The number of "smeared" nearest neighbors in this area is

$$c \cdot \frac{2\pi a^2 \sin\theta \, d\theta}{4\pi a^2} = \frac{c}{2} \sin\theta \, d\theta$$

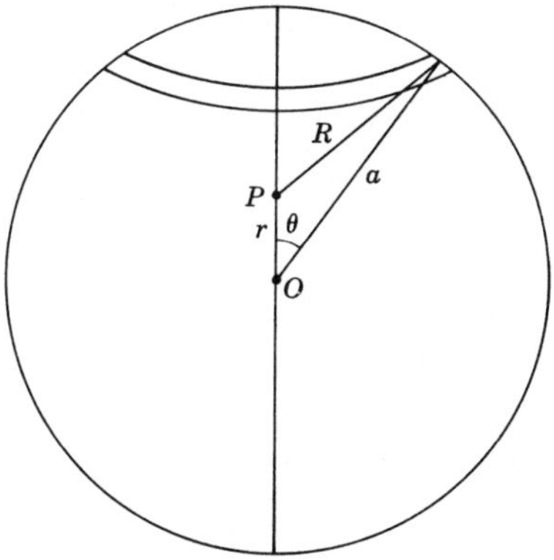

FIG. 61. Geometry within cell.

and the potential energy of interaction between the molecule at P and the neighbors in the ring is

$$u(R) \cdot \frac{c}{2} \sin\theta \, d\theta$$

where
$$R^2 = r^2 + a^2 - 2ar\cos\theta$$

Hence the total energy of interaction between the molecule at P and all of its c neighbors is

$$E(r) = \frac{c}{2} \int_0^{\pi} u(R) \sin\theta \, d\theta \tag{49.4}$$

where we take [see Eq. (27.29)]

$$u(R) = -2\varepsilon \left(\frac{r^*}{R}\right)^6 + \varepsilon \left(\frac{r^*}{R}\right)^{12} \tag{49.5}$$

We substitute Eq. (49.5) into Eq. (49.4), and obtain, on carrying out the integration,

$$\psi_1(r) = E(r) - E(0) = c\varepsilon \left[\left(\frac{r^*}{a}\right)^{12} l\left(\frac{r^2}{a^2}\right) - 2\left(\frac{r^*}{a}\right)^6 m\left(\frac{r^2}{a^2}\right) \right] \tag{49.6}$$

where $$E(0) = c\varepsilon\left[-2\left(\frac{r^*}{a}\right)^6 + \left(\frac{r^*}{a}\right)^{12}\right] \tag{49.7}$$

$$l(x) = (1 + 12x + 25.2x^2 + 12x^3 + x^4)(1 - x)^{-10} - 1 \tag{49.8}$$

$$m(x) = (1 + x)(1 - x)^{-4} - 1 \tag{49.9}$$

If we define

$$\Lambda^* = c\varepsilon \tag{49.10}$$

$$v^* = \frac{v}{a^3}r^{*3} = \frac{r^{*3}}{\gamma} \tag{49.11}$$

Equations (49.6) and (49.7) become

$$\psi_1(r) = E(r) - E(0) = \Lambda^*\left[\left(\frac{v^*}{v}\right)^4 l\left(\frac{r^2}{a^2}\right) - 2\left(\frac{v^*}{v}\right)^2 m\left(\frac{r^2}{a^2}\right)\right] \tag{49.12}$$

$$E(0) = \Lambda^*\left[-2\left(\frac{v^*}{v}\right)^2 + \left(\frac{v^*}{v}\right)^4\right] \tag{49.13}$$

When v/v^* is small ($<\sim 1.6$), $\psi_1(r)$ has a minimum at $r = 0$ and rises rapidly as r increases. For large v/v^*, as should be expected from the physical model, $\psi_1(r)$ has a low maximum at $r = 0$, a minimum near $r = a - r^*$, and rises rapidly when $r > a - r^*$.

The potential $\psi_1(r)$ in Eq. (49.12) is substituted in Eq. (49.2), and we find (putting $y = r^2/a^2$)

$$v_{f(1)} = 2\pi a^3 g \tag{49.14}$$

$$g = \int_0^s \exp\left\{-\frac{\Lambda^*}{\mathbf{k}T}\left[\left(\frac{v^*}{v}\right)^4 l(y) - 2\left(\frac{v^*}{v}\right)^2 m(y)\right]\right\} y^{1/2}\,dy$$

The upper limit s in the integral is determined by the consideration that, when $\Lambda^*/\mathbf{k}T = 0$ (perfect gas),

$$v_{f(1)} = v = 2\pi a^3 \int_0^s y^{1/2}\,dy$$

or $$s = \left(\frac{3}{4\pi\gamma}\right)^{2/3}$$

The Helmholtz free energy of the system is

$$A = -N\mathbf{k}T \ln \frac{v_{f(1)}\sigma}{\Lambda^3} + \frac{NE(0)}{2} \tag{49.15}$$

and all thermodynamic properties may be found from this equation provided some decision is made about the value of σ. In the original LJD theory (of the liquid state) $\sigma = e$ was chosen, but there is no known justification for this. If $\sigma =$ constant, the particular choice of the constant does not affect the equation of state, the internal energy, etc., but it does affect the chemical potential, vapor pressure, boiling point, etc. Instead of $\sigma =$ constant one might consider using the expression given in Eq. (48.100), so that, in Eq. (49.15),

$$v_{f(1)}\sigma = v_{f(1)} + \sqrt{2v^*_{f(1)}v_{f(1)}} \tag{49.16}$$

Unfortunately, $v^*_{f(1)}$ has not been calculated for the potential of Eq. (49.5). But a more serious objection is that a nonconstant σ is logically inconsistent with Kirkwood's derivation of the LJD equations (see Sec. 48).

If $\sigma =$ constant, the equation of state, from Eqs. (48.68) and (49.15), is

$$\frac{pv}{\mathbf{k}T} = 1 - \frac{2\Lambda^*}{\mathbf{k}T}\left[\left(\frac{v^*}{v}\right)^2 - \left(\frac{v^*}{v}\right)^4\right] + 4\frac{\Lambda^*}{\mathbf{k}T}\left[\left(\frac{v^*}{v}\right)^4\frac{g_l}{g} - \left(\frac{v^*}{v}\right)^2\frac{g_m}{g}\right] \tag{49.17}$$

where

$$g_l = \int_0^s \exp\left\{-\frac{\Lambda^*}{\mathbf{k}T}\left[\left(\frac{v^*}{v}\right)^4 l(y) - 2\left(\frac{v^*}{v}\right)^2 m(y)\right]\right\} y^{1/2} l(y)\, dy \tag{49.18}$$

$$g_m = \int_0^s \exp\left\{-\frac{\Lambda^*}{\mathbf{k}T}\left[\left(\frac{v^*}{v}\right)^4 l(y) - 2\left(\frac{v^*}{v}\right)^2 m(y)\right]\right\} y^{1/2} m(y)\, dy \tag{49.19}$$

The quantities g, g_l, and g_m have to be calculated by numerical integration for each pair of values of v^*/v and $\Lambda^*/\mathbf{k}T$. Extensive tables of g, g_l, g_m and several thermodynamic properties have been published by Wentorf et al.[1] Some of these results will be included in the comparisons of different theories made in Sec. 50. We might mention here, though, that (1) $pv/\mathbf{k}T$ is a function of $\Lambda^*/\mathbf{k}T$ and v^*/v only, and hence the model obeys the law of corresponding states [as it must, since we have adopted the two-parameter potential, Eq. (49.5)]; (2) Eqs. (49.17) to (49.19) give a loop in $pv^*/\mathbf{k}T$ versus v/v^* at low enough temperatures, thus predicting a first-order phase transition (actually, a sublimation); and (3) the expansion of $pv/\mathbf{k}T$ in powers of v^*/v has no first power term, so that the second virial coefficient is zero. This is a consequence of the single-occupation restriction, for at low densities (large cells) the frequency of binary collisions at the surfaces of the cells becomes of a negligible order.

In view of the single-occupation assumption, the LJD theory can obviously be taken very seriously only for a crystal. To be strictly applicable to the liquid state, the theory must be modified (see Sec. 48) to remove the long-range (periodic) order associated with the lattice structure.

[1] R. H. Wentorf, R. J. Buehler, J. O. Hirschfelder, and C. F. Curtiss, *J. Chem. Phys.*, **18**, 1484 (1950). The tables given actually include the contribution of second and third neighbor shells, but this refinement turns out to have little effect on thermodynamic properties.

50. Hole Theories of the Liquid and Solid States

In so-called "hole theories" of liquids, a lattice approach is again used but the cells are chosen small enough so that double occupation can be neglected because of intermolecular repulsion at small separations. Suppose the lattice geometry is chosen to conform with the crystal geometry (this is not necessary in a completely rigorous treatment but it is helpful when approximations are to be introduced), and that the volume V is divided up into $\mathscr{B}$ cells, where $\mathscr{B} \geqslant N$. Then Eq. (48.3) becomes

$$Z = \sum_{l_1=1}^{\mathscr{B}} \cdots \sum_{l_N=1}^{\mathscr{B}} \int_{\Delta_{l_1}} \cdots \int_{\Delta_{l_N}} e^{-U/\mathbf{k}T} \, d\mathbf{r}_1 \cdots d\mathbf{r}_N \tag{50.1}$$

There are $\mathscr{B}^N$ terms in this sum but only $\mathscr{B}!/(\mathscr{B} - N)!$ of these avoid multiple occupation in all cells and are therefore nonzero. Let $Z^{(m_1 \cdots m_{\mathscr{B}})}$ be the partition function of the system when there are m_1 molecules in cell 1, m_2 in cell 2, etc., where m_i can be zero or unity. Then, as in Eq. (48.5),

$$Z = N! \sum_{\substack{m_1, \ldots, m_{\mathscr{B}}=0 \\ (\Sigma m_s = N)}}^{1} Z^{(m_1 \cdots m_{\mathscr{B}})} \tag{50.2}$$

where there are $\mathscr{B}!/N!(\mathscr{B} - N)!$ terms in this sum.

Equation (50.2) is exact (for small enough cells) and applies equally well to a gas, liquid, or solid. The same free energy will of course be obtained for different (sufficiently small) cell sizes. Every configuration or term (choice of $m_1, \ldots, m_{\mathscr{B}}$) in Eq. (50.2) refers to a system with N occupied cells and $\mathscr{B} - N$ vacant cells or holes; there is no natural separation into free volume and communal free energy as in Eqs. (48.7) to (48.21). The consideration of communal free energy is thus completely avoided here; it is a well-defined [Eq. (48.8)] property of the system, but obviously not one that it is essential to discuss in order to derive thermodynamic quantities.[1]

Let us rewrite Eq. (50.2) as

$$\frac{Z}{N!} = \sum_{\mathbf{l}} Z^{(\mathbf{l})} \tag{50.3}$$

where $\mathbf{l}$ represents a "configuration" $l_1, l_2, \ldots, l_N$ and l_i signifies that molecule i is in cell Δ_{l_i}. That is,

$$Z^{(\mathbf{l})} = \int_{\Delta_{l_1}} \cdots \int_{\Delta_{l_N}} e^{-U/\mathbf{k}T} \, d\mathbf{r}_1 \cdots d\mathbf{r}_N \tag{50.4}$$

The sum in Eq. (50.3) is over the $\mathscr{B}!/N!(\mathscr{B} - N)!$ different ways of placing N molecules in $\mathscr{B}$ cells. The actual numbering of the molecules in any

[1] Of course, the same remark applies as well to the concept of "holes."

configuration of filled and empty cells is arbitrary, but some definite choice is made. We define the probability density $P^{(\mathbf{l})}$ for the configuration $\mathbf{l}$ by

$$P^{(\mathbf{l})} = \frac{e^{-U/\mathbf{k}T}}{Z^{(\mathbf{l})}} \tag{50.5}$$

We introduce the approximation[1] for each configuration, that

$$P^{(\mathbf{l})}(\mathbf{r}_1, \ldots, \mathbf{r}_N) = \prod_{s=1}^{N} \varphi_s(\mathbf{r}_s,\mathbf{l}) \tag{50.6}$$

$$\int_\Delta \varphi_s(\mathbf{r},\mathbf{l})\, d\mathbf{r} = 1 \tag{50.7}$$

We now wish to maximize $Z/N!$ (minimize A) with respect to the functions $\varphi_s(\mathbf{r}_s,\mathbf{l})$. Since different configurations $\mathbf{l}$ in the sum in Eq. (50.3) do not "interact," the sum is maximized when each $Z^{(\mathbf{l})}$ is maximized separately. Hence we define, for an arbitrary configuration $\mathbf{l}$,

$$A^{(\mathbf{l})} = -\mathbf{k}T \ln Z^{(\mathbf{l})} \tag{50.8}$$

$$= E^{(\mathbf{l})} - TS^{(\mathbf{l})} \tag{50.9}$$

where

$$E^{(\mathbf{l})} = \int_{\Delta_{l_1}} \cdots \int_{\Delta_{l_N}} UP^{(\mathbf{l})} d\mathbf{r}_1 \ldots d\mathbf{r}_N \tag{50.10}$$

$$S^{(\mathbf{l})} = -\mathbf{k} \int_{\Delta_{l_1}} \cdots \int_{\Delta_{l_N}} P^{(\mathbf{l})} \ln P^{(\mathbf{l})}\, d\mathbf{r}_1 \cdots d\mathbf{r}_N \tag{50.11}$$

and proceed below to minimize $A^{(\mathbf{l})}$. Equations (50.9) to (50.11) follow from Eq. (50.5). We note that, using Eq. (50.6),

$$S^{(\mathbf{l})} = -\mathbf{k} \sum_{i=1}^{N} \int_\Delta \varphi_i(\mathbf{r},\mathbf{l}) \ln \varphi_i(\mathbf{r},\mathbf{l})\, d\mathbf{r} \tag{50.12}$$

and

$$E^{(\mathbf{l})} = \sum_{1 \leqslant i<j \leqslant N} \int_{\Delta_{l_i}} \int_{\Delta_{l_j}} \varphi_i(\mathbf{r}_i,\mathbf{l})\varphi_j(\mathbf{r}_j,\mathbf{l}) u(r_{ij})\, d\mathbf{r}_i\, d\mathbf{r}_j$$

$$= \int_\Delta \int_{\Delta'} [\sum_{1 \leqslant i<j \leqslant N} \varphi_i(\mathbf{r},\mathbf{l})\varphi_j(\mathbf{r}',\mathbf{l}) u(\mathbf{R}_{ij} + \mathbf{r} - \mathbf{r}')]\, d\mathbf{r}\, d\mathbf{r}' \tag{50.13}$$

It is convenient to use $\alpha_i(\mathbf{l}) + (N-1)\bar{\bar{\mathrm{E}}}(\mathbf{l})$ as undetermined multipliers in minimizing $A^{(\mathbf{l})}$ subject to Eq. (50.7), where

$$\bar{\bar{\mathrm{E}}}(\mathbf{l}) \equiv \frac{2}{N(N-1)} E^{(\mathbf{l})} \tag{50.14}$$

[1] The general argument from this point on was suggested by the papers of Z. W. Salsburg and J. G. Kirkwood [*J. Chem. Phys.*, **20**, 1538 (1952); **21**, 2169 (1953)] on a cell theory for mixtures. Here holes are the second component.

Then, from Eqs. (50.7), (50.9), (50.12), and (50.13), we find

$$\sum_{i=1}^{N} \int_{\Delta} \left\{ \int_{\Delta'} [\sum_{\substack{j=1 \\ j \neq i}}^{N} \varphi_j(\mathbf{r}',\mathbf{l}) u(\mathbf{R}_{ij} + \mathbf{r} - \mathbf{r}')] \, d\mathbf{r}' + \mathbf{k}T \ln \varphi_i(\mathbf{r},\mathbf{l}) - \alpha_i(\mathbf{l}) - (N-1)\bar{\bar{\mathrm{E}}}(\mathbf{l}) \right\} \delta\varphi_i(\mathbf{r}) \, d\mathbf{r} = 0$$

or

$$\ln \varphi_i(\mathbf{r},\mathbf{l}) + \frac{1}{\mathbf{k}T} \int_{\Delta'} [\sum_{\substack{j=1 \\ j \neq i}}^{N} \varphi_j(\mathbf{r}',\mathbf{l}) u(\mathbf{R}_{ij} + \mathbf{r} - \mathbf{r}')] \, d\mathbf{r}' = \frac{\alpha_i(\mathbf{l})}{\mathbf{k}T} + \frac{(N-1)\bar{\bar{\mathrm{E}}}(\mathbf{l})}{\mathbf{k}T} \qquad i = 1, 2, \ldots, N \tag{50.15}$$

If we define $\psi_i(\mathbf{r},\mathbf{l})$ by

$$\varphi_i(\mathbf{r},\mathbf{l}) = \exp\left[\frac{\alpha_i(\mathbf{l}) - \psi_i(\mathbf{r},\mathbf{l})}{\mathbf{k}T}\right]$$
$$\exp\left[-\frac{\alpha_i(\mathbf{l})}{\mathbf{k}T}\right] = \int_{\Delta} \exp\left[-\frac{\psi_i(\mathbf{r},\mathbf{l})}{\mathbf{k}T}\right] d\mathbf{r} \tag{50.16}$$

Eq. (50.15) becomes

$$\psi_i(\mathbf{r},\mathbf{l}) = \int_{\Delta'} \left\{ \sum_{\substack{j=1 \\ j \neq i}}^{N} \exp\left[\frac{\alpha_j(\mathbf{l}) - \psi_j(\mathbf{r}',\mathbf{l})}{\mathbf{k}T}\right] [u(\mathbf{R}_{ij} + \mathbf{r} - \mathbf{r}') - \bar{\bar{\mathrm{E}}}(\mathbf{l})] \right\} d\mathbf{r}' \qquad i = 1, 2, \ldots, N \tag{50.17}$$

This is a system of N interconnected integral equations, and there is one such system for each configuration $\mathbf{l}$.

Let us define the free volume $v_i(\mathbf{l})$ by

$$v_i(\mathbf{l}) = \int_{\Delta} \exp\left[-\frac{\psi_i(\mathbf{r},\mathbf{l})}{\mathbf{k}T}\right] d\mathbf{r} \tag{50.18}$$

Then, from Eqs. (50.12), (50.16), (50.17), and (50.18),

$$S^{(\mathbf{l})} = \mathbf{k} \ln [\prod_{i=1}^{N} v_i(\mathbf{l})] \tag{50.19}$$

and

$$Z^{(\mathbf{l})} = \exp\left[-\frac{A^{(\mathbf{l})}}{\mathbf{k}T}\right] = \exp\left[-\frac{N(N-1)}{2}\bar{\bar{\mathrm{E}}}(\mathbf{l})\right] \prod_{i=1}^{N} v_i(\mathbf{l}) \tag{50.20}$$

As a zeroth approximation, in the solution of Eq. (50.17), we put

$$\varphi_{j(0)}(\mathbf{r},\mathbf{l}) = \delta(\mathbf{r})$$

which leads to the first approximation

$$\psi_{i(1)}(\mathbf{r},\mathbf{l}) = \sum_{\substack{j=1 \\ j\neq i}}^{N} u(\mathbf{R}_{ij} + \mathbf{r}) - (N-1)\overline{\overline{\mathrm{E}}}_{(0)}(\mathbf{l}) \tag{50.21}$$

$$(N-1)\overline{\overline{\mathrm{E}}}_{(0)}(\mathbf{l}) = \frac{2}{N} \sum_{1\leqslant i<j\leqslant N} u(\mathbf{R}_{ij}) \tag{50.22}$$

$$v_{i(1)}(\mathbf{l}) = \int_{\Delta} \exp\left[-\frac{\psi_{i(1)}(\mathbf{r},\mathbf{l})}{\mathbf{k}T}\right] d\mathbf{r} \tag{50.23}$$

As a further approximation, we assume that the cell size (which is fixed by the distance a' between centers of nearest-neighbor cells) can be chosen in such a way that not only double occupation of a cell is excluded but also only interactions between molecules in nearest-neighbor cells need be taken into account. In general, these requirements cannot both be satisfied without some error. With this nearest-neighbor assumption introduced, the problem begins to take on some aspects of the nearest-neighbor problems discussed in Chap. 7.

Eq. (50.21) becomes, then,

$$\psi_{i(1)}(\mathbf{r},\mathbf{l}) = \sum_{\substack{\text{n.n. to} \\ \Delta_{l_i} \text{ in } \mathbf{l}}} u(\mathbf{R}_{ij} + \mathbf{r}) - (N-1)\overline{\overline{\mathrm{E}}}_0(\mathbf{l}) \tag{50.24}$$

$$(N-1)\overline{\overline{\mathrm{E}}}_0(\mathbf{l}) = \frac{2N_{AA}(\mathbf{l})u(a')}{N} \tag{50.25}$$

where the sum is over those cells nearest neighbor to Δ_{l_i} which are occupied in the configuration **l**, and $N_{AA}(\mathbf{l})$ is the number of nearest-neighbor pairs of cells both occupied, in the configuration **l**.

Let $\sum_i^{(1)} u$ represent the nearest-neighbor sum in Eq. (50.24). Then, in Eq. (50.20),

$$\prod_{i=1}^{N} v_{i(1)}(\mathbf{l}) = \exp\left[\frac{2N_{AA}(\mathbf{l})u(a')}{\mathbf{k}T}\right] \prod_{i=1}^{N} \int_{\Delta} \exp\left[-\frac{\Sigma_i^{(1)}u}{\mathbf{k}T}\right] d\mathbf{r} \tag{50.26}$$

Now suppose, in the configuration **l**, that there are $h_j(\mathbf{l})$ occupied cells each of which has j vacant nearest-neighbor cells. Clearly,

$$\sum_{j=0}^{c} h_j(\mathbf{l}) = N \tag{50.27}$$

$$\sum_{j=1}^{c} jh_j(\mathbf{l}) = N_{AB}(\mathbf{l}) \tag{50.28}$$

where $N_{AB}(\mathbf{l})$ is the number of nearest-neighbor pairs of cells with one cell occupied and the other vacant. Now define an average nearest-neighbor interaction energy $\overline{U(\mathbf{r},j,\mathbf{l})}$ for those molecules with j nearest-neighbor vacant cells by

$$\left\{\int_\Delta \exp\left[-\frac{\overline{U(\mathbf{r},j,\mathbf{l})}}{\mathbf{k}T}\right] d\mathbf{r}\right\}^{h_j(\mathbf{l})} = \prod_{\substack{i \\ (j\text{ n.n. holes})}} \int_\Delta \exp\left[-\frac{\sum_i^{(\mathbf{l})} u}{\mathbf{k}T}\right] d\mathbf{r} \tag{50.29}$$

Then, since [see Eq. (41.17)]

$$2N_{AA}(\mathbf{l}) = cN - N_{AB}(\mathbf{l}) = \sum_{j=0}^{c} (c - j) h_j(\mathbf{l})$$

Eq. (50.26) can be written

$$\prod_{i=1}^{N} v_{i(\mathbf{l})}(\mathbf{l}) = \prod_{j=0}^{c} (v_{[j]}(\mathbf{l}))^{h_j(\mathbf{l})}$$

where
$$v_{[j]}(\mathbf{l}) = \int_\Delta \exp\left\{-\frac{[\overline{U(\mathbf{r},j,\mathbf{l})} - (c-j)u(a')]}{\mathbf{k}T}\right\} d\mathbf{r} \tag{50.30}$$

and Eq. (50.20) becomes

$$Z^{(\mathbf{l})} = \exp\left\{-\frac{[cN - N_{AB}(\mathbf{l})]u(a')}{2\mathbf{k}T}\right\} \prod_{j=0}^{c} (v_{[j]}(\mathbf{l}))^{h_j(\mathbf{l})} \tag{50.31}$$

The quantity $v_{[j]}(\mathbf{l})$ is an average free volume for molecules with j vacant nearest-neighbor cells in the configuration $\mathbf{l}$.

We next introduce an approximation for $\overline{U(\mathbf{r},j,\mathbf{l})}$. If we smear the $c - j$ nearest-neighbor molecules over a sphere of radius a', we have

$$\overline{U(\mathbf{r},j,\mathbf{l})} = \frac{c-j}{c} E(\mathbf{r}) \tag{50.32}$$

and
$$\overline{U(\mathbf{r},j,\mathbf{l})} - (c-j)u(a') = \frac{c-j}{c}[E(\mathbf{r}) - E(0)]$$

$$= \left(1 - \frac{j}{c}\right) \psi_1(\mathbf{r}) \tag{50.33}$$

where $E(\mathbf{r})$, $E(0)$, and $\psi_1(\mathbf{r})$ are the LJD functions of Sec. 49 except that a is replaced by a'. Then

$$v_{[j]} = \int_\Delta e^{-[1-(j/c)]\psi_1(\mathbf{r})/\mathbf{k}T}\, d\mathbf{r} \tag{50.34}$$

which is the same for all $\mathbf{l}$, and the volume of the cell Δ is $V/\mathscr{B} = a'^3/\gamma$ ($\gamma = \sqrt{2}$ if $c = 12$). This smearing of molecules is not serious in the LJD theory where there are no neighboring holes and hence where the symmetry is practically spherical. But, with neighboring holes, spherical symmetry is lacking. The largest factors in the product on the right-hand side of Eq. (50.29) will, in fact, be associated with unsymmetrical arrangements of neighboring holes which allow the central molecule to move relatively freely away from the center of its cell and in the direction of the vacant neighboring cells. These large factors in Eq. (50.29) are not given proper weight in carrying over the "smearing" technique to the present problem.

Now let $\Gamma(h_0, \ldots, h_c, \mathscr{B})$ be the number of configurations **1** with h_j occupied cells each of which has j ($j = 0, 1, \ldots, c$) vacant nearest-neighbor cells. Then

$$\sum_{j=0}^{c} h_j = N \tag{50.35}$$

$$\sum_{j=1}^{c} jh_j = N_{AB} \tag{50.36}$$

$$\sum_{\substack{h_0, \ldots, h_c \\ (\Sigma h_j = N) \\ (\Sigma jh_j = N_{AB})}} \Gamma(h_0, \ldots, h_c, \mathscr{B}) = g(N, N_{AB}, \mathscr{B}) \tag{50.37}$$

where $g(N, N_{AB}, \mathscr{B})$ is the function already introduced in Chap. 7. That is,

$$\sum_{N_{AB}} g(N, N_{AB}, \mathscr{B}) = \frac{\mathscr{B}!}{N!(\mathscr{B} - N)!} \tag{50.38}$$

If, in Eq. (50.31), we replace $v_{[j]}(\mathbf{1})$ by $v_{[j]}$ [Eq. (50.34)], $Z^{(\mathbf{1})}$ obviously has the same value for all of the Γ configurations belonging to the set $h_0, \ldots, h_c$. Let us denote this value by $Z^{[\mathbf{h}]}$. Then, from Eqs. (50.3) and (50.31),

$$\frac{Z}{N!} = \sum_{\substack{h_0, \ldots, h_c \\ (\Sigma h_j = N)}} \Gamma(h_0, \ldots, h_c, \mathscr{B}) Z^{[\mathbf{h}]} \tag{50.39}$$

$$= \sum_{N_{AB}} \sum_{\substack{h_0, \ldots, h_c \\ (\Sigma h_j = N,\ \Sigma jh_j = N_{AB})}} \Gamma(h_0, \ldots, h_c, \mathscr{B}) Z^{[\mathbf{h}]} \tag{50.40}$$

$$= e^{-cNu(a')/2\mathbf{k}T} \sum_{N_{AB}} e^{N_{AB}u(a')/2\mathbf{k}T} \sum_{\substack{h_0, \ldots, h_c \\ (\Sigma h_j = N,\ \Sigma jh_j = N_{AB})}} \Gamma(h_0, \ldots, h_c, \mathscr{B}) \prod_{j=0}^{c} v_{[j]}{}^{h_j} \tag{50.41}$$

With some approximation[1] for $\Gamma(h_0, \ldots, h_c, \mathscr{B})$, Eq. (50.41) can be used to calculate thermodynamic properties. Actually, no such calculations have yet been carried out. Instead, a further approximation has been introduced

[1] The exact expression for Γ is unknown for two- and three-dimensional lattices. See Eq. (50.37) and Chap. 7.

to simplify $v_{[j]}$. Figure 62 shows a typical computed curve of ln $v_{[j]}$ versus j/c. When $j/c = 0$, $v_{[0]}$ is just the LJD free volume (with a' in place of a). When $j/c = 1$, $v_{[c]} = V/\mathscr{B} = a'^3/\gamma$. The approximation is to replace the actual dependence of ln $v_{[j]}$ on j/c [Eq. (50.34)] by a linear relationship:

$$\ln v_{[j]} = \left(1 - \frac{j}{c}\right) \ln \alpha_0 + \frac{j}{c} \ln \alpha_1 \tag{50.42}$$

where α_0 and α_1 are constants. Equation (50.42) leads to a considerable simplification in Eq. (50.41), for

$$\prod_{j=0}^{c} v_{[j]}{}^{h_j} = \prod_{j=0}^{c} \left[\alpha_0 \left(\frac{\alpha_1}{\alpha_0}\right)^{j/c}\right]^{h_j} = \alpha_0{}^N \left(\frac{\alpha_1}{\alpha_0}\right)^{N_{AB}/c}$$

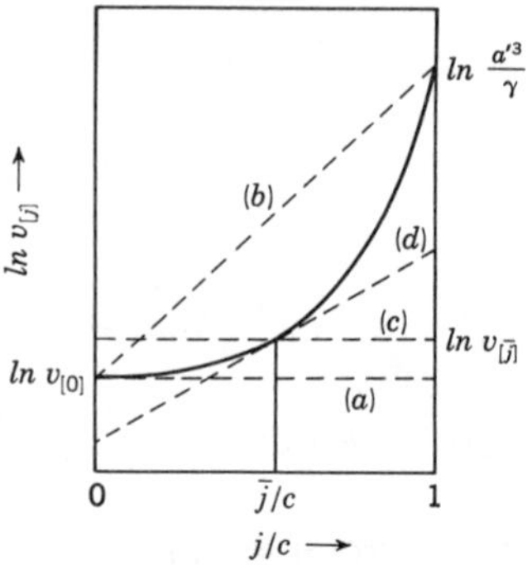

FIG. 62. Four approximations: (*a*) Cernuschi and Eyring; (*b*) Ono; (*c*) Peek and Hill; (*d*) Rowlinson and Curtiss.

Using Eq. (50.37), we find

$$\frac{Z}{N!} = e^{-cNu(a')/2\mathbf{k}T} \alpha_0{}^N \sum_{N_{AB}} g(N, N_{AB}, \mathscr{B}) \left[e^{u(a')/2\mathbf{k}T} \left(\frac{\alpha_1}{\alpha_0}\right)^{1/c}\right]^{N_{AB}} \tag{50.43}$$

If $\alpha_1 = \alpha_0$ (that is, $v_{[j]} =$ constant), Eq. (50.43) gives

$$Q = \left(\frac{\alpha_0}{\Lambda^3}\right)^N e^{-cNu(a')/2\mathbf{k}T} \sum_{N_{AB}} g(N, N_{AB}, \mathscr{B}) e^{u(a')N_{AB}/2\mathbf{k}T} \tag{50.44}$$

which is identical with Eqs. (41.18), (41.19), and (41.28) for a lattice gas, with

$$\varepsilon_{AB} = \varepsilon_{BB} = 0$$

$$j \text{ (lattice gas)} \leftrightarrow \frac{\alpha_0}{\Lambda^3}$$

$$\varepsilon_{AA} \text{ (lattice gas)} \leftrightarrow u(a')$$

The quasi-chemical approximation (Sec. 46) can now be introduced for $g(N,N_{AB},\mathscr{B})$ in Eq. (50.43), leading to equations for which numerical computations are practical. We first deduce, from Eqs. (41.18), (41.19), (41.28), (46.25), (46.26), and (in the notation of Chap. 7)

$$A = -\mathbf{k}T \ln Q = N\mu - p\mathscr{B}$$

that

$$\sum_{N_{AB}} gx^{N_{AB}} = \left\{\left[\frac{(\beta+1)(1-\rho)}{\beta+1-2\rho}\right]^{c/2}\frac{1}{1-\rho}\right\}^{\mathscr{B}}\left\{\left[\frac{(\beta+1-2\rho)\rho}{(\beta-1+2\rho)(1-\rho)}\right]^{c/2}\frac{1-\rho}{\rho}\right\}^{N} \tag{50.45}$$

where
$$\rho = \frac{N}{\mathscr{B}}$$

$$\beta = [1 - 4\rho(1-\rho)(1-x^{-2})]^{1/2}$$

Equation (50.45) may be substituted for the sum in Eq. (50.43), if we put

$$x = e^{u(a')/2\mathbf{k}T}\left(\frac{\alpha_1}{\alpha_0}\right)^{1/c} \tag{50.46}$$

Numerical results, based on Eqs. (50.43), (50.45), and (50.46), have been obtained in four different "linear" cases,[1] as indicated in Fig. 62. Rowlinson and Curtiss have given a detailed and critical analysis of these cases. We shall confine ourselves to a rather brief discussion. Cernuschi and Eyring used approximation (*a*), Fig. 62, $v_{[j]} = v_{[0]}$ (= constant). As we have seen in Eq. (50.44), this means that the Cernuschi and Eyring treatment is identical with the quasi-chemical approximation to a lattice gas. The equation of state and critical properties[2] have already been deduced in Chap. 7. In Ono's approximation (*b*), Fig. 62, the same formal expressions for the equation of state and critical properties are obtained as in case (*a*), but here x is given by [Eq. (50.46)]

$$x = e^{u(a')/2\mathbf{k}T}\left(\frac{v_{[c]}}{v_{[0]}}\right)^{1/c}$$

We have already mentioned that in an exact calculation of Z, using a lattice and Eq. (50.2), the same thermodynamic properties would have to be obtained with any choice of cell size, provided only that the cells are small enough so that double occupation can be neglected. But as soon as approximations are introduced, the results become dependent on the size of the cell.

[1] F. Cernuschi and H. Eyring, *J. Chem. Phys.*, **7**, 547 (1939); S. Ono, *Mem. Fac. Eng. Kyushu Univ.*, **10**, 190 (1947); H. M. Peek and T. L. Hill, *J. Chem. Phys.*, **18**, 1252 (1950); J. S. Rowlinson and C. F. Curtiss, *J. Chem. Phys.*, **19**, 1519 (1951).

[2] Equations (46.26) and (46.28) with $x = \exp[u(a')/2\mathbf{k}T]$.

Perhaps the most natural choice of a' is $a' = r^*$, for this would be the nearest-neighbor intermolecular distance in the crystalline state (with nearest-neighbor interactions only) at zero pressure and $T = 0°K$. This is the value of a' used in obtaining the numerical results mentioned below for cases (*a*) and (*b*).

Instead of choosing a definite value of a', in approximate evaluations of Z, a more complicated but more refined procedure is to select that value of a' which minimizes A for given values of N, V, and T. Of course the value of a' adopted from this procedure for each N, V, and T should be consistent, to a reasonable approximation, with our neglect of multiple occupation of cells and of higher neighbor interactions. Peek and Hill [(*c*) in Fig. 62] and Rowlinson and Curtiss [(*d*) in Fig. 62] used this minimization procedure. Incidentally, no simple expression can be derived for the critical properties when this principle of free-energy minimization is employed. Peek and Hill chose $v_{[j]} = v_{[\bar{j}]}$ (= constant), where $\bar{j}$ is the mean value of j as deduced from the quasi-chemical approximation. This was improved by Rowlinson and Curtiss who selected their straight line (Fig. 62) not only to pass through the point $v_{[j]} = v_{[\bar{j}]}$, $j = \bar{j}$ but also to have the correct slope at this point.

Numerical Results. We summarize here[1] some of the properties of the four special cases described above. The results are not especially encouraging. However, we mention again that, in comparing these results with experimental properties of gases and liquids, we must keep in mind the fact that the equations are strictly suitable only for a system with periodic order (crystal or lattice gas); also, *many* approximations have been made (see above).

a. Second virial coefficient. The second virial coefficient B_2 is defined by

$$\frac{pV}{N\mathbf{k}T} = 1 + \frac{B_2}{v} + \frac{B_3}{v^2} + \cdots$$

Figure 63 presents B_2/b_0 versus the reduced temperature $\mathbf{k}T/\varepsilon$, where $b_0 = \sqrt{2}\pi r^{*3}/3$. The curve labeled "exact" has been computed from Eqs. (22.26), (25.9), (25.31), and (49.5). Approximations (*a*) and (*b*) are seen to be much worse than (*c*) and (*d*), while (*d*) is somewhat better than (*c*). The LJD theory, it will be recalled, gives $B_2 = 0$.

b. Critical constants. The critical constants are presented in Table 18 (see also Table 11, Chap. 6), in which $v^* = r^{*3}/\sqrt{2}$. None of the theories is particularly good and not much improvement over LJD is shown. The experimental value of $\bar{j}_{crit}/c$ in the table is a rough estimate from X-ray

[1] From Rowlinson and Curtiss, *loc. cit.* See this paper for further details. The LJD results are from Wentorf et al., *loc. cit.*

TABLE 18. CRITICAL CONSTANTS

	$\frac{\mathbf{k}T_{\text{crit}}}{\varepsilon}$	$\frac{p_{\text{crit}}v^*}{\varepsilon}$	$\frac{v_{\text{crit}}}{v^*}$	$\frac{p_{\text{crit}}v_{\text{crit}}}{\mathbf{k}T_{\text{crit}}}$	$\frac{\bar{\jmath}_{\text{crit}}}{c}$
Mean value for:					
Ne, N_2, A, CH_4	1.277	.121	3.09	.292	(.6–.7)
LJD	1.30	.434	1.77	.591	0
(*a*) Cernuschi and Eyring	2.74	.469	2.00	.342	.455
(*b*) Ono	0.75	.128	2.00	.342	.455
(*c*) Peek and Hill ..	1.18	.261	3.25	.719	.175
Mayer and Careri ..	1.42	.414	2.32	.676	~ .23

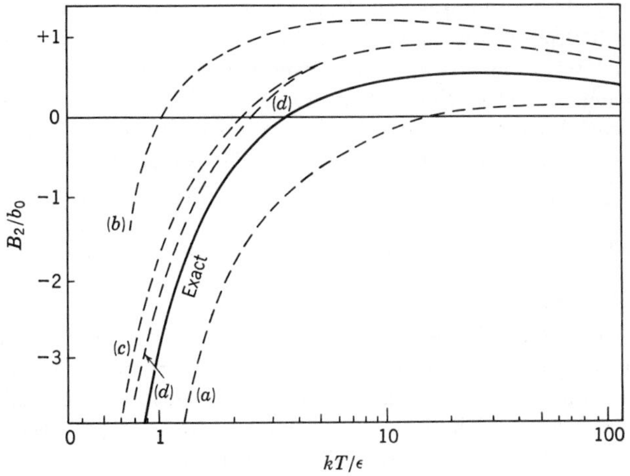

FIG. 63. Second virial coefficient against temperature according to the four approximations.

diffraction measurements on argon.[1] The number of holes in the more refined case (*c*) is much too small; in fact, at the boiling point of the liquid, Peek and Hill found $\bar{\jmath}/c = 0.004$, whereas the experimental value for argon is about 0.125. It seems very likely that this scarcity of holes is related (*a*) to the fact that in the theory the "liquid" is not a liquid but a crystal and (*b*) to the "smearing" approximation, Eq. (50.32). A modification (see below) to ensure a uniform $\rho^{(1)}$ for the liquid state would presumably introduce many more holes.

[1] See Peek and Hill, *loc. cit.*

c. Vapor pressure. Both the LJD and Ono theories predict that $\ln p_0$ (where p_0 is the vapor pressure) varies linearly with $1/T$, in qualitative agreement with experiment. The lines may be represented by:

$$\text{Experimental: } \ln\left(\frac{p_0 v^*}{\varepsilon}\right) = 2.97 - \frac{6.53\varepsilon}{\mathbf{k}T} \tag{50.47}$$

$$\text{LJD: } \ln \frac{p_0 v^*}{\varepsilon} = 4.41 - \frac{8.14\varepsilon}{\mathbf{k}T} \tag{50.48}$$

$$\text{Ono: } \ln \frac{p_0 v^*}{\varepsilon} = 5.91 - \frac{5.99\varepsilon}{\mathbf{k}T} \tag{50.49}$$

However, quantitative agreement is definitely lacking.

Theory of Mayer and Careri. Mayer and Careri[1] have made rather closely related computations with roughly the same degree of success as in the theories just discussed. A number of the features of the Mayer and Careri treatment are different from those embodied in Eqs. (50.43), (50.45), and (50.46):

1. The Bragg-Williams approximation instead of the quasi-chemical approximation is used for the configurational part of the problem.

2. Essentially the choice $v_{[j]} = \alpha_0 =$ independent of j is made, but the argument is developed in terms of the probability function $\varphi = \exp(-\psi/\mathbf{k}T)/\int_\Delta \exp(-\psi/\mathbf{k}T)\, d\mathbf{r}$ instead of ψ itself. Rather than the LJD expression for φ in a cell, Mayer and Careri used for φ the Gaussian function

$$\varphi(r) = \text{const} \times e^{-\delta r^2} \tag{50.50}$$

where the constant can be determined by normalization, and δ (independent of j) is found by minimization of the free energy with respect to this parameter.

3. In computing the potential energy, the molecules in cells nearest neighbor to the cell of interest are not fixed at the centers of their cells, as in the LJD theory, but are also allowed to move in accordance with Eq. (50.50).

4. The size (or number) of cells is determined by minimizing the free energy with respect to this parameter (a' or $\mathscr{B}$) also.

5. A Morse function is used for $u(r)$, with a minimum at $u = -\varepsilon$, $r = r^*$ and curvature at this point the same as in Eq. (49.5).

The Mayer and Careri critical constants are included in Table 18. The value of $\bar{\jmath}_{\text{crit}}/c$ has been estimated from Mayer and Careri's Table I (in the Bragg-Williams approximation, $\bar{\jmath}/c = 1 - \rho$). Other thermodynamic properties are discussed in the original paper.

Mayer and Careri include in their paper a careful discussion of the self-consistency (or lack thereof) of approximate evaluations of the configuration integral Z.

[1] J. E. Mayer and G. Careri, *J. Chem. Phys.*, **20**, 1001 (1952). See also G. Careri, *J. Chem. Phys.*, **20**, 1114 (1952).

Modification for Fluids. The situation here is very similar to that in free-volume theory (Sec. 48). Equation (50.3), if used without approximations, will automatically yield $\rho^{(1)} = \text{constant} = N/V$ for values of N, V, and T appropriate to liquid and gaseous states. This would again be achieved through a sensitive balancing of different probability densities. Consider, for example, an arbitrary cell γ in the lattice. Let $\varphi_\gamma(\mathbf{r},\mathbf{l})$ be the singlet probability density in cell γ in the configuration $\mathbf{l}$ (a configuration in which cell γ is occupied), normalized according to

$$\int_\Delta \varphi_\gamma(\mathbf{r},\mathbf{l})\, d\mathbf{r} = 1 \tag{50.51}$$

Then the total or net $\rho^{(1)}$ in cell γ is given by

$$\rho^{(1)} = \frac{\sum_{\mathbf{l}}' \varphi_\gamma(\mathbf{r},\mathbf{l}) Z^{(\mathbf{l})}}{\sum_{\mathbf{l}} Z^{(\mathbf{l})}} \qquad \gamma = 1, 2, \ldots, \mathscr{B} \cdot \tag{50.52}$$

where the sum Σ' is over those configurations $\mathbf{l}$ in which cell γ is occupied. In a fluid, we will have $\rho^{(1)} = \text{constant} = N/V$, but in a crystal $\rho^{(1)}$ will not be constant. In physical terms, this balancing of the various $\varphi_\gamma(\mathbf{r},\mathbf{l})$ to give $\rho^{(1)} = \text{constant}$ in a fluid arises as follows (using the liquid state as an illustration). In those configurations in which all cells nearest neighbor to cell γ are occupied, φ_γ will have a maximum at $\mathbf{r} = 0$ and fall off rapidly toward the periphery of the cell. But this behavior will be counterbalanced by those φ_γ's associated with configurations in which cell γ has nearest-neighbor holes. For, when a nearest-neighbor cell is vacant, the molecule in cell γ will have relatively high probability of being found near the periphery of the cell in a region adjacent to the vacant neighboring cell. Since all neighboring cells are equally likely to be vacant, neighboring holes will tend to increase the probability of finding the central molecule near the periphery of its cell (γ), with spherical symmetry.

As soon as approximations are introduced [e.g., Eqs. (50.6), (50.21), and (50.32)], Eq. (50.52) will presumably not lead to $\rho^{(1)} = \text{constant}$ at any density (except zero) and hence fluid states, strictly speaking, are excluded. The "smearing" approximation [Eq. (50.32)] is especially serious in this connection for the uniformizing effect of neighboring holes on $\rho^{(1)}$ is completely lost when this approximation is employed.

To discuss fluid states using approximate hole theories, the additional restraint $\rho^{(1)} = N/V$ in Eq. (50.52) should be imposed. For example, instead of maximizing $Z^{(\mathbf{l})}$ as in Eqs. (50.15) to (50.20), one should maximize $\sum_{\mathbf{l}} Z^{(\mathbf{l})}$ subject to the restraints Eq. (50.7) and $\rho^{(1)} = N/V$ in Eq. (50.52). The $Z^{(\mathbf{l})}$'s cannot be maximized separately here because Eq. (50.52) introduces an interconnection between the different configurations. The complete program would include, finally, maximizing $\sum_{l} Z^{(l)}$ with respect to a'.

The remarks at the end of Sec. 48 concerning gas-liquid and liquid-solid transitions apply here as well.

GENERAL REFERENCES

DeBoer, J., *Proc. Roy. Soc. (London)*, **A215**, 4 (1952).
Fowler, R. H., and E. A. Guggenheim, "Statistical Thermodynamics" (Cambridge, London, 1939).
Hirschfelder, J. O., C. F. Curtiss, and R. B. Bird, "Molecular Theory of Gases and Liquids" (Wiley, New York, 1954).
Kimball, G. E., "A Treatise on Physical Chemistry" (Van Nostrand, New York, 1951), edited by H. S. Taylor and S. Glasstone.

APPENDIX I

NATURAL CONSTANTS

In Sec. 14 a universal constant $\mathbf{k}$ was introduced whose numerical value was not specified. The value of $\mathbf{k}$ can be found from a special case, for example, from a comparison of the experimental equation of state of a very dilute monatomic gas with the theoretical equation of state as given by statistical mechanics (including the as yet unspecified $\mathbf{k}$). The theoretical equation of state is (Chap. 5)

$$pV = N\mathbf{k}T \tag{A1.1}$$

while the experimental equation of state is usually written

$$pV = nRT = \frac{N}{N_A} RT \tag{A1.2}$$

where n is the number of moles, N_A is Avogadro's number, and R is the empirical gas constant per mole. Therefore the constant $\mathbf{k}$ should be identified with the empirical gas constant per molecule, R/N_A (Boltzmann's constant).

We list below the values of a few important natural constants:[1]

Avogadro's number, N_A	6.02544×10^{23} mole^{-1}
Velocity of light, c	2.997902×10^{10} cm-sec^{-1}
Electronic charge, e	4.80223×10^{-10} esu
Electron rest mass, m	9.10721×10^{-28} g
Planck's constant, h	6.62377×10^{-27} erg-sec
Mass of hydrogen atom	1.67312×10^{-24} g
Mass of unit atomic weight	1.65963×10^{-24} g
Mass of proton	1.67221×10^{-24} g
Boltzmann's constant, $\mathbf{k}$	1.38026×10^{-16} erg-deg^{-1}
Temperature scales	$0°C = 273.16°K$

[1] J. W. M. Du Mond and E. R. Cohen, *Phys. Rev.*, **82**, 555 (1951).

APPENDIX 2

ONE-COMPONENT, PERFECT, MONATOMIC GAS

We derive here the thermodynamic properties of a perfect, classical, one-component, monatomic gas using several different ensembles.

Canonical Ensemble. We have from Eq. (16.56)

$$Q(N,V,T) = \frac{1}{h^{3N}N!}\int \exp\left[-\frac{\sum_{i=1}^{N}(p_{xi}^2 + p_{yi}^2 + p_{zi}^2)}{2m\mathbf{k}T}\right]\prod_{i=1}^{N} dx_i\, dy_i\, dz_i\, dp_{xi}\, dp_{yi}\, dp_{zi} \tag{A2.1}$$

$$= \frac{V^N}{N!\Lambda^{3N}} \tag{A2.2}$$

$$\ln Q = N \ln V - N \ln N + N - N \ln \Lambda^3 \tag{A2.3}$$

From Eqs. (14.85) to (14.90), we obtain

$$A = -\mathbf{k}T \ln Q$$

$$= N\mathbf{k}T\left(\ln \frac{N\Lambda^3}{V} - 1\right) \tag{A2.4}$$

$$\bar{p} = \mathbf{k}T\left(\frac{\partial \ln Q}{\partial V}\right)_{T,N} = \frac{N\mathbf{k}T}{V} \tag{A2.5}$$

$$\mu = -\mathbf{k}T\left(\frac{\partial \ln Q}{\partial N}\right)_{T,V} = \mathbf{k}T \ln\left(\frac{N\Lambda^3}{V}\right) \tag{A2.6}$$

$$= \mathbf{k}T \ln\left(\frac{\bar{p}\Lambda^3}{\mathbf{k}T}\right) \tag{A2.7}$$

$$= \mu^0(T) + \mathbf{k}T \ln \bar{p} \tag{A2.8}$$

where

$$\mu^0(T) = \mathbf{k}T \ln\left(\frac{\Lambda^3}{\mathbf{k}T}\right)$$

Also

$$S = \mathbf{k} \ln Q + \mathbf{k}T\left(\frac{\partial \ln Q}{\partial T}\right)_{V,N}$$

$$= -N\mathbf{k}\left(\ln \frac{N\Lambda^3}{V} - \frac{5}{2}\right) \tag{A2.9}$$

Grand Canonical Ensemble. In this case

$$\Xi(V,\mu,T) = \sum_{N=0}^{\infty} e^{N\mu/\mathbf{k}T} Q(N,V,T)$$

$$= \sum_{N=0}^{\infty} \frac{a^N}{N!} = e^a \tag{A2.10}$$

where
$$a = \frac{Ve^{\mu/\mathbf{k}T}}{\Lambda^3} = \ln \Xi \tag{A2.11}$$

Then Eqs. (14.46) to (14.50) lead to

$$\bar{p}V = \mathbf{k}T \ln \Xi = \frac{\mathbf{k}TVe^{\mu/\mathbf{k}T}}{\Lambda^3} \tag{A2.12}$$

$$\bar{N} = \mathbf{k}T \left(\frac{\partial \ln \Xi}{\partial \mu}\right)_{T,V} = \frac{Ve^{\mu/\mathbf{k}T}}{\Lambda^3} \tag{A2.13}$$

so that

$$\bar{p}V = \bar{N}\mathbf{k}T \tag{A2.14}$$

Also
$$S = \mathbf{k} \ln \Xi + \mathbf{k}T \left(\frac{\partial \ln \Xi}{\partial T}\right)_{V,\mu}$$

$$= \bar{N}\mathbf{k} + \bar{N}\mathbf{k}\left(\frac{3}{2} - \frac{\mu}{\mathbf{k}T}\right) \tag{A2.15}$$

which agrees with Eq. (A2.9).

Isothermal-Isobaric Ensemble. From Eqs. (A2.2) and (14.71),

$$\Delta(N,p,T) = \int_0^\infty d\,\frac{V}{V^*}\, e^{-pV/\mathbf{k}T}\, \frac{V^N}{N!\Lambda^{3N}}$$

$$= \frac{\mathbf{k}T}{pV^*}\left(\frac{\mathbf{k}T}{p\Lambda^3}\right)^N \tag{A2.16}$$

and
$$\ln \Delta = N \ln \frac{\mathbf{k}T}{p\Lambda^3} \tag{A2.17}$$

where we have omitted a thermodynamically negligible term in Eq. (A2.17). Using Eqs. (14.65) to (14.70), we find

$$F = N\mu = -\mathbf{k}T \ln \Delta = -N\mathbf{k}T \ln \frac{\mathbf{k}T}{p\Lambda^3} \tag{A2.18}$$

$$\bar{V} = -\mathbf{k}T\left(\frac{\partial \ln \Delta}{\partial p}\right)_{T,N} = \frac{N\mathbf{k}T}{p} \tag{A2.19}$$

$$S = \mathbf{k} \ln \Delta + \mathbf{k}T\left(\frac{\partial \ln \Delta}{\partial T}\right)_{p,N}$$

$$= N\mathbf{k} \ln \frac{\mathbf{k}T}{p\Lambda^3} + \frac{5}{2} N\mathbf{k} \tag{A2.20}$$

Generalized Ensemble. We start with[1] Eqs. (14.101), (14.109), and (A2.10):

$$\Upsilon = \int_0^{2\bar{V}/V^*} d\,\frac{V}{V^*}\, e^{-pV/\mathbf{k}T} e^{a}$$

$$= \int_0^{u_m} e^{-\alpha u}\, du \tag{A2.21}$$

where
$$u = \frac{V}{V^*} \qquad u_m = \frac{2\bar{V}}{V^*}$$

$$\alpha = \frac{pV^*}{\mathbf{k}T}\left(1 - \frac{\mathbf{k}Te^{\mu/\mathbf{k}T}}{p\Lambda^3}\right) \tag{A2.22}$$

Then
$$\Upsilon = \frac{1}{\alpha}\,(1 - e^{-\alpha u_m}) \tag{A2.23}$$

We remark in passing that, since u_m is of the order of $\bar{N}$, $\ln \Upsilon$ is not negligible compared to $\bar{N}$ for negative α but is negligible for $\alpha \geqslant 0$.

To find $f(T,p,\mu) = 0$ we use Eq. (14.21) and obtain

$$\mathbf{k}T\left(\frac{\partial \ln \Upsilon}{\partial p}\right)_{T,\mu} = \frac{V^*(\alpha u_m e^{-\alpha u_m} - 1 + e^{-\alpha u_m})}{\alpha(1 - e^{-\alpha u_m})} \tag{A2.24}$$

The right-hand side of Eq. (A2.24) reduces to $-\bar{V}$, as required by Eq. (14.21), if and only if $\alpha \to 0$. Thus we find

$$f(T,p,\mu) = 1 - \frac{\mathbf{k}Te^{\mu/\mathbf{k}T}}{p\Lambda^3} = 0 \tag{A2.25}$$

or
$$e^{\mu/\mathbf{k}T} = \frac{p\Lambda^3}{\mathbf{k}T} \tag{A2.26}$$

in agreement with Eq. (A2.7).

Equations (14.22) and (A2.23) then give

$$\lim_{\alpha\to 0} \mathbf{k}T\left(\frac{\partial \ln \Upsilon}{\partial \mu}\right)_{T,p} = \frac{p\bar{V}}{\mathbf{k}T} = \bar{N} \tag{A2.27}$$

which is the equation of state. Also, from Eq. (14.20),

$$\lim_{\alpha\to 0} \mathbf{k}T\left(\frac{\partial \ln \Upsilon}{\partial T}\right)_{p,\mu} = -\mathbf{k}T\bar{V}\left(-\frac{p}{\mathbf{k}T^2} + \frac{\mu}{\mathbf{k}T^2}\,\frac{e^{\mu/\mathbf{k}T}}{\Lambda^3} - \frac{3e^{\mu/\mathbf{k}T}}{2T\Lambda^3}\right)$$

$$= S = \frac{p\bar{V}}{T}\left(\frac{5}{2} - \ln\frac{p\Lambda^3}{\mathbf{k}T}\right)$$

$$= \bar{N}\mathbf{k}\left(\frac{5}{2} - \ln\frac{p\Lambda^3}{\mathbf{k}T}\right) \tag{A2.28}$$

as in Eq. (A2.20).

---

[1] In a one-component system with one external variable (V), note that $\Gamma_V = \Xi$ and $\Gamma_N = \Delta$.

The same results can of course be found from the alternative expression

$$\Upsilon = \sum_{N=0}^{2\tilde{N}} e^{N\mu/\mathbf{k}T}\Delta$$

$$= \frac{\mathbf{k}T}{pV^*} \frac{1 - (\mathbf{k}Te^{\mu/\mathbf{k}T}/p\Lambda^3)^{2\tilde{N}+1}}{1 - (\mathbf{k}Te^{\mu/\mathbf{k}T}/p\Lambda^3)} \tag{A2.29}$$

Equation (A2.23) leads to

$$\lim_{\alpha \to 0} \Upsilon = \frac{2\tilde{V}}{V^*} \tag{A2.30}$$

This agrees with Eqs. (14.110) and (A2.10). That is, $C_V = 1$. From Eq. (A2.29),

$$\lim_{\alpha \to 0} \Upsilon = \frac{2\tilde{N}\mathbf{k}T}{pV^*} \tag{A2.31}$$

This result is in agreement with Eqs. (14.112) and (A2.16) since $C_N = \mathbf{k}T/pV^*$. Note that Eqs. (A2.30) and (A2.31) are self-consistent; the equation of state follows on combining the two equations.

The quantity

$$P(N,V) = e^{N\mu/\mathbf{k}T}e^{-pV/\mathbf{k}T}Q(N,V,T) \tag{A2.32}$$

is proportional to the probability of observing N and V in a system in thermal, mechanical and material equilibrium. [That is, $P(N,V)$ is proportional to $p_k(N,V)$ in Sec. 14 summed over all k.] For a monatomic perfect gas, using Eq. (A2.2) for Q and Eq. (A2.26) for the relation of μ to p and T, we have

$$P(N,V) = \frac{1}{N!}\left(\frac{pV}{\mathbf{k}T}\right)^N e^{-pV/\mathbf{k}T} \tag{A2.33}$$

or

$$\ln P(N,V) = N \ln \left(\frac{pV}{\mathbf{k}T}\right) - \frac{pV}{\mathbf{k}T} - N \ln N + N - \tfrac{1}{2} \ln 2\pi N \tag{A2.34}$$

where the last term in Eq. (A2.34) is unimportant for most but not all purposes (see below). It is instructive to examine the behavior of $P(N,V)$. To do this, we note that

$$\left(\frac{\partial \ln P}{\partial V}\right)_N = \frac{N}{V} - \frac{p}{\mathbf{k}T} \tag{A2.35}$$

$$\left(\frac{\partial \ln P}{\partial N}\right)_V = \ln \left(\frac{pV}{\mathbf{k}T}\right) - \ln N \tag{A2.36}$$

and

$$\frac{\partial^2 \ln P}{\partial V^2} = -\frac{N}{V^2} \qquad \frac{\partial^2 \ln P}{\partial N \partial V} = \frac{1}{V} \qquad \frac{\partial^2 \ln P}{\partial N^2} = -\frac{1}{N} \tag{A2.37}$$

From these derivatives we see that $P(N,V)$ has a maximum with respect to both N = constant and V = constant along the line in the N,V plane, $N = \left(\frac{p}{\mathbf{k}T}\right) V$. The value of P on this line, $P_{\max}$, is seen from Eq. (A2.34) to be

$$P_{\max} = \frac{1}{(2\pi N)^{1/2}} \tag{A2.38}$$

The dependence of P on N along the line $V = V' =$ constant, and expanded about $N = pV'/\mathbf{k}T \equiv N'$, is

$$\ln P(N,V') = \ln P_{\max} + \frac{1}{2}\left(\frac{\partial^2 \ln P}{\partial N^2}\right)_{N=N'}(N - N')^2 + \cdots \tag{A2.39}$$

or

$$P(N,V') = \frac{1}{(2\pi N')^{1/2}} \exp\left[-\frac{N'}{2}\left(\frac{N - N'}{N'}\right)^2\right] \tag{A2.40}$$

This is the normal dependence on or fluctuation in N (see Chap. 4). The integral of $P(N,V')$ over N gives a result which is independent of V', as expected from

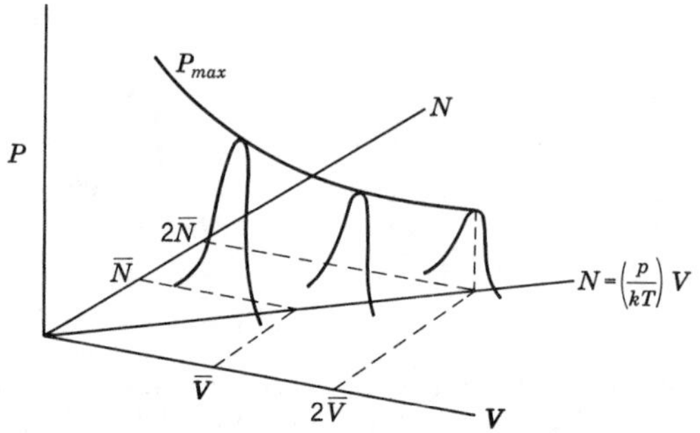

FIG. 64. The probability function $P(N,V)$ for a perfect monatomic gas.

Eq. (14.107). Similarly, along the line $N = N' =$ constant, and defining $V' = (\mathbf{k}T/p)N'$ here,

$$P(N',V) = \frac{1}{(2\pi N')^{1/2}} \exp\left[-\frac{N'}{2}\left(\frac{V - V'}{V'}\right)^2\right] \tag{A2.41}$$

This is the normal fluctuation in V, and the integral of $P(N',V)$ over V is independent of N', as predicted by Eq. (14.108).

$P(N,V)$ can be described (Fig. 64) as a "Gaussian mountain range," along the line $N = (p/\mathbf{k}T)V$ in the N,V plane, which becomes somewhat narrower and taller as the origin is approached. If $\bar{N}$ and $\bar{V}$ are the mean values of N and V, then because of the integral properties of $P(N,V')$ and $P(N',V)$ mentioned above, $P(N,V)$ must be cut off at $2\bar{N}$, $2\bar{V}$, as shown in the figure.

APPENDIX 3

BINARY PERFECT-GAS MIXTURE

Here we extend the treatment of the generalized ensemble in Appendix 2 to a binary mixture. We give first, however, expressions for the other partition functions. For Q we find easily

$$Q(N_1,N_2,V,T) = \frac{V^{N_1+N_2}}{N_1!N_2!\Lambda_1^{3N_1}\Lambda_2^{3N_2}} \tag{A3.1}$$

Then

$$\begin{aligned} \Xi(V,\mu_1,\mu_2,T) &= \sum_{N_1,N_2=0}^{\infty} e^{(N_1\mu_1+N_2\mu_2)/\mathbf{k}T}Q \\ &= e^{a_1+a_2} \end{aligned} \tag{A3.2}$$

where

$$a_i = \frac{Ve^{\mu_i/\mathbf{k}T}}{\Lambda_i^3} \qquad i = 1, 2$$

For Δ, we find

$$\begin{aligned} \Delta &= \int_0^\infty d\,\frac{V}{V^*}\, e^{-pV/\mathbf{k}T}Q \\ &= \frac{\mathbf{k}T}{pV^*}\left(\frac{\mathbf{k}T}{p\Lambda_1^3}\right)^{N_1}\left(\frac{\mathbf{k}T}{p\Lambda_2^3}\right)^{N_2}\frac{(N_1+N_2)!}{N_1!N_2!} \end{aligned} \tag{A3.3}$$

Turning now to Υ,

$$\begin{aligned} \Upsilon &= \int_0^{2\bar{V}/V^*} d\,\frac{V}{V^*}\, e^{-pV/\mathbf{k}T}\,\Xi \\ &= \frac{1}{\alpha'}(1 - e^{-\alpha' u_m}) \end{aligned} \tag{A3.4}$$

where

$$\alpha' = \frac{pV^*}{\mathbf{k}T}\left(1 - \frac{\mathbf{k}Te^{\mu_1/\mathbf{k}T}}{p\Lambda_1^3} - \frac{\mathbf{k}Te^{\mu_2/\mathbf{k}T}}{p\Lambda_2^3}\right) \tag{A3.5}$$

Then

$$\mathbf{k}T\left(\frac{\partial \ln \Upsilon}{\partial p}\right)_{T,\mu_1,\mu_2} = \frac{V^*(\alpha' u_m e^{-\alpha' u_m} - 1 + e^{-\alpha' u_m})}{\alpha'(1 - e^{-\alpha' u_m})}$$

This reduces to $-\bar{V}$ if and only if $\alpha' \to 0$. Thus the thermodynamic relation between T, p, μ_1, and μ_2 is

$$\frac{p}{\mathbf{k}T} = \frac{e^{\mu_1/\mathbf{k}T}}{\Lambda_1^3} + \frac{e^{\mu_2/\mathbf{k}T}}{\Lambda_2^3} \tag{A3.6}$$

Equations (14.22) and (A3.4) give

$$\lim_{\alpha'\to 0} \mathbf{k}T\left(\frac{\partial \ln \Upsilon}{\partial \mu_i}\right)_{T,\mu_j,p} = \frac{\bar{V}e^{\mu_i/\mathbf{k}T}}{\Lambda_i^3}$$

$$= \bar{N}_i \qquad i = 1, 2 \tag{A3.7}$$

When combined with Eq. (A3.6), this results in the equation of state

$$\frac{p\bar{V}}{\mathbf{k}T} = \bar{N}_1 + \bar{N}_2 \tag{A3.8}$$

The entropy can be found from Eq. (14.20):

$$\lim_{\alpha'\to 0} \mathbf{k}T\left(\frac{\partial \ln \Upsilon}{\partial T}\right)_{\mu_1,\mu_2,p} = \frac{p\bar{V}}{T}\left(1 + \frac{3}{2}n_1 - n_1 \ln \frac{\bar{N}_1\Lambda_1^3}{\bar{V}} + \frac{3}{2}n_2 - n_2 \ln \frac{\bar{N}_2\Lambda_2^3}{\bar{V}}\right)$$

$$= \bar{N}_1\mathbf{k}\left(\frac{5}{2} - \ln \frac{\bar{N}_1\Lambda_1^3}{\bar{V}}\right) + \bar{N}_2\mathbf{k}\left(\frac{5}{2} - \ln \frac{\bar{N}_2\Lambda_2^3}{\bar{V}}\right) \tag{A3.9}$$

where n_1 and n_2 are mole fractions.

Alternative expressions for Υ arise if the last summation is over N_1 or N_2 instead of V. For example,

$$\Upsilon = \sum_{N_2=0}^{2\bar{N}_2} e^{N_2\mu_2/\mathbf{k}T}\Gamma_{N_2} \tag{A3.10}$$

where

$$\Gamma_{N_2} = \sum_{N_1=0}^{\infty} e^{N_1\mu_1/\mathbf{k}T}\Delta$$

$$= \frac{\mathbf{k}T}{pV^*}\left(\frac{\mathbf{k}T}{p\Lambda_2^3}\right)^{N_2}\frac{1}{N_2!}\sum_{N_1=0}^{\infty}\frac{(N_1+N_2)!}{N_1!}x_1^{N_1} \tag{A3.11}$$

where

$$x_1 = \frac{e^{\mu_1/\mathbf{k}T}\mathbf{k}T}{p\Lambda_1^3}$$

The sum, Σ, in Eq. (A3.11) is

$$\Sigma = \sum_{N_1=0}^{\infty}\frac{d^{N_2}}{dx_1^{N_2}}x_1^{N_1+N_2} = \frac{d^{N_2}}{dx_1^{N_2}}x_1^{N_2}\sum_{N_1=0}^{\infty}x_1^{N_1}$$

$$= \frac{d^{N_2}}{dx_1^{N_2}}\frac{x_1^{N_2}}{1-x_1} = \frac{N_2!}{(1-x_1)^{N_2+1}} \qquad x_1 < 1 \tag{A3.12}$$

Hence

$$\Gamma_{N_2} = \frac{\mathbf{k}T}{pV^*}\frac{1}{(1-x_1)}\left[\frac{\mathbf{k}T}{p\Lambda_2^3(1-x_1)}\right]^{N_2} \tag{A3.13}$$

and

$$\Upsilon = \frac{\mathbf{k}T}{pV^*}\frac{1}{(1-x_1)}\sum_{N_2=0}^{2\bar{N}_2}\left(\frac{x_2}{1-x_1}\right)^{N_2}$$

$$= \frac{\mathbf{k}T}{pV^*}\frac{1}{(1-x_1)}\frac{1-[x_2/(1-x_1)]^{2\bar{N}_2+1}}{1-[x_2/(1-x_1)]} \tag{A3.14}$$

The limit $\alpha' \to 0$ corresponds to $x_2/(1-x_1) \to 1$ [see Eq. (A3.6)], so that

$$\lim_{\alpha'\to 0}\Upsilon = \frac{\mathbf{k}T}{pV^*}\frac{2\bar{N}_2}{(1-x_1)} = \frac{\mathbf{k}T}{pV^*}\frac{2\bar{N}_2}{x_2} \tag{A3.15}$$

From Eq. (A3.4),

$$\lim_{\alpha' \to 0} \Upsilon = \frac{2\bar{V}}{V^*} \tag{A3.16}$$

That is [Eqs. (14.105), (14.106), (A3.2), and (A3.13)],

$$C_V = 1 \qquad C_{N_2} = \frac{\mathbf{k}T}{pV^*} \frac{1}{(1 - x_1)}$$

Equations (A3.15) and (A3.16) give

$$x_2 = \frac{\bar{N}_2 \mathbf{k} T}{p\bar{V}} = \frac{\bar{N}_2}{\bar{N}_1 + \bar{N}_2} = n_2 \tag{A3.17}$$

APPENDIX 4

ONE-COMPONENT PERFECT LATTICE GAS

Suppose we have a collection of B equivalent sites in a lattice. Each site may be empty or occupied by one molecule. Let the (canonical ensemble) partition function for the motion of a single molecule in the neighborhood of its site be $j(T)$. Then if the system consists of N molecules distributed among the B sites,

$$Q(N,B,T) = \frac{B!j^N}{N!(B-N)!} \tag{A4.1}$$

provided there are no interactions between the molecules ("perfect" lattice gas). The lattice may be one-, two-, or three-dimensional. We shall use here language appropriate to three dimensions. The two-dimensional case is essentially the Langmuir adsorption problem. Let the relation between V and B be $V = B\alpha$, where α is a constant.

Canonical Ensemble. From Eq. (A4.1),

$$\ln Q = B \ln B - N \ln N - (B - N) \ln (B - N) + N \ln j \tag{A4.2}$$

Then

$$\frac{\mu}{\mathbf{k}T} = -\left(\frac{\partial \ln Q}{\partial N}\right)_{B,T} = \ln \frac{\rho}{(1-\rho)j} \tag{A4.3}$$

and

$$\frac{\bar{p}\alpha}{\mathbf{k}T} = \left(\frac{\partial \ln Q}{\partial B}\right)_{N,T} = \ln \frac{1}{1-\rho} \tag{A4.4}$$

where $\rho = N/B$. Incidentally, as $\rho \to 0$, $\bar{p} \to N\mathbf{k}T/V$. Combining Eqs. (A4.3) and (A4.4),

$$\frac{\mu}{\mathbf{k}T} = \ln \frac{e^{\bar{p}\alpha/\mathbf{k}T} - 1}{j(T)} \tag{A4.5}$$

which is the thermodynamic relation between μ, p and T in this case. Also, the entropy [Eq. (14.89)] can be written in the form

$$S = \mathbf{k} \ln \frac{B!}{N!(B-N)!} + S_i \tag{A4.6}$$

$$S_i = N\mathbf{k}\left(\ln j + T\frac{\partial \ln j}{\partial T}\right) \tag{A4.7}$$

where S–S_i is the "configurational" entropy and S_i is the entropy associated with $j(T)$.

Isothermal-Isobaric Ensemble. The partition function is

$$\Delta(N,p,T) = \sum_{B=N}^{\infty} e^{-pB\alpha/\mathbf{k}T} \frac{B!j^N}{N!(B-N)!} \tag{A4.8}$$

Summation over B is appropriate here instead of integration over V, on account of the "discrete" nature of the lattice. If we write $M = B - N$, Eq. (A4.8) becomes

$$\Delta = \frac{e^{-p\alpha N/\mathbf{k}T}j^N}{N!} \sum_{M=0}^{\infty} \frac{(M+N)!}{M!} x^M \tag{A4.9}$$

where $x = e^{-p\alpha/\mathbf{k}T}$

But, from Eq. (A3.12),

$$\Delta = \left(\frac{j}{e^{p\alpha/\mathbf{k}T} - 1}\right)^N \frac{1}{1 - e^{-p\alpha/\mathbf{k}T}} \tag{A4.10}$$

and

$$\ln \Delta = N \ln \frac{j}{e^{p\alpha/\mathbf{k}T} - 1} \tag{A4.11}$$

where a negligible term has been dropped in Eq. (A4.11).

From $N\mu = -\mathbf{k}T \ln \Delta$

we again find Eq. (A4.5) for μ, and from

$$-\bar{B}\alpha = \mathbf{k}T \left(\frac{\partial \ln \Delta}{\partial p}\right)_{T,N}$$

we obtain

$$\rho = 1 - e^{-p\alpha/\mathbf{k}T} \tag{A4.12}$$

in agreement with Eq. (A4.4). Equation (A4.6) may also be verified for S.

Grand Canonical Ensemble. We have

$$\Xi = \sum_{N=0}^{B} e^{N\mu/\mathbf{k}T} \frac{B!j^N}{N!(B-N)!}$$

$$= (1 + je^{\mu/\mathbf{k}T})^B \tag{A4.13}$$

and

$$\ln \Xi = B \ln (1 + je^{\mu/\mathbf{k}T})$$

$$= \frac{\bar{p}B\alpha}{\mathbf{k}T} \tag{A4.14}$$

Equation (A4.14) is the same as Eq. (A4.5). Also, using

$$\bar{N} = \mathbf{k}T \left(\frac{\partial \ln \Xi}{\partial \mu}\right)_{T,B}$$

we get

$$\rho = \frac{je^{\mu/\mathbf{k}T}}{1 + je^{\mu/\mathbf{k}T}} \tag{A4.15}$$

which agrees with Eq. (A4.3). Equation (A4.15) is the adsorption isotherm in the Langmuir case mentioned above. We omit the verification of Eq. (A4.6) for S.

Generalized Ensemble. Let us use the form

$$\Upsilon = \sum_{B=0}^{2\bar{B}} e^{-pB\alpha/\mathbf{k}T}\Xi$$

$$= \sum_{B=0}^{2\bar{B}} y^B \tag{A4.16}$$

where

$$y = e^{-p\alpha/\mathbf{k}T}(1 + je^{\mu/\mathbf{k}T})$$

Then

$$\Upsilon = \frac{1 - y^{2\bar{B}+1}}{1 - y} \tag{A4.17}$$

We find on differentiating Eq. (A4.17),

$$\mathbf{k}T\left(\frac{\partial \ln \Upsilon}{\partial p}\right)_{\mu,T} = -y\alpha\left[\frac{2\bar{B}y^{2\bar{B}+1} - (2\bar{B}+1)y^{2\bar{B}} + 1}{(1-y)(1-y^{2\bar{B}+1})}\right] \tag{A4.18}$$

The limit $y \to 1$ gives $-\alpha\bar{B}$ and therefore

$$e^{p\alpha/\mathbf{k}T} = 1 + je^{\mu/\mathbf{k}T} \tag{A4.19}$$

is the relation between p, T, and μ. Then

$$\lim_{y\to 1} \mathbf{k}T\left(\frac{\partial \ln \Upsilon}{\partial \mu}\right)_{p,T} = (1 - e^{-p\alpha/\mathbf{k}T})\bar{B} = \bar{N}$$

or

$$\rho = 1 - e^{-p\alpha/\mathbf{k}T} \tag{A4.20}$$

Finally, Eq. (A4.6) is found again from

$$\lim_{y\to 1} \mathbf{k}T\left(\frac{\partial \ln \Upsilon}{\partial T}\right)_{p,\mu} = S \tag{A4.21}$$

We can also write

$$\Upsilon = \sum_{N=0}^{2\bar{N}} e^{N\mu/\mathbf{k}T}\Delta$$

$$= \frac{1}{1 - e^{-p\alpha/\mathbf{k}T}} \sum_{N=0}^{2\bar{N}} w^N$$

$$= \frac{1}{1 - e^{-p\alpha/\mathbf{k}T}} \frac{1 - w^{2\bar{N}+1}}{1 - w} \tag{A4.22}$$

where

$$w = \frac{je^{\mu/\mathbf{k}T}}{(e^{p\alpha/\mathbf{k}T} - 1)}$$

The limit $y \to 1$ corresponds to $w \to 1$; thus we have, from Eq. (A4.22),

$$\lim_{y\to 1} \Upsilon = \frac{2\bar{N}}{1 - e^{-p\alpha/\mathbf{k}T}} \tag{A4.23}$$

and, from Eq. (A4.17),

$$\lim_{y\to 1} \Upsilon = 2\bar{B} \tag{A4.24}$$

Also

$$C_N = \frac{1}{(1 - e^{-p\alpha/\mathbf{k}T})} \qquad C_B = 1$$

Equations (A4.23) and (A4.24) are consistent with Eq. (A4.20).

APPENDIX 5

MULTILAYER GAS ADSORPTION

A typical example of the use of the grand ensemble to avoid the restraint $N =$ constant (see Sec. 20) in the canonical ensemble is the following. Consider a two-dimensional lattice of B equivalent sites. One gas molecule can be adsorbed on each site, with partition function j_1 (see Appendix 4). In addition assume that each of these "first layer" adsorbed molecules is a "site" for a "second layer" adsorbed molecule, etc., and that the partition function for all second and higher layer adsorbed molecules is j_∞. If a total of N molecules are adsorbed, of which N_1 are in the first layer and N^* in higher layers, then[1]

$$Q(N,B,T) = \sum_{N_1} \sum_{N^*} \left[\frac{B!j_1^{N_1}}{N_1!(B-N_1)!} \right] \left[\frac{(N_1+N^*-1)!j_\infty^{N^*}}{N^*!(N_1-1)!} \right] \tag{A5.1}$$

where the sums are over values of N_1 and N^* such that

$$\begin{array}{llll} N_1 + N^* = N & N \geqslant N_1 \geqslant 1 & N-1 \geqslant N^* \geqslant 0 & \text{if } N \leqslant B \\ N_1 + N^* = N & B \geqslant N_1 \geqslant 1 & N-1 \geqslant N^* \geqslant N-B & \text{if } N > B \end{array} \tag{A5.2}$$

The first combinatorial expression in Eq. (A5.1) is the number of ways of putting N_1 identical objects in B boxes, not more than one to a box. The second expression is the number of ways of putting N^* identical objects into N_1 boxes, with no restriction on the number per box.

The complicated restraints (A5.2) can be avoided by use of the grand partition function. We have

$$\begin{aligned} \Xi(B,\mu,T) &= \sum_{N \geqslant 0} e^{N\mu/\mathbf{k}T} Q(N,B,T) \\ &= 1 + \sum_{N_1=1}^{B} \frac{B!(j_1 e^{\mu/\mathbf{k}T})^{N_1}}{N_1!(B-N_1)!(N_1-1)!} \sum_{N^*=0}^{\infty} \frac{(N_1+N^*-1)!(j_\infty e^{\mu/\mathbf{k}T})^{N^*}}{N^*!} \end{aligned} \tag{A5.3}$$

From Eq. (A3.12) we see that the sum over N^* is

$$\sum_{N^*} = \frac{(N_1-1)!}{(1-j_\infty e^{\mu/\mathbf{k}T})^{N_1}}$$

Hence

$$\Xi = \sum_{N_1=0}^{B} \frac{B!y^{N_1}}{N_1!(B-N_1)!} = (1+y)^B \tag{A5.4}$$

[1] T. L. Hill, *J. Chem. Phys.*, **14**, 263 (1946).

where
$$y = \frac{j_1 e^{\mu/\mathbf{k}T}}{1 - j_\infty e^{\mu/\mathbf{k}T}} \tag{A5.5}$$

From
$$\bar{N} = \mathbf{k}T \left(\frac{\partial \ln \Xi}{\partial \mu}\right)_{T,B} \tag{A5.6}$$

and Eq. (A5.4) we find
$$\frac{\bar{N}}{B} = \frac{cx}{(1 - x + cx)(1 - x)} \tag{A5.7}$$

where
$$c = \frac{j_1}{j_\infty} \qquad x = j_\infty e^{\mu/\mathbf{k}T} \tag{A5.8}$$

Equation (A5.7) is the well-known Brunauer-Emmett-Teller adsorption isotherm equation.[1]

The "spreading" or surface pressure $\bar{p}$ follows immediately from Eq. (A5.4):
$$\frac{\bar{p}\alpha}{\mathbf{k}T} = \frac{1}{B} \ln \Xi = \ln \left(\frac{1 - x + cx}{1 - x}\right) \tag{A5.9}$$

Here $\alpha = \mathscr{A}/B$, where $\mathscr{A}$ is the surface area.

If adsorption is restricted to n layers,
$$\Xi = \sum_{N_1=0}^{B} \sum_{N_2=0}^{N_1} \sum_{N_3=0}^{N_2} \cdots \sum_{N_n=0}^{N_{n-1}} \frac{B!}{(B - N_1)!N_1!} \frac{N_1!}{(N_1 - N_2)!N_2!} \cdots \frac{N_{n-1}!(cx)^{N_1} x^{N_2 + \cdots + N_n}}{(N_{n-1} - N_n)!N_n!} \tag{A5.10}$$

Summing in turn over $N_n, N_{n-1}, \ldots, N_1$ we find
$$\begin{aligned}\Xi &= [1 + cx(1 + x + x^2 + \cdots + x^{n-1})]^B \\ &= \left[1 + cx\left(\frac{1 - x^n}{1 - x}\right)\right]^B\end{aligned} \tag{A5.11}$$

From Eqs. (A5.6) and (A5.11) we obtain the Brunauer-Emmett-Teller result for restricted adsorption
$$\frac{\bar{N}}{B} = \frac{cx[1 - (n + 1)x^n + nx^{n+1}]}{(1 - x)(1 - x + cx - cx^{n+1})} \tag{A5.12}$$

The limits on the sums in Eq. (A5.10) would be very complicated in the canonical ensemble. Equation (A5.12) reduces to Eq. (A5.7) when $n \to \infty$ and to Eq. (A4.15) when $n = 1$.

The spreading pressure $\bar{p}$ is, from Eq. (A5.11),
$$\frac{\bar{p}\alpha}{\mathbf{k}T} = \ln \left[\frac{1 - x + cx(1 - x^n)}{1 - x}\right] \tag{A5.13}$$

See also Appendix 10.

[1] Essentially the same derivation has been given independently by T. Keii, *J. Chem. Phys.*, **22**, 1617 (1954).

APPENDIX 6

QUANTUM AND CLASSICAL LIMITS

In a system with no intermolecular forces ($U = 0$), it is clear from the discussion of Eq. (16.55) that quantum (symmetry) effects in translational degrees of freedom occur only when two particles are within a distance of the order of Λ (the mean de Broglie wavelength at T) from each other. The average distance between nearest-neighbor molecules is of the order of $(V/N)^{1/3}$; thus the condition for the absence of quantum effects is

$$\frac{\Lambda}{(V/N)^{1/3}} \ll 1$$

or

$$\frac{\Lambda^3 N}{V} \ll 1 \tag{A6.1}$$

Quite aside from symmetry considerations, the discussion of Eq. (16.29) ($U \neq 0$) shows that quantum effects associated with molecular interactions depend on the ratio of Λ to the range of intermolecular forces. In a liquid, $(V/N)^{1/3}$ and the range of the intermolecular forces are of the same order of magnitude so that the two quantum effects mentioned are mixed and both can be related to (A6.1) for order of magnitude arguments.

For computational purposes,

$$\frac{\Lambda^3 N}{V} = \frac{3206\rho}{M^{5/2}T^{3/2}} \tag{A6.2}$$

where ρ is the density in grams per cubic centimeter, M is the molecular weight, and T the absolute temperature. Systems containing light molecules at low temperatures can therefore be expected to furnish the best illustrations of quantum effects. For example, for the liquid phase at the normal boiling point, we have, from Eq. (A6.2),

$$\text{He},\ 4.2°\text{K},\ \frac{\Lambda^3 N}{V} = 1.5$$

$$H_2,\ 20.4°\text{K},\ \frac{\Lambda^3 N}{V} = 0.44$$

$$\text{Ne},\ 27.2°\text{K},\ \frac{\Lambda^3 N}{V} = 0.015$$

$$\text{A},\ 87.4°\text{K},\ \frac{\Lambda^3 N}{V} = 0.00054$$

In these examples, quantum statistics must obviously be used for the translational degrees of freedom of helium and hydrogen, while classical statistics will provide a good approximation for neon and an excellent approximation for argon.

APPENDIX 7

NORMALIZATION OF RADIAL DISTRIBUTION FUNCTION

We give a few more details here concerning the asymptotic behavior of distribution functions in the canonical and grand canonical ensembles. Expansions in powers of the activity are employed so that the results are strictly proved only for gases.

Canonical Ensemble. DeBoer[1] has shown, from a modified cluster expansion of Eq. (29.5), that

$$\begin{aligned}V^2P_N^{(2)}(r_{12}) = 1 &- (2\cdot 1Vb_2)\frac{Z_{N-2}}{Z_N} - (N-2)(3\cdot 2Vb_3)\frac{Z_{N-3}}{Z_N}\\ &- (N-2)(N-3)(4\cdot 3Vb_4)\frac{Z_{N-4}}{Z_N} - \cdots + V^2h_2(r_{12})\frac{Z_{N-2}}{Z_N}\\ &+ (N-2)V^2h_3(r_{12})\frac{Z_{N-3}}{Z_N}\\ &+ (N-2)(N-3)V^2h_4(r_{12})\frac{Z_{N-4}}{Z_N} + \cdots \end{aligned} \tag{A7.1}$$

where $h_2(r_{12}) = e^{-u(r_{12})/\mathbf{k}T} - 1$ (A7.2)

$$\begin{aligned}h_3(r_{12}) = \int_V \{e^{-[u(r_{12})+u(r_{13})+u(r_{23})]/\mathbf{k}T} &- e^{-u(r_{12})/\mathbf{k}T}\\ &- e^{-u(r_{13})/\mathbf{k}T} - e^{-u(r_{23})/\mathbf{k}T} + 2\}\,d\mathbf{r}_3\\ &\text{etc.}\end{aligned} \tag{A7.3}$$

All the $h_l \to 0$ as $r_{12} \to \infty$. Now, we define z_l as

$$z_l = \frac{Z_{N-l}N!}{Z_N(N-l)!} \tag{A7.4}$$

so that

$$V^2P_N^{(2)}(r_{12}) = 1 + \frac{V^2\sum_{l\geqslant 2} h_l(r_{12})z_l}{N(N-1)} - \frac{\sum_{l\geqslant 2} l(l-1)Vb_lz_l}{N(N-1)} \tag{A7.5}$$

[1] J. DeBoer, Thesis, Amsterdam, 1940.

In the limit $\rho \to 0$ (or $V \to \infty$ with N constant),

$$z_l \to \frac{1}{V^l}\frac{N!}{(N-l)!} = \rho^l \left[1 + O\left(\frac{1}{N}\right)\right] \to 0$$

In particular,

$$z_2 \to \frac{N(N-1)}{V^2}$$

Then

$$V^2 P_N^{(2)}(r_{12}) = e^{-w_N^{(2)}(r_{12})/\mathbf{k}T}$$
$$\to 1 + h_2(r_{12}) = e^{-u(r_{12})/\mathbf{k}T} \tag{A7.6}$$

or[1]

$$\frac{\rho_N^{(2)}(r_{12})}{\rho^2} = g_N^{(2)}(r_{12}) = \frac{N(N-1)V^2P_N^{(2)}(r_{12})}{N^2}$$
$$\to \left(1 - \frac{1}{N}\right) e^{-u(r_{12})/\mathbf{k}T} \tag{A7.7}$$

Equation (A7.7) agrees with Eqs. (32.17) and (31.7).

In the limit $r_{12} \to \infty$, $h_l \to 0$ and

$$V^2 P_N^{(2)} \to 1 - \frac{\sum_{l \geqslant 2} l(l-1)Vb_l z_l}{N(N-1)} \tag{A7.8}$$

This result can be expressed in thermodynamic terms by recalling that DeBoer[2] also deduced from Z_N that

$$\rho = \sum_{j \geqslant 1} j b_j z_j \tag{A7.9}$$

On comparing Eq. (A7.9) with Eq. (23.22), we conclude that for thermodynamic purposes, we may write $z_j = z^j$, where z is the activity. Then using

$$\kappa = -\frac{1}{V}\left(\frac{\partial V}{\partial p}\right)_{N,T} = \frac{1}{\rho}\frac{\partial \rho}{\partial z}\frac{\partial z}{\partial p} = \frac{z}{\rho^2 \mathbf{k}T}\frac{\partial \rho}{\partial z}$$

and

$$\sum_{l \geqslant 2} l(l-1)b_l z^l = z^2 \frac{\partial(\rho/z)}{\partial z} = z\frac{\partial \rho}{\partial z} - \rho$$

we find, as $r_{12} \to \infty$,

$$V^2 P_N^{(2)} \to 1 + \frac{1 - \rho \mathbf{k}T\kappa}{N-1} = 1 + O\left(\frac{1}{N}\right)$$
$$g_N^{(2)} \to \frac{N-1}{N} + \frac{1 - \rho\mathbf{k}T\kappa}{N} = 1 + O\left(\frac{1}{N}\right) \tag{A7.10}$$

Grand Canonical Ensemble. We first show the implications here of Eqs. (A7.6) and (A7.10) above, using

$$\rho^{(2)} = \sum_{N \geqslant 2} P_N \rho_N^{(2)}$$

---

[1] Recall that the discussion refers to a fluid (gas).

[2] J. DeBoer, *loc. cit.*, Eq. (51), p. 21. See also Eq. (48), p. 20.

to establish the connection between the grand canonical and canonical ensembles. In the limit $\rho \to 0$,

$$\rho^{(2)}(r_{12}) \to \sum_{N\geqslant 2} P_N \frac{N(N-1)}{V^2} e^{-u(r_{12})/\mathbf{k}T}$$

$$= \frac{(\overline{N^2} - \bar{N})}{V^2} e^{-u(r_{12})/\mathbf{k}T}$$

But for a perfect gas ($\rho \to 0$) $\overline{N^2} - (\bar{N})^2 = \bar{N}$ [Eq. (19.39)] so that

$$\rho^{(2)}(r_{12}) \to \left(\frac{\bar{N}}{V}\right)^2 e^{-u(r_{12})/\mathbf{k}T} \tag{A7.11}$$

or

$$g^{(2)}(r_{12}) = e^{-W^{(2)}(r_{12})/\mathbf{k}T} = \frac{\rho^{(2)}(r_{12})}{\rho^2} \to e^{-u(r_{12})/\mathbf{k}T} \tag{A7.12}$$

which verifies Eq. (37.15).

In the limit $r_{12} \to \infty$, treating $\rho\mathbf{k}T\kappa$ as a thermodynamic quantity,

$$\rho^{(2)} \to \sum_{N\geqslant 2} P_N \left[1 + \frac{1 - \rho\mathbf{k}T\kappa}{N-1}\right] \frac{N(N-1)}{V^2}$$

$$= \frac{\overline{N^2} - \bar{N}}{V^2} + \frac{(1 - \rho\mathbf{k}T\kappa)\bar{N}}{V^2}$$

$$= \frac{\overline{N^2} - \bar{N}\rho\mathbf{k}T\kappa}{V^2} \tag{A7.13}$$

But, from Eq. (19.39),

$$\overline{N^2} = \bar{N}\rho\mathbf{k}T\kappa + (\bar{N})^2$$

Therefore Eq. (A7.13) becomes ($r_{12} \to \infty$)

$$\rho^{(2)} \to \left(\frac{\bar{N}}{V}\right)^2 \tag{A7.14}$$

or

$$g^{(2)} = e^{-W^{(2)}/\mathbf{k}T} \to 1 \tag{A7.15}$$

Turning now to the direct use of the definition of distribution functions in the grand canonical ensemble, it has already been pointed out [Eq. (37.15)] that Eq. (37.4) gives, as $\rho \to 0$,

$$\rho^{(n)} \to \rho^n e^{-U_n/\mathbf{k}T}$$

or

$$g^{(n)} \to e^{-U_n/\mathbf{k}T} \tag{A7.16}$$

To examine the behavior as all r_{ij} ($1 \leqslant i < j \leqslant n$) $\to \infty$, we rewrite Eq. (37.79) as

$$\left(\frac{\rho}{z}\right)^n g^{(n)}(\infty) = 1 + z[I_1^{(n)}(\infty) - Z_1]$$

$$+ z^2\left[\frac{I_2^{(n)}(\infty)}{2} - Z_1 I_1^{(n)}(\infty) + \left(Z_1^2 - \frac{Z_2}{2}\right)\right] + \cdots \tag{A7.17}$$

where
$$I_1^{(n)}(\infty) = \int_V e^{-[u(r_{1,n+1}) + \cdots + u(r_{n,n+1})]/kT} \, d\mathbf{r}_{n+1}$$
$$= \int_V (1 + f_{1,n+1}) \cdots (1 + f_{n,n+1}) \, d\mathbf{r}_{n+1}$$
$$= V + 2nb_2 \qquad \text{(A7.18)}$$

since the product of two or more of these f's is zero in view of r_{ij} $(1 \leqslant i < j \leqslant n) \to \infty$. Similarly,

$$I_2^{(n)}(\infty) = \iint_V (1 + f_{1,n+1}) \cdots (1 + f_{n,n+1})(1 + f_{1,n+2}) \cdots \times (1 + f_{n,n+2})(1 + f_{n+1,n+2}) \, d\mathbf{r}_{n+1} \, d\mathbf{r}_{n+2}$$
$$= V^2 + 2(2n + 1)b_2V + 4n(n - 1)b_2^2 + 6nb_3 \qquad \text{(A7.19)}$$

etc.

When these expressions for $I_1^{(n)}(\infty)$, $I_2^{(n)}(\infty)$, . . . are introduced in Eq. (A7.17), the right-hand side becomes identical with the expansion of $(\rho/z)^n$ in Eq. (37.80). That is,

$$g^{(n)}(\infty) = 1 \qquad \text{(A7.20)}$$

Incidentally, with $n = 2$, Eq. (37.79) can be written

$$g^{(2)}(r_{12}) = 1 + \left(\frac{z}{\rho}\right)^2 \left\{ e^{-u(r_{12})/kT} - \left(\frac{\rho}{z}\right)^2 + z[I_1^{(2)} - Z_1 e^{-u(r_{12})/kT}] + \cdots \right\}$$

Using Eq. (37.80) for $(\rho/z)^2$, this becomes

$$g^{(2)}(r_{12}) = 1 + \frac{1}{\rho^2} \sum_{l \geqslant 2} h_l(r_{12}) z^l \qquad \text{(A7.21)}$$

where the h_l are the same as those in Eq. (A7.5). On comparing Eq. (A7.21) with Eq. (A7.5), it should be remembered that

$$g_N^{(2)} = \frac{N - 1}{N} V^2 P_N^{(2)}$$

APPENDIX 8

GLOSSARY OF CERTAIN DEFINITIONS IN CHAPTER 7

1. $\rho = \dfrac{N}{\mathscr{B}} \qquad \mathscr{I} = \mathscr{B} - 2N \qquad I = \dfrac{\mathscr{I}}{\mathscr{B}} = 1 - 2\rho$

$$M = \lim_{\mathscr{H}\to 0} I(\mathscr{H}) = 1 - 2\rho^0$$

2. $2\mathscr{H} = 2H + \dfrac{\mathbf{c}}{2}(\varepsilon_{AA} - \varepsilon_{BB}) = 2H + \mathbf{k}T \ln \sigma$

$$\sigma = e^{c(\varepsilon_{AA} - \varepsilon_{BB})/2\mathbf{k}T} \qquad y = \exp\left(-\frac{2\mathscr{H}}{\mathbf{k}T}\right) = \frac{e^{-2H/\mathbf{k}T}}{\sigma} = \frac{z}{\sigma}$$

$$z = je^{\mu/\mathbf{k}T}$$

3. $w = \varepsilon_{AB} - \dfrac{\varepsilon_{AA}}{2} - \dfrac{\varepsilon_{BB}}{2} \qquad J = \dfrac{w}{2} \qquad K = \dfrac{J}{\mathbf{k}T}$

$$u = \tanh K \qquad x = e^{-2K} = e^{-2J/\mathbf{k}T} = e^{-w/\mathbf{k}T} \qquad t = 1 - x^2$$

4. $G(N_{AB},\mathscr{B}) = \sum_N g(N,N_{AB},\mathscr{B})$

$$\Omega = \frac{\mathscr{B}!}{N!(\mathscr{B} - N)!} = \sum_{N_{AB}} g(N,N_{AB},\mathscr{B})$$

5. $\Xi = \exp\left(\dfrac{p\mathscr{B}}{\mathbf{k}T}\right) = \exp\left(-\dfrac{c\mathscr{B}\varepsilon_{BB}}{2\mathbf{k}T\cdot}\right)\exp\left(-\dfrac{c\mathscr{B}K}{2}\right)\mathscr{S}$

$$= \exp\left(-\frac{c\mathscr{B}\varepsilon_{BB}}{2\mathbf{k}T}\right)\Lambda^{\mathscr{B}}$$

$$\Lambda = \mathscr{S}^{1/\mathscr{B}} e^{-cK/2} = e^{p'/\mathbf{k}T}$$

$$p' = p + \frac{c\varepsilon_{BB}}{2}$$

APPENDIX 9

FIRST-ORDER PHASE TRANSITIONS

In this appendix we consider the question of whether or not a "loop" should be expected in the μ–N/V or p–V/N curves for an exact theory of a first-order phase transition, using the canonical ensemble. Certain related topics are also discussed. Some of the conclusions reached here are utilized in the various chapters (omitting many details).

The material included is based largely on the papers of Hill[1] and Katsura.[2]

The Argument of van Hove. A careful argument has been given by van Hove[3] showing that no loop is obtained from the complete evaluation of the canonical ensemble partition function of a fluid. The argument is based on the division of the system into sufficiently large cells and neglect (which is shown to be justified) of molecular interactions between cells in the limit of an infinite system. Katsura's conjecture[2] concerning a van der Waals loop, referred to further below, contradicts the above result of van Hove, and hence Katsura[2] questions the validity of van Hove's reasoning. It will be pointed out in the later discussion that it is in fact not necessary for Katsura to invoke a van der Waals type of loop in order to explain his results, and hence his reason for questioning van Hove's argument is eliminated.

It may, however, be helpful here to digress and supplement van Hove's discussion by the following alternative but essentially equivalent considerations.

Let Q be the canonical ensemble partition function for a system of N molecules and volume V. Divide V into M cells of equal volume. V/M is taken large enough so that molecular interactions between molecules in different cells can be neglected. Then

$$Q = \Sigma \frac{M!Q_1^{n_1}Q_2^{n_2}\cdots}{n_1!n_2!\cdots} \tag{A9.1}$$

where $Q_m(V/M,T)$ is the canonical ensemble partition function for a system of m molecules and volume V/M, and n_m is the number of cells containing m molecules. The sum is over all distributions of the N molecules among cells such that

$$\begin{aligned}\sum_m n_m &= M\\ \sum_m mn_m &= N\end{aligned} \tag{A9.2}$$

[1] T. L. Hill, *J. Phys. Chem.*, **57**, 324 (1953); *J. Chem. Phys.*, **23**, 812 (1955).

[2] S. Katsura, *J. Chem. Phys.*, **22**, 1277 (1954); *Progr. Theoret. Phys.* (*Japan*), **11**, 476 (1954).

[3] L. van Hove, *Physica*, **15**, 951 (1949).

The most probable distribution [maximize the summand in Eq. (A9.1) subject to Eqs. (A9.2)] is easily found by Lagrange's method to be

$$\frac{n_m}{M} = \frac{e^{-\gamma m}Q_m}{\sum_m e^{-\gamma m}Q_m} \tag{A9.3}$$

where γ is an undetermined multiplier. If we replace $\ln Q$ by the logarithm of the largest term in Eq. (A9.1) and use Eq. (A9.3) to eliminate the n_m, we obtain

$$\ln Q = M \ln \sum_m e^{-\gamma m}Q_m + \gamma N \tag{A9.4}$$

Now we know that

$$\left(\frac{\partial \ln Q}{\partial N}\right)_{V,T} = -\frac{\mu}{\mathbf{k}T} \tag{A9.5}$$

Taking this same derivative in Eq. (A9.4), keeping M constant, we find

$$\left(\frac{\partial \ln Q}{\partial N}\right)_{V,T,M} = \frac{-M\dfrac{\partial \gamma}{\partial N}\sum_m e^{-\gamma m} m Q_m}{\sum_m e^{-\gamma m}Q_m} + \gamma + N\frac{\partial \gamma}{\partial N} = \gamma \tag{A9.6}$$

From Eqs. (A9.5) and (A9.6) we conclude that

$$\gamma = -\frac{\mu}{\mathbf{k}T} \tag{A9.7}$$

Equations (A9.4) and (A9.7) give

$$A = \mu N - pV = -\mathbf{k}T \ln Q$$

$$= -\mathbf{k}TM \ln \sum_m e^{\mu m/\mathbf{k}T}Q_m + \mu N$$

or

$$p\left(\frac{V}{M}\right) = \mathbf{k}T \ln \sum_m e^{\mu m/\mathbf{k}T}Q_m\left(\frac{V}{M}, T\right)$$

$$= \mathbf{k}T \ln \Xi \tag{A9.8}$$

where Ξ is the grand partition function for a system of volume V/M. Equation (A9.8) is the conventional relationship for the pressure using a grand ensemble.

Now a pressure-volume loop can never be obtained from a grand partition function (see below); therefore we can conclude from Eq. (A9.8) that whenever the canonical ensemble partition function Q can be written in the form of Eq. (A9.1), a loop will not be found. It should be noted that Eq. (A9.1) will be valid for any theory (regardless of how exactly the Q_m are evaluated) (1) which allows all possible fluctuations in the number of molecules in each cell and (2) with cells chosen large enough (infinite system, strictly) so that interactions between molecules in different cells can be neglected. Approximate theories of the van der Waals, Bragg-Williams, etc., type, which contain the implicit or explicit restriction of the same fixed density in every cell, are excluded because of condition (1) above, and, as is well known, lead to loops from the canonical ensemble.

General Relations. In the canonical ensemble we find μ and p from

$$\frac{\mu}{\mathbf{k}T} = -\left(\frac{\partial \ln Q}{\partial N}\right)_{V,T} \tag{A9.9}$$

$$\frac{p}{\mathbf{k}T} = \left(\frac{\partial \ln Q}{\partial V}\right)_{N,T} \tag{A9.10}$$

There is also the thermodynamic relation [Eq. (19.38)],

$$-\left(\frac{N}{V}\right)^2 \left(\frac{\partial \mu}{\partial N}\right)_{V,T} = \left(\frac{\partial p}{\partial V}\right)_{N,T} \tag{A9.11}$$

which also follows from Eqs. (A9.9) and (A9.10) on taking cross derivatives. If the μ–N/V curve from Eq. (A9.9) has a loop, then the p–V/N curve from Eq. (A9.10) will also have a loop, and vice versa. This follows from Eq. (A9.11): in the thermodynamically unstable region where $\partial\mu/\partial N$ is negative, $\partial p/\partial V$ is positive.

In the grand ensemble, let us define $P(N)$ here by

$$P(N) = Q(N,V,T)\lambda^N \tag{A9.12}$$

$$\lambda = e^{\mu/\mathbf{k}T}$$

where V, T, and μ are constants. That is, $P(N)$ is proportional to the probability that a system of the grand ensemble will contain N molecules. To locate the maxima and minima in $P(N)$, we write

$$\frac{\partial P}{\partial N} = 0 = \left(\frac{\partial Q}{\partial N}\right)_{V,T} \lambda^N + Q\lambda^N \frac{\mu}{\mathbf{k}T} \tag{A9.13}$$

Now Eq. (A9.13) is the same as Eq. (A9.9). Thus we see that the μ–N/V curve deduced from the canonical ensemble is the locus of the maxima and minima in the grand ensemble.[1] It is easy to show that the maxima are associated with regions where $\partial\mu/\partial N > 0$ and minima with $\partial\mu/\partial N < 0$. Numerical examples are presented below, but we may outline the qualitative situation further here. For a value of μ not in the neighborhood of a loop, only one value of N satisfies Eq. (A9.9) or (A9.13), and this corresponds to an extremely sharp maximum in $P(N)$ for large V, as we have essentially already found in Chap. 4. In the region of a loop, three values of N are found for a given μ from Eq. (A9.9). The middle value of N is associated with the unstable part of the μ–N/V curve and with a minimum in $P(N)$; the other two values of N correspond to maxima in $P(N)$. One of these maxima is very high (the stable equilibrium portion of the μ–N/V curve) while the other is extremely small (the metastable equilibrium part of the μ–N/V curve). At the value[2] of μ at which the phase transition occurs, the two maxima reverse their roles.

In the isothermal-isobaric ensemble, we define $P(V)$ by

$$P(V) = Q(N,V,T)e^{-pV/\mathbf{k}T} \tag{A9.14}$$

---

[1] We have already pointed out in Sec. 20 that picking out the maximum term in the grand partition function is equivalent to using the canonical ensemble.

[2] Actually, this occurs over a small range of values of μ, which vanishes as $V \to \infty$. In this range the maxima have the same order of magnitude.

where N, T and p are constants. The maxima and minima in $P(V)$ are located by

$$\frac{\partial P}{\partial V} = 0 = \left(\frac{\partial Q}{\partial V}\right)_{N,T} e^{-pV/\mathbf{k}T} - Qe^{-pV/\mathbf{k}T}\frac{p}{\mathbf{k}T} \qquad \text{(A9.15)}$$

This is the same as Eq. (A9.10). Hence the p–V/N curve from the canonical ensemble is the locus of the maxima and minima in $P(V)$. The situation parallels exactly that described above for $P(N)$ and need not be discussed further.

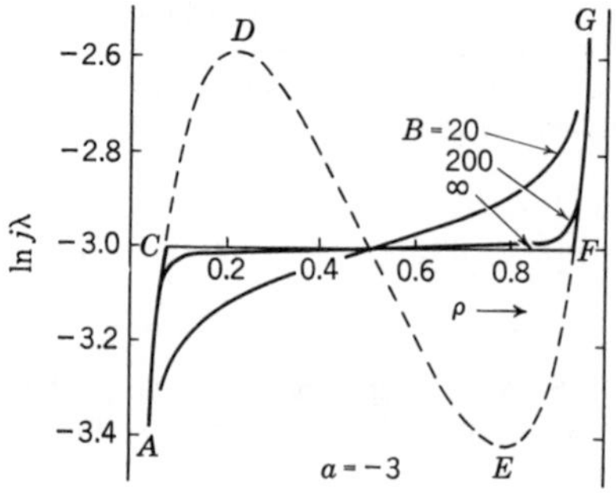

FIG. 65. Curves of μ versus ρ below the critical temperature from the Bragg-Williams approximation using the canonical ensemble (dashed curve) and grand ensemble (solid curves).

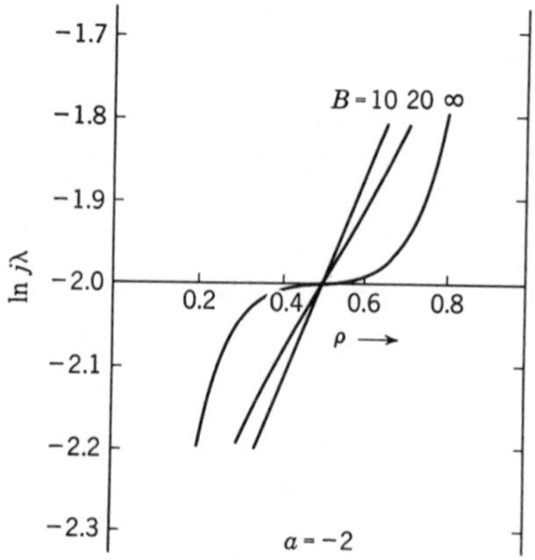

FIG. 66. Curves of μ versus ρ at the critical temperature from the Bragg-Williams approximation. $B = 10$ and 20 from the grand ensemble; $B = \infty$ from grand and canonical ensembles.

Though a loop in the μ–N/V curve may be found from the canonical ensemble [Eq. (A9.9)], it is impossible to get a loop by using the grand ensemble. This follows on differentiating

$$\bar{N}\sum_N Q\lambda^N = \sum_N QN\lambda^N \qquad \text{(A9.16)}$$

with respect to μ. The result is Eq. (19.33) for any Q, approximate or exact. Hence $\partial\bar{N}/\partial\mu$ is necessarily positive.[1]

Similarly, though a loop in the p–V/N curve may be encountered through the use of Eq. (A9.10), it is not possible to obtain such a loop from the isothermal-isobaric ensemble. One finds $\partial p/\partial\bar{V} < 0$ from Eqs. (A9.14) and (19.11).

Approximate Theories. We use the Bragg-Williams approximation of a lattice gas to illustrate qualitative features which would also be found from the van der Waals equation, quasi-chemical approximation, etc. From Eq. (46.3) and Appendix 8, putting $\varepsilon_{AB} = \varepsilon_{BB} = 0$, we have

$$\Xi = \sum_N Q\lambda^N \qquad \text{(A9.17)}$$

[1] See also Eqs. (28.6) to (28.10).

where
$$Q = \frac{B!j^N}{N!(B-N)!}e^{-\alpha N^2/B} \tag{A9.18}$$

and
$$\alpha = \frac{c\varepsilon_{AA}}{2\mathbf{k}T}$$

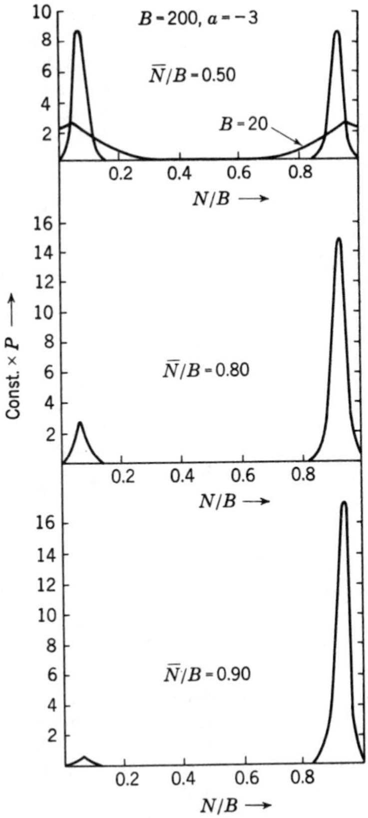

Fig. 67. $P(N)$ curves in the Bragg-Williams approximation showing a phase transition.

The canonical ensemble Eqs. (A9.9) and (A9.18) lead to

$$j\lambda = \frac{\rho}{1-\rho}e^{2\alpha\rho} \tag{A9.19}$$

where $\rho = N/B$. Equation (A9.19) gives the characteristic loop $ACDEFG$ in Fig. 65 for $\alpha = -3$ and the critical curve in Fig. 66 for $\alpha = -2$.

Numerical calculations[1] for the grand ensemble have also been carried out, employing Eqs. (A9.16), (A9.17), and (A9.18). Figure 65 shows ($B = 20, 200, \infty$) that, with the use of Ξ, $ACFG$ is approached rather than $ACDEFG$ as $B \to \infty$. Analogous curves are given in Fig. 66 at the critical ($B = \infty$) temperature, $\alpha = -2$. Figure 67 illustrates the transition, mentioned above, between the

[1] T. L. Hill, *J. Phys. Chem.*, **57**, 324 (1953).

two maxima in $P(N)$ in the two-phase region (Fig. 65) for $\alpha = -3$, $B = 200$ and $\rho = \bar{N}/B = 0.50$, 0.80, and 0.90. The curves shown are normalized to unity. The $\rho = 0.50$ case includes P for $B = 20$. Figure 68 gives $P(N)$, normalized to unity, at the critical temperature ($\alpha = -2$) and critical ρ ($\rho = 0.50$), for $B = 10$, 20, 100, and 200.

For practical purposes, Fig. 67 shows "phase separation" even with B only 200. As $B \to \infty$, the behavior of $P(N)$ is the same except that the peaks get extremely sharp. Any approximate theory with a Q leading to loops will give $P(N)$ curves of this type for reasons already discussed.

Exact Theories. We use the Ising model of a lattice gas to be specific. Instead of Eq. (A9.18) we have, from Eq. (42.26), Appendix 8, and Eq. (41.17),

$$Q = j^N \sum_{N_{AB}} g(N, N_{AB}, B) e^{-\varepsilon_{AA} N_{AA}/\mathbf{k}T} \tag{A9.20}$$

Let $\lambda_0 = \exp(\mu_0/\mathbf{k}T)$ be the value of λ for which $\bar{N} = B\,2$ in the grand ensemble,[1] and $P_0 \equiv Q\lambda_0^N$. Then $P_0(N)$ is symmetrical about $N = B/2$.

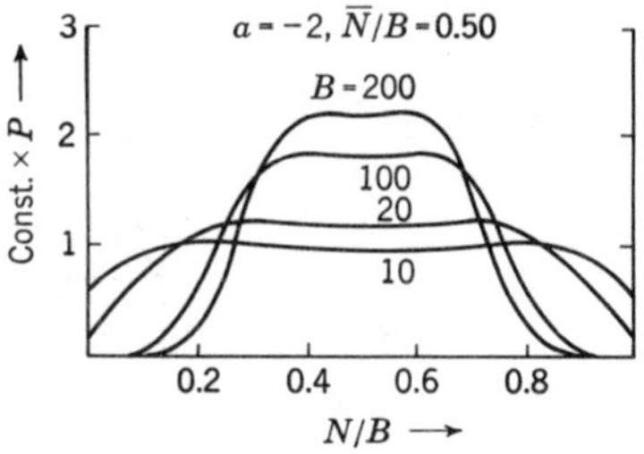

FIG. 68. $P(N)$ curves in the Bragg-Williams approximation at the critical temperature.

All configurations, for given N and B, are included in the sum in Eq. (A9.20). The conventional view (supported by the work of van Hove) is that, below the critical temperature and for a value of N in the two-phase region, configurations with separated phases will predominate in the sum relative to, for example, configurations with an essentially uniform macroscopic density throughout V. If we accept this point of view, we can write (ignoring interfacial effects temporarily)

$$\begin{aligned} A &= A_1 + A_2 = (N_1\mu_0 - pV_1) + (N_2\mu_0 - pV_2) \\ Q(N,V,T) &= Ce^{-(N_1\mu_0 - pV_1)/\mathbf{k}T} e^{-(N_2\mu_0 - pV_2)/\mathbf{k}T} \\ &= Ce^{-(N\mu_0 - pV)/\mathbf{k}T} \end{aligned} \tag{A9.21}$$

where A, Q, N, and V refer to the complete system, while A_1, Q_1, N_1, V_1 and A_2, Q_2, N_2, V_2 refer to the separate phases. V is proportional to B. The

---

[1] This value, for both Eqs. (A9.18) and (A9.20), is given by $j\lambda_0 = \exp(c\varepsilon_{AA}/2\mathbf{k}T)$.

constant[1] $\ln C$ may be of order $\ln N$, etc., but is thermodynamically negligible. From Eq. (A9.21),

$$P_0(N) = Q(N)\lambda_0^N = Ce^{pV/\mathbf{k}T} \qquad \text{(A9.22)}$$

which is independent[1] of N. That is, $P_0(N)$ will behave as in Fig. 69 instead of as in Fig. 70 (see Fig. 67, $\rho = 0.50$). Correspondingly, a flat μ–ρ curve in the two-phase region would be found from the canonical ensemble.

On the other hand, if we should restrict the sum in Eq. (A9.20) to configurations with uniform macroscopic density throughout V (one phase), we would expect to obtain metastable states and a van der Waals type of loop from the canonical ensemble and $P_0(N)$, as in Fig. 70, from the grand ensemble.

Actually, Katsura and Hill have found that $P_0(N)$ behaves as in Fig. 70 in exact calculations for finite B, and not as in Fig. 69. This is illustrated in Fig. 71 for 12 sites on the surface of a sphere. Figure 72 gives $P(N)$ at $\lambda/\lambda_0 = 0.9$ for the same system. Katsura has also noted that the maxima in $P_0(N)$ become

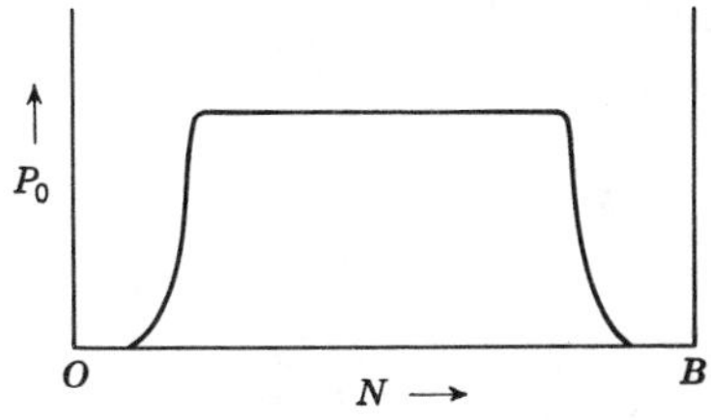

FIG. 69. Schematic $P_0(N)$ curve.

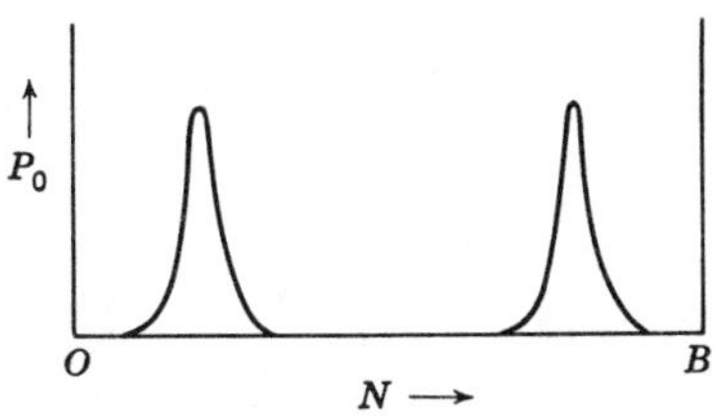

FIG. 70. Schematic $P_0(N)$ curve.

sharper as B increases. Offhand, then, it might appear that a van der Waals type of loop must also be expected from an exact theory and Katsura so conjectures.

However, it is generally believed that such a loop could only follow from the restraint of one phase or uniform density in V; the complete sum in Eq. (A9.20) is not subject to this restraint and should not give a van der Waals loop. As an alternative to Katsura's conjecture, there is presented below an argument showing that *interfacial effects* [omitted in Eqs. (A9.21) and (A9.22)] can account for $P_0(N)$ as in Fig. 70, for an exact theory, and will lead to a corresponding loop which, however, unlike a van der Waals loop, vanishes as $B \to \infty$. Interfacial contributions are automatically taken care of, of course, in Eq. (A9.20); the object here is to estimate their order of magnitude.

[1] This is the quantity $C_{N,V}$ in the notation of Eqs. (14.105), (14.106), and (21.21). For simplicity, and because precise information is lacking in the two-phase region, we treat $C_{N,V}$ as a constant. Actually, it can in general vary slowly with N and V. For example, from Appendix 2, we have for a perfect gas $C_V = 1$, $C_N = \mathbf{k}T/pV^*$ and $C_{N,V} = (2\pi N)^{-1/2}$ [compare Eqs. (A2.2) and (A2.4), using $N! = (N/e)^N(2\pi N)^{1/2}$]. Similarly, for a perfect lattice gas (Appendix 4), $C_B = 1$, $C_N = (1 - e^{-p\alpha/\mathbf{k}T})^{-1}$ and $C_{N,B} = [B/2\pi N(B - N)]^{1/2}$. In the Ising problem, $C_{N,B}$ has to be symmetrical about $N = B/2$ so that Eq. (A9.23) is correct even taking this complication into account. Throughout the book no qualitative conclusions are affected by treating these "constants" as constant.

Let N_1' be the number of molecules in V when phase 1 (the less dense phase) occupies the entire volume, and similarly for $N_2' = B - N_1'$ (phase 2, the more dense phase). For simplicity, instead of considering all values of N between

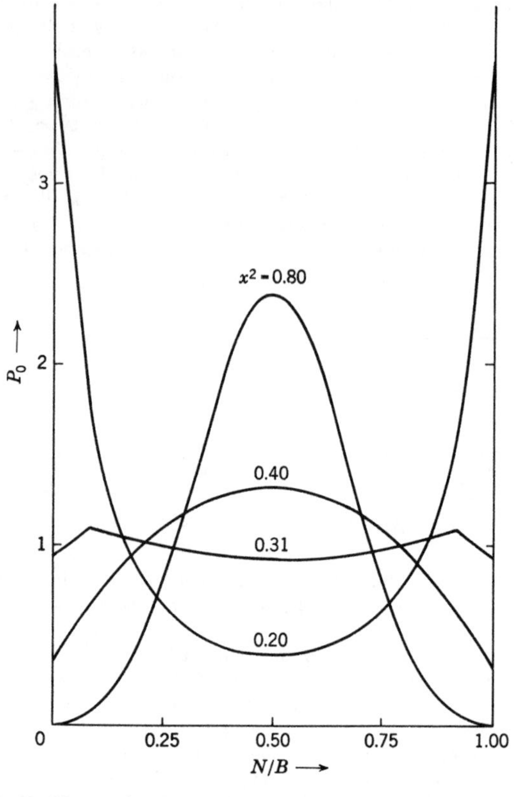

FIG. 71. Exact $P_0(N)$ curves for twelve sites on the surface of a sphere (cubic close packing). $x^2 = \exp(\varepsilon_{AA}/\mathbf{k}T)$. $x^2 \to 1$ at high temperatures ("one phase") and $x^2 \to 0$ at low temperatures ("two phases").

N_1' and N_2', let us discuss $N = N_1'$, $B/2$ and N_2'. At N_1' and N_2' a single phase exists and

$$\begin{aligned} P_0(N_1') = P_0(N_2') &= Q(N_1')\lambda_0^{N_1'} = Q(N_2')\lambda_0^{N_2'} \\ &= Ce^{-(N_1'\mu_0 - pV)/\mathbf{k}T}\lambda_0^{N_1'} \\ &= Ce^{pV/\mathbf{k}T} = C' \end{aligned} \tag{A9.23}$$

At $N = B/2$, two phases exist, but A is not quite additive as assumed in Eq. (A9.21). There will be of the order of $B^{2/3}$ molecules at the interface, which will result in of the order of $B^{2/3}$ "missing" nearest-neighbor interactions. The total surface free energy, to be added to A, will then be of the order $\mathbf{k}TB^{2/3}$. Thus

$$Q\left(\frac{B}{2}\right) = Ce^{-\left(\frac{B}{2}\mu_0 - pV\right)/\mathbf{k}T}e^{-aB^{2/3}} \tag{A9.24}$$

$$P_0\left(\frac{B}{2}\right) = C'e^{-aB^{2/3}} \tag{A9.25}$$

where a is a positive constant of order unity. The term in $B^{2/3}$ in Eq. (A9.24) is negligible thermodynamically (that is, in A) but is not negligible in Eq. (A9.25).

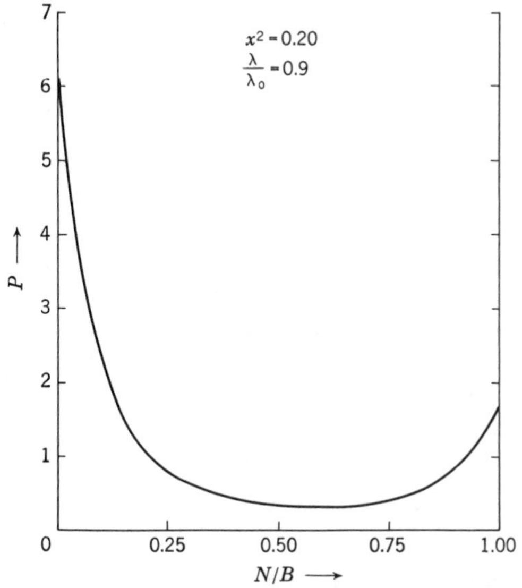

FIG. 72. Same system as in Fig. 71. $P(N)$ at $x^2 = 0.20$ and $\lambda/\lambda_0 = 0.9$.

In fact, from Eqs. (A9.23) and (A9.25) and the symmetry of $P_0(N)$, it is clear that $P_0(N)$ has maxima at N_1' and N_2' and a minimum at $B/2$. The ratio

$$\frac{P_0(N_1')}{P_0(B/2)} = e^{aB^{2/3}} \tag{A9.26}$$

is very large for large B and increases as B increases. Thus the behavior found in Fig. 70 by Katsura and Hill can be accounted for as a surface phenomenon.[1] In two dimensions, the ratio in Eq. (A9.26) is exp $(aB^{1/2})$; in a theory with the restraint of uniform density (including Bragg-Williams, etc.), it is rather obviously exp (aB).

We now wish to consider the extent of the loop to be expected in a μ–ρ curve (from the canonical ensemble), as compared to the loop in an approximate theory

[1] "Surface effects" will play a role when B is not large but the $B^{2/3}$ behavior is asymptotic for large B.

(see Fig. 73). As pointed out above, a loop in Fig. 73 for the canonical ensemble is the locus of the two maxima and one minimum in Fig. 70 for the grand ensemble. The value of $\ln(\lambda_1/\lambda_0)$, which locates the extent of the loop (see Fig. 73), can thus be determined as the value of $\ln(\lambda/\lambda_0)$ at which the maximum at $N = N_1'$ disappears. This will occur essentially when

$$\ln P(N_1') = \ln P\left(\frac{B}{2}\right) \tag{A9.27}$$

Now we have

$$\begin{aligned} P(N) &= Q(N)\lambda^N \\ &= P_0(N)\left(\frac{\lambda}{\lambda_0}\right)^N \end{aligned}$$

$$\ln P = \ln P_0 + N \ln \frac{\lambda}{\lambda_0} \tag{A9.28}$$

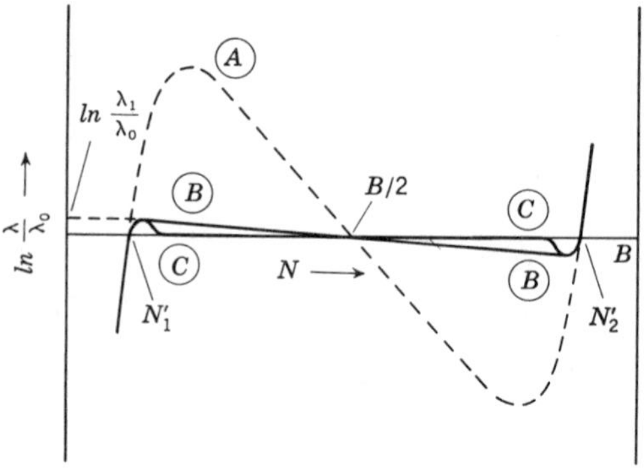

FIG. 73. Schematic μ versus ρ curves. Curve A: approximate theory or model restrained to a single phase. Curves B and C: loops arising in an exact theory from interfacial effects.

Then $\ln(\lambda_1/\lambda_0)$ is given by

$$\begin{aligned} \ln P(N_1') &= \ln C' + N_1' \ln \frac{\lambda_1}{\lambda_0} \\ &= \ln P\left(\frac{B}{2}\right) \\ &= \ln C' - aB^{2/3} + \frac{B}{2} \ln \frac{\lambda_1}{\lambda_0} \end{aligned}$$

or

$$\ln \frac{\lambda_1}{\lambda_0} = \frac{aB^{2/3}}{(B/2) - N_1'} = O(B^{-1/3}) \tag{A9.29}$$

As $B \to \infty$, $\ln(\lambda_1/\lambda_0) \to 0$. Thus the canonical ensemble loop (solid curves B and C in Fig. 73) vanishes as $B \to \infty$ in an exact theory, as found by van Hove. In two dimensions, $\ln(\lambda_1/\lambda_0) = O(B^{-1/2})$. In an approximate theory (or with the restraint of uniform density), $\ln(\lambda_1/\lambda_0) = O(1) =$ *constant* as $B \to \infty$; the loop (dotted curve A in Fig. 73) in this case does *not* vanish as $B \to \infty$.

Whether an exact theory will give a loop of type B or C in Fig. 73 depends on the geometry of the system.

Figure 70 implies that an *open* system at λ_0 will practically always be found to have $N = N_1'$ or N_2' but not intermediate values of N (except for very small fluctuations in the neighborhood of N_1' and N_2'). We should expect this result from the grand ensemble point of view, for the total free energy (see above) of an ensemble of open systems will be lower if (1) half of the systems have $N = N_1'$ and half $N = N_2'$ rather than if (2) the systems have all values of N from N_1' to N_2' with an average N of $B/2$. This follows because there will be an interface with an extra surface free energy in the systems of (2) but there will be no surface contributions in (1) since interactions between molecules in different systems of the ensemble are not counted. These remarks apply to idealized systems, without "edges" (e.g., the surface of a torus or a sphere in two dimensions) or interacting containing walls, such as are usually employed for theoretical calculations. Of course in a *closed* system (canonical ensemble), both phases will in general be present at $\lambda = \lambda_0$. The interface cannot be avoided as in an open system.

A completely analogous argument relating the p–V/N curve to $P(V)$ can be carried through, and with the same results.

Our conclusion is then that, because of interfacial contributions, a loop will be encountered in μ–N/V and p–V/N curves obtained from an exact application of the canonical ensemble to a system with a first order phase transition. However, the range in μ or p covered by the loop approaches zero as the system becomes larger, in contrast to a van der Waals loop. Within this narrowing range, the peaks in $P(N)$ or $P(V)$ become more pronounced as the system becomes larger.

System-Subsystem Relation. We have had occasion to emphasize above that interactions between molecules in different systems of a grand ensemble are not counted. If a subvolume of a large system is considered an open "system" with the remainder of the large system its "surroundings," a grand ensemble would appear to be the representative ensemble to use in this case. This conclusion is legitimate *except* in a two-phase or critical region where the interactions of the "system" with its "surroundings" cannot be neglected. See Sec. 18 in this connection.

APPENDIX 10

GAS ADSORPTION ON A SOLID SURFACE

The treatment of the general imperfect gas in Sec. 23 can easily be extended to the present problem.[1] Consider first a one component gas (at μ and T) which does not interact with the walls of the container. This case was treated in Sec. 23, but we change the notation here to read

$$\Xi^0 = \sum_{N \geqslant 0} \frac{Z_N{}^0 z^N}{N!} \tag{A10.1}$$

$$Z_N{}^0 = \left(\frac{V}{Q_1{}^0}\right)^N N!Q_N{}^0 \tag{A10.2}$$

$$z = \left(\frac{Q_1{}^0}{V}\right)\lambda \tag{A10.3}$$

Now consider the same gas (still at μ and T) and container except that the gas molecules interact with one of the walls (the adsorbent) as well as with each other. We write here

$$\Xi = \sum_{N \geqslant 0} \frac{Z_N z^N}{N!} \tag{A10.4}$$

$$Z_N = \left(\frac{V}{Q_1{}^0}\right)^N N!Q_N \tag{A10.5}$$

In the special case of a monatomic, classical gas with pair interactions,

$$Z_N{}^0 = \int_V \cdots \int e^{-U_N{}^0/\mathbf{k}T}\, d\mathbf{r}_1 \cdots d\mathbf{r}_N \tag{A10.6}$$

$$Z_N = \int_V \cdots \int e^{-U_N/\mathbf{k}T}\, d\mathbf{r}_1 \cdots d\mathbf{r}_N \tag{A10.7}$$

---

[1] For other derivations, see A. Wheeler, American Chemical Society Meeting, Atlantic City, 1949; S. Ono, *J. Chem. Phys.*, **18**, 397 (1950); *Mem. Fac. Kyushu Univ.*, **12**, 9 (1950); *J. Phys. Soc. Japan*, **6**, 10 (1951); M. P. Freeman and G. D. Halsey, *J. Phys. Chem.*, **59**, 181 (1955).

$$U_N{}^0 = \sum_{1 \leqslant i < j \leqslant N} u(r_{ij}) \tag{A10.8}$$

$$U_N = U_N{}^0 + \sum_{i=1}^{N} \bar{u}(\mathbf{r}_i) \tag{A10.9}$$

where $U_N{}^0$ is the total intermolecular potential energy and $\bar{u}(\mathbf{r}_i)$ is the potential energy of interaction of molecule i at $\mathbf{r}_i$ with the adsorbent.

From Eqs. (A10.1) and (A10.4) we have

$$\ln \Xi^0 = \sum_{j \geqslant 1} V b_j{}^0 z^j \tag{A10.10}$$

$$\ln \Xi = \sum_{j \geqslant 1} V b_j z^j \tag{A10.11}$$

where

$$\begin{aligned} 1!Vb_1{}^0 &= Z_1{}^0 = V \\ 2!Vb_2{}^0 &= Z_2{}^0 - Z_1{}^{02} \\ &\text{etc.} \end{aligned} \tag{A10.12}$$

and

$$\begin{aligned} 1!Vb_1 &= Z_1 \\ 2!Vb_2 &= Z_2 - Z_1{}^2 \\ &\text{etc.} \end{aligned} \tag{A10.13}$$

as in Eqs. (23.43). The pressure of the gas far from the surface is

$$\frac{p}{\mathbf{k}T} = \sum_{j \geqslant 1} b_j{}^0 z^j \tag{A10.14}$$

The mean numbers of molecules in V in the two cases are

$$\bar{N}^0 = z \frac{\partial \ln \Xi^0}{\partial z} = \sum_{j \geqslant 1} V j b_j{}^0 z^j \tag{A10.15}$$

$$\bar{N} = z \frac{\partial \ln \Xi}{\partial z} = \sum_{j \geqslant 1} V j b_j z^j \tag{A10.16}$$

The amount of gas adsorbed, as a function of the activity of the gas,[1] is then

$$\bar{N} - \bar{N}^0 = \sum_{j \geqslant 1} V j (b_j - b_j{}^0) z^j \tag{A10.17}$$

That is, the number of molecules "adsorbed" is defined as the mean excess of molecules in V owing to the presence of the adsorbent. If desired, the activity (or fugacity) of the gas can be replaced as independent variable in Eq. (A10.17) by the pressure, through the use of Eq. (A10.14).

In the special case of Eqs. (A10.6) to (A10.9), the leading terms are

$$\begin{aligned} N - \bar{N}^0 = {} & z \int_V [e^{-\bar{u}(\mathbf{r}_1)/\mathbf{k}T} - 1]\, d\mathbf{r}_1 + z^2 \iint_V \{e^{-[u(r_{12}) + \bar{u}(\mathbf{r}_1) + \bar{u}(\mathbf{r}_2)]/\mathbf{k}T} \\ & - e^{-[\bar{u}(\mathbf{r}_1) + \bar{u}(\mathbf{r}_2)]/\mathbf{k}T} - e^{-u(r_{12})/\mathbf{k}T} + 1\}\, d\mathbf{r}_1\, d\mathbf{r}_2 + \cdots \end{aligned} \tag{A10.18}$$

$$\begin{aligned} = {} & \frac{p}{\mathbf{k}T} \int_V [e^{-\bar{u}(\mathbf{r}_1)/\mathbf{k}T} - 1]\, d\mathbf{r}_1 + \left(\frac{p}{\mathbf{k}T}\right)^2 \iint_V \{e^{-[u(r_{12}) + \bar{u}(\mathbf{r}_1) + \bar{u}(\mathbf{r}_2)]/\mathbf{k}T} \\ & - e^{-[\bar{u}(\mathbf{r}_1) + \bar{u}(\mathbf{r}_2)]/\mathbf{k}T} - e^{-u(r_{12})/\mathbf{k}T} + 1 \\ & - \tfrac{1}{2} [e^{-u(r_{12})/\mathbf{k}T} - 1][e^{-\bar{u}(\mathbf{r}_1)/\mathbf{k}T} - 1]\}\, d\mathbf{r}_1\, d\mathbf{r}_2 + \cdots \end{aligned} \tag{A10.19}$$

---

[1] Recall that $z = f/\mathbf{k}T$, where f is the fugacity of the gas.

INDEX

A CATALOG OF SELECTED

DOVER BOOKS

IN SCIENCE AND MATHEMATICS